W9-CKR-784

*The Eye of the Lynx*

DAVID
FREEDBERG

# *The Eye of the Lynx*

## GALILEO, HIS FRIENDS, AND THE BEGINNINGS OF MODERN NATURAL HISTORY

*The University of Chicago Press ⁂ Chicago and London*

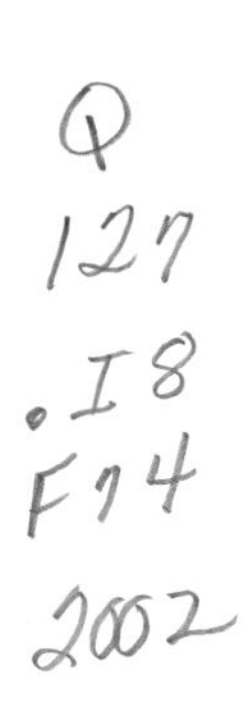


DAVID FREEDBERG is professor of art history and director of the Italian Academy for Advanced Studies in America at Columbia University. His books include *The Power of Images; Dutch Landscape Prints of the Seventeenth Century; Rubens: The Life of Christ after the Passion;* and *Iconoclasts and Their Motives.*

The University of Chicago Press, Chicago 60637
The University of Chicago Press, Ltd., London

Printed in Italy
Published with the assistance of the Getty Grant Program.


11 10 09 08 07 06 05 04 03 02    1 2 3 4 5

ISBN: 0-226-26147-6 (cloth)

Library of Congress Cataloging-in-Publication Data

Freedberg, David.
The eye of the lynx: Galileo, his friends, and the beginnings of modern natural history/ David Freedberg.
p. cm.
Includes bibliographical references and index.
ISBN 0-226-26147-6 (cloth: alk. paper)
1. Accademia nazionale dei Lincei—History.
2. Science—Italy—History. I. Title.
Q127.I8 F74 2002
509.45—dc21

2002000361

This book is printed on acid-free paper.

FOR HANNAH AND WILLIAM

*Filiis Dilectis*

*Contents*

# *Acknowledgments*

The bulk of this book was written while I had the privilege of being Andrew W. Mellon Professor at the Center for Advanced Study in the Visual Arts at the National Gallery of Art, Washington, D.C., in 1996–98. I am grateful to all those at the Center and at the Gallery who patiently put up with my endless requests for scholarly and administrative help. Above all I must thank Henry Millon, then director of CASVA, for his support at every stage. Of all those who helped with the research for this book, four must be singled out here: Irina Oryshkevich, my research assistant of many years standing, who could probably have written this book better than I; Henrietta Ryan, who worked with me from the very beginning on the drawings at Windsor Castle, and who willingly supplied me with every new form of information that emerged about them; Alberta Campitelli, who assisted me with a thousand requests on the Roman front; and Francesco Solinas, with whom I embarked on this extraordinary journey through the histories of science and art.

But very many other colleagues and friends helped me over the fifteen-year period in which this book was being prepared and written. In addition to Simon Varey, who generously allowed me to read a copy of the two-volume collection of essays on Francisco Hernández he has been editing with Rafael Chabrán and others, I would especially like to record my gratitude to Enrico Baldini, Ugo Baldini, Noit Banai, Cecilia Bartoli, Gabriella Becchina, Mario Biagioli, Akeel Bilgrami, Howard Bloch, Horst Bredekamp, Jamie Brindley, Eric Brothers, Anna Maria Capecchi, Patrizia Cavazzini, Joseph Connors, Hubert Damisch, Charles Dempsey, Silvia De Renzi, Oliver Everett, Seth Fagen, Claire Flemming, Marc Fumaroli, Sven Gahlin, Bella Galil, Fabio Garbari, Robert and Pippa Gerard, Carlo Ginzburg, Anthony Grafton, David Helfand, Ingo Herklotz, David Jaffé, Deborah Kahn, Eileen Kinghan, Rosalind Krauss, Claudia Kryza-Gersch, Paola Lanzara, Evonne Levy, Ross MacPhee, Sarah McPhee, Polly Maguire, Paula Mikkelsen, Sara Morasch, Caterina Napoleone, Anna Nicolò, Therese O'Malley, Mireille Pastoureau, David Pegler, Eileen

Reeves, Jane Roberts, Jerome Rozen, Enrica Schettini Piazza, Jeffrey Schnapp, Sebastian Schütze, Andrew Scott, John Beldon Scott, Ursula Sdunnus, Ada Segre, William Stearn, Claudia Swan, Jo Taylor, Lucia Tongiorgi Tomasi, Andrea Ubriszy, Chris Vogel, and the late and very much missed Onno Wijnands. I am especially grateful to all those scientists, especially the botanists, who unstintingly shared their expert knowledge with me, in areas far beyond my own competence. Inevitably, the names of some who helped me will have escaped my memory; and to them I record my gratitude and apologies. The staffs of many libraries were patient with my endless requests, most notably those of the Royal Library, Windsor Castle, the Biblioteca Apostolica Vaticana, the Biblioteca dell' Accademia Nazionale dei Lincei, the National Gallery of Art, the Biblioteca Angelica, the Biblioteca Casanatense, the Archivium Storicum Societatis Jesu, the British Library, the Bibliothèque de l'Institut de France, the Warburg Institute and, of course, the Rare Book Room, Butler Library and Avery Library, all at Columbia University.

# *A Note to Historians of Science*

No assessment of the achievement of Federico Cesi and his fellow Linceans could be complete without an account of their involvement in the discoveries and publications of Galileo in the critical years between the appearance of the *Starry Messenger* in 1610 and the *Assayer* of 1624. Chapters 4 and 5 bring together the evidence for the fundamental role played by Cesi and the Academy of Linceans in the controversies that raged round Galileo's cosmological theories in these years. These chapters also set out the strategies, both worldly and personal, that the Linceans devised in order to support and encourage him. While the general outlines of this story are well-known, a considerable amount of new material is offered in both these chapters; they also provide a critical transition between chapter 3 (on Cesi's, Heckius's, and Galileo's concern with the supernova of 1604), and chapter 6, which outlines the Linceans' pathfinding and as yet largely unknown work with the microscope.

This book covers a great deal of ground. At an early stage of its writing I decided not to enter into a number of current controversies and issues in the history of science, simply because the quantity of new material on Cesi and the early history of the Academy he founded was so substantial. For this reason, then, I chose not to engage explicitly with the debates on the nature and definition of the "scientific revolution of the seventeenth century," on the role of courtly behavior and civility in scientific discovery, or on the phenomenon of collecting in late sixteenth and early seventeenth century Europe. Rich discussions are now available in each of these areas; I hope that the present book contributes to each of them. In some cases my views on the more controversial aspects of these topics will be clear; in others I would simply refer readers to the first-rate literature on each of them.

On the problem of the scientific revolution see now the nuanced summary and comprehensive bibliography by Steven Shapin ("there was no such thing as the scientific revolution, and this is a book about it").[1] For the boldest statement of the influence and pressures of courtly behavior and patronage on scientific discovery, see the work of Mario Biagioli, especially his *Galileo Courtier* (1993). For the whole phenomenon of collecting and the related issues of monstrosity, anomaly, and the whole distinction between *naturalia* and *artificialia*, see now the excellent summation by Lorraine Daston and Katherine Park (1998). While this last subject has never really been ignored (see, for example, the famous work by Schlosser of 1908), it received a major shot in the arm when Impey and MacGregor published their collection of essays in 1985 (Impey and MacGregor now also edit *The Journal of the History of Collections*). Then followed a remarkable series of articles by Giuseppe Olmi and the rich book on the subject by Paula Findlen (1994). Findlen's work often has a direct bearing on the researches, collections, and broad scientific approaches of the Linceans.

In all my work I have inevitably been influenced by Alexandre Koyré, Thomas Kuhn, Michel Foucault, Paul Feyerabend, and Steven Shapin. But in the end, as readers will judge for themselves, I am nothing like as relativist as Feyerabend, not as much a social constructionist as Thomas Kuhn or his more radical followers, slightly less a believer in definitive epistemic ruptures than Foucault, and not quite so committed to the decisiveness of Galileo's role in the geometricization of the universe (or the shift from a closed world to an infinite cosmos) as Koyré.

Had I had all the time in the world (and readers all the patience), I would have liked to have written more specifically than I have here about the relations between truth and trust, between truths and dissimulation, and above all about not telling the truth in order to tell the truth. But that would have meant writing another book entirely—and this one is long enough.

# *Introduction*

## ACROSS THE DISCIPLINES

This book tells of a forgotten yet exalted episode in the history of science and art. At its center stands the figure of Federico Cesi (1585–1630), Galileo's most devoted and ardent supporter. Alongside Cesi are the brilliant and often troubled members of the Academy he founded in 1603 at the age of eighteen. Their story has neither been fully told nor properly appreciated.[1] As it unfolds, it will seem to be divided within itself. It will seem, as has been said of the great astronomer Kepler and his contemporaries in Prague, to reveal

> two world views in collision. One is empirical, turned toward the direct study of nature, open to imaginative speculations that could go wherever the facts might lead. The other is literary, bounded by vast authoritative texts that made speculation difficult.[2]

In the surviving work of Cesi and his friends, much of what we would now call scientific discovery is embedded in the apparently conservative fields of archaeology, philology, and theology. New and direct observations of nature appear in the context of arcane antiquarian researches. Firsthand observation goes alongside the repetition of stale and obscure passages from classical antiquity. Fresh and startling discoveries about the world of nature seem to be accorded no more merit—and sometimes considerably less—than the excavation and explication of some recondite text. Given the importance of so many of these discoveries, such an outlook often appears perplexing. What we would now regard as experimental and empirical activity is frequently accompanied by occult explanations from the fields of astrology and alchemy. Such explanations might seem like holdovers from the sixteenth century; what is new is precisely the emphasis on observation and experiment. Even so, the old disciplines, such as physiognomy, phytognomy, and chiromancy, where cause is fundamentally predicated on the reading of external appearance, are retained. And the old magical explanations never seem to be entirely renounced.

To us, all of this may seem like a fundamental tension. But at the beginning of the seventeenth century that tension was felt much less keenly than it is now. Scholarly and scientific practices, as everyone knows, have become much more fragmented than they were in the late sixteenth and seventeenth centuries, when the division between "humanists" and "scientists" was far less marked. Indeed, "the ability of seventeenth-century scholars to combine scientific and humanistic interests, to use Near Eastern languages as well as Western ones, to move with obvious intellectual comfort from history to law to moral philosophy, is more likely to inspire bewilderment than admiration in the modern reader."[3]

Nevertheless, it would be wrong to suggest that for Cesi and his friends the movement between disciplines was always comfortable. Even in the sixteenth century the gaps between new evidence and old texts were by no means always bridged. Things were being discovered in nature that could not be found in any of the ancient authorities. Once you actually went out into the meadows and woods, you soon began to suspect the reliability and exhaustiveness of the classical handbooks. With Cesi the problem became acute. However easily he and his friends may have moved from one field to another, the tensions between what we now call the humanistic and the scientific fields began to emerge with great force—especially where they intersected in the domain of theology. At first, the members of the Lincean Academy,[4] just like their hero Kepler and so many other distinguished intellectuals of the time, proceeded in the optimistic view that archaeology and theology could at least sometimes be the handmaidens of discovery and progress in natural history, and assist in the firsthand exploration of the world around them. They seemed to enjoy an equally fluent command of every literary and classical field. For a long time, the path to saving the phenomena seemed to be discernible through the refracting glass of Christianity, antiquity, and the ancient forms of rhetoric; but there were other occasions when they could not be saved by these means. Because the Linceans' discoveries were so often buried within texts and practices that were so clearly stamped with the mark of traditional classical scholarship, particular care has to be taken in order to discern them.

Very much the same applies to the juxtaposition of science—or what was then often called philosophy—and poetry. When Giovanni Battista Ferrari, Jesuit father and one of the great seventeenth-century students of flowers and fruit, felt that certain problems regarding the monstrous generation of fruit could not be resolved by any "philosophical" explanation, he turned to poetry and poetic narrative as an alternative.[5] This is not an approach that Galileo would have tolerated. But it often went hand in hand with some of the most important empirical discoveries and speculations of the age. We have to learn to take seriously the overtly poetic language in which scientific and observational breakthroughs (by whatever reckoning) are frequently couched. And not just the language, but also the form. *We* may not expect our biology to come in the form of panegyric or poetry; but once it did. *We* may resist the idea that if we could reconcile the views of Plato and Pythagoras with the evidence of God's operations then we might somehow come closer to the logic of Creation. Yet this is exactly what the great precursors of Galileo's major empirical and theoretical achievements did habitually.

## THE EVIDENCE OF THE EYES

The scientific revolution of the seventeenth century—as it has long been called—has primarily been associated with the great strides in astronomy, physics, and mathematics made by Galileo.[6] But it has not generally been considered in terms of the extraordinary work in the field of natural history that forms the central focus of these pages. Cesi and the men—and the few women—he gathered round him took an active interest in the cosmological, mathematical, and physical researches led by Galileo. They supported him and believed passionately in his theories. At the same time, they studied and collected fossils and mushrooms and plants, whether local or exotic; and they made pioneering contributions to the understanding of them all. But their efforts and the techniques they used to describe all the world of nature in pictorial or graphic form were unprecedented. They believed that no overtly theoretical step could be taken prior to assembling as complete a visual record of nature as possible.

It would, of course, be naive to think that the Linceans drew no theoretical conclusions before that task was concluded. On the contrary. But for many years they proceeded in their belief that the material evidence of their researches had first to be observed, collected, identified, and disseminated by means of pictures and drawings. At the same time, overwhelmed by the multiplicity, variety, and dense texture of the things around them, and the zoological and botanical specimens that were being brought back from the New World in ever greater quantities, Cesi sought to define ways of ordering the abundance of nature. And then, rapidly, he seems to have realized that when it came to the requirements and exigencies of order, pictures were only of limited use.[7]

The first great age of visual encyclopedias was the sixteenth century. It is true that throughout the Middle Ages attempts had indeed been made to assemble compendia of visual information about the world of nature, but they were mostly sporadic and scant in comparison with those that appeared in the wake of the printing revolution. Printing—and the associated arts of woodcut and engraving—enabled the easy reproduction and dissemination of visual information, and students of the natural world were not slow to exploit it. The first to make extensive use of illustration in their works were the great botanists—Brunfels, Bock, and Fuchs, above all. They were followed by a host of others, including Mattioli in Italy and Dodoens, De L'Obel, and Clusius in the north. Then came the zoologists of every stripe, and the anatomists. In Switzerland, Conrad

Gesner published his seemingly all-encompassing works on plants, animals, and minerals, with unprecedented quantities of illustrations. In Bologna, the doctor, antiquarian, and naturalist Ulisse Aldrovandi assembled a body of drawings of the natural world that was so extensive that he only succeeded in having a fraction published in his lifetime. Realizing the importance for science of disseminating the information he had gathered, he directed a vast program of woodcut manufacture. But what happened between the belief in pictures of men like Aldrovandi and Gesner on the one hand, and the reliance on geometry and mathematics of less than a century later? A new commitment to order and classification supervened.

To some extent this part of the story has been told by Michel Foucault in *Les mots et les choses*, appropriately translated into English as *The Order of Things*. Foucault's view was that in the sixteenth and early seventeenth centuries, resemblance and similitude provided the basis for understanding the relations between things; in the seventeenth century, roughly from Descartes onward (in what Foucault called the Classical Age), difference and identity did. From then on geometry and number became still more critical. But in describing the transition from one *episteme* to another, Foucault not only postulated too clear a rupture, he also omitted the crucial role played in this development by Cesi and the Academy he founded. His omission was not willful; he was simply unaware of the newly discovered material to be described in these pages. When we survey this material, Cesi too seems to have started out believing that everything could be made to resemble everything else, in one great seamless fabric of knowledge. Much of what he wrote suggests that for him too, just like the great sixteenth-century naturalists, history itself was implicated in this sense of universal continuity. But as soon as the great mass of material he collected made him aware of the need for more efficient principles of ordering the world, and for better taxonomies, the old schemes failed. Difference and identity, to use Foucault's terms, replaced similitude and resemblance as the chief motors for classification. The graphic description of the surfaces of things could not yield the principles of order; these could only be achieved by penetrating beneath the surface, by counting, and by reducing the fullness of pictorial description to their essential geometrical abstractions. In this respect Cesi's work in natural history marks the same rupture with sixteenth-century procedures as does the geometrization of the world view described by Alexandre Koyré in the case of Galileo's physics.[8]

Like Aldrovandi before him, Cesi set out to document everything around him in visual form. But the drawings Cesi commissioned were far superior in terms of their technical refinement, precision, and attention to detail. Where Aldrovandi had had many drawings of impossible or improbable subjects, Cesi's drawings, with a few significant exceptions, were of actual specimens—however problematic they might be. Often they were almost unidentifiable;[9] but they always presented a higher degree of fidelity to nature than those of almost every one of Aldrovandi's many artists.[10]

In terms of range and scale, too, Cesi's collection of drawings was even more comprehensive than Aldrovandi's. On the whole, they are more careful and precise in the rendering of surface detail, and much more analytic in their attentiveness to the parts of things—as if the aim was to provide new and more substantial bases for their taxonomy. Where Aldrovandi and his followers rushed to provide definitive names for the things they had drawn, Cesi and his friends, more aware of the problems of classification, were much more cautious. And whenever they could they pried and cut things open, in order to draw what they discovered inside.

The new evidence of the eyes was gathered not only on the basis of the huge variety of species daily flooding in from the New World, but also on that of the material collected with new and passionate intensity in the local environment. Increasingly, this evidence stood at odds with the very authorities upon whom most natural historians still relied—Aristotle, Dioscorides, and Galen. Either the Linceans found themselves struggling to reconcile their discoveries in the field with the descriptions in such writers, or they realized that the new species were simply absent from pages that had once been regarded as authoritative and complete.

The crisis in natural history was as great as the crisis in cosmology. Ever since Copernicus had suggested that the earth did not stand at the center of the universe, the old Aristotelian and Ptolemaic systems had been breaking down with ever greater speed. To this momentum was now added the pressure coming from new terrestrial discoveries as well. Apart from their strong personal affinity, nothing joined Cesi to Galileo as closely as Cesi's ever more strongly held conviction that Aristotle and the whole Peripatetic school could no longer be relied upon to provide definitive solutions—either to the structure of the heavens or to the contents and order of life on earth.

But for the students of terrestrial life there was a further dimension to the crisis. Even though he went too far and declared the orbits to be circular, Galileo had shown the logic in simplifying the movements of the heavenly spheres and the bodies contained in them. To Cesi and his followers it became clear that the multitudinous variety of life on earth ought also to be simplified. But how? This was the struggle they never resolved; it dogged them until the day each one of them died.

For one thing, they came to realize that the simple collecting of pictures could never suffice in this endeavor. Picture making, they began to understand, was fundamentally descriptive and synthetic; it stood at odds with order and analysis. Like the old forms of rhetoric it could not completely satisfy the exigencies of logic,[11] nor convey an adequate—or even plausible—idea of the logical and systematic construction of the world. Rhetoric, as it was understood in the classical period and in the sixteenth century, was too dependent on its own devices of color, shade, and nuance—the tricks, in short, of the speaker. So too with pictures: they could not provide a sufficiently impersonal—and therefore natural—reflection of the organization and systematization of content. The point was to discover some sort of order in nature itself—free, it was hoped, of the intervention of ever labile subjectivities. As soon as Cesi set out to order the world in what he thought of as a more logical way, and to examine nature for the traces of what could be taken to be its *own* order, he realized the insufficiency—indeed the essential inadequacy—of pictures.

## TO ORDER THE WORLD

It is no accident that in the history of science, the striving for order is often associated with the geometrical diagram, the table, and the grid. The order of things cannot, it is felt, be conveyed by pictures; but it can by diagrams. When Copernicus set out to simplify the dense and disorderly systems of the universe he placed the ancient motto (Platonic, not Aristotelian) "Let no one unskilled in geometry enter here" on the title page of his *De revolutionibus orbium coelestium* of 1543.[12] In the outline of his system of plant classification of almost two centuries later, Linnaeus wrote a sharp tirade against the use of pictures (or "icons," as he called them);[13] while his distinguished but too often neglected precursor Joachim Jungius insisted that one could only order the data of sensual experience by reducing them to their geometrical essentials. For Jungius, it was only after having found and defined the lowest levels of being that one could constitute an axiomatic science of nature, "in the manner of geometry that starts from the point, the line, the circle, or the parallels."[14]

It is not as if there were no precedents for the resistance to pictures. Ancient medical and natural historical writers like Galen and Pliny were both seriously skeptical about their value. For them, pictures were misleading and seductive; words provided a more adequate, accurate, and concise way of conveying natural historical information. But with the advent of printing such misgivings became ever more muted; and men like Aldrovandi and Cesi—to say nothing of the hosts of lesser figures in the history of natural history—seem to have become altogether carried away by the ever more attractive and useful possibilities of visual description and reproduction.

But not, in Cesi's case, for too long. The more drawings he collected the more acute the problem of order became. They were of limited use when it came to the fundamental problems of classification. To some extent, this was a problem that was anticipated by the sixteenth-century naturalists. But they could not have anticipated the crisis provoked by the use of the microscope. Although its principles had been discovered a few years before they actually acquired one, Cesi and his closest friends were the first to make illustrations of what they saw with its aid. Inspired by the example of Galileo, they were the first to put the microscope to systematic use in the observation and recording of natural specimens.

The use of the microscope was precipitated by the simple discovery that by inverting the telescope, one could see the very smallest of things as if they were large (instead of the most distant things as if nearby). Now one could penetrate to the hearts of the things themselves; and this, especially in the light of the old traditions, posed a simple problem. One might have thought that the rich, dense, and complex surfaces of things—such as the microscope now revealed—only masked equally rich, dense, and complex interiors. But that easy equivalence was swiftly undermined. It became clear that within those interiors there were elements that were common to species whose exterior appearances differed widely. The old idea, held so tenaciously by the occult scientists and magicians of the sixteenth century, was that with the proper training one could accurately learn to read the souls of things, whatever they were, from their external appearances. This hoary notion could now be discarded (though it was by no means always).

All this offered unheard-of possibilities for classification. One could now group species (however so understood) on the basis of a logic that depended on what one discovered on the insides of things, rather than on their surfaces and outward appearances. For example, when one concentrated on the reproductive organs of plants and animals—or on their eggs and spores—as a basis for determining the just class to which a specimen belonged, it became clear that superficial appearance could not possibly be an adequate guide to the establishment of secure and what soon came to be regarded as natural relationships. Even more significantly, examination by microscope immediately began to reveal inner structures whose regularity could never have been predicted on the grounds of their irregular, inconsistent, multitextured, and multicolored outward forms. Such structures, as Kepler had already noted, were apparently geometrical in their formations; and it was these that came to be so fetishized in the microscopic investigations from the middle of the century onward and were celebrated in the great publications of 1665 and 1682 by Robert Hooke and Nathaniel Grew.

In short, by enabling investigators to penetrate into the depths of the specimens they examined, the microscope revealed relationships that could never have been derived from examination of their more or less attractive surfaces. But it was just these surfaces that were so precisely recorded in the drawings assembled by Cesi and his friends. Even when they cut things up into sections, the drawings never went much below the inner surfaces. In the course of assembling this material the members of the young Academy came to realize, along with their leader, that pictures could neither provide sufficient clues to the essential relations between the objects of their studies nor offer evidence of their internal structures. It was these, however, that increasingly provided the key to primary qualities, rather than to the passing, fading, and deliquescent colors and variable textures of their surfaces. Whatever pleasure he took in the first illustrations of zoological and botanical specimens to be seen under the microscope, Cesi came to realize that progress in the field of natural history could no longer come through visual record and description alone but rather through the plotting of the relationships and networks that he was now beginning to discover in the things of nature. This entailed the deployment of number and the spatial systems of geometry, rather than the descriptive processes of picturing. It is true that the further one plumbed the depths of things, there always seemed to be yet another surface for description; but with the new optical instruments, the geometrical principles of the organization of forms now seemed very much clearer. Cesi began to discover what Kepler had deduced from the hexagonal snowflake that fell on him when he crossed the Charles Bridge in Prague one day in the winter of 1611: the geometrical logic of even the most impermanent of forms.

On the one hand, the new technologies of vision made a whole new world available for description; on the other, they contained within them the seeds of destruction of visual description itself. That destruction, however, would be a long time in coming. It would entail much clearer views than Cesi ever had about the ordered classification of the contents of the world. In the meantime, as he struggled with the realization of the need for order, the desire for pictures abated. But it could never be entirely quenched because of his pride in the results of his first microscopic investigations: they had to be announced, published, and communicated to others—even in triumph.

Astonishingly, this will to make known the results of the most secret investigations of his Academy ran exactly contrary to Cesi's earlier commitment (and it was fierce) not to communicate too much to the enemies of the new empiricism and the sycophants of old authority. But now there were additional motives; and scientific pictures were enlisted in the service of propaganda. It was precisely this aspect of the need for pictures that sustained them in the same years that Cesi realized he had to renounce them, and turn his back on the descriptive density that was their most constitutive part of all.

## TRADITION, OPPOSITION, AND CHANGE

All this was played out against a background of intense personal drama. The formation of the Academy of Linceans—which has often been called the first modern scientific academy—was attended by difficulties of every kind. It was at once too ambitious and too exclusive, thwarted as much by the high and rigorous aims of its founder as by its differences from the common science of the age. Cesi's own work was unremitting in the face of all kinds of opposition, beginning with that of his loutish and ignorant father. The lives of his friends were fraught by shifting allegiances and by unusual personal fragilities. Their support for Galileo was constantly threatened by the pressures of the Church, of their weightiest patrons, and finally of the pope himself. Galileo even wavered at times, and at several crucial moments could not have gone on had it not been for the encouragement of his fellow Linceans and their work on his behalf. They continued to believe in him when almost everyone began to have doubts about the motives if not the work of the irascible and headstrong mathematician from Tuscany. His work was fortified by his contact with the Linceans and by their own researches into natural history. He kept abreast of their work on American plants, local fossils, and geological and marine anomalies, and he helped them to their conclusions. The methodological strategies of the Linceans' work were slowly purified as a result of his input—even though at first such purification was impeded by the fierceness of their desire to accumulate and compare information, and then cut short by death.

But in what ways and to what extent did the Linceans free themselves from the traditional forms and practices of science? How are we to reconcile their study of antiquity and their engagement in more purely literary activity with their empirical researches into physics and natural history? To what extent were their results hampered and obscured by their forms of presentation, and how are we to assess such results?

In order to answer such questions we must take every aspect of their work as seriously as they did. We shall have to free ourselves from modern presuppositions about the nature of science and of scientific activity, and to remain attentive to different and to shifting paradigms. We cannot dismiss as mere antiquarianism their interest in archaeology or in ancient texts, and we must respect the ideological pressures on the ways in which they published—or suppressed—their discoveries. Above all we must acknowledge that science for them covered a much larger field than it does now. It ranged from what we call the humanities to fundamental physical phenomena and to mathematics. It included the occult sciences too, and it should not surprise us to find in them, as in Kepler, a faith in astrology that often seems as strong as the commitment to the new astronomy.

But central to the work of the Linceans were their unceasing researches in the field of what we still call natural history, and it is this that will form the central focus of *The Eye of the Lynx*. While the evidence for much of these researches has long been known, much of it has remained unpublished. Since the early 1980s a great deal of new material, both pictorial and documentary, has become available. As a result it is now possible to gauge much more accurately than ever before the full extent of the Linceans' contribution to the history of science. It is a heroic and magnificent one, worthy of being set beside the achievement of Galileo.

The Linceans did not, of course, transform natural history and biology in the way that Galileo revolutionized astronomy, mathematics, and physics; but if they had been less concerned with secrecy, if they had published more, and if the fates had allowed Cesi to live beyond the age of forty-five, who knows what they might finally have achieved? Even now, with the vastly increased knowledge we have of their activities, their researches seem tragically fragmentary. This is less because of the lacunae of the historical record than because of the ultimate impossibility of their ambitions. When we look into that record we discover a small group of men overwhelmed by the multiplicity of things not just in the New World, but in a new universe. They began with a commitment to the local and, never relinquishing that, ended by reaching across oceans and into the starry heavens. Not once did they abandon their commitment to the need for direct observation or for the dissemination and sharing of knowledge—even if, at first, that sharing was only among themselves and the few whose judgment they trusted and respected. Since they hardly ever seemed daunted by the sheer quantity of what they discovered and learned, perhaps it was inevitable that their work should often have stopped short of all but the most perfunctory of theories. Cesi alone seems to have come to the realization that the need for order, in the end, was at least as urgent as the inevitably haphazard accumulation of data. The *Eye of the Lynx*, then, will be the story of their struggle to comprehend the world.

*The Eye of the Lynx*

# PART ONE

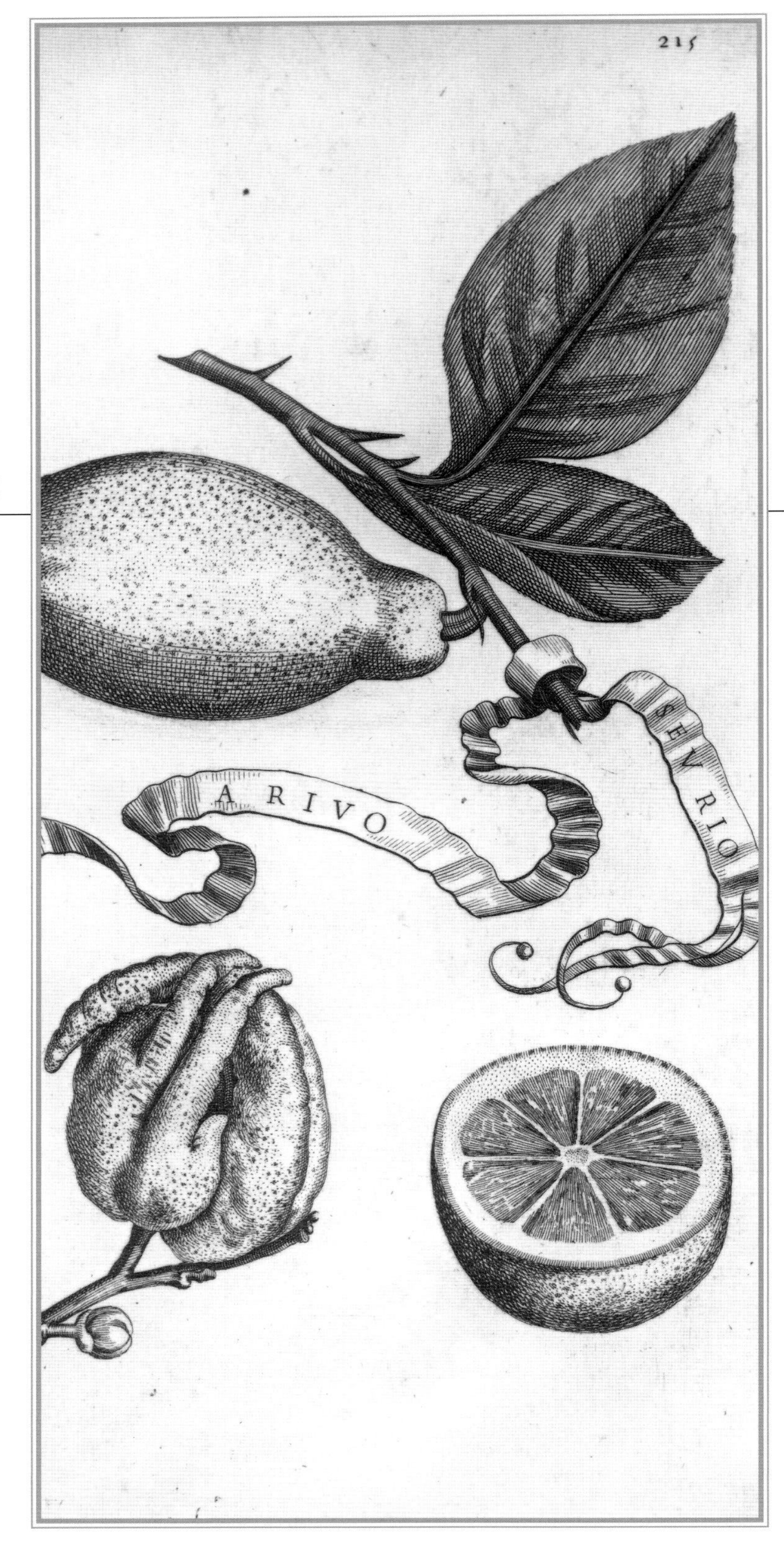

# *Background*

# ONE *The Paper Museum*

## THE DISCOVERY

In 1986, in a cupboard in Windsor Castle, I came across hundreds of the finest natural historical drawings I had ever seen (e.g., figs. 1.1–29). They showed a spellbinding variety of zoological, botanical, ornithological, mycological, and geological specimens. There were truffles, tubers, and sponges; minerals, gems, and fossil woods; flowers, animals, fruits, grasses, and vegetables of the Old World and New; an aviary, so to speak, of magnificently depicted birds, and a cornucopia of different kinds of citrus fruit, some ordinary, and some monstrous and misshapen. There were sheets that showed the details of animals, such as the claws, snout, and quill of a porcupine, the feather of a stork, and the webbed foot of a duck; other illustrations included a giant broccoli, a cluster of insignificant mushrooms, corals, shells, a page of worked asbestos, a frog in amber. There were nestlings, fetuses, bottle-imps,[1] and a number of things I—and no one I consulted—could identify.

Most of the drawings were in color and masterfully precise and detailed. They seemed to come from the nineteenth century, but these were from the seventeenth. Who commissioned them and why? For what purposes were they made? What did the person—or persons—who commissioned them think to achieve? It all seemed to be a kind of megalomaniacal effort to document as much of nature as possible in visual form; but did the project these drawings entailed have some more specific limits, and was there some particular ordering principle at stake? Whatever the case, the drawings as a whole seemed to bear witness to an almost limitless faith in the possibilities of visual representation as an aid to understanding the world around us.

The first question was not difficult to answer. The drawings turned out to have belonged to Cassiano dal Pozzo, one of the best known antiquarians and collectors of paintings in Rome during the prestigious papacy of Urban VIII (1623–1646).[2] French painter Nicolas Poussin's earliest supporter in Rome, Cassiano eventually owned more than fifty paintings by him.[3] In his efforts to record all the surviving remains of classical antiquity, Cassiano commissioned many hundreds of drawings of ancient objects, fragments, and statues.[4] These drawings he kept in twenty-three large vellum-bound volumes in what he called his "*museo cartaceo*" (paper museum).[5] His residence in the Via dei Chiavari in Rome also housed a small museum of natural specimens and art objects, as well as a laboratory in which he made anatomies of animals and conducted chemical and other scientific experiments.

By 1985 most people had forgotten that Cassiano ever owned natural historical drawings and that he had a laboratory in which he conducted his researches into nature. His palazzo seemed to contain little more than the kind of cabinet of curiosity that was common at the time.[6] Galileo, for example, used the image of a typical small collector's cabinet in order to illustrate how the apparently disjunct and piecemeal poems of Tasso were inferior to the grandly conceptual work of Ariosto. Tasso's works were strangely like the collection of an *ometto curioso*,

> who has taken a delight in adorning it with things that have something exotic about them, either because of age or of rarity or some other reason, but are in effect bric-a-brac—a petrified crayfish; a dried-up chameleon, a fly, and a spider embedded in a piece of amber, some of those little clay figures said to be found in the ancient tombs of Egypt, and (when it comes to painting) a sketch or two by Bandinelli and Parmigianino and other similar trifles.[7]

Was this really what Cassiano's collection was like? And was it in fact as aimless and piecemeal as all that?

To judge from two vivid seventeenth-century accounts of visits to Cassiano's house, this is just what one might have thought. John Evelyn, the English diarist, antiquarian, and student of nature, visited Cassiano on November 21, 1644:

> On the 21 I was carried to a great virtuoso, one Cavalliero Pozzo, who showed us a rare collection of all kind of antiquities, a choice library, over which are the effigies of most of our late men of polite literature: That which was most new to me was his rare collection of the antique bassirelievos about Rome, which this curious man[8] had caused to be designed in diverse folios: he showed us also many fine medals and among other curiosities a pretty folding ladder, . . . and a number of choice designs and drawings. He also showed us that stone Pliny calls *Enhydrus* of the bigness of a walnut: it had plainly in it the quantity of half a spoonful of water, of a yellow pebble color, and another in a ring, paler than amethyst, which yet he affirmed to be the true *carbuncle* and harder than diamond . . .[9]

The antiquities were only to be expected, but for the rest—what a jumble! And so much was omitted, not least the great collection of paintings by

Nicolas Poussin. Only the reference to the stones offers a hint of something more portentous.[10]

The second description comes from 1663, six years after Cassiano's death. His collections had been preserved—and augmented—by his devoted brother and partner in collecting, Carlo Antonio dal Pozzo.[11] In that year Philip Skippon visited their house, and it is from him, if we read carefully, that we gain a sense of some of the priorities of the collection. At first Skippon seems to convey an impression that is no less confusing than Evelyn's. In addition to the statues, medals, lamps, and other small antique objects (including several of the kinds of priapic lamps, small bronzes, and terra-cottas so beloved of antiquarians at the time), he also saw a large number of drawings. These were an extraordinary mix, and included a "picture of a boy that defended philosophical theses when but ten years old, now grown a most ignorant man" and "of a stone Priapus, the lower parts like a lion; figures of animals &c. hung round the glans."

But there was another group of drawings entirely. It contained

> four folios pictured with plants, well done. Many pictures of birds &c. in loose papers. The picture of an onocrotalus [pelican], phaenicopterus [flamingo]. . . . In a book of birds, the picture of a white parrot . . .

There were many more bird drawings too—another pelican, a toucan, an owl, and so on—as well as one of

> a dolphin brought to the fish market in Rome. . . . Sagovius, a sort of jack-an-ape, with large white ears. An Egyptian mouse with long hind legs, and very short ones before. The plant that budded out of a man's side in Spain in 1626. . . . A little embryo about an inch and a half long fully shaped, which was observed to pant *in menstruis*. Seven books of John Heckius a German, wrote in his travels; he observed plants, insects, &c and was one of the *Accademici Lyncei illum*. . . . A chopping knife and a saw the martyrs were put to death with, were found in church-yards. . . . Ancient brass armour, very light, easy to be worn, and fitted. . . . The pictures of three mummies, which were in Pietro della Valle's possession. . . . The picture of the mummied leg at Cavaliero Corvino's. Matthiolus curiously painted. These books are painted very exactly, the heads, legs, and other parts of animals being distinctly drawn. The picture of Sada, Petrarch's mistress.[12]

Here was work to be done. Where were all these drawings? Could they have had anything to do with those in Windsor? At least Skippon's description gave some idea of the substantial place held by drawings after nature in Cassiano's collection—even amid all those strange and seemingly random objects. Perhaps there was indeed some overarching idea behind them, rather more like Galileo's description of Ariosto than that of Tasso, whom he clearly preferred less.[13]

I kept finding more drawings at Windsor; and later on in private and public collections elsewhere. Though the original group I came across in Windsor consisted of loose and sometimes rather dusty sheets, there were many more, in finer condition yet, bound into splendid late eighteenth-century volumes. These volumes turned out to have made for King George III after his acquisition of all of Cassiano's drawings in 1763.[14] In addition there were also a few more modest volumes, clearly older than those bound for George III, in rather plain vellum bindings.

Four of the large volumes, each inscribed *Natural History: Fossils*, contained more than two hundred of the most precise drawings, mostly in pen and ink, of pieces of fossilized wood and other fossil phenomena;[15] a fifth showed a wide range of gems, marbles, mineralogical specimens and geological curiosities painted in brilliant watercolors (fig. 1.18).[16] There was also a splendid volume of bird drawings, of a similarly high quality, and clearly related to the surviving loose sheets (e.g., figs. 1.3–8; 1.44, 1.46).[17]

Of the two vellum-bound volumes, one was an illustrated herbal inscribed "*Erbario Miniato*" on its spine;[18] the other, still more modest, showed a large quantity of Central American plants, accompanied by their names in Nahuatl and charming descriptions of their medicinal uses. This was a copy of the famous work painted by a Nahuatl Indian for Philip II of Spain in 1563 and now known as the Codex Badianus (cf. fig. 9.3).[19]

Slowly Skippon's description began to make more sense. Since many of the drawings in the *Erbario miniato* were accompanied not only by long texts explaining their names and outlining their medicinal qualities, but also by references to a particular edition of the famous sixteenth-century herbalist Mattioli,[20] this was surely the "Matthiolus curiously Painted" recorded by Skippon. The volume of bird drawings was presumably one of those referred to by him; and so things gradually began to fall into place.

The loose drawings, at Windsor Castle and elsewhere, were generally of the highest quality, meticulously drawn in a sensitive and subtle range of watercolors and often varnished in gum arabic. Many seemed to be by a distinctive hand. Quite a few could be correlated with particular drawings mentioned by Skippon. There was a beautiful drawing of a dolphin, for example, still enclosed in the distinctive wash mount prepared for these drawings just after they entered the collection of George III (fig. 1.11). This was surely the specimen "brought to the fish market in Rome" before being drawn for Cassiano.[21] Skippon referred to two drawings of a pelican: both reappeared at a 1988 sale of drawings that had once belonged to Windsor Castle (fig. 1.5).[22] The drawing of the "phenicopterus" he mentioned is presumably one of several drawings of flamingos surviv-

Fig. 1.1. Vincenzo Leonardi, porcupine (*Hystrix cristata*). Watercolor and bodycolor heightened with gum over black chalk, 341 x 481 mm. London, collection Sven Gahlin.

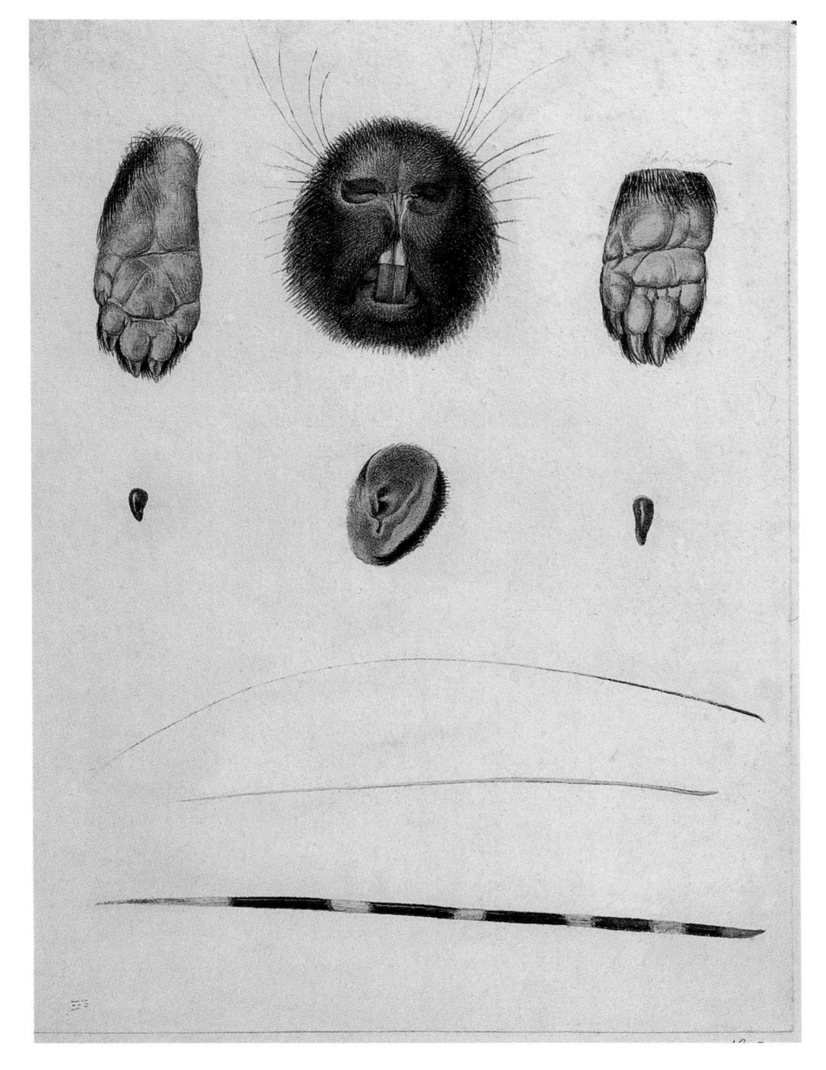

Fig. 1.2. Vincenzo Leonardi, anatomical details of the common or crested porcupine (*Hystrix cristata*). Watercolor and bodycolor heightened with gum over black chalk, 411 x 218 mm (inscribed, in Cassiano's hand, *Zampa/ l'altra zama/unghia*). Windsor, RL 19438. The Royal Collection © 2000, Her Majesty Queen Elizabeth II.

Fig. 1.3. Vincenzo Leonardi, head of a European white stork (*Ciconia ciconia*). Watercolor and bodycolor over black chalk, 208 x 271 mm. Windsor, RL 28740. The Royal Collection © 2000, Her Majesty Queen Elizabeth II.

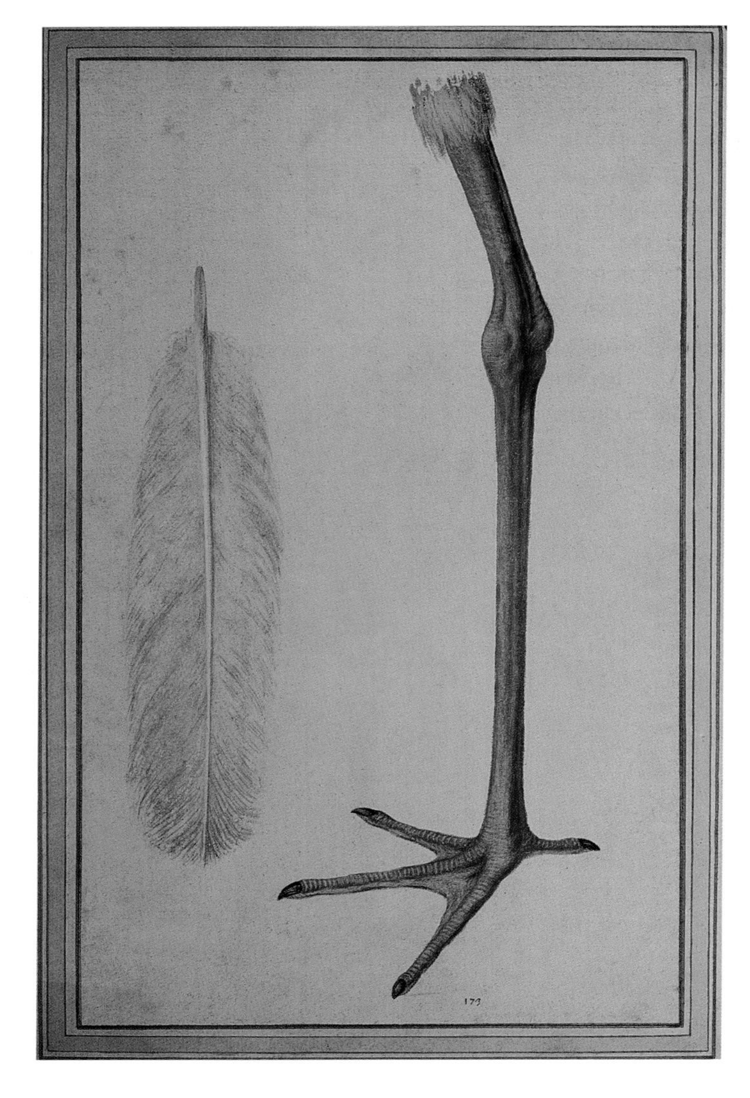

Fig. 1.4. Vincenzo Leonardi, leg and feather of a European white stork (*Ciconia ciconia*). Watercolor and bodycolor with touches of gum over black chalk, 381 x 227 mm. Windsor, RL 28739. The Royal Collection © 2000, Her Majesty Queen Elizabeth II.

Fig. 1.5. Vincenzo Leonardi, European or great white pelican (*Pelecanus onocrotalus*). Watercolor and bodycolor heightened with gum over black chalk, 364 x 452 mm. Windsor, RL 28746. The Royal Collection © 2000, Her Majesty Queen Elizabeth II.

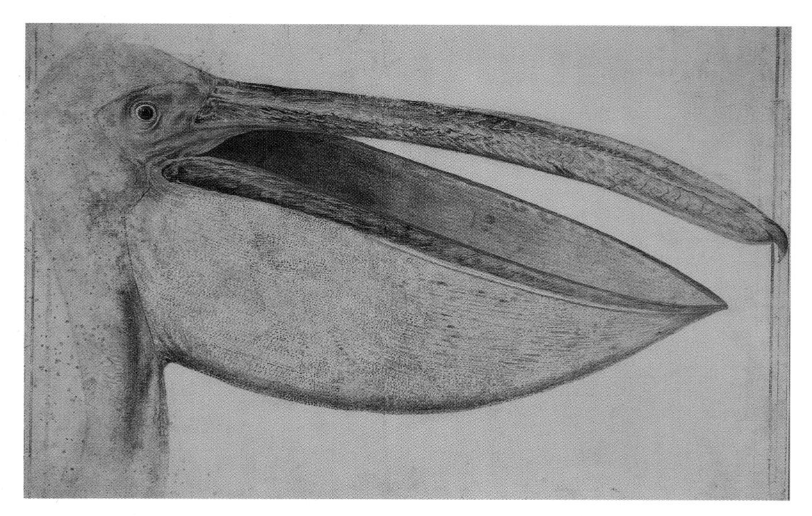

Fig. 1.6. Vincenzo Leonardi, head of European or great white pelican. Watercolor and bodycolor heightened with gum over black chalk, 355 x 544 mm. Windsor, RL 19437. The Royal Collection © 2000, Her Majesty Queen Elizabeth II.

ing in private collections (e.g., fig. 1.7).[23] The sprightly owl, having been dispersed from the Royal Collection, was rediscovered; and the "little embryo about an inch and a half long which was observed to pant *in menstruis*" may well be one or the other of several such drawings still at Windsor Castle (cf. fig. 1.20).

Skippon was especially struck by the fact that in many of the illustrations the "heads, legs and other parts of animals were "very distinctly drawn"; and indeed, as one goes through the drawings, one cannot help but notice this concentration on the parts of animals (as well, of course, as their presentation as a whole). One drawing vividly portrays the common or crested porcupine (fig. 1.1), another shows its snout, paws, ear, quills, and other anatomical details (fig. 1.2). One sheet depicts a Euro-

Fig. 1.7. Vincenzo Leonardi (?), three studies of the head, beak, and tongue of a flamingo (*Phoenicopterus ruber*). Watercolor and bodycolor over pen and ink, 533 x 402 mm. London, collection Sven Gahlin.

pean white stork (fig. 1.3); another its head; another a leg and a feather, both minutely detailed and with careful attention, as always, to color (fig. 1.4). One of the pelicans—the great white pelican—had its head portrayed with care (figs. 1.5–6); the flamingo, its legs, head, wings, and tongue (e.g., fig. 1.7).[24] It is hard to think of any precedents—either for the sustained analysis of parts, or for the extraordinarily attentive care devoted to portraying them.

Just as Skippon's account suggested, there were many more drawings of birds, most of the same high standard of draftsmanship (e.g., fig. 1.8); of unusual animals such as the civet (fig. 1.9), the oryx (fig. 1.10) and the sloth; of fishes and other aquatic creatures (figs. 1.11–12); of local as well as exotic plants and fruits in great abundance; of leaves, grasses, and vegetables, including what is surely the most spectacular representation of a broccoli in the history of art (fig. 1.13). A water lily was shown in its various phases of transformation from flower to fruit (in fact two drawings now pasted onto the same sheet, fig. 1.14), while

Fig. 1.8. Vincenzo Leonardi, Squacco heron (*Ardeola ralloides*). Watercolor and bodycolor heightened with gum over black chalk; pentimenti in black chalk, 475 x 343 mm. Windsor, RL 28741. The Royal Collection © 2000, Her Majesty Queen Elizabeth II.

Fig. 1.9. Vincenzo Leonardi, African civet (*Viverra civetta*). Watercolor and bodycolor heightened with gum over black chalk, 344 x 476 mm. Windsor, RL 21145. The Royal Collection © 2000, Her Majesty Queen Elizabeth II.

Fig. 1.10. Vincenzo Leonardi, Arabian oryx (*Oryx gaziella*). Watercolor and bodycolor heightened with gum over black chalk, 440 x 327 mm. London, private collection.

Fig. 1.11. Vincenzo Leonardi, dolphin (*Delphinus delphis*). Watercolor and bodycolor with touches of silver, heightened with gum over black chalk, 338 x 538 mm. Windsor, RL 28735. The Royal Collection © 2000, Her Majesty Queen Elizabeth II.

another sheet presented one of the earliest depictions of a healthy pineapple (fig. 1.15).

Who could not admire the subtlety and sheer beauty of the many drawings—several finely depicted on blue paper—of corals, sponges, fossils, truffles, tubers, and fungi (e.g., figs. 1.16–17); of fossil woods, concretions, aetites, bezoars, and mineralogical specimens of many kinds (fig. 1.18; cf. figs. 11.5–23)? If you needed a representation of how asbestos could be woven into fabric, you could find just such a thing here (fig. 1.19). There were pictures of deformed nestlings, sacs, fetuses, and egglike things that for a long time resisted identification (figs. 1.20–21). Many of the drawings were actually the first such representations ever,

Fig. 1.12. Vincenzo Leonardi, Mediterranean slipper lobster (*Scyllarus latus*). Watercolor and bodycolor heightened with gum arabic over black chalk, 292 x 445 mm. London, collection Sven Gahlin.

Fig. 1.13. Vincenzo Leonardi, head of broccoli (*Brassica oleracea, Italica group*). Watercolor and bodycolor over black chalk, 342 x 476 mm. Windsor, RL 21143.

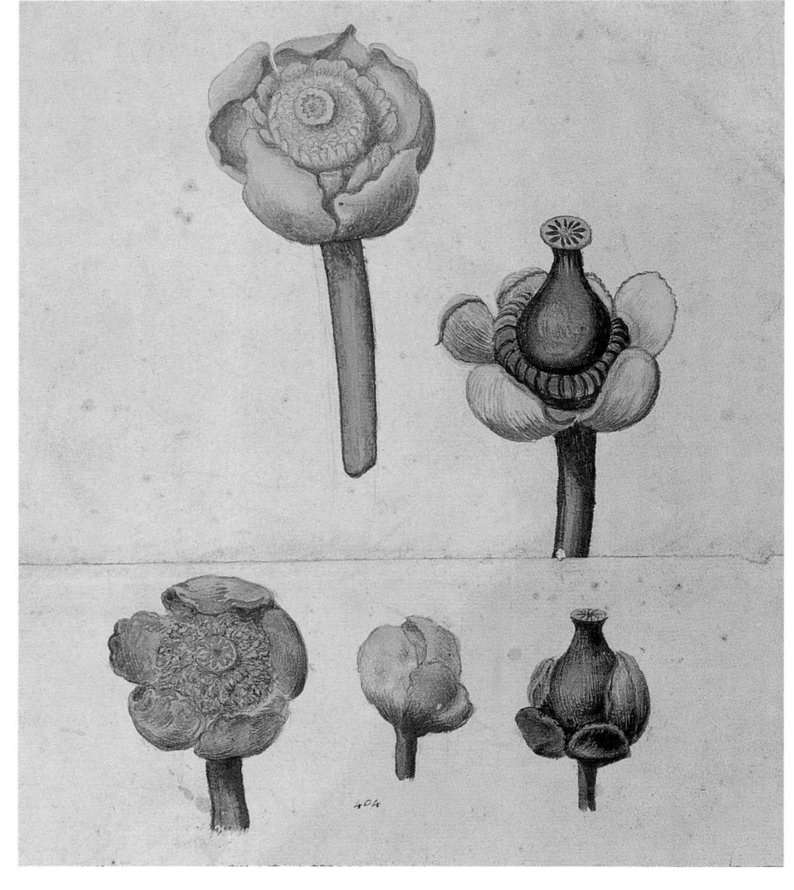

Fig. 1.14. Stages in the growth of a yellow water lily (*Nuphar lutea* L). Watercolor and bodycolor over black chalk, 278 x 252 mm. Windsor, RL 19397. Both the Royal Collection © 2000, Her Majesty Queen Elizabeth II.

Fig. 1.15 Vincenzo Leonardi, pineapple (*Ananas comosus*). Watercolor and bodycolor over black chalk, 381 x 284 mm. Windsor, RL 28738. The Royal Collection © 2000, Her Majesty Queen Elizabeth II.

as, most notably, the drawing of little frogs in a piece of old American amber (fig. 1.22), shown with a kind of precision that still excites zoologists and paleontologists today.[25]

A fair number of drawings clearly showed anomalous specimens; but was it only the elements of gigantism and monstrosity they displayed that inspired their production? This question seemed especially pressing in the case of one of the largest single groups of drawings, namely the spectacular series of drawings of citrus fruit. Here were oranges, lemons, and citrons in great abundance. Many seemed ordinary enough specimens; but then there were hybrids and monstrosities, elephantine citrons with phallic growths, wrinkly and rugged lemons, and oranges with tuberous and tumorous excrescences. One prodigious specimen followed another, some with fingerlike projections, others with one fruit enclosed by another, as if pregnant. Some did not look like members of the citrus family at all, while others seemed sooner to belong in a museum of marvels (figs. 1.23–29).[26]

Fig. 1.16. Vincenzo Leonardi two outgrowths of an unidentified coral (?). Watercolor and bodycolor with touches of gum on blue paper, 222 x 178 mm. Windsor, RL 19346. The Royal Collection © 2000, Her Majesty Queen Elizabeth II.

Fig. 1.17. Vincenzo Leonardi (?), views of two stages of growth on a fungus, probably the spring agaric (*Agrocybe praecox* (Pers.: Fr.) Fayod). Watercolor and bodycolor over black chalk, 215 x 171 mm. Windsor, RL 19369. The Royal Collection © 2000, Her Majesty Queen Elizabeth II.

Fig. 1.18. An assortment of corals, patterned stones, asbestos, fossils, ivory, ammonite, and semiprecious stones. Watercolor and bodycolor over black chalk, 604 x 465 mm. Windsor, RL 25497. The Royal Collection © 2000, Her Majesty Queen Elizabeth II.

All in all there were no fewer than twenty-seven hundred drawings of natural historical subjects, some loose, others bound together. With the exception of the great efforts at visual documentation by the Bolognese doctor Ulisse Aldrovandi, there was no other collection remotely like Cassiano's from either the sixteenth or the seventeenth centuries.[27]

Particular groups of drawings—especially the citrus fruit—raised further issues. For example: What lay behind so concentrated an effort at recording such a large number of what we would now regard simply as varieties of single species? How much of the precision in representing surface appearance was driven by the need to find an adequate basis for

Fig. 1.19. Vincenzo Leonardi (?), natural and worked specimens of asbestos. Watercolor and white heightening with touches of gum on blue paper, 184 x 225 mm. Windsor, RL 25522. The Royal Collection © 2000, Her Majesty Queen Elizabeth II.

the classification of one or another puzzling specimen, or to separate one specimen from another that seemed closely to resemble it? The general classificatory thrust of these drawings could not have been clearer—and whatever system of classification there was seemed less encumbered with irrelevance than the rambling natural historical textbooks produced by Aldrovandi and his editors.[28]

Questions such as these only became more urgent when, following the rediscovery of the Windsor drawings, several other of the groups of drawings mentioned by Skippon began to reemerge. For a start, four of the "books of John Heckius," the "German" member of the Academy of the

Fig. 1.20. Deformed nestling: front and back views. Watercolor and bodycolor over black chalk, (*a*) 126 x 68 mm; (*b*) 126 x 85 mm. Windsor, RL 19431–32. The Royal Collection © 2000, Her Majesty Queen Elizabeth II.

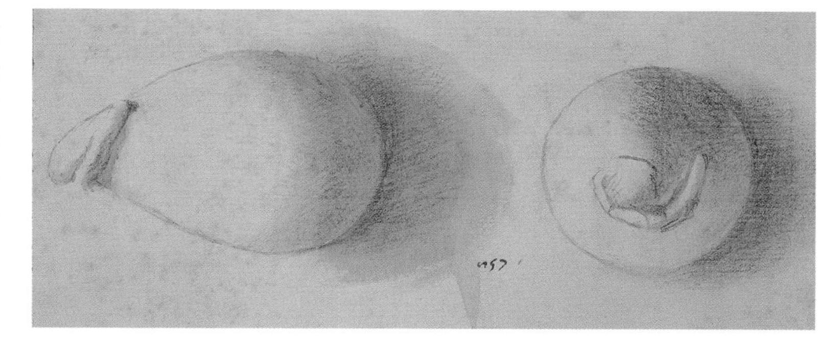

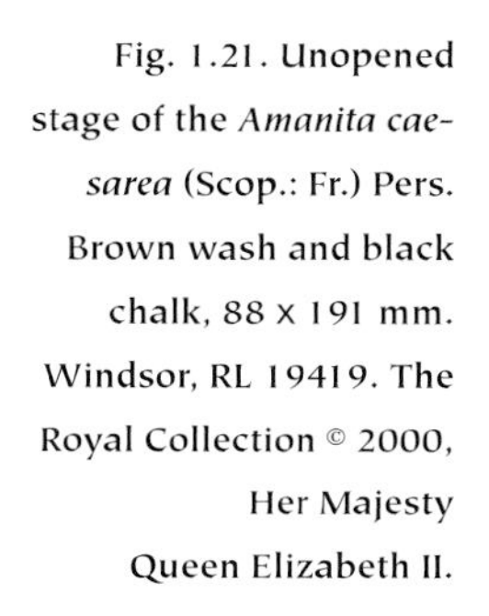

Fig. 1.21. Unopened stage of the *Amanita caesarea* (Scop.: Fr.) Pers. Brown wash and black chalk, 88 x 191 mm. Windsor, RL 19419. The Royal Collection © 2000, Her Majesty Queen Elizabeth II.

Lincei, turned out to be preserved in the library of the old medical school in Montpellier.[29] Just as Skippon noted, they contained a record of the "plants, insects, &c" that Heckius had seen in the course of his travels in northern Europe—the "&c" presumably referring to the many different kinds of butterflies, serpents, and spiders that drew his attention and that he chose to draw on the pages of his notebooks.[30] But these were very crude drawings in comparison with those in Windsor.[31]

In the early 1980s attention was drawn to a vastly more important group of drawings than those in Montpellier.[32] Eight extraordinary volumes of plants and fungi, also from the collections of Cassiano and containing more than two thousand drawings, were found in the library of the Institut de France in Paris (figs. 8.20–40). These too had been ac-

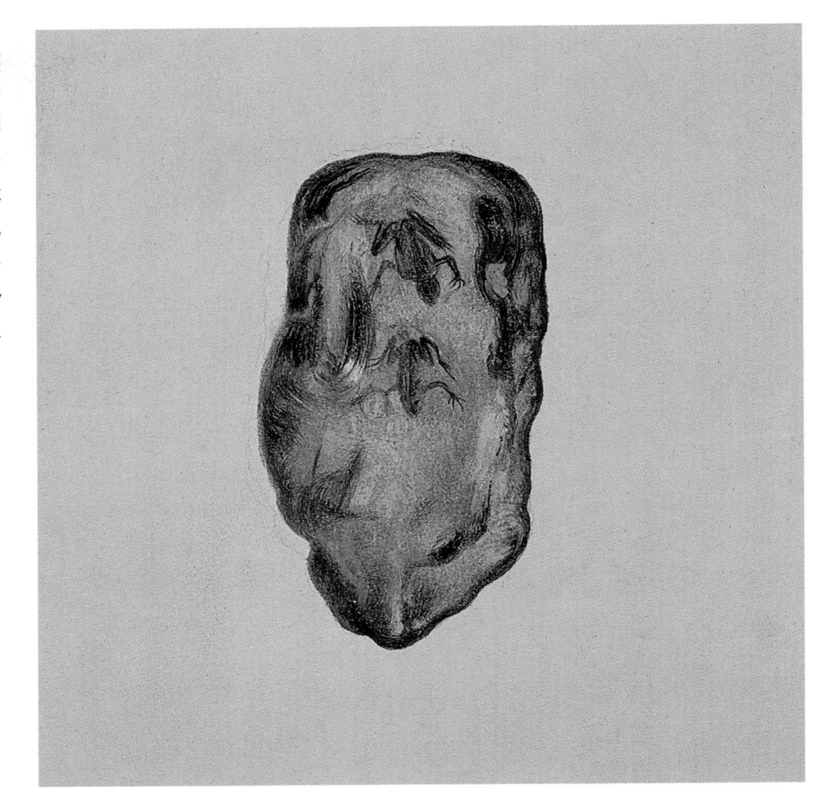

Fig. 1.22. Vincenzo Leonardi (?), frogs in Amber. Watercolor and bodycolor with gum arabic over black chalk, 150 x 206 mm. Windsor, RL 25481. The Royal Collection © 2000, Her Majesty Queen Elizabeth II.

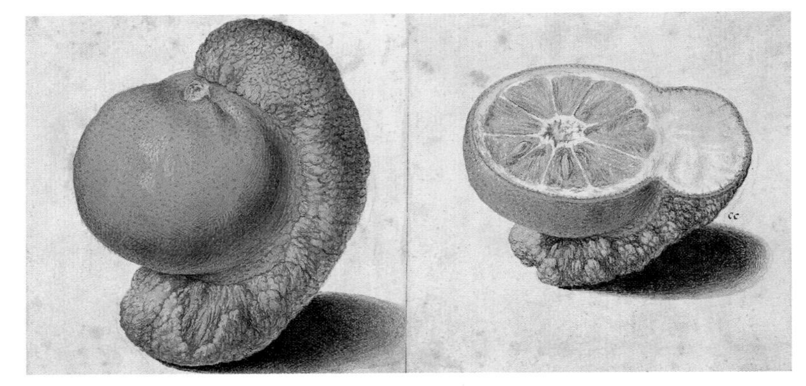

Fig. 1.23. Vincenzo Leonardi, chimeric citron sour orange (?) with excrescence. Watercolor and bodycolor with gum heightening over black chalk, 129 x 130 mm. Windsor, RL 1929 and 19330 (mounted together). The Royal Collection © 2000, Her Majesty Queen Elizabeth II.

quired—perhaps simply confiscated—during the French occupation of Rome, this time by an officer of the revolutionary army in 1798.[33]

Three of the volumes in the Institut de France were labeled *Fungorum genera et species* and five *Plantae et flores*. The first three contained roughly six hundred pages of colored drawings of fungi (often several to a page), observed with extraordinary closeness and intensity. No such close examination of the fungi of a particular region (southern Umbria and the vicinity of Rome, as the inscriptions on many pages revealed) had ever been made before.[34] It was soon apparent that their place in the history of mycology was similar to that of the drawings of fossils in Windsor Cas-

Fig. 1.24. Vincenzo Leonardi, misshapen whole citron. Watercolor and bodycolor over black chalk, 310 x 217 mm. England, private collection.

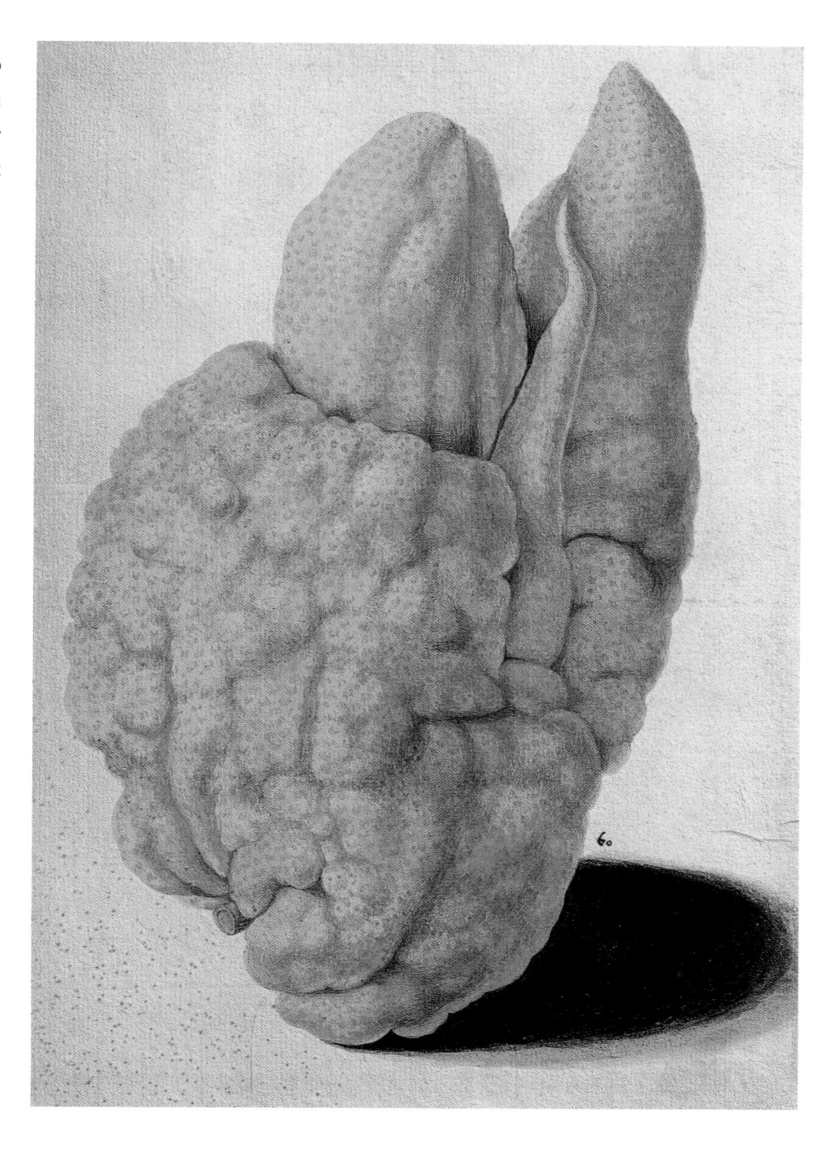

tle in the history of paleontology: both were the first field surveys of their kind in visual form.[35]

The second set of volumes in Paris contained well over eight hundred pages of drawings of plants and cryptogams. They ranged from exotic plants from the Americas to page after page of mosses, lichens, liverworts, algae, and ferns. All were shown in exceptional detail. None were dull; most were remarkable for their delicate attentiveness to color, morphology, and texture. Most importantly, however, a large number of drawings in both volumes turned out to be the earliest surviving drawings ever made with the aid of a microscope.[36] Since the inscriptions on several of these drawings noted that they were made between 1623 and 1628,[37] they were done about forty years before the pioneering work by van Leeuwenhoek and the well-known microscopic illustrations of the

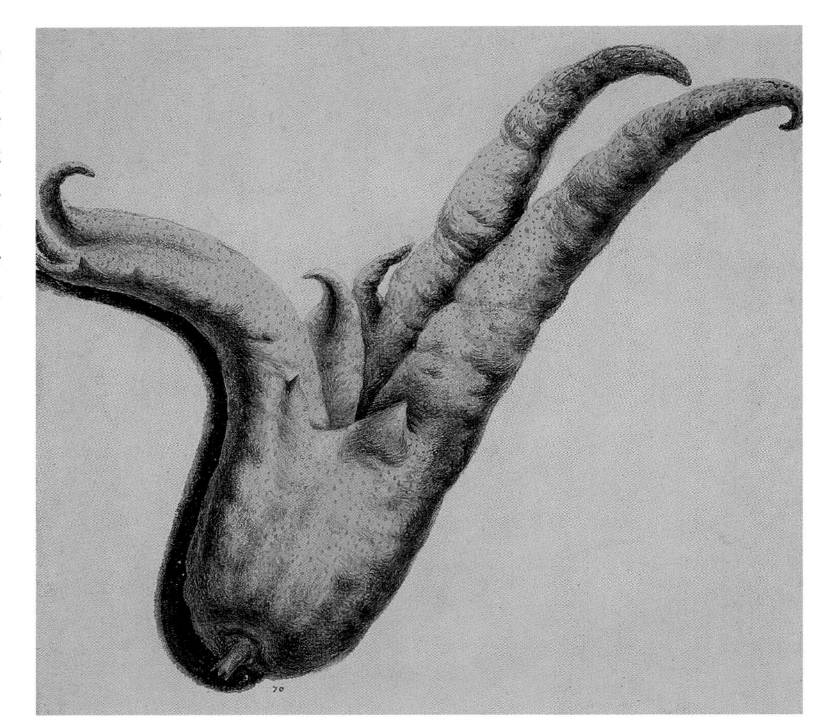

Fig. 1.25. Vincenzo Leonardi, fingered lemon. Watercolor and bodycolor over black chalk, 248 x 255 mm. Windsor, RL 19358. The Royal Collection © 2000, Her Majesty Queen Elizabeth II.

Fig. 1.26. Vincenzo Leonardi, fingered lemon. Watercolor and bodycolor over black chalk, 117 x 139 mm. Windsor, RL 19379. The Royal Collection © 2000, Her Majesty Queen Elizabeth II.

details and innards of things published by Hooke in his famous *Micrographia* of 1665 and Grew in the *Anatomy of Plants* of 1682.

One did not have to look far to see how much the drawings in this corpus differed from one another, sometimes very substantially. Some, like the majority of the drawings at or originally from Windsor, were of great

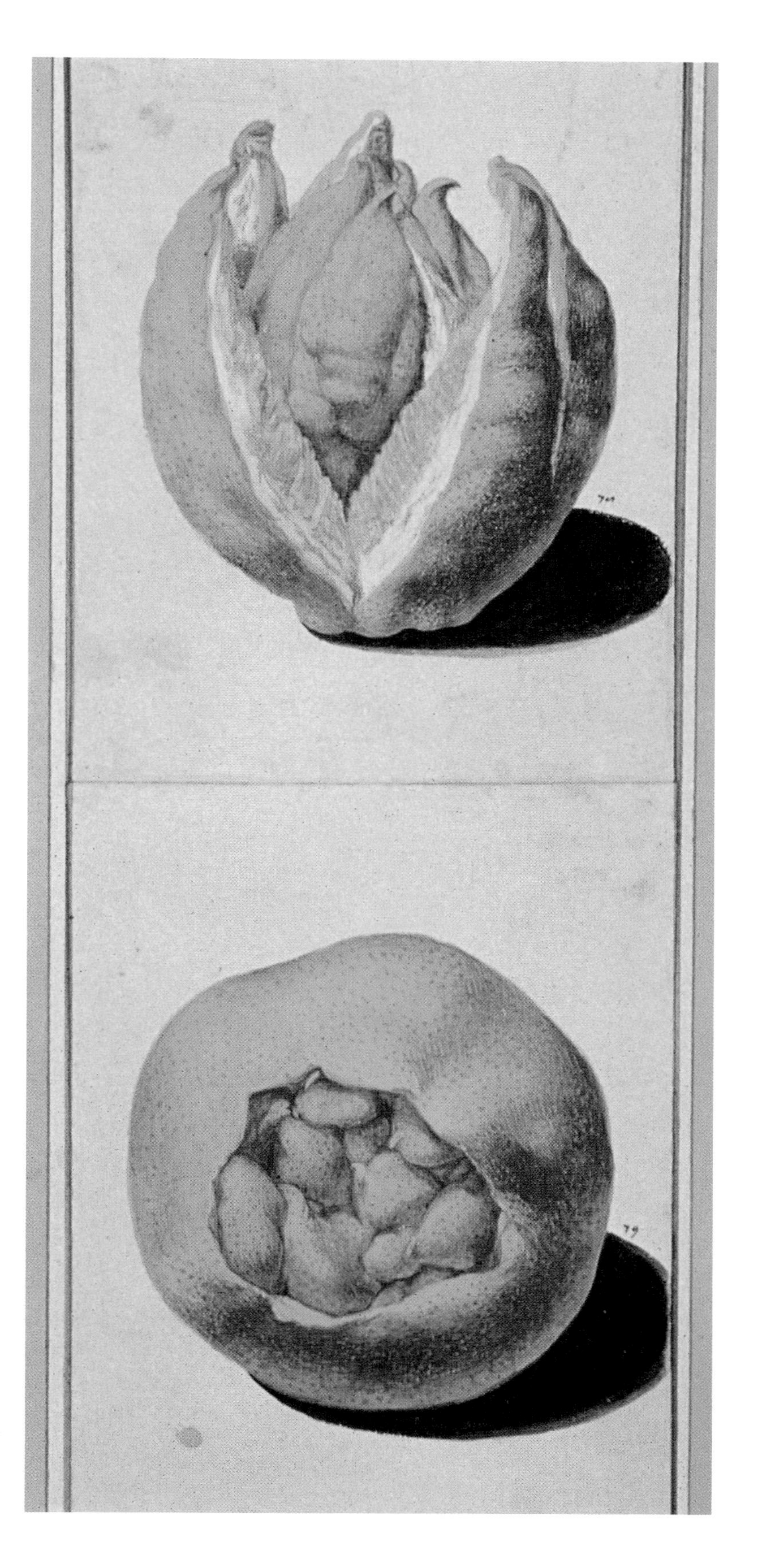

Fig. 1.27. Vincenzo Leonardi, pregnant whole fruit of a citron-lemon. Watercolor and bodycolor with gum heightening and blackened lead white highlights over black chalk, 374 x 141 mm. Sale, Sotheby's, New York, September 16, 1988 (property from the estate of James R. Herbert Boone), lot 149.

Fig. 1.28. Vincenzo Leonardi, flowering citron-lemon (*Limon citratus silvestris*). Watercolor and bodycolor with gum heightening over graphite, 258 x 174 mm. Windsor, RL 19372. The Royal Collection © 2000, Her Majesty Queen Elizabeth II.

beauty and displayed the closest attention to details of color, texture, and modeling of form. Others, such as those in the Institut de France, were of a lower level of artistic and technical skill, even though they too varied considerably in their visual effect. But at the same time they devoted considerably more attention to habitat, environment, growth, and reproduction than anything in Windsor, even the so-called *Erbario miniato* (e.g., figs. 8.27–35).

All in all, the scale, range, accuracy, and careful detail of the drawings that once formed part of Cassiano dal Pozzo's *museo cartaceo* opened a new chapter in the history of science. I had come across well over six thousand drawings of nature never previously studied, let alone photographed. Foucault himself could not have asked for a better corpus to

Fig. 1.29. Vincenzo Leonardi, sweet "pregnant" orange. Watercolor and bodycolor with gum heightening over black chalk, 320 x 202 mm. Windsor, RL 21146. The Royal Collection © 2000, Her Majesty Queen Elizabeth II.

reveal the role of the visual in the study of nature, the changing bases for natural historical classification in the Renaissance and the Baroque periods, and the relationship between observing, naming, and classification.

But there was much work to be done. Here were drawings not only of local Italian specimens, but also of the newest species of animals and plants that were just then being imported from the Indies and elsewhere. There were drawings that examined both the interior and exterior of things, with an intense and analytical detail that seemed quite new.

Why were the drawings made, and what was the driving force behind the whole effort? What criteria determined the representation of one particular specimen and not another; and why the specific concentration on coloristic precision, almost above all else? Here, in short, was an extraordinary attempt to document all of nature—both of the Old World and the New—in visual form; and to classify it. Could it be that just as Cassiano had undertaken the huge task of recording all traces of Roman civilization, from the smallest and most battered objects to the grandest and most famous, he also attempted to do the same for nature? This, after all, had to be an even larger enterprise.

## FERRARI, FLOWERS, AND THE MICROSCOPE

I had come across the drawings serendipitously. For several years I had been writing a book about a seventeenth-century Jesuit priest named Giovanni Battista Ferrari, who was born in Siena in 1583 and died there in 1655.[38] Most of his adult life he spent at the newly built Jesuit college in Rome, in the shadow of much greater men than he.[39] His story is a poignant one, and his career remarkable and unusual—even in those days of unusual Jesuit careers. During the early years of the century he was a classmate of Orazio Grassi, who would soon become one of Galileo's most famous adversaries. When Maffeo Barberini was elevated to the papacy as Urban VIII in 1623, Ferrari had just been elected to the post of professor of Hebrew and Syriac at the College.[40] At the same time, and for many years after, Ferrari also served as professor of rhetoric, to instruct the Jesuit novices and the young noblemen at the College in that still indispensable field.[41] But shortly after Urban became pope (and Cassiano set up his collections in the Via dei Chiavari), Ferrari's career suddenly changed direction. He became a gardener and acted as chief horticultural consultant to the Barberini family in their magnificently rebuilt palazzo on the Quirinal (appropriately sited on the ancient circus of Flora).[42] Barely missing a beat, he designed parterres,[43] introduced and grew exotic plants for the first time in Rome, and turned his rhetorical skills to horticultural purposes.

Here I must ask the reader to permit a digression—but a critical and relevant one. In 1633, the very year of Galileo's trial, Ferrari published his *De Florum Cultura*, the first book ever to be devoted to the cultivation of flowers for solely ornamental and horticultural purposes. It eschewed the pharmacological concerns of the traditional illustrated herbals. Inspired by the gardening craze that just then was sweeping Rome,[44] it was magnificently illustrated with etchings and engravings, mostly of stories made up by Ferrari himself (e.g., fig. 1.30), as well as a number of exotic flowers supposedly from the Indies. Interspersed among these were engravings of garden plans, garden implements, bouquets, and a number of vases and flower pots, one even designed by Galileo's famous Jesuit opponent, Orazio Grassi. The allegorical designs were engraved by that favorite of the Barberini family, Johann Friedrich Greuter,[45] while the carefully descriptive engravings of flowers were almost all by the Dutchman resident in Rome, Cornelis Bloemaert.[46]

One of the chief concerns of the *De Florum Cultura* was its attempt to

Fig. 1.30. Johann Friedrich Greuter after Guido Reni, *The Indies present Neptune with seeds of Indian flowers to take back to the Barberini gaardens in Rome.* Engraving, Ferrari, *De Florum Cultura,* p. 377.

establish firm bases for the taxonomy of the multitude of plants now pouring into Rome from the New World. The old criteria had begun to seem inadequate to this task. For Ferrari, for example, color remained a clear criterion—as in the apparently strange decision to include two plates of what are self-evidently no more than two very similar varieties of what most botanists today regard as the same species (the *Amaryllis belladonna*). One he called *Narcissus indicus liliaceus saturo colore purpurascens* ("a dark purple lily-like Indian narcissus") and the other a *Narcissus indicus liliaceus diluto colore purpurascens* ("a light purple lily-like Indian narcissus") (figs. 1.31–32).[47] For Ferrari these were different species.

Texture played a similar role in helping to determine the differences between things;[48] but even as Ferrari wrote his book, he seemed to become ever more aware of the problem of using such external subjective appearances as the basis for making species distinctions.[49] And so he turned to that brand-new instrument, the microscope, only recently given

Fig. 1.31. Cornelis Bloemaert, *Narcissus indicus liliaceus saturo colore purpurascens* (*Amaryllis belladonna*, Masson). Engraving, Ferrari, *De Florum Cultura*, p. 119.

by Galileo to some of his closest friends, to look within and to examine the reproductive elements of plants.[50]

Throughout his book, Ferrari never gave up his efforts to sort out the names and species of plants that to us post-Linnaeans would sometimes simply seem different *varieties* of the same plant. One of the most striking cases—and certainly the one closest to Ferrari's heart (because, as he claimed, he was the first to grow it in Rome)—was that of the Chinese rose, the modern *Hibiscus mutabilis*. Perhaps misled by its greatly varying number of petals, Ferrari gave it no less than three separate illustrations. One he titled "the leafy Chinese rose"(1.33), the second "the leafier Chinese rose," and the third "the five-leafed Chinese rose" (fig. 1.34). Such were the limits of botanical classification in those days.

To these illustrations, however, Ferrari added two more; and they are

Fig. 1.32. Cornelis Bloemaert, *Narcissus indicus liliaceus diluto colore purpurascens* (*Amaryllis belladonna*, Masson). Engraving, Ferrari, *De Florum Cultura*, p. 121.

of signal importance in the history of science. Realizing that color, texture, and general morphology could not, in the end, be sufficient bases for making distinctions into species, Ferrari decided, rather as Linnaeus would many years later, to cut the fruit into a section (fig. 1.35)—and then examine the seed with a microscope (fig. 1.36). This was the first botanical illustration ever to be made with the aid of a microscope[51]—and it was done in the very year in which Galileo paid the price for the conclusions he drew from his use of the telescope.

These two plates are of different order altogether from the others in the book. In them the focus is narrowed, the presentation much sparer and more overtly didactic than anything else there.[52] They reveal a concern with the insides of things, rather than with surface alone.

In the first (fig. 1.35), a straight section of a fruit is presented beside a

Fig. 1.33. Cornelis Bloemaert, *Rosa Sinensis Foliosa* (*Hibiscus mutabilis* L.). Engraving, Ferrari, *De Florum Cultura*, p. 491.

three-quarter view, seen slightly from above. This is the only illustration in the book to show different viewpoints—let alone a section of so small a part of a plant. But the second plate (fig. 1.36) is even more atypical, dramatically so. It represents a dark seed, shown not just from two but from three viewpoints, represented almost as slightly sexualized Rorschach blots, the seed's tufts of hair shown as spiky excrescences neatly radiating from each image. There is a kind of geometry here, a blatant spareness absent from every other illustration in the book. And the scroll proclaims its secret: "the same seed presented in threefold view beneath a microscope."

The chapter preceding this illustration is titled *A wonder of Nature greater than those of art*[53] and in it Ferrari explains why he chose to exam-

Fig. 1.34. Cornelis Bloemaert, *Rosa Sinensis Foliosior* (*Hibiscus mutabilis* L.). Engraving, Ferrari, *De Florum Cultura*, p. 489.

ine the Chinese rose. Not only had he been the first to grow it successfully in Rome,[54] it also presented a critical problem in classification. As he knew only too well, this was a plant that changed color from white to flaming red to purple in the course of a single day. It also occurred in a puzzling variety of forms. In both ways, Ferrari felt, nature rose to the level of art, and to its challenge.

The chapter begins typically enough. It starts with the "sweet"—what all the rhetoricians then called the *dulcis*—before proceeding to the "useful," the *utilis*. This will be followed by hard investigation. First Ferrari makes up pleasant narrative (a poetic fable, as he calls it), set in the Barberini gardens, as one way of accounting for the coloristic changes of the Chinese rose, which he describes at extraordinary length.[55] But then he

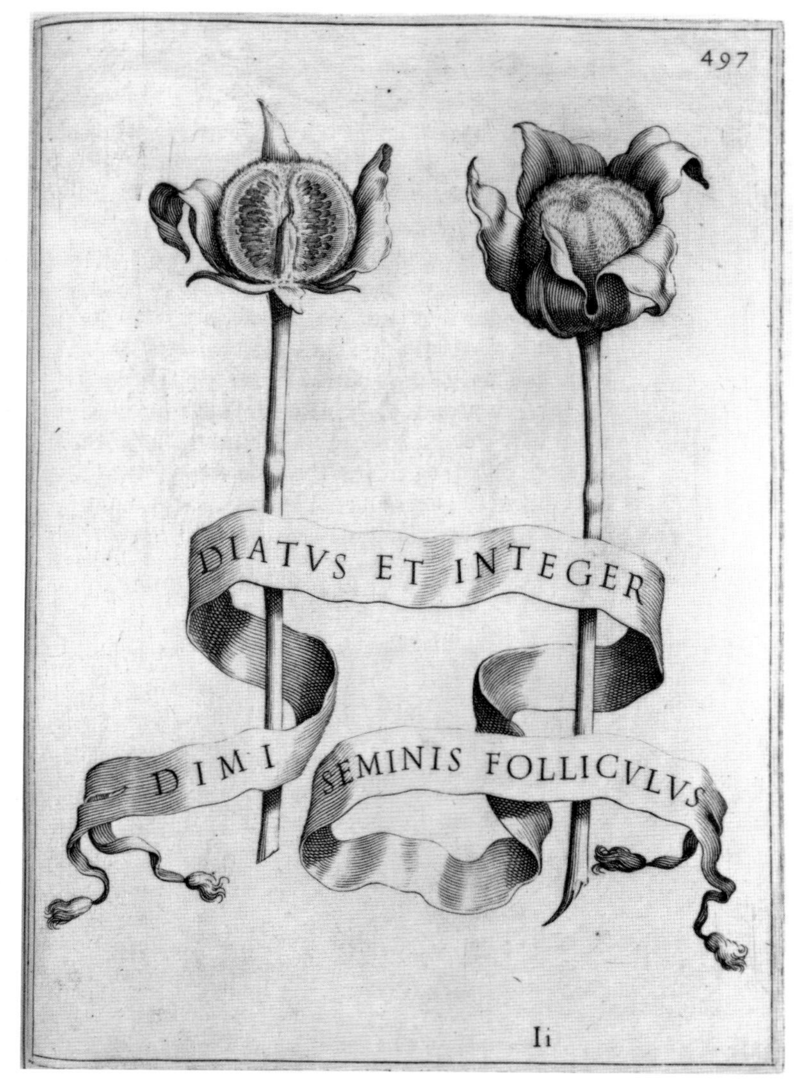


Fig. 1.35. Cornelis Bloemaert, fruit and section of fruit of a Chinese rose (*Hibiscus mutabilis* L.). Engraving, Ferrari, *De Florum Cultura*, p. 497.

realizes the inadequacy of this strategy. As if sharpened by his awareness that a piece of rhetorical invention, however charming, could not satisfy the need for an explanation of something so problematic, Ferrari discards his narrative and sets aside his panegyric on his favorite flower. Suddenly he abandons sentiment in favor of analysis, in a more much forthright way than one might have expected from a professor of rhetoric. After all, how can a "poetical" explanation really account for such strange and apparently inexplicable things as the coloristic changes and structure of the Chinese rose?[56]

And so Ferrari has to adopt some other approach. Texture, for a start, will no longer do. "Now," he declares, "in order to examine the very fine natural roughness of the petals, we were not content solely to use our hands; and so, to make it capable of being minutely investigated with the

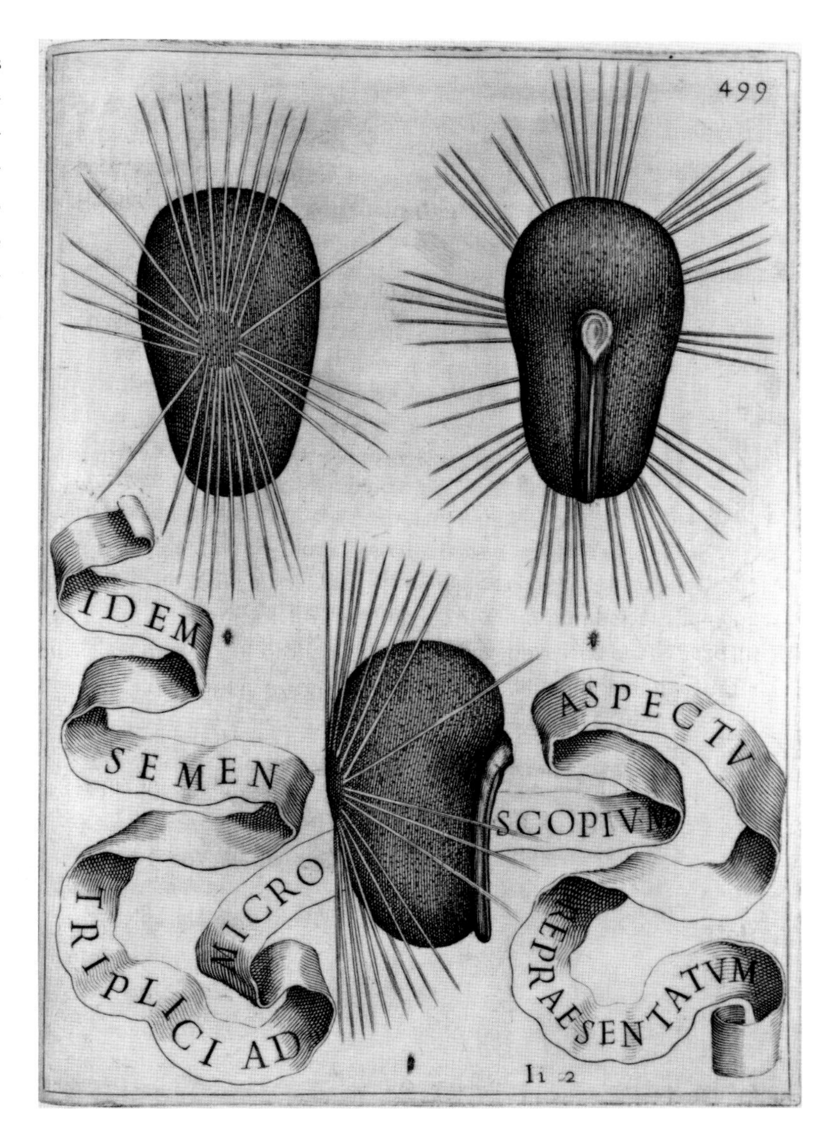


Fig. 1.36. Cornelis Bloemaert, seed of a Chinese rose (*Hibiscus mutabilis* L.) seen under a microscope. Engraving, Ferrari, *De Florum Cultura*, p. 499.

eyes, we applied the Lincean explorer, that is to say, the microscope, a small viewing glass enclosed in a tube."[57]

In the fine 1638 translation of the *De Florum Cultura*, this passage runs slightly differently. Not only is it less precise, all mention of the society that had supported Galileo through some of his most difficult struggles (and of which he was himself a proud member) is expunged. Only one subtle recollection of that association is preserved by the translator's clever allusion to the famous phrase that Galileo himself had applied to the microscope, namely "that kind of lens in a tube *which makes very small bodies look very large*, and can show each part distinctly."[58]

Having realized that he could not rely on pure externals for his classification, Ferrari had decided to look deeper within; and so, putting his microscope to use, turns to discuss its seeds. Here his precision takes one by

Fig. 1.37. Cornelis Bloemaert after Nicolas Poussin, *Nymphs of the Hesperides Presenting Citrons to the Deities of Lake Garda*. Engraving, Ferrari, *Hesperides* (1646), p. 97.

surprise. As if perfectly exemplifying what Foucault called the age of *mathesis*, Ferrari begins to count. He opens a seed-pod of the Chinese rose, and counts its seeds. He finds exactly 163, divided into six compartments.[59] And with the patient task of counting over, Ferrari realizes that he has to look more closely at the structure of each seed, in the hope that somehow the seed itself might contain the secret of the flower's mutability. It would be left to the society he was never asked to join, the Academy of Linceans, to pursue the problem to its limits.

## ORANGES AND LEMONS

For all the importance of his book on flowers, the work by Ferrari that interested me most was his *summa*, namely the *Hesperides, or On the Cultivation of the Golden Apples*, of 1646. This was (and for a long time re-

Fig. 1.38. Cornelis bloemaert after Andrea Sacchi, *The Transformation of Harmonillus*. Engraving, Ferrari, *Hesperides* (1646), p. 74.

mained) the largest work ever to have been devoted to citrus fruit, and certainly the most prodigally illustrated. Like the *De Florum Cultura*, it was adorned with engravings after some of the finest artists of the day. Pietro da Cortona, Francesco Albani, Andrea Sacchi, Francesco Romanelli, Giovanni Lanfranco, Domenichino, and Poussin each provided an allegorical plate either to show the arrival of different kinds of citrus fruits in Italy (e.g., fig. 1.37) or to account for the origins of digitated and "pregnant" specimens (fruits enclosed one within the other; cf. figs. 1.38–39). But my main concern was with the botanical plates. Whereas the *De Florum Cultura* had only eighteen pages of flowers, the *Hesperides* had no fewer than eighty-two such plates, often with several specimens or parts of specimens to a page. With high scrupulousness, the etchings seemed to convey every textural element of skin and pulp, every lump, rugosity, ridge,

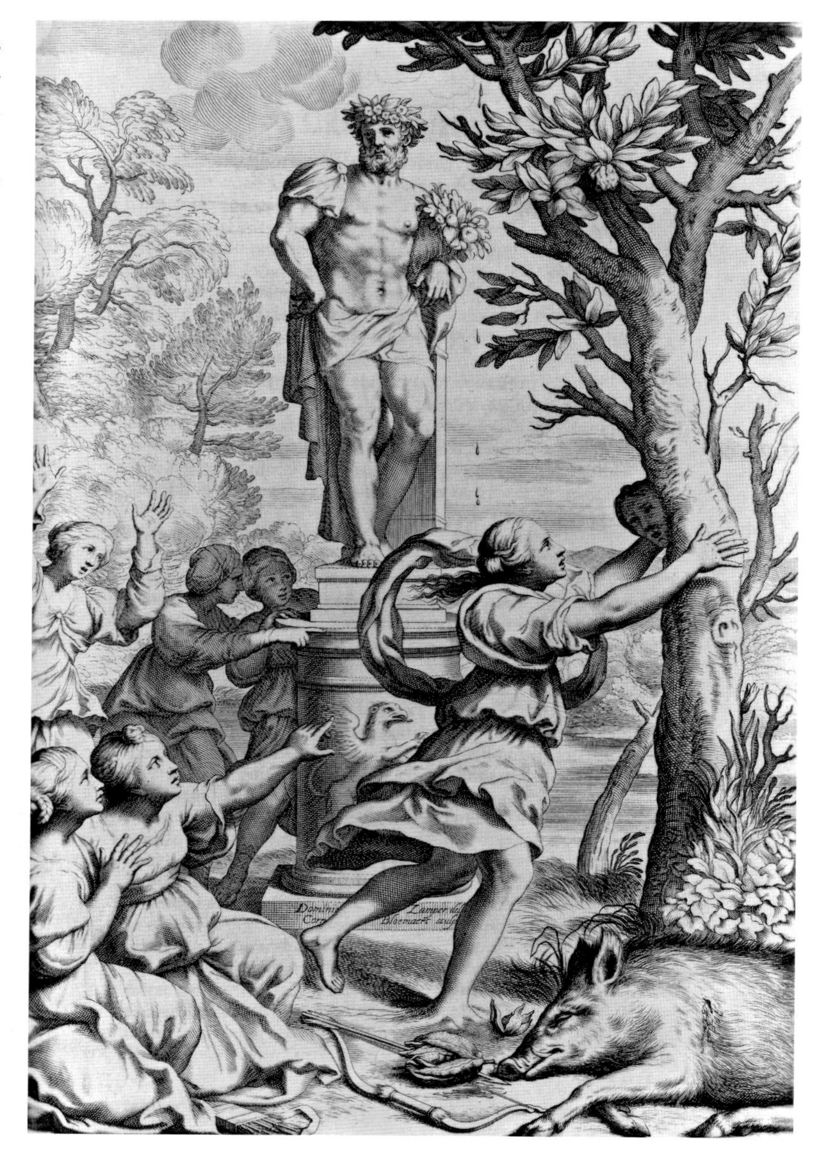

Fig. 1.39. Cornelis Bloemaert after Domenichino, *The Transformation of Leonilla*. Engraving, Ferrari, *Hesperides* (1646), p. 418.

and furrow. It is impossible to think of a single other book in which the members of a single genus are represented with such care in differentiating one from the other, or with such careful attention to the feel of things. Equal attention was bestowed on leaves and flowers, on sections, segments, seeds, and stalks. Often the specimens seem to be anatomized, dissected, and shown from several viewpoints, as if presented for scientific and classificatory inspection. Above and around the engravings of fruits, beside and below them, are graceful banderoles, just as in the book on flowers, containing the specific name of each fruit. No such detailed taxonomy of citrus fruit had even remotely been produced before—and has hardly been equaled since.[60]

Once more the relationship between illustration and classification was

Fig. 1.40. Cornelis Bloemaert after Vincenzo Leonardi, chimeric citron and sour orange with excrescence (*Aurantium callosum*). Engraving, Ferrari, *Hesperides* (1646), p. 411.

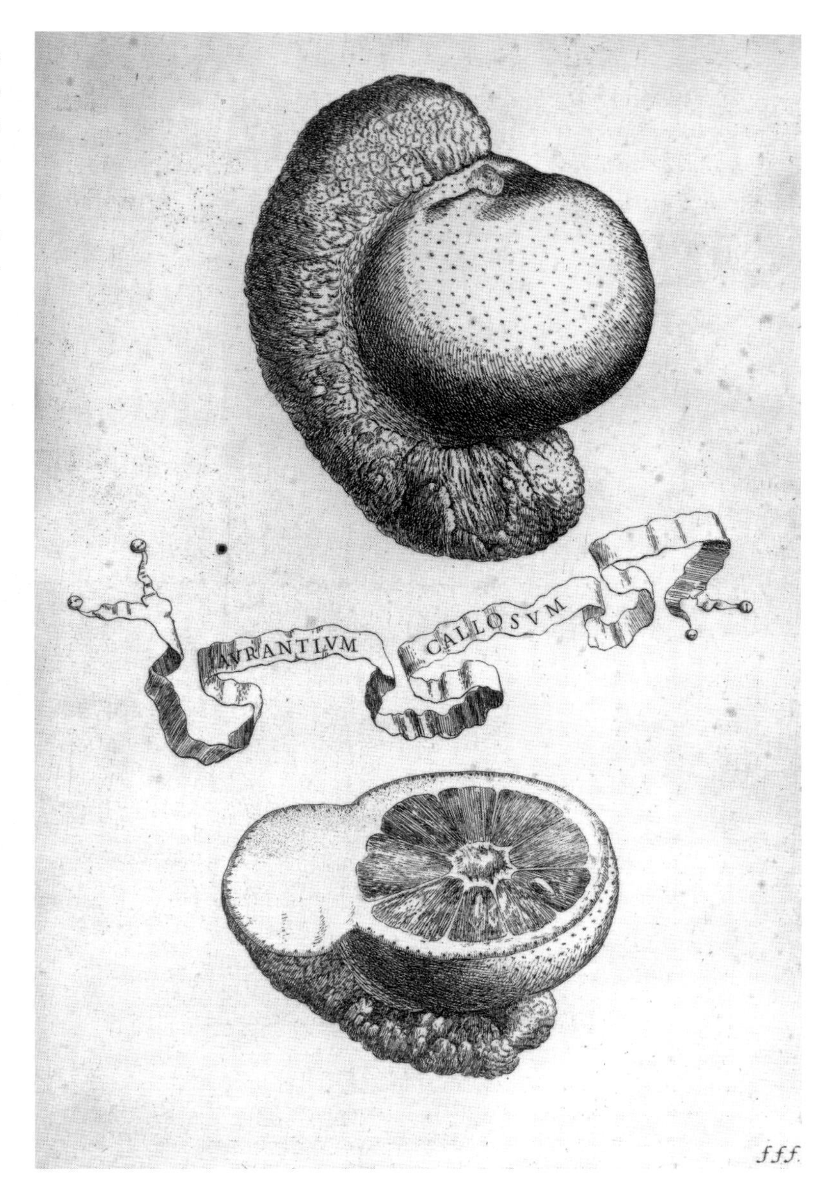


very close. But what impelled the classificatory and taxonomic drive, and how was it aided by visual reproduction? The real problems arose with specimens that seemed to stand on the borderlines between species—and sometimes even genera.[61] How did Ferrari know that this specimen was just an ordinary lemon and this one a citrated lemon (to say nothing of what he dubbed to be the pseudo-citrated lemon)? How did he distinguish between borderline species or establish a classification for fruits that seemed to partake of more than one class, whether species or genus?[62] These were questions that haunted me. But first I resolved to find the preparatory drawings for these plates.[63]

A recollection of a discussion I once had with Anthony Blunt, the fa-

Fig. 1.41. Cornelis Bloemaert after Vincenzo Leonardi, misshapen citron (*Malum citreum multiforme*). Engraving, Ferrari, *Hesperides* (1646), p. 77.

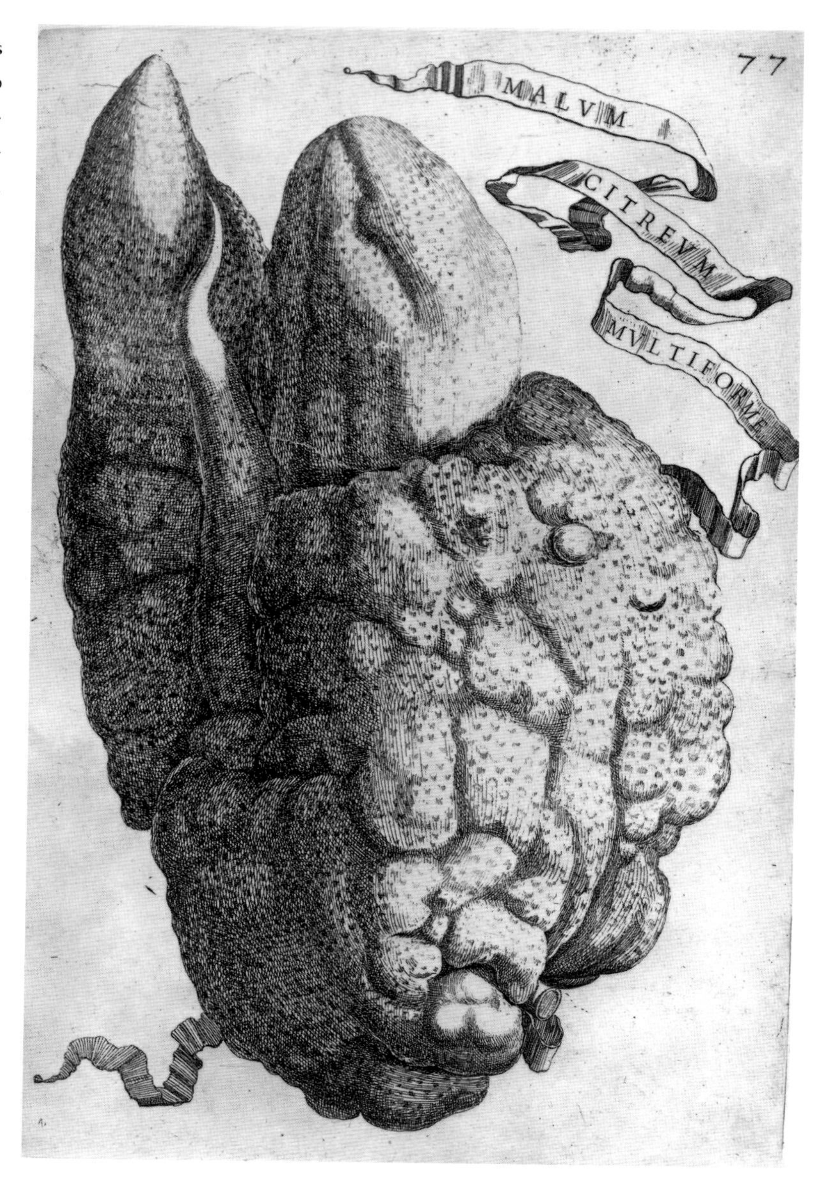


Fig. 1.42. Cornelis Bloemaert after Vincenzo Leonardi, fingered lemon (*Limon a rio seu rivo*). Engraving, Ferrari, *Hesperides* (1646), p. 215.

mous art historian, disgraced spy, and former Surveyor of the Queen's Pictures, led me to Windsor Castle. And it was there, after some searching, that I found the citrus drawings (figs. 1.23–29). They were not in black chalk or in pen and ink, as I expected, but rather in brilliant watercolor, finished with white gouache heightening and delicate varnishes. Not all of them seemed to me to relate to the plates in the *Hesperides*, but many undoubtedly did. Some were more or less ordinary specimens, but quite a few were breathtaking: prodigious lemons with long fingerlike extrusions, and an extraordinary orange with a strange growth on it. Its warty excrescence, turning from orange to green and various nacreous colors, looked so familiar that I realized immediately that it must have

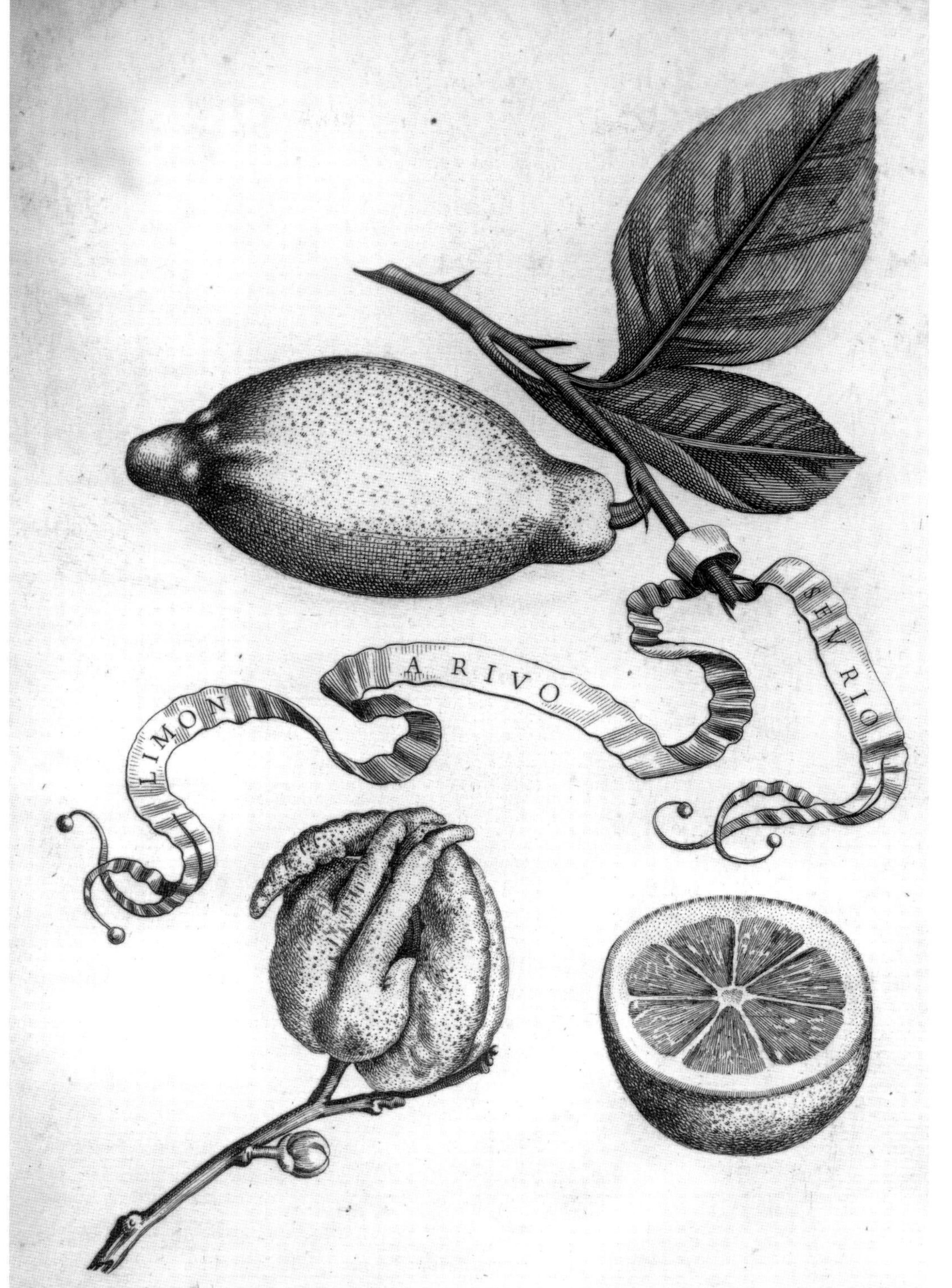
LIMON A RIVO SEU RIO

LIÆ FORMÆ
CITRATI
LIMONIS
ALIOS
INCLVDENTIS

been the preparatory design for one of the most striking plates in the *Hesperides* (fig. 1.40). Thus it became possible to correlate most (though not all) of the citrus drawings with the relevant illustrations in Ferrari's chef d'oeuvre (cf. figs. 1.24, 1.26-27 and 1.41–43; as well as 1.39).

## THE ARTIST: VINCENZO LEONARDI

But who did these drawings—to say nothing of the others in Cassiano's paper museum? For many years inconclusive arguments had raged about the possible authors of the antiquarian drawings. It seemed even less likely that the natural historical drawings would yield a name. But they did.

Early on in the *Hesperides*, Ferrari composed a paean to the painter who had provided the botanical drawings for his book. This man had wrought a miracle, Ferrari rhapsodized, "because the same things are born on these pages as in the soil"; and since "what he painted for this

Fig. 1.43. Cornelis Bloemaert after Vincenzo Leonardi, "pregnant" citron-lemon (*Aliae formae citrati limonis alios includentis*). Engraving, Ferrari, *Hesperides* (1646), p. 271.

Fig. 1.44. Etching after Vincenzo Leonardi, *Pettirosso*. Olina, *Uccelliera* (1623), p. 15.

Fig. 1.45. Vincenzo Leonardi, robin (*Erithacus rubecula*). Watercolor, bodycolor, pen and brown ink over black and red chalk, 135 x 192 mm. Windsor, RL 27626. The Royal Collection © 2000, Her Majesty Queen Elizabeth II.

Fig. 1.46. Vincenzo Leonardi, female Italian sparrow (*Passer Italiae*). Watercolor and bodycolor over black chalk, 185 x 206 mm. Signed on the verso: *Vin. Leonardi F. 1629*. Windsor, RL 27628.The Royal Collection © 2000, Her Majesty Queen Elizabeth II.

volume had brought forth real fruit."[64] And in the margin, Ferrari named the artist as Vincenzo Leonardi. At last one had a name.

We know nothing else about Leonardi (except that he was perhaps Sienese, like Ferrari himself);[65] but his name came up twice again, in unexpected places.

In his preface to an exceedingly discursive book on birds titled the *Uccelliera* or *Aviary*, which appeared in Rome in 1626 under the name of a certain Giovanni Pietro Olina, the publisher recorded that the drawings for the plates "were most diligently done by Vincenzo Leonardi." Subtitled *Discourse on the Nature and Distinctive Characteristics of diverse birds, and in particular of those which sing, together with the way of catching them, recognizing them, raising them and maintaining them,*[66] and illustrated with engravings by Antonio Tempesta and Francesco Villamena, the book included at least one illustration of a bird that was clearly based on a drawing in Windsor Castle—the charming portrait of a robin (figs. 1.44 and 1.45). Could it be that the same Leonardi who did the citrus fruit was also the author of at least some of Cassiano's splendid ornithological drawings?

This question was soon answered. Shortly after the rediscovery of Cassiano's drawings in Windsor, a decision was made to lift a small drawing of a female Italian sparrow from its mount (fig. 1.46).[67] On its verso was Leonardi's signature and a date: *Vin. Leonardi F[ecit] 1629.*[68] Though the drawing was not related to any illustration in Olina, it was by the same hand as the other bird drawings in Windsor Castle. Finally there was a name to attach to two of the finest groups of natural history drawings in Windsor Castle. The best artist in the whole natural historical corpus had emerged. There could now be no question that Leonardi was the author of many of the other natural history drawings in Cassiano's paper museum, including the magnificent representations of the pineapple, the broccoli, the dolphin, the porcupine, and so on and so forth, all done with the same sensitivity to color and texture (and the same technical skill in watercolor) as the birds and citrus fruits.[69]

## CASSIANO'S ROLE

The circle, it seemed, was closing. The books by Olina and Ferrari had provided the name of the finest of Cassiano's artists; and at the same time they cast a sharp light on Cassiano's role as a natural historian and as a scientific facilitator,[70] as well as on at least some of his purpose in collecting the drawings. They also led me to the heart of this story.

First Ferrari. He will not be not be a major player in what I have to tell. He stood on the periphery of the dramatic events of his lifetime, and his chief work, the *Hesperides,* appeared long after the commotions around Galileo had died down. But Ferrari knew all the players, on both sides, and at many levels. He shared several of the preoccupations of the main figures in this narrative: the preoccupation with taxonomy and classification, an obsession with how biological reproduction works or goes astray, a commitment to the use of visual reproduction as a critical aid to the study and ordering of nature.

Cassiano is mentioned only twice in the book on flowers. And yet without him it probably would not have existed. The first time Cassiano is referred to is in connection with one of Ferrari's favorite flowers, the Indian jasmine. This flower, Ferrari wrote, was best to be seen in the Barberini gardens, "upon which the Urban Sun shines so felicitously, the same that shines so majestically and resplendently in the Vatican."[71] And who was the first to introduce so lovely a plant to the Barberini gardens? Cassiano dal Pozzo, of course, "a person abundantly endowed with the gifts of body, soul, and fortune, prerogatives that otherwise are rarely united together." And who sent the plant to Cassiano? None other than his learned friend and correspondent in France, the famous antiquarian from Aix, Nicolas Claude Fabri de Peiresc, whom Ferrari briefly praised too.[72] Cassiano and Peiresc were united in their natural historical as well as their antiquarian interests, and we know from their correspondence that he sent several other drawings to Cassiano as well, including the drawings of a flamingo he had made in 1627–28.[73]

But was the connection with the Indian jasmine the only reason to bring in Cassiano? No. The second reference to Cassiano was in the context of a gilded painting made up of flower petals and painted colors, apparently so well done that one could not make out whether it was a production of nature or art—the very dilemma that animated much of his discussion in this book:

> Such a painting I have seen at the house of Cassiano dal Pozzo, a man notable not only for the purity of his blood, but for his choice and well-stocked library, and for his museum full of rare things; and above all for his rare gifts of character, and for the sweetness of his soul, with which he is able to captivate everyone. No surprise, therefore, that such sweetness should be welcomed and loved by the Barberini Bees. . . . Thanks to him, this work of mine, that is, the flowers of the garden, can now acknowledge the benevolence with which they have been welcomed and honoured by the reigning Bees.[74]

This is a courteous acknowledgment of a famously courteous man, in terms that are typical of Barberini rhetoric (the bees, the Barberini symbol, will return again and again). Even so, the relatively limited mention of Cassiano in the book is surprising. We can be almost certain that it was he who introduced the otherwise obscure Jesuit father to the Barberini family, in particular to the man who actually seems to have paid for most of the book, Francesco Barberini. And it was Francesco who would soon turn against Galileo with even more vigor than his uncle.[75]

When it came to the *Hesperides*, Cassiano played an even more direct role in the preparation and funding of the work. But still he is only mentioned twice, with no reference whatsoever to his critical role in producing it. The first time is in connection with the plate designed by Poussin showing the Hesperides presenting citrons to Lake Garda (fig. 1.37). Just as in the case of the other allegorical plates in the book, Ferrari offers an expansive passage on the talents of the artist who designed it. And it is here that he first mentions Cassiano, not only for his generosity but also for his keen recognition of artistic talent.[76] The second time Ferrari refers to Cassiano is in the course of a longish chapter on lemon sherbet. Two recipes for it had caught Ferrari's fancy: one by that intrepid early seventeenth-century traveler to Turkey, Persia, and Mesopotamia, Pietro della Valle, and the other, needless to say, by Cassiano himself.[77]

This is all very well (and diverting too); but given Cassiano's central and intense involvement in the production of the book, the meagerness of these references is strange. Not only had Cassiano helped fund the book; he had gathered much of the information in it too. A thick volume of notes on citrus fruit was put together by Cassiano, and it survives in the library of the Accademia dei Lincei in Rome.[78] It reveals that Cassiano himself wrote to friends and correspondents all over Italy, sometimes in the most out-of-the-way places, asking for information about every aspect of the culture and cultivation of the distinctive citrus fruit of each region. This information was presented under a series of fixed headings—or "capi d'informatione" as they were called—prepared by Cassiano and Peiresc in consultation with the well-read Jesuit father himself. From the same set of notes (and from Ferrari's letters to Cassiano, also preserved in Rome),[79] we learn that Cassiano assembled many of the citrus drawings, as well as several of the designs for the allegorical plates, including the drawing by Poussin—for which, of course, he paid himself.[80] The whole conception of the work, as well as the classificatory thrust and much of the actual research, was Cassiano's. By 1646, when the *Hesperides* appeared, Urban was dead, the Barberini family in disgrace, and their protégé Ferrari exiled to Siena. There could be no question any more of a subvention from his old employers and supporters; but even then Cassiano tried to help Ferrari, now exiled to Siena, to find a sponsor. Their chief hope lay with Louis XIII of France (Poussin spoke with his closest counselors on behalf of Cassiano and Ferrari);[81] but in vain. In the end, we do not know who provided the final subvention for the book. Perhaps it was someone else prevailed upon by Cassiano; perhaps it was Ferrari's own hometown of Siena, where documents still survive recording his gratitude to the city.[82]

But why was Cassiano's role in producing these books suppressed? The question is even more puzzling in the case of Olina's *Aviary* of 1622. This otherwise obscure work leads us on to the major player in our story. The book is even more directly the result of Cassiano's own researches than the *Hesperides;* and in it he is even more concerned with the quality of the illustrations. This, in short, is the work that Cassiano offered as his presentation piece (and as evidence of his commitment to pictures as an aid to scientific research) when Prince Federico Cesi inscribed him as a member of the Academy of Linceans in 1622.[83] For several years before, Cassiano had worked hard with Cesi and three other members of the Academy—Giovanni Ciampoli, Johannes Faber, and Virginio Cesarini—in editing the manuscript of Galileo's *Assayer* (*Il Saggiatore*) and preparing it for publication,[84] and Cesi valued his work so much that he asked him to join their still small Academy.

Cesi was delighted with the *Uccelliera*. In a letter of October 4, 1622, he thanked Cassiano for the book and said that he believed that it would serve as a great aid to the study of natural history. This, after all, was itself the basis of good science: "la buona e non mascherata filosofia" is how Cesi puts it in a significant phrase—in other words, good and not fake ("masked") science. The engravings in the book would appeal to all students of natural history, but especially to their own colleagues.[85]

Once more, however, the question arises: why did Cassiano want his name to be suppressed in the *Uccelliera*, if not as its actual writer, then certainly as its compiler? What was being masked? Was it simply that this kind of modesty was becoming to a gentleman-virtuoso—or were there other reasons? As in all cases of dissimulation, the matter is not easy to resolve; but the context is clear enough.

The *Uccelliera* is something of a farrago of all kinds of ornithological information, presented in a most unsystematic way. Much more than in the case of Ferrari's *Hesperides*, it veers from the promiscuously encyclopedic to the taxonomic. As in the *Hesperides*, there is a great deal on habitat. Bird behavior receives a good amount of attention. There are chapters on everything from methods of hunting and trapping to culinary and musical details (including recipes for nightingale soup and clues on how to identify birds by their song). It is as much a book on bird lore and bird myth as it is on anything that we might now call ornithology.

But contact with Cesi and the other Linceans helped Cassiano trim away some of the incidental aspects of his subsequent ornithological research—in other words, some of the folkloric, culinary, and fabulous and emblematic material of the kind that characterizes the work of writers like Aldrovandi and is still present in the *Uccelliera*. With the occasional help of his brother Carlo Antonio, Cassiano went on to prepare at least four immensely detailed ornithological treatises, or *discorsi*.[86] One was about a toucan he saw in the collections of Louis XIII when he visited Paris in 1625 (in the course of diplomatic legation led by Francesco Barberini to France and Spain).[87] It was originally accompanied by a drawing of the bird seen by Skippon in 1663.[88]

The other three *discorsi* are on the bearded vulture, a pair of ruby-throated hummingbirds sent by a Jesuit from Canada to Rome, and the Dalmatian and European pelicans (cf. fig. 1.5).[89] On the basis of specimens sent to him from Ostia in 1635, Cassiano made a point of distinguishing between two species of pelicans (*onocrotalus*) that had generally been conflated. Three years later he dissected one of the white pelicans sent to him from Ostia in order to determine its sex.[90]

The detailed drawings of a white stork and its beak, leg, and feathers that survive in Cassiano's collection (figs. 1.3–4)[91] must also have been intended to accompany a treatise on that bird, while the drawings of a flamingo (*phoenicopterus*)—also mentioned by Skippon—turn out to have been commissioned by Peiresc and was later given to Cassiano when he visited France in 1625.[92] Typically, they show both the whole animal and its parts. The flamingo is shown both at rest and in flight; while its head, beak, webbed feet, wings, feathers, and both sides of the tongue are represented with the usual high attentiveness to form and to color (cf. fig. 1.7).[93] When we consider Cassiano's interests in color and in matters such as the flight of birds, it does not come at all as a surprise to discover that from the early 1630s on he was actively engaged in a plan to gather and publish an edition of the manuscripts of Leonardo da Vinci.[94]

Cassiano's work in the field of ornithology (perhaps the natural historical area closest to his heart) offers considerable insight into his approach to natural historical problems and into his working methods. It makes very plain his profound commitment to the use of images in the study of nature.[95] However much apparently irrelevant matters—about hunting or cooking or clothesmaking, for example—sometimes come to the fore, his descriptions are always firmly rooted in firsthand knowledge;[96] and in this they differ significantly from the earlier ornithological compendia that were at least as dependent on lore and on written authority as on empirical study. In the *Uccelliera* to a certain degree, but above all in the series of minutely detailed and intensely observed *discorsi*, the emphases are on direct observation, habitat, behavior, and the colors and coloristic nuances of the birds. All this may not in the end have resulted in any particularly fruitful conclusions; but it certainly produced some of the finest images in the whole corpus of drawings collected by Cassiano.

What, however, of the other drawings, the drawings clearly *not* by Leonardi? After all, there are hundreds of others in the corpus, including the two herbals in Windsor, the loose sheets associated with them, the fossil drawings in Windsor, and the whole body of material in the Institut de France. All these are of a considerably lower technical and artistic level and by a number of different hands. Some are rather amateurish, others more skilled, but almost all lack the scale and sure finesse of the certain drawings by Leonardi. Nevertheless, they are arguably of much greater significance in the history of science.

## CASSIANO'S DRAWINGS AND BELIEFS

Cassiano set his paper museum on secure foundations when he moved from the house he shared with his brother on the Via della Croce to the larger one down the road from Sant' Andrea della Valle in the Via dei Chiavari in 1626.[97] In 1630 his dear friend Federico Cesi died unexpectedly. After three years of negotiations with his widow, Cassiano finally acquired all the books, manuscripts, drawings, and scientific instruments that constituted Cesi's library and museum.[98] He made the purchase not just to help Isabella Salviati, the widow, but above all to ensure that the

Lincean heritage—now under great threat because of Galileo's trial—would not be lost.

At the same time Cassiano made the single largest addition to the natural history section of his own museum on paper. He acquired from Cesi's estate the volumes on fossil woods, the two illustrated herbals, and the eight volumes of fungi and plants now in the Institut de France.

Cesi's hand may still be recognized in the annotations to the fossil volumes in Windsor and those in the Institut de France. As soon as he acquired them, Cassiano set about having a scribe provide all the volumes with alphabetical indices of what they contained and with transcriptions of Cesi's brief annotations on many of the drawings.[99]

When Cassiano died in 1657, the Florentine antiquarian Carlo Dati wrote a long eulogy for him, and it is this (published in 1664) that offers us the fullest contemporary insight into Cassiano and his collections. But it is not always entirely straightforward. In it, Dati confirmed the cold documentary evidence for the acquisition in these warm terms:

> Nor is this the only debt which the Academy of Linceans owes to Cassiano, for after the sad death of their great Founder, he gathered together all Cesi's records and writings in his museum, without any regard to cost; and in his heart he held all the plans and thoughts of so learned a group, the life of which, already pitifully languishing, he thus prolonged. Even more than this, he saved it from future accident, and by his own virtue made it immortal.[100]

Thus Dati recorded Cassiano's devotion to Cesi and the Academy he founded. But what was the "other debt" the Academy had to Cassiano? Dati mentioned it next:

> A noble trophy of Cassiano's generosity that will last through the centuries is the precise Compendium which he had made on his journey to Spain of the Natural History of Mexico, gathered in eighteen volumes by Francisco Hernández at truly royal expense and, I would say, almost superhuman foresight by Philip II of Spain. Everyone knows how much light those erudite Academicians derived from this, when they published in Rome an illustrated version of the epitome made by Nardo Antonio Recchi of this same natural history.[101]

When the Linceans were not engaged in supporting Galileo, or involved in one of their own natural historical projects, the primary task they set themselves was the completion of a vast illustrated natural history of Central America. It was based on notes and illustrations collected in Mexico between 1571 and 1578 by the personal physician of Philip II of Spain, Francisco Hernández. Some of this material—including the Aztec herbal in Windsor Castle—Cassiano had ordered to be copied when he went to

Spain in 1626. A whole series of further copies of the original drawings had already been acquired by Cesi from the descendants of Recchi on a visit to Naples in 1610—just before he met Galileo for the first time.

But Cesi had a further influence on Cassiano. Not only did "The Duke of Acquasparta, the clearest light of our century, and the founder of the Accademia dei Lincei, choose Cassiano as one of his first companions, to help him in his aim of compiling the story of nature," he also helped him "to examine the purpose and composition of every created thing by means of physical and chemical experiments." One wonders indeed where Cassiano found the time to write so many letters,[102] organize his museum, conduct his natural historical researches, and do scientific experiments in his laboratory. As Dati noted, "Cassiano deeply immersed himself in these curious and subtle speculations, and did not pass over any of those means by which he might be arrive at so noble an end."[103] But what did Dati really mean by this? Was he *only* referring to the intensity of Cassiano's devotion to the Lincean project, or did it imply something more?

Probably the latter. Throughout his tribute to Cassiano, Dati went to some length to insist that Cassiano was *not* engaged in any kind of heterodox science, that he was *not* an atheist or heretic, and that his researches were purely motivated by the desire for knowledge.

> He who enters into the viscera of the earth to see the metals, minerals, stones and gems; who observes the plants, quadrupeds and serpents on the surface of the earth; he who immerses himself in the sea in order to contemplate the fish and other marine things; who lifts himself into the skies to wonder at so many different kinds of apparitions and generations, the birds and so many flying insects; and then perceives in them the mechanisms and the harmony of each smallest particle so well adapted to the whole—how could he then, seeing everything so infused with Providence and Divinity, how could he not detest Atheism, and inebriated by the grandeur and bounty of the Highest Power, not preach it constantly, and proclaim it, just as did our Cavaliere Cassiano?[104]

All this is just a little too strenuous. Dati seems to be trying too hard to clear Cassiano's name from the taint of heterodoxy. Why drag in atheism, for example? Certainly, Cassiano was on the closest terms with many of the well-known group of intellectual libertines in Rome;[105] yet at every stage of his biography, Dati insisted on the orthodoxy of all his activities, especially those on behalf of Cesi and his Lincean colleagues. The eulogistic tone of Dati's biography—if it can be called that—belies his awareness of the subversive nature of many of Cassiano's activities, both of the experiments he did in his own laboratory and the work he did with and on behalf of the Linceans.

But first we must ask what it was about Cesi that inspired the devotion apparent from Dati's *Life* of Cassiano. What were the aims of the Academy Cesi founded so hopefully in 1603, and how did it develop? The story has still not been fully told. While Galileo's contributions to the history of science have long been appreciated, the Linceans' have not. By the time Galileo was defeated, they too were spent, and Cesi was dead. Only with the rediscovery of the drawings they made can we finally begin to recover the full measure of the Lincean achievement.

## APPENDIX: A BRIEF NOTE ON THE PROVENANCE OF CASSIANO'S DRAWINGS

For the most part, the drawings after antiquities in Cassiano's paper museum ended up in Britain. Having passed from the Dal Pozzo family to that of the Albani in 1703, they were bought by the architect brothers Robert and James Adam for George III of England in 1762, from an ever more impecunious and spendthrift Cardinal Albani. Most then entered the British Royal Collection, and there, for the most part (having been rebound and remounted), they remained. Several hundred drawings (of antiquarian subjects) ended up in the British Museum, while four volumes of architectural drawings, kept by the Adam Brothers for themselves, were bought by Sir John Soane and kept in his own museum, where they survive to this day.[106] All these drawings were known to scholars, and had not, for the most part, gone unnoticed.[107]

But what of the natural history drawings mentioned by Skippon? Though also mentioned by Cassiano's friend and biographer, Carlo Dati, as early as 1664,[108] no one seemed to have studied them at all, let alone attempted to find out where they were.[109] And, when they started to reappear, their history turned out to be a little more complicated than that of Cassiano's drawings of classical art and antiquities.

Very many had indeed entered the British Royal Collection, along with the drawings of antiquities purchased from the Albani family. They were looked down upon from the start. Just before his purchase of the Albani volumes for George III, James Adam admitted to his brother Robert

> I am far from saying that these volumes are all interesting. This is not the case. In such a vast collection much rubbish must be expected. The mosaics, paintings, and bas-reliefs of the primitive church or the first stages of Christianity will entertain you little, though there is some curiosity even in this subject. But . . . there is a collection of drawings after antiquities, bas-reliefs, altars, tripods, cenarie [sic] and urns that I believe you will own to be most valuable.[110]

The natural history drawings were not even *mentioned* by Adam. Could it be that they came under the category of "rubbish," as he put it?

In fact, this seems to have been very much the case when it came to the judgment of Sir John Fortescue, Royal Librarian in the teens and twenties of this century. Apparently he felt that fifteen of the volumes of natural history and one volume of antiquities were either "uninteresting" or "unimportant," and so he sold them in the course of a few years just after the First World War. Wishing to raise money to finance the upkeep of the Royal Library, he did not think that drawings of plants and animals were quite as prestigious as drawings of antiquities (to say nothing of the great treasures of the Royal Collections, including the staggering assemblage of Old Master drawings).

Most of the natural history volumes were bought—either directly or indirectly, we do not know—by a London book dealer, Jacob Mendelson. Mendelson then broke up the historic folios and sold the drawings as loose sheets at his shops on the Tottenham Court Road and on the King's Road in the 1930s and for years afterward. Thus they entered the collections of a number of subsequently well-known figures in the London art historical world, like Anthony Blunt and John Pope-Hennessy.

At the very end of the 1970s a large group of these drawings was returned to the Royal Library by an art dealer, Rex Nan Kivell.[111] More than any other single event it was probably this that led to a renewed interest in Cassiano's drawings of nature (although the Surveyor of the Queen's Pictures, Sir Anthony Blunt, had mounted a small exhibition of them at the Courtauld Institute in London in 1968). In 1988, 107 drawings that had once belonged to the Royal Collection but had been bought by a Baltimore collector, Herbert Boone, were auctioned in New York.[112] By then a project to catalog all of drawings that had once come from Cassiano's paper museum was already under way. At this point many of the other drawings that had been sold from the Royal Library and were still in the hands of private collectors began to resurface as well.

Some sense of the exceptional quantity and range of Cassiano's collection of natural history drawings may be obtained from an inventory—the so-called Inventory A—made shortly after their arrival in England in 1763. This inventory lists five folio volumes of drawings of birds, six of plants, five of fossils, and one each of animals, fishes, fruits, and corals; an illustrated herbal titled "Erbario miniato" containing no fewer than 209 drawings; a small book of Mexican plants; and a large group of loose sheets—ninety-two of birds, sixteen of shells, twenty-five of quadrupeds, and twenty-nine more plants. All in all there seemed to no fewer than twenty-seven hundred drawings of natural historical subjects, some loose, others bound together.

Fortunately enough, eight of the volumes listed in Inventory A were not sold by Fortescue, and they survive intact in the Royal Library in Windsor Castle, still shelved alongside the much better-known volumes of drawings after the antique. These are the five volumes with the words

*Natural History: Fossils* stamped on their spines and still in the splendid gold-blocked calf bindings made for them at the end of the eighteenth century; the single volume titled *Natural History: Birds;* and the two vellum-bound volumes inscribed "Erbario Miniato" and "Erbe Mexicane," that is, the little Aztec herbal, both referred to above.[113]

# TWO *Lynxes*

## THE WANDERER OF THE HEAVENS

They called him *Coelivagus*, or "wanderer of the heavens." Though born to a family that might have prepared him for the world of politics, he did not participate in political and worldly affairs. Even had he wished to do so, his unceasing investigations into the natural world and the resolve with which he sought to order its fractious multiplicity would have left him no time for such things. "I hate the court and courtiers like the plague," he once wrote to his closest friend, "they are all traitors and I trust none of them."[1] He tasted no passing fame, nor seemed much concerned about it.

When he died, his oldest and closest surviving friend wrote to Galileo:

> with trembling hand and eyes full of tears I write to tell you of the sad news of the death of my Lord, the Prince of Acquasparta. An acute fever came and took him from us yesterday. The damage to the Republic of Letters caused by the loss of so many of his fine studies, all of which he left incomplete at the time of his death, is inestimable, and it leaves me unimaginably sad.[2]

"In its relatively short duration," wrote Giuseppe Gabrieli, the historian who best knew Cesi and the work of the Academy he founded, "his life was wholly interior. It was philosophical in its aspirations, full of setbacks and troubles of every kind, but of an unalterable rectitude, humanity, and nobility."[3] It is an admiring but just tribute.

Cesi bore misfortune and personal travail with a stoicism that exceeded the standards of the fashionable stoicism of his day. His father hated everything he undertook, and set whatever obstacle he could in his path; Cesi's two sons for whom he hoped so much died in childbirth or within a few days thereafter; four daughters did not survive infancy. These and many other of his setbacks and misfortunes are occasionally mentioned in his correspondence, but always with reticence—even when it comes to seeking their causes. When his children died, his grief may be discerned not from any complaint, but from the characteristic perseverance with which he made yet another attempt to find a medical reason for the repeated tragedy of their deaths, or to seek out one more doctor for a possible explanation. He passed away, Gabrieli recorded, "mourned only by the few who remained faithful to him. The world was indifferent to him, the silence almost general. Swiftly forgotten by his contemporaries, he was almost totally ignored by immediate posterity."[4]

Not, perhaps, so unusual a fate for many whose greatness comes to be acknowledged long after their deaths; but the achievements of Federico Cesi are still far from having received their due. Fate has not been generous in according him the privilege of fame. Born on March 13, 1585, in the family palace that still stands in the Via Maschera d'Oro in Rome, he came from an old and well-connected Umbrian family. His grandfather had married into the famous and ambitious clan of the Caietani; his mother was from the ancient, rich, and reckless family of the Orsini. This was a lineage of consequence and power. His uncle Bartolomeo, whom he much admired, became a cardinal in 1596. His pleasure-loving, spendthrift, and deeply anti-intellectual father carried the titles of Marquis of Monticelli and Duke of Acquasparta; while Paul V would bestow on our Federico—partly in compensation for the fact that his disapproving father withdrew from him the right of primogeniture—the title of Prince of Sant'Angelo and San Polo.[5]

The family's main country estate lay in the small town of Acquasparta, on a promontory facing the foothills of the Apennines in a part of Umbria, not more than fifty miles from Rome, which Federico loved more than any other place. Soon, exploring its hills, collecting its plants, searching out its animals, and digging up its abundant fossil remains, he came to know every inch of its terrain.[6]

In August 1603 the eighteen year-old youth made the decision that would determine the course of his life and change the face of science. On the seventeenth of that month, under the combined influence of Jupiter, Saturn, and Mercury, he called together three of his friends to his Roman home in order to found his Academy. He called it the Academy of Linceans, naming it not simply after Lyncaeus, that most keen-eyed of the Argonauts, but above all after the lynx, the small, wily, and intensely sharp-eyed animal that could still occasionally be found in the hills around his Umbrian home.[7]

The story of his Academy, at least in outline, has often been told in Italian, but only superficially in other tongues.[8] Although often called the world's first modern scientific academy (it preceded such later academies as the *Accademia del Cimento* in Florence, the Royal Society in Britain, and the *Académie des sciences* in France), its true achievements—chiefly those of Cesi himself—have remained obscure. They seem to have been lost in the minute and often overwhelming detail of the researches of the group itself, which even at its height never numbered more than twenty living members (only thirty-one were ever enrolled).

Unlike the work of Galileo, the sixth member of the Academy, whose profile is so easily determined thanks to the sharpness and brilliance of his controlling insights, their labor was more diffuse. Its detail can be distracting, its high and often unrelenting empiricism seemingly devoid of the kinds of discrimination necessary for theory and a more systematic view of the world. But the drive to order that characterizes the later work of Cesi, struggling to make sense of the great mass of information that he and his colleagues collected untiringly over many years, merits the closest attention. The members of the Academy did so much sheerly practical work in the interests of gathering specific data about the world of nature that it is all too easy to overlook the conceptual bases that may have informed their investigations. This, it is true, is often the problem with old science; but as one surveys the immense body of material left by Cesi and his most active colleagues—much of it still lying unpublished in the archives—it is hard to discern any broader vision.

At times the work of the Linceans can seem cripplingly old-fashioned and derivative, far too deferential to ancient authority and much too uncritical of what we would now call "magical" explanation. They could never hope to free themselves altogether from Aristotelian natural history, and often they were less advanced they thought themselves to be. Early on they described themselves as "most sagacious investigators of the arcanities of nature and dedicated to the Paracelsan disciplines,"[9] in an allusion to the influential Swiss doctor known as Paracelsus. Paracelsus had detached alchemy from its old mystical connotations and pushed it well along the road to something more closely approaching what we would now call chemistry; but at the same time, he continued to believe in such doctrines as the influence of the planets and stars on the lives of men and women as well as on other natural relations. Nevertheless, one often finds that it is precisely when the promise of the Linceans' work seemed most skewed by its allegiance to the manners of the past that it made its most remarkable advances. Only by relinquishing our modern attachment to the divorce between the technological and the theoretical, and by seeking to disentangle the visionary from the practical, the courtesies to authority from the revolutionary, can we arrive at some measure of what was begun at Palazzo Cesi on that day in 1603 when the stars were so fortunately and so favorably conjoined.

## FOUR FRIENDS

None of the four original members were older than twenty-six.[10] Johannes Heckius (van Heeck), an obsessive Dutch doctor from the town of Deventer, was chosen as master of their small Academy. Recognizing his dedication to the study of nature, Cesi had seen to it that he was freed from a Roman prison, where he had been sent just over a month earlier for the murder of an apothecary with whom he had a long-standing dispute.[11] Cesi and his devoted friend Francesco Stelluti from Fabriano in the Marches were declared the "chief counselors," while Anastasio de Filiis, a young nobleman from Terni, near to Acquasparta, took on the task of secretary.

As was the custom, they referred to one another by academic names: Cesi was called *Coelivagus*; Heckius, *Illuminatus*; Stelluti, lovingly, *Tardigradus* or "slow-paced"; and De Filiis, *Eclipsatus*, or "eclipsed." Because of their interest in oriental languages, Heckius and Cesi also took the Arabic (but latinized) equivalents of these names, *Sammavius* for Cesi, *Monourus* for Heckius. The lynx they chose as their emblem could not have been more appropriate for a group that would from then on devote itself to the keen exploration of the minutiae of nature, whether on earth or in heaven. "Of all animals they see the most clearly," noted Pliny the Elder in his *Natural History*.[12]

The four young men wasted no time in beginning their work. They went outdoors and collected specimens, examining, recording, and naming whatever they saw, whether animal, plant, or mineral. They charted the heavens and discussed the intricacies of natural philosophy and metaphysics. They built scientific instruments, including a complex astrolabe, or a large celestial planisphere, where they marked the constellations and the planets. They studied Arabic, in order to read the manuscript treatises of the great Arab scientists of the Middle Ages. Whenever they could, they met to discuss their ideas and to share the abundant information they were gathering about the phenomena of nature, anomalies and all.

At their meeting on Christmas Day 1603, Cesi was declared *Princeps* of the Academy. A spirit of frank criticism reigned at all their meetings. They established a rigorous code of intellectual and personal conduct, and began to correspond with doctors, scientists, and students of natural history all over Europe. Each of the four was supposed to have his own academic specialty—Heckius was expected to lead their discussions in philosophy and metaphysics, Cesi in botany, De Filiis in history, and Stelluti in astronomy and mathematics; but those were times when the boundaries between disciplines were porous. Each worked assiduously in every one of these fields, and they shared their results at every juncture. They brought strange plants and minerals to their meetings, devised new ways of examining them, and discussed theories of sidereal motion and the principles of geometry, mathematics, and physics.

Above all they came to realize that instead of relying on the authority of the ancients—and Aristotle in particular—it was crucial to make practical experiments and firsthand observations, directly from nature. Whenever possible, what they saw they had drawn by their draftsman. They employed rhizotomes to dig up plants by their roots and they dissected the animals and insects they brought into their simple laboratories, searching, above all, for the secrets of reproduction. Theory would never be divorced from the object itself, and they realized that understanding could only come if they penetrated beneath the surfaces of things.

Almost every day from 1603 on the four Linceans went out into the countryside around Cesi's family home in Acquasparta. There, in the hills and valleys of southern Umbria, they examined and collected whatever they could. The drawings they had made supplemented in fuller and more satisfactory ways what they did not or could not draw themselves. In doing all this they stumbled, quite literally, upon the two classes of terrestrial phenomena that formed the center of their work in natural history from its very beginnings until its end.

In the hills around the small Cesi castle and especially in the neighboring region of Dunarobba and Sismano, one may still see the extraordinary remains of fossil lignites that first gripped Cesi's attention. Perhaps even as a child he had wandered among them. Embedded with other fossils in the ground, or still standing as strange and forlorn trunks, these pyritic and lignitic remains seemed to offer the key to two of the central problems of their scientific lives: the classification of the natural world and the nature of light and heat. What exactly were these fossils—animal, vegetable, or mineral? Why did they glow in the dark, or heat up when dipped in cold water, and what was the nature of the subterranean fires that burned beneath them?

There was hardly a free moment, it seemed, when they were not agonizing about these questions, or discussing them (once they met him) with Galileo. But at the same time Cesi and Heckius were just as involved in the study of the fungi and ferns they found growing among the fossil remains—and in many other places too. The cryptogams, as we now call them, raised very similar questions to those posed by the fossils. They too raised difficult problems of classification; but even more than the fossils, they forced Cesi to confront one of the central and most critical mysteries of nature: that of generation and reproduction.

## SEPARATION AND EXILE: PRAGUE, SPAIN, AND NAPLES

Already by the end of 1603 Federico's father had begun his machinations against his son and the Academy. The elder Cesi did everything he could to separate the four young men and subvert their researches. He convinced himself that Heckius was trying to seduce his impressionable young son to become a Protestant and return with him to Holland. Heckius was thus set up on charges of heresy—of all things, since the rabidly Catholic Dutchman had come to Italy precisely in order to get away from the "injuries of the heretics," as he put it.[13] Moreover, Cesi's father and several members of his household had grown paranoid about the cryptographic writing the Linceans had devised to communicate with one another (and which further complicates the reading of their often dense manuscripts). Within a few months, he had the unstable Heckius and the well-adjusted Stelluti expelled from Rome. Stelluti went no farther than Parma, but for Heckius the consequences were both more significant and more unsettling.

First he traveled to Siena, and then to Florence, Pisa, Genoa, Milan, and Turin. Everywhere Heckius sought to meet the most prominent doctors and scholars. He propagandized on behalf of the Academy, and diligently studied every plant, animal, and insect he encountered.[14] In his spare time he continued to write his notes and treatises on everything from the plague to the Protestant heresy, and from mechanics to every form of natural history. From Italy Heckius went to Lyons, Paris, and Dieppe (where he again attempted to stab someone who disagreed with him about religion); and then across the Channel to the British Isles. Then he returned to his hometown of Deventer, apparently in order to reclaim some family property. There too he made trouble, always arguing with the Protestant "heretics," and the exasperated town authorities soon had him expelled.

Heckius headed north for Norway, Sweden, and Denmark, where he traveled and observed for several months. Then, restless as ever, but faithfully keeping his notebooks, and drawing as much as he could, he went south to Flanders again, and east into Germany, Bohemia, and Poland.[15]

By the end of 1604 Heckius had settled in Prague, where—perhaps fearing persecution, perhaps because he felt his natural historical researches would have aroused superstition—he lived under the assumed name of Gisberto Tacconi (sometimes Tassoni).[16] This was typical of the veil of secrecy that often enveloped the activities of the Linceans.

There was much to detain Heckius in Prague. The Habsburg emperor Rudolf II had set up a glittering and strange court there, and surrounded himself with all kinds of intellectuals and pseudo-intellectuals—from magicians to mathematicians, from experts in botany and pharmacology to astronomers and astrologers. Heckius visited the collection of natural and artificial curiosities amassed by Rudolf, and studied the exotic plants from the Americas and Asia growing in the gardens that had been established under the inspiration of Charles de L'Escluse (Clusius), the distinguished botanist of Leiden.

In that intense and slightly overheated intellectual atmosphere, where science and magic went even more closely together than usual, he met the astronomers Johannes Kepler and Franciscus Tengnagel (son-in-law of the famous Danish astronomer Tycho Brahe).[17] With Kepler he discussed the significance of a supernova that had appeared over Prague in October 1604, rousing the attention of astronomers all over Europe.[18]

Back in Acquasparta and Rome, Cesi and Stelluti continued their botanical and geological studies in the field. Every day they seemed to discover new plants and animals not to be found in the ancient authorities Aristotle, Dioscorides, and Pliny; and every day they heard of new and unknown plants from the Americas. Soon their correspondents and friends began sending them drawings and actual specimens too. The old edifices were beginning to crumble; but the new ones could only be set up on the basis of direct observation—or, failing that, with the aid of pictures and drawings of what they and others saw with their own eyes.

When Cesi began to realize that the fossil woods around his home posed some of the most fundamental problems about the organization of nature—Were they vegetable or mineral remains, for example, and how were they formed? Could they be not of one class of things but of two?—he made a point of having them meticulously drawn, from different angles, down to their most minute details.[19]

The problem of classification now began to obsess him and grew ever stronger. Indeed, it lay at the basis of the attempt he began at this time to systematize all of nature, and to put it into some kind of logical and comprehensive order. One had to know where to put individual species, and how to categorize and to group them. Swiftly Cesi became aware that these were not easy matters to resolve. If you were going to order nature, or discover its own order (for the two were by no means always regarded as synonymous),[20] you had to be able to classify rigorously and correctly. And this, Cesi realized long before Linnaeus, could not depend on the changing external characteristics of things, but rather on some more essential characteristic—an organism's means of reproduction, for instance.

Almost from the beginning of Cesi's researches, issues of biological reproduction were bound up with those of classification. Already in 1603 he gave a "phytosophical lecture," in which he defined the plant as a "vegetative animal."[21] The twin issues of definition and generation would haunt him for the rest of his life.

In order to discover the secrets of reproduction you could no longer be content with the surfaces of things; you had to penetrate within them as far as you could possibly go. Only then could you discover what lay at the basis of the exceptions that so bedeviled every classificatory attempt—monsters, anomalies, and prodigies, for example—and begin to grapple with the persistently vexed issue of borderline cases and intermediate and overlapping classes. For both these reasons, perhaps, but also out of a sense that it was precisely the lowliest and most ill-defined things that provided the best clues to the order of nature—Cesi and Stelluti began to turn their attention to another class of things where these problems were intensely at issue: fungi (or imperfect plants, as Cesi would soon be calling them). How did they reproduce? And what class of things did they belong to exactly?[22]

Cesi moved quickly. Finding Rome so hostile an environment, and knowing of the resources of Naples, he went down to that city, which he loved more than any other besides Rome. He thought of setting up his society—or at least a branch of it—there.[23] In Naples Cesi visited (or tried to visit) almost everyone who counted in the fields that were now absorbing him, from the brilliant young Fabio Colonna whose botanical researches had been stimulated by his search for the Dioscoridean *phu* as a cure for his own epilepsy (cf. fig. 9.1),[24] to the famous collector, museum owner, and natural history writer on everything from fossils to fish, Ferrante Imperato. Cesi also went to examine the summary and copies of parts of an extraordinary collection of material about Aztec plants and animals assembled by the *protomedicus* of the Indies (or the *Pliny of the Indies* as he was also sometimes called), Francisco Hernández;[25] and he began negotiations to acquire this material, made and put together by the very Nardo Antonio Recchi mentioned by Dati in his life of Cassiano.[26]

But the figure whom Cesi was keenest of all to meet was the man then probably regarded as the greatest natural scientist of his age, the seventy-year-old Giovanni Battista Della Porta. From our modern vantage point he was much more a representative of the old science than the new. Della Porta was possessive and jealous of his multifold researches, but he was known throughout Europe for his work in a multitude of fields—astrology, alchemy, magic, mathematics, optics, meteorology, chemistry, distillation, numerology, ciphers, physiognomy, and phytognomy.[27]

The latter two were altogether typical of Della Porta's science. You read from the surface what lay beneath—the hidden secrets and essences of things. In phytognomy the medicinal powers of plants were taken to be directly related to the way they looked. This, like the widespread doctrine of signatures that it perfectly embodied, is perhaps the most extreme form of what Foucault called the doctrine of similitude, the leading *episteme* of science in the preclassical era. The doctrine of signatures is that notion whereby plants are assigned their characters on the basis of their apparent similarity to human anatomical features. Thus, a plant like the walnut was good for the cure of a whole range of ailments in the head (because of the similarity between the kernel itself and the brain), and plants with hairy roots served to alleviate baldness and other conditions affecting one's body hair.

Della Porta's immensely popular books on physiognomy and phytognomy (as well as his work on "celestial physiognomy" and chiromancy, also, in a sense, a science of signatures)[28] embody the essence of an approach to the study of the natural world predicated on the view that the surface appearances of things provide crucial clues to their inner natures, to their very character and powers. Indeed, in his terms, the word "character" (or often also *nota*) meant the outward sign of inner essence, describable by analogy with other visible phenomena in the natural world. Since the inner secrets of nature were always inscribed on the outward face of nature and its beings, only by ever closer and more detailed examination could one hope to come closer to what lay within.

To turn from Della Porta's publications on optics and mathematics to his massive tomes on natural magic, physiognomy, and phytognomy is to be amazed, even in an age of intellectual omnivores, by the range of his work and interests. No wonder Cesi felt he had to go to meet him.

The irascible old sage immediately took to the nineteen-year-old prince. In Cesi he seems to have recognized the potential for changing the face of science. From then on Della Porta adored him. Already on June 25, 1604, he wrote to Cesi and the other Linceans (presumably in response to a letter of theirs), assuring them of his willingness to cooperate with them. "In the meantime, however little I am worth, I shall devote myself to you; I shall be deserving if not by learning, then at least by affection."[29] Less than a month later Della Porta prepared the long dedication to a treatise on distillation, in which he praised both the *adolescens illustrissimus*, "blessed with a new virtue and an even greater hope for the future," and the Academy he had founded.[30] When the *De distillatione* finally appeared in 1608, it was published under the auspices of that promising new society, the Academy of the Lynx-eyed.[31]

It is as if the researches of Della Porta's old age were rejuvenated by his contact with the young men; and in the next few years he dedicated two more books to Cesi, first a book on curvilinear geometry and in particular on squaring the circle, the *Elementa curvilinea*;[32] and then another called *The transformations of the atmosphere*, on meteorites and other atmospheric and celestial phenomena.[33]

In 1610 Cesi went back to Naples and invited Della Porta to become a member of the Academy. By now there were only two of them left: Cesi and the faithful Stelluti—for De Filiis had died two years earlier, and Heckius, off again on his pan-European wanderings, was losing his mind. The time had come to start building anew—and more intensely than before. With the inscription of Della Porta into the old album that Cesi kept expressly for the purpose of recording new enrollments (and which he and Stelluti had brought down to Naples with them), it was as if a block had been overcome.

## GALILEO AND THE LINCEI: PROLOGUE

If Della Porta's acceptance of the invitation to join the apparently faltering Academy was testimony to the seriousness with which he regarded the young men's researches and the promise of their investigations into the natural world, then the acceptance of their next invitation was even more significant. It would change their world—and ours—forever. In the following year, 1611, Galileo became the sixth member of the Academy of Linceans. It is an indication of his pride in their society that from then on his works would carry the emblem of the lynx on their frontispieces, and that even before his other titles "Florentine nobleman, Philosopher, and First Mathematician of the Grand Duke of Tuscany," he would proudly add "Lyncaeus" or "Linceo" to his name.

Galileo's inscription into the Academy followed hard on the heels of his triumphant visit to Rome in April 1611, when he lectured on his new discoveries in the heavens—the flawed and cratered surface of the moon, the satellites that revolved round Jupiter, and the multitude of hitherto unseen stars—and demonstrated the use of the still-new instrument that had enabled him to make these discoveries.[34]

From then onward the Academy expanded its membership and its activities with a speed that could not have been predicted. In October and

December 1611 two Germans who had been present at Galileo's demonstration of his telescope were elected into the Academy. Both were experts in the field of botany. One, Johannes Schreck, had to resign a few years later. Contrary to the Academy's rules forbidding members to join a religious order, he became a Jesuit; but even after being sent as a missionary to China, Schreck continued to send back the results of his researches into the flora and fauna of Asia.[35] The other German inscription was that of a doctor from Bamberg, Johannes Faber, who not only worked at the hospital of Santo Spirito but was also professor of medicine and lecturer in simples at the University of Rome, and superintendent of the pope's garden.[36] His correspondence and collaborations with Cesi reveal his skills both in botany and in zoology, as well as his full awareness of the importance of Galileo's progressive dismantling of the old Ptolemaic cosmology. But it was to a third German, Markus Welser of Augsburg (who became a member in 1612), that Galileo addressed his *Letter on the Sunspots* of 1613. This was the work in which he proved that the sun was no perfect orb of light, but was flawed too, like the other celestial bodies, and was a fixed star, like all the others. The battle lines were drawn, and the Jesuits, who had at first received Galileo with a striking mixture of enthusiasm and caution, thereafter yielded some of his most hostile opponents.

No one could fail to detect the ironies in the story that now unfolded. Among Galileo's most fervent supporters was the young, eloquent, and sickly Roman nobleman, Virginio Cesarini, elected to the Academy in 1618.[37] The more strongly Cesarini campaigned on behalf of Galileo, the closer he drew to the Jesuits. His campaign came to a head in the years between 1620 and 1623, when he and his friend and fellow Lincean Giovanni Ciampoli (also a high official of the papal court), along with Faber and Cassiano dal Pozzo, worked under the direction of Cesi on the preparation and edition of one of the most robust of all Galileo's works, the *Assayer* of 1623.[38]

It was at this point that Cassiano appeared for the first time in his role as scientific enabler. It was he who bound the small group of four editors together, and he, probably, who determined the complex strategy of publication of the work that put a further nail in the coffin of Ptolemaic astronomy, made daring hypotheses about the nature of the heavenly bodies, and attempted to put the rising numbers of Galileo's Jesuit opponents to flight. Lacking the scientific skills of Faber, and less eloquent than Cesarini or Ciampoli, Cassiano ensured the unity of the group that worked so assiduously—and by now so riskily—on behalf of a sick and dispirited Galileo.

In 1622, as we have seen, Cassiano was himself rewarded with election to the Academy. And then, in August of the following year, Maffeo Barberini became Pope Urban VIII. Already in 1612 the talented and deeply intellectual Barberini had befriended Galileo in Florence. There they discussed poetry and topics such as the problem of tides and the behavior of frozen bodies in water. But by the time of Maffeo's elevation, the

Holy Office of the Inquisition had condemned Galileo's Copernican views of the movement of the earth and the stability of the sun. Galileo was forbidden from putting forward or even defending such views again. How then to get the provocative *Assayer* past the papal censor?

Cassiano and his friends Cesi, Faber, and Ciampoli took an extraordinary step. They made Galileo's subversive text take the form of a letter to Cesarini, whom Urban loved and had just appointed as his *maestro di camera*. It was a masterful psychological move. What could the pope and his family do against a text that was addressed so sympathetically to one who was so close to them? And how discomfited the Jesuits must have been by the close bond that existed between Galileo and the very same young man who supported their order and was thinking of joining it!

The *Assayer* was cleared by the censors and received a laudatory *imprimatur*. It was reported that Urban himself enjoyed reading it.[39] Indeed, Galileo's robust style makes it one of the most readable of all revolutionary books in the history of science. Not for the first time—nor the last—were the Linceans two-faced, all in the interests of the highest moral purpose (or so they must have thought to themselves). This was not, to be sure, the year in which to act slowly or timorously. The results of both Galileo's and the Linceans' researches flew ever more blatantly in the face of received doctrine. Their own anti-Aristotelianism, just like Galileo's, stood at complete odds with the official teachings of the Church on any number of matters. Once more they needed support from those who stood closest to the pope. And there was an ideal candidate to hand—the pope's favorite nephew, Francesco, just elevated to the cardinalate. Here was someone who could not have been closer to the pope, who was deeply interested in most of the subjects that the Linceans were investigating, and who, best of all, was endowed with an ever-growing but already vast fortune. So they invited him to become a member. He gladly accepted. Their hopes were not misplaced (for the time being at least), since the new cardinal did indeed take a deep interest in their studies and helped finance a number of their most important projects.

From then onward things moved forward with a speed that not one of them could have foreseen.

## THE LEGATION TO FRANCE AND SPAIN

In 1625–26 Francesco Barberini was sent by Urban VIII on a pair of diplomatic missions to France and Spain.[40] With him he took a glittering entourage, including Cassiano. Everywhere Cassiano noted matters of botanical, ornithological, zoological, geological, archaeological, and artistic interest. He sent constant reports back to Cesi, for future use in their publications. From France Cassiano wrote to Cesi that although he had not found a soul there worthy of election to their Academy—the French were too interested in women and the court—there was nevertheless one Englishman whom he wanted to mention. This was Francis

Bacon. Cassiano reminded Cesi of the *Moral Essays* (which had first appeared in Italian in 1618)[41] and other important scientific works of his. "If he weren't living in England, I'd want us to make every effort to make him one of us." Bacon's recent *De dignitate et augmentis scientiarum* contained a quantity of material of benefit to all the sciences, Cassiano noted, and had brought many important issues to the fore.[42] In general, the name of Francis Bacon occurs very much less in the Lincean correspondence than one might have expected;[43] but it is significant that his was the name mentioned by Cassiano amid all the material of scientific and natural historical interest that he found in the course of his trip to France and communicated back to Cesi.

When the legation went to Spain, Cassiano ensured that both the text and the illustrations of several of the manuscripts that Hernández had brought back from Mexico were copied. It was on this trip too that he commissioned the Lincean copy of the Aztec herbal made for Philip II in the middle of the previous century (fig. 9.3). Soon the edition of Hernández's manuscripts, largely based on copies acquired by Cesi, as well as the notes on them by Faber, Schreck, and Colonna, would be ready. They would call this work their *Tesoro Messicano*, or "Mexican Treasury."[44]

By 1630 Galileo's old friend Maffeo Barberini was seriously irritated with him. Though still lying quiet after the publication of the *Assayer*, Galileo continued to make plain his adherence to the doctrines of Copernicus. In that year Cesi died. He was only forty-five years old, worn out by his labors and the constant opposition to his efforts. In the moving letter he wrote to Galileo the day after Cesi's death, Stelluti expressed his sadness that their leader had not made arrangements for the things belonging to the Academy, "to which he wanted to leave his whole library, his museum, his manuscripts and other beautiful things, and I don't know in what hands they will end up."[45]

Stelluti need not have worried; there was always Cassiano to turn to. Encouraged by Stelluti, Cassiano realized the value of Cesi's laboratory and library and bought all his books, manuscripts, and scientific instruments in 1633[46]—the very year of Galileo's trial, and one year after the publication of his most systematic and compelling book, the *Dialogue on the Two Great World Systems*.

It was in this brilliant and often scathing work that Galileo cogently put an end to the old Aristotelian and Ptolemaic views of things. The book seemed to make fools of many of his opponents. Cesarini and Faber were dead, and Ciampoli had been sent into a retirement of disgrace. Francesco Barberini, obliged to take up a hostile position to Galileo, withdrew from the domain of the scientific endeavors initiated by the Linceans and which he had so often and so generously sponsored. And in his late opposition to Florentine plans for the monument to the man he and his uncle once so admired—even loved—no one could have been colder than he.[47]

## CASSIANO TO THE END

From then onward matters were in the hands of Cassiano and, to a lesser extent, the less resourceful but ever-devoted Stelluti. But the atmosphere had turned against them. Their association with Galileo had made them all suspect, and the intellectual climate was far less favorable than it had once so briefly been. They did not dare to publish the more radical of their findings. Besides, Cesi had left too much undone, too much in an incipient state.[48] Often, in reading his manuscripts, one senses that he came to realize how fraught were the implications of his investigations into the nature of the heavens and the animals and vegetables buried in the earth. He could go no further.

Aside from a brief and seemingly innocuous treatise prepared by Stelluti on the basis of Cesi's work on fossils,[49] nothing more appeared from the Linceans—with one notable exception. This was the definitive edition of their Mexican Treasury in 1651.[50] It would never have appeared had Cassiano not encouraged Stelluti to keep working on its enormous text, and had he not found a sponsor for it (in the surprising form of a Spanish aristocrat living in Rome).[51] Two years later Stelluti died, and Cassiano himself four years after that. The time of the Linceans was over.

Without Cassiano, the achievements of one of the great heroes of modern science and of the Academy he founded would have been obscured for ever. Not only was Cassiano one of the firmest of both Galileo's and Cesi's supporters, it was also he who kept the group together through its most difficult and controversial times, who arranged for periodic injections of money, and who kept both Colonna and Stelluti going in the almost intolerable few years after Cesi's death. And by incorporating Cesi's drawings into his own paper museum, Cassiano had preserved them for posterity.

But what of the more crucial cover-up, the one hinted at by Dati in his life of Cassiano? What lay behind the relentless insistence on Cassiano's orthodoxy, and how much might it have to do with the persistent suppression of Cassiano's name and of the credit he deserved for his natural historical researches? There seemed to be nothing very unusual—let alone unorthodox—in his ornithological or botanical work, aside, perhaps, from the remarkable density of description and the strong faith in illustration as an adjunct to the study of nature. But could the trouble have been the friends with whom he associated? After all, Cassiano had played a seminal role in keeping the Linceans on course through some of their most difficult times, and he had consistently sought to facilitate their often heretical researches and publications. In the years that Cesi was alive, the Linceans pursued topics that threatened to undermine both the stability of the Church and the authoritative classical traditions on which this depended. And one of them, more brilliant and significant than any of the others, had to be published at all costs—if only because his opponents so persisted in forcing his hand. This, of course, was Galileo.

# PART TWO

# *Astronomy*

# THREE *The New Star*

## THE SUPERNOVA OF 1604

In October 1604 a new star—what astronomers now call a supernova—appeared in the heavens above central and southern Europe.[1] For a few days it burned brilliantly, and then it gradually faded from view.[2] All over Europe one astronomer after another claimed to have been the first to spot it. Among them was Cesi's sad friend Heckius, now an exile in Prague. Within a few months he had written a short treatise on it. Published soon afterward, it would be the very first of the Lincean publications. We will examine it shortly. Everyone asked themselves what the nova meant. Observing it in the same part of the heavens as the exceptional conjunction of Mars and Jupiter in Sagittarius, the astrologers quickly interpreted it as some devastating portent of doom—of famine, plague, war, or political upheaval. In Hungary, they noted, the revolts had already started.[3]

Soon the most distinguished astronomer in all of Europe, Johannes Kepler, began composing his refutation of the astrologers' views. His *De Stella Nova* of 1606 remains one of the basic books on the subject, and astronomers still refer to the nova of 1604 as "Kepler's nova."[4] For him, as for the other astronomers and mathematicians of the day, the issues were very different from those of the astrologers. Such men were all too aware of the fact that the sudden appearance of the nova required explanations of a kind that threatened to shake the very bases of the traditional Christian view of the universe and the Aristotelian doctrines on which it depended.

To begin with, the nova called into question the standard Christian and Aristotelian view that the heavens were eternal and incorruptible. According to the then orthodox combination of Aristotelian and Ptolemaic cosmology, the earth stood motionless at the center of the universe. Around it revolved a succession of ever-widening spheres, carrying the moon, the planets (including the sun), and the stars. The first, or the terrestrial, sphere was that "beneath" the moon. Everything within it was subject to change and decay. But the celestial spheres beyond the moon, which carried Mercury, Venus, the sun, Mars, Jupiter, Saturn, and finally the fixed stars (in that order), were regarded as solid as crystal or diamond, and incorruptible. This was neatly consistent with the orthodox Christian view of a perfectly created universe, with the earth at its center.

The nova thus raised a series of truly critical questions about these orthodoxies. It did indeed seem to be new, but could it really have appeared from nothing, ex nihilo? Why had it not been known to the ancients? Where had it come from and of what was it composed? The astronomers tied themselves into knots over all this. They fretted over the question of how the star had only *then* become visible, if it had somehow *always* been there. And how was it possible to account for the fact that it burned so brilliantly for only a few days, before gradually disappearing from view altogether?

These were not the only problems that provoked consternation among the astronomers and theologians. There was another, equally crucial one. It had to do with the difficulty of determining exactly where the nova was located. This was not just a technical matter. Everyone argued about how best to measure the distance of the nova from the earth; but the implications of the arguments about measurement ran much deeper than we would now suppose. If the nova belonged not to the sublunary sphere between the earth and the moon, but was instead a fixed star in the solid firmament above it, then the most basic element in the whole edifice of Ptolemaic cosmology would be threatened: the notion—or rather the doctrine—of the crystalline spheres, which carried the planets in their orbits round the earth. To claim that the nova belonged to the firmament meant that the spheres were permeable and neither perfect nor immutable.

Immediately astronomers (and many others) rushed to enter the fray. Galileo gave lectures on it in Padua,[5] the brilliant young Flemish Jesuit father Odo van Maelcote lectured on it in the Jesuit college in Rome,[6] and publishers in Padua, Rome, Frankfurt, and Prague were kept busy as they rushed into print the latest opinions on the nova.[7] The matter was immensely fraught. How could one draw the obvious conclusions without offending the Church, and in particular those most dogged proponents of the unity of Christian doctrine and Aristotelian science, the Jesuits? Compromises were necessary on every side.

Not that the problems surrounding the nova were entirely new. Already in the case of the still well-remembered nova of 1572, and the equally famous comet of 1577 (which posed very similar questions), the great Danish astronomer Tycho Brahe had revealed just to what extent it was possible to trim one's sails. On the one hand, Brahe's meticulous observations of both phenomena had led him to conclude that comets, like the nova, were located beyond the moon. This, along with his opinion that comets moved in circular orbits around the sun in the vicinity of Venus, seriously impugned the doctrine of the solidity of the spheres—for the comet would necessarily have penetrated them.[8] What was at stake had exceptional implications. Suddenly the old Ptolemaic cosmos lost all plausibility. And yet Brahe could not bring himself to take the further, wholly Copernican step of affirming that the sun stood at the center of the universe. Instead, as Stillman Drake has succinctly put it, "he proposed a third alternative, in which the earth remained fixed at the centre of the universe while the moon and sun went around it, the planets revolving about the sun as it went round the earth."[9]

No wonder that Galileo would never have much patience with Brahe's work! It must all have seemed too weak and too obliging to those who (like the Jesuits) could see the problems with the fully fledged Ptolemaic system, but were forbidden to support the Copernican one.[10]

## BRUNO

On a cold winter's morning in February 1600 Giordano Bruno was led to the stake. The Inquisition had handed him over to the secular authorities to pay the ultimate price for his many heresies. Everyone remembered the spectacle with horror. Bruno was stripped, tied to a stake, and burned alive at the flower market just a few hundred yards from the Jesuit college in Rome. Every one of his many published works had already been consigned to the pyre. Even the Calvinist authorities in Geneva and the Lutheran ones in Helmstedt had long since excommunicated him from their respective faiths. (He had repeatedly flirted with these in the course of his pan-European travels.)

For almost eight years the lapsed Dominican philosopher with the phenomenal memory techniques had been subjected to interrogation and torture by the Inquisitional courts of Venice and Rome. Accused of atheism, blasphemy, sorcery, and heresy, he expressed doubts about some of the most fundamental aspects of Catholic doctrine: the nature of the Trinity, the personhood of Christ, and the status of the Eucharistic doctrine of transubstantiation, to take just the most salient examples.

But almost as dangerous, and much more prophetic, were his views about the nature of the world and the universe. Bruno took the Copernican position that the earth was not the center of the universe but moved around the sun. He maintained that there was a plurality of worlds (as many as there were stars).[11] He believed in the immobility of the heavenly spheres. All this also stood in direct opposition to the traditional teachings of the Church. He had even suggested that the smallest particles of the universe were infinite quantities of atoms, and that these were not subject to an external force (such as God), but rather to their own internal laws![12]

Bruno's cosmological and atomist views were by no means the chief objections the Church had against him (they were much more concerned with his heretical theological opinions), but each one of these profoundly anti-Aristotelian positions was enough to set the teeth of the orthodox on edge. Hearing the sentence of the Inquisition read to him, Bruno rose from his knees and exclaimed to his judges: "You probably deliver this sentence against me with greater fear than that with which I receive it."[13] Did the great Cardinal Robert Bellarmine, chief among his judges, shiver just a little when he heard this? Although there is nothing in Bellarmine's character to suggest that he would have been shaken by such a challenge, little can he have dreamed how soon he would have to respond to cosmological views not very different from those of Giordano Bruno, the archheretic himself.

## ABOVE THE MOON—OR BELOW?

At a time when no intellectual could conceivably have forgotten the pyre in the very heart of old Rome, the nova of 1604 raised equally dangerous questions about the nature of the universe. But Galileo was not one to hold back. On the face of it, the lectures he delivered in November 1604 to a packed lecture hall at the University of Padua (more than a thousand are said to have attended) do not seem to have been particularly provocative.

The first lecture began innocuously enough. It started with a description of what was empirically observable about the new star:

> A strange and unexpected star was first seen in the heavens on 10 October 1604. Initially it seemed to be of small bulk, but after a few days, it so greatly increased in size, that with the sole exception of Venus, it appeared to be larger than all the other stars, both fixed and wandering. It glittered and shone very brilliantly with a reddish glow, to such a degree that the vibration of its light seemed at one moment to be extinguished and then immediately to light up again. Its ruddy brilliance exceeded that of all the fixed stars, including the Dog star itself; and the colour of its light imitated the golden splendor of Jupiter and the yellow fire of Mars.

Galileo was clearly fascinated by the nova's incandescence. "At one moment," he observed, "it withdrew its tremendous rays and seemed to extinguish itself, becoming almost reddish white; at another, as if reviving itself, it poured forth wider rays and shone with Jovian brilliance."[14] Then he outlined his position that the fiery star was not combustible, as the Aristotelians had held, but instead was composed of vaporous masses or "exhalations"—a view that would have an interesting future.[15] In suggesting that the new star had actually originated in a mass of vapor arising from the earth,[16] Galileo was already exploring issues that would form—mistakenly as it turned out—the basis of his speculations about the even more controversial comet of 1618.

But these were not matters that could be resolved yet. Their resonance would only become clear later. For the moment, Galileo moved on to the larger issues that were claiming everyone's attention. To judge from the surviving fragments of these lectures (which were not published until the nineteenth century),[17] he swiftly turned to emphasize that the nova could not possibly be so close to the earth as to be on *this* side of the moon (that is, in the sublunary sphere). He firmly agreed with those who claimed it was to be placed in the sphere of the fixed stars far beyond it. Next, he moved on to the technical details of the measurement of parallax as a way of determining the distance of the star from earth. Parallax is the phenomenon that refers to the shift in view that occurs when an object is seen from different positions. If one holds a pencil in front of one's face and then looks at it with one eye closed, the pencil will seem to have moved; but the greater the distance of the pencil, the smaller the shift or parallax. When the moon is observed from two points on the surface of the earth, it shows a small but definite parallactic displacement against the background of distant stars.[18] And since, according to Galileo and several other observers at the time, the nova showed hardly any parallax at all, it had to be well above the moon—and not, as the orthodox Ptolemaic and Aristotelian view held, in the sublunary sphere below it.

No wonder the views proposed by Galileo and the other *novatores* were such a threat to the established order of things. The orthodox astronomers, though certainly struck by the nova, had little real difficulty explaining it. In the first place, they maintained that the nova was a sublunary phenomenon, to be located in the one sphere subject to change and decay—that is, the terrestrial sphere between the earth and the moon. Second, they invented a variety of reasons to explain why it had somehow escaped the attention of the ancient authorities (as the *novatores* loved to point out). And third, they devised one argument after another to account for the ways in which it might have been obscured from their eyes.

If, however, one maintained (as Galileo and his allies did), that the nova was indeed a new star in the heavens, that it had never been seen before, that it came and went, and that it was to be placed not in the sublunary sphere but far beyond it—even in the remote sphere of the fixed stars—then the twin doctrines of incorruptibility of the spheres and their adamantine impenetrability, those mainstays of Aristotelian-Ptolemaic cosmology, had to come tumbling down.

Of course, Brahe had taken very similar positions in the case of the nova of 1572; but you could always say, first, that he was a Protestant, and second, that he never asserted that the sun was the center of the universe. The official Church thus remained comparatively unconcerned about the few astronomers who, like Galileo, were beginning to believe in a heliocentric view of the universe. Copernicus himself had only proposed it as hypothesis, rather than declaring it outright. Nor were the authorities too much perturbed—yet—by the implications of the supralunary view of the nova. Even van Maelcote took this position in his lecture on the subject at that bastion of Aristotelianism, the Roman Jesuit College.[19] The old cosmos might seem to be disintegrating before the very eyes of its beholders, but still the Church stood firm, confident in its authority and reach. Galileo, however, was not one to compromise.

## COSMOS AND COUNTRYSIDE

Following his lectures in Padua, Galileo pondered his next step, as if waiting for the response of his audience and his colleagues. It soon came. In February he published his first major statement on the subject of astronomy. It was written in reply to a booklet by his local colleague, Antonio Cremonini, and an elderly ally of Cremonini's, Antonio Lorenzini, who had both been provoked by his lectures. In their booklet (with a foreword dated January 15, 1605), they defended the incorruptibility of the spheres, assigned the nova to the sublunary world, and denied the use of parallax to establish its astronomical position.[20]

Galileo's response took the form of an anonymous dialogue. It was written not in the sober scientific Latin of the day but in a particularly robust form of the local Paduan dialect.[21] Later on he would compose his works in the hard-edged Tuscan for which he became renowned, but for the moment he decided to use the language of the people of the Paduan plain.[22] In this Galileo was probably following the example of the rustic dialogues of the sixteenth-century Paduan poet Angelo Beolco, known as Ruzzante.[23] At least part of the appeal and effect of Ruzzante's works lay precisely in the use of this kind of language—just as it did for Galileo.

In any case, the dialogue he now wrote was both brilliant and rebarbative, as all his subsequent statements in the field would be. Nowhere in it does his name appear, but its mordant style—as well as its theoretically rigorous content—is unmistakably Galilean.[24] Probably written with the help of a friendly colleague, it fiercely satirized the conservative positions of his local colleagues.

In putting his own ideas into the mouths of two seemingly simple and plain-speaking *contadini*, Galileo could hardly have expressed himself more directly. The typical Galilean strategy was now taking shape. He realized that the vernacular was a more efficient vehicle for the dissemination of his ideas than the old language of authority. Later he might write in a higher vernacular, but now he seemed to be going out of his way to prove his point that science should be accessible to all, and not just to those trained in Latin and Greek. Galileo understood that there could be no more effective way of dismantling the status of the traditional cosmological positions than by making the counterpositions available to everyone. Authority, even in the form of language, would be stripped of the last shreds of its exclusivity.

Anyone with a modicum of sophistication would have recognized Galileo's views in the new dialogue; but how were his opponents to make them stick to him? Not for the last time would he show himself to be a master of dissimulation.

And so Galileo presents some of his earliest and most crucial thoughts on the structure and phenomena of the cosmos by way of an engaging discussion between two smart locals. In their simple and pungent way they take apart almost every one of the Aristotelian positions held by the philosophers (in those days, it should be remembered, "philosophy" was commonly used to describe what we would now call science, though it was clearly distinguished from the more precise judgments of mathematics). And they make a meal of those who insist on the sublunary position of the nova, who deny the use of parallax in determining its distance from the earth, and who assert the incorruptibility of the celestial spheres. "What would you say, most Reverend Sir," asks the preface,

> if you saw a poor servant of yours, who'd never done anything else except work in the cowshed, and who'd never measured anything but his fields, now picking an argument with one of the Professors at Padua? Wouldn't you split a gut laughing?

In this manner the countrymen proceed to demolish the professors, piece by piece. It is a virtuoso performance.

> I wouldn't have known what to say about the stars if I hadn't heard you Professors say one thing one day and another thing the next about all this, at least a thousand times over, no a million. It's the same with this new star which has so amazed the whole world. . . .
>
> Your faithful servant
> Cecco di Ronchitti,
> in Padua, the first of February 1605.[25]

Of course there was no such person as Cecco di Ronchitti. Galileo made him up for the occasion. "Blind man with the coughs" would be a good way to translate his name.

The two protagonists of the dialogue are called Matteo and Natale. They may speak earthily, but they succeed in discussing almost all the important issues raised by the appearance of the nova, even more competently than their opponents Cremonini and Lorenzini. And their skeptical assault on Aristotle and the Aristotelians is all the more effective for the barnyard language in which it is expressed.

First they lay to rest the astrological notion that the new star was somehow responsible for earthly vicissitudes. The lack of rain has nothing to do with it. But what about the claim that it had never been seen before? Surely this could not mean, as the professor was claiming, that it never *existed* before. "Let the roach get back to his Paduan roaches' nest," exclaims Matteo. After all, "in all my experience I've never seen a troop of German lansquenets—but still I know they exist."[26]

The satire continues vigorously. Unable to resist one final swipe at the astrologers, Galileo has Natale say this: "Yes, but since the new star's so far away, and since we can't even tell where it really is, how can we say that it's the reason that there's no rain?" And so he introduces the question of parallax: "Far away—for God's sake! It's not even beyond the moon, at least that's what the book said."[27]

> But what would the Professor really know?
> —What's he do? Is he a field-surveyor?
> —No, he's a philosopher.
> —Don't you know that a cobbler should stick to his last?[28]

Little could Galileo then have suspected, when he and his anonymous friend put together their dialogue, how loudly that particular metaphor would continue to resonate for the next quarter century![29]

And so, after Matteo's colorful dismissal of the philosophers (and he does so again and again), he returns to vindicate the mathematicians. They are the only ones, according to him, who are truly capable of judging the issues at stake.

> —One has to believe the mathematicians, who are measurers of the air. Just as I'm able to measure the fields, and can measure their length and breadth, so they're the ones who can measure the heavens.[30]

Eventually the countrymen return to the fact that Lorenzini's pamphlet had attacked the mathematicians for claiming that the nova was very high in the spheres. It had made the usual Aristotelian case for a sublunary position of the nova. "The great Aristotelian philosopher," as Matteo describes the author at one point, had even claimed that the new star was no more than ten miles above the earth. But what did he know? If he had been skilled and competent enough to know how to use the method of parallax correctly he couldn't possibly have made such a mistake.[31]

But there was also another, equally critical matter: that of the alleged incorruptibility of the heavens. Even though the mathematicians were only supposed to deal with measurement and geometry, as Natale observed, they were also claiming not only that the heavens were corruptible but that new things could actually be generated in them from nothing, and that this could take place spontaneously.

"Where're the mathematicians who argue in this way?" asks Matteo, now playing the part of devil's advocate,

> —if they're so busy *measuring*, what's it to them if the heavens can generate new things or not? I mean, even if it were made of polenta, these guys would have something to say about that too.

They continue arguing for a while about whether or not a star really could have been generated totally anew;[32] and then Galileo (speaking as "Matteo") goes on to make a number of remarkably prescient statements:

> —And does he think that one can see every one of the stars in the sky—go tell this to a priest! It's just not possible. You aren't going to tell me, are you, that there couldn't be at least three or four stars, maybe even more, which you can't see—especially if they're smaller ones.

Natale makes a last-ditch attempt to defend the professor:

—But he says that these are the very muscles and sinews of Aristotle's reasoning!
—Forget it—if the muscles and sinews are so pathetic, you might as well chuck all his generation and corruption into the soup.[33]

Matteo and Natale continue squabbling about the matter of the incorruptible heavens. Barely containing their derision, they return once more to Lorenzini:

—He says that if this star were in the heavens, the whole of natural philosophy would be a joke; and that Aristotle thought that if one star were added to the heavens, they wouldn't be able to move all.
—What a plague! This star has really made a big mistake. It's gone and ruined all their philosophy.[34]

The tone is so engaging, especially when the agricultural metaphors begin to flow, that it is easy to miss the allusions—sometimes breathtaking in their apparent foresight—to one or another crucial element of the new cosmology. Among the many prophetic moments in the dialogue, there can hardly be one more so than when Matteo says in passing that there are even people who believe that the world turns round and round like the wheel of a mill.[35]

Few were better at making things plain than Galileo, or more easily comprehensible; and if he had to use such shrewd but plain-speaking rustics to express his ideas he both could and would. Down to the last expletive:

onions, boils, and balls to all my opponents. Let Lorenzini's star lick the ass of the moon.[36]

It is easy to understand how this kind of approach might have alienated his opponents. The Jesuits must have been particularly irritated. Already in 1593–4 and again in 1599 the fundamental outline of the syllabus for instruction in their colleges—the *Ratio Studiorum*—had clearly asserted the primacy of Aristotelian science in their teaching.[37]

But Galileo's style was set. The equation between the new science and the language of the people could not have been clearer than in this dialogue. The old authorities had to be demystified. In the same way the German reformers of the previous century had demystified Sacred Scripture (and religion itself) by the simple device of making it available in the vernacular. And if there was anyone whose command of the vernacular was as robust as Galileo's, it was none other than Martin Luther himself. As long as one wrote in Latin, debate was just about tolerable, but if one made heretical ideas available to all, the threat to orthodoxy became more substantial. The authorities can hardly have been unworried, as has sometimes been claimed. A new period of censorship was about to begin; and every tactic had to be used to avoid it.

Although not many people are likely to have read the dialogue by

"Cecco di Ronchitti," and although Galileo may have cloaked his ideas under the veil of anonymity, he knew he was playing with fire. He was expressing the most dangerous and heterodox ideas about the nature of the cosmos in the most seductive and plainspoken way. Whatever precautions he took to pretend that he was not actually saying what he seemed to be saying, or that someone else was saying them, the appealing directness with which he expressed his heretical views was unmistakable. His friends, upon whom he depended so much not only for moral support but also for the dissemination of his ideas, had to take more elaborate precautions.

## HAWKS AND DOVES

In none of the controversy about the nova is there anything remotely as engaging as this dialogue. There is little amusement to be found in the two little books that antedate Galileo's pamphlet—either Lorenzini's *Discorso* or Baldassare Capra's *Astronomical consideration of the new and portentous star*. Capra claimed to have been the first to see the nova in Padua along with his teacher Simon Mayr on October 10, 1604, and in *his* book he claimed that though the nova was indeed above the moon, Galileo had not properly applied the system of parallax to measuring its distance from the earth. Such was the pettiness, the envy, and the hostility that the mathematician from Pisa raised from the start.

Two days before Christmas 1604, when van Maelcote gave his lecture on the nova to the Jesuit college in Rome, he was inhibited from drawing the full anti-Aristotelian conclusions of his views by the rules of his order;[38] otherwise he too might have been counted among those who took the supralunary position. So also might the otherwise obscure Raffaelle Gualterotti, whose discourse on the new star appeared in Florence in 1605, and, of course, Johannes Kepler himself, who published his *De Stella Nova* in Prague the following year.

Most of the others, however, roused Galileo's ire—and none so much as the man who would soon become one of the most irritating thorns in his side, Ludovico delle Colombe. The title of his book, published in Florence in 1606, says it all: *Discourse in which it is demonstrated that the new star which appeared last October 1604 in Sagittarius is neither a Comet, nor a [newly] generated star, nor created anew, nor apparent; but is rather one of those stars that were always in the heavens from the beginning; and as such conforms to both to true Philosophy and Theology, as well as to astronomical demonstrations*. No mutable heavens for him! Galileo was incensed, and wrote yet another pseudonymous response in defense of his own views under the name of one "Alimberto Mauri."[39] This work, just as sarcastic but much less flavorful than the one by "Cecco di Ronchitti," concluded that Colombe had better stick to his philosophical astronomy—which was bad enough—and leave mathematical astronomy out of it altogether.[40]

Then, in April 1606, as if exhausted by the nova controversy, Galileo did something that must have caused more than a few raised eyebrows. While waiting for his appointment as professor of mathematics at Padua to be renewed, he went to nearby Venice. This was a risky move, guaranteed to rouse suspicion, for it was at this very moment that Cardinal Bellarmine, instructed by Paul V, placed Venice and its chief theologian and politician, Paolo Sarpi, under the interdict. Sarpi's heretical—some said directly Protestant views—had brought this on himself. He responded to the interdict by having the Jesuits, whom he thought lay behind it, expelled from Venice.

And with whom had Galileo long been in touch concerning almost all his discoveries and theories?[41] None other than Sarpi himself. More than one Jesuit must already have taken note—to say nothing of the brilliant and ever-vigilant Bellarmine himself.

## CESI INTERVENES

The controversy about the nova of 1604 thus played a fundamental role in the formation of Galileo's polemical style and laid the foundations for his eventual rupture with the Church. It also marks the first serious entry of the Linceans onto the astronomical stage, indeed onto the broader scientific stage as a whole. Cesi's burgeoning anti-Aristotelianism[42] immediately engaged with the new controversies about the nature and constitution of the heavens. It must have seemed as if every issue raised by the new star was relevant to him and to his young Academy. The Linceans were familiar with the discussions about whether or not the sublunary atmosphere and bodies within it could be composed of fiery substances (Galileo, it will be remembered, claimed that it was simply a reflection of light on gases rising from the earth).[43] Cesi, of course, would immediately have grasped the threatening implications for the Aristotelian view of the cosmos.

Sometime in the first quarter of 1605, Cesi sponsored the publication of a treatise by Heckius on the subject of the nova. With the proud ascription "Lyncaeus" attached to its author's name, this was the first work ever to appear under the aegis of the Linceans.[44] It stood at the very beginning of the confluence of interests that soon brought Galileo and the Linceans together, with profound consequences for both sides. The book is so rare and apparently so obscure that it was never seriously studied by any scholar until 1988.[45] Galileo himself does not even seem to have known it, and even if he had, it is unlikely that he would have had much patience with it. It was too equivocal—and sometimes too reactionary—about the implications of the nova. On the other hand, a copy that recently appeared on the market in New York carries on its owner's binding the arms of none other than Cardinal Bellarmine.[46] Bellarmine knew about it from the outset.

By the end of 1604, Heckius was already in Prague. Fraternizing with the leading astronomers and mathematicians at the strange court of Rudolf II, he urged them to correspond with the Linceans. But Heckius felt his travails all too keenly. Probably the most diligent and intense of the original group, he had had enough of the lonely, vagabond life he had been forced to lead. His letters to Cesi are full of plaintive declarations about how much he longed to return to Italy and how desperately he missed his friends. He was in constant need of money and worried about sex.[47] Often these letters take on such paranoid tones that they seem to foretell the madness that would eventually be the end of him: Why has Cesi or the others not replied to him? he asks. Surely it cannot only be a matter of the post. Where is the money he has been promised? When can he return to Italy? There is not a place on earth that is safe for him, thanks to the influence of Cesi's father and his other enemies in Rome. Should he go on to Lithuania or Constantinople? But he is tired of traveling in the lands of Protestants and barbarians.[48] And so on and so forth.

The tone of these letters is so intimate, so fraught with the pressures of personal love and hostility, that one can only assume the bond between the two young men was sexual as well as intellectual.[49] And Cesi does indeed respond in the most loving and consolatory of tones, always assuring Heckius that he will eventually be able to return to Italy (incognito) and that he should continue the valuable work he is doing on behalf of the Academy. Stelluti and de Filiis love him too, he is the adamantine pillar of their society, they depend on him, and he should certainly not even think of leaving them, or marrying in order to drown his sorrows.[50] But Heckius is inconsolable.

Then, in October 1604, he sees the nova, perhaps even a week before Kepler did.[50] Soon it becomes the talk of Prague, and briefly it provides Heckius with a new focus in his life. Had he not been one of the first, he claims, to see the new star burning so brilliantly in the skies?[52]

For a few months Heckius worked and wrote on the problem with passionate intensity. Cesi immediately understood its scientific importance and its relevance to many of the Linceans' other projects, astronomical as well as geological (for how was one to account for brilliant and heated bodies on earth as much as in the heavens?). All these were beginning to disclose ever greater cracks in the whole Aristotelian scheme of things, and make Cesi and his friends doubt even further the authority of the ancients. And so, for the moment, he encouraged Heckius not to abandon his astronomical researches.

When Cesi founded his Academy in the previous year, Francesco Stelluti was the one who had been assigned responsibility for astronomical and mathematical matters. Heckius was allotted the whole broad area of philosophy and metaphysics, while Cesi took on botany, and De Filiis, never a very significant player in the group, history. But boundaries such as these fell away almost instantly. Like Cesi, Heckius had always been passionately interested in botany, while Stelluti helped Cesi in collecting and studying the fossils—and especially the fossil woods and underground fires—of Umbria. The notebooks Heckius kept on his journeys, as we have seen, are full of his studies of butterflies, serpents, mushrooms, and other things of the earth;[53] but even before his exile from Italy he (and Cesi too) had been concerned with cosmological matters. By 1604 Heckius had already written a number of manuscripts (most of them now lost) about astrological and astronomical problems, including one called *On the neglected science of the stars*. Here—to judge from the few comments on it in the Lincean papers—he took a wholly traditional view of cosmological matters; he asserted the validity of the Aristotelian point of view against those who believed in the existence of an infinite universe beyond the realm of the celestial spheres; and he attacked those who, like Bruno, argued for a plurality of worlds.[54] To have taken any other position would have been too radical, indeed too heretical, for someone who so hated Protestantism.

No wonder, then, that Heckius found himself caught up in the newest astronomical controversy. Almost from the outset, the nova presented major difficulties to the Aristotelian scheme of things. And how could Heckius have resisted the influence of Kepler, whom he was asking to write to the Linceans, and who saw in the new star further proof of the fact that the orbs were not solid, impermeable, or incorruptible?

On January 5, 1605, Heckius wrote to Cesi about the nova. His letter testifies both to his excitement and to his discomfort. Beginning in a typically plaintive way, he accuses the Linceans of having failed to write to him "at a time when fate has forced me to wander about the earth like the perpetual movement of a wandering star."[55] The issue struck a deep chord within the lonely young man and had already led to larger reflections on the profound instability of things on the earth and in the heavens.

With his complaints out of the way, Heckius could proceed. He described to Cesi how there were objects appearing in the sky that seemed to break every law of mathematics and were unsettling the foundations of traditional science ("antiquissimam disciplinam"). Recently, he wrote, some new movements had been observed in the skies that were keeping the astrologers very busy: the new star seen in October. Its beginnings had been clearly observed in the eighteenth grade of Sagittarius. There it could be seen before sunrise, glittering with reddish-white rays. It continued to burn even during daylight.

Was this a new addition to the skies, Heckius asked, or rather some variation of an existing heavenly body? The whole idea of the perfection and stability of the universe was being called into question. Could even the heavenly bodies themselves be subject to change and variation? The ancient truths seemed no longer to hold. It was hard to believe that something could arise in the heavens entirely anew, *de novo*. And then he paused for reflection: The law of necessity was a marvel. A man rises and falls; he rules and serves; he hides and comes forth into the open. So with the whole of history and the whole of the world; even this sublunary world was subject to change, in ways that confounded the wisest of men.[56]

Heckius had his treatise on the nova ready as early as January 24, 1605. On this day he sent a draft to Cesi for his approval.[57] In the accompanying letter he noted that it had been written in great haste, *raptim et ex tempore*.[58] Cesi must have been delighted. Their most talented member—however erratic he may sometimes have been—had produced a treatise on one of the most critical cosmological issues of the day. Now they could show to the world the true sharpness of their Lincean eyes. Cesi immediately set about finding a publisher for the treatise.

But the more he read of Heckius's manuscript, the more he became dismayed. His initial pleasure at receiving the fruit of Heckius's labors rapidly diminished at the evidence of his friend's habitual prejudices. All too clearly motivated by his animosity toward Protestants and frightened by the anti-Aristotelian implications of the nova, Heckius had attacked the very astronomer upon whom his conclusions depended—Tycho Brahe; and set upon everyone whom he regarded as an anti-Aristotelian.

Cesi realized that this was no way to proceed. The Linceans both needed and wanted the support of the astronomical *novatores*. Cesi himself had not the slightest inclination to renounce the Brahean view of the cosmos, not only because he believed in most of it, but also because he had no wish to alienate Kepler. The most advanced astronomical school of the time *was* the Copernican and Keplerian one, as Cesi well knew; and he wanted to be part of it—just as Galileo already was (whatever he may have said to the contrary). The painful truth was that Heckius's treatise, in the form he sent it to Cesi, was intemperate and unnecessarily provocative. Cesi could not let it go ahead as it stood. He had to alter it before publication; but he could not afford to derail his unstable friend even further. Heckius's professed Aristotelianism in cosmological matters was just what Cesi did *not* wish his young Academy to stand for. What should have been a calm and intelligent demonstration of Lincean skills had turned into a violently polemical tract.

Yet there was a great deal to applaud in the text. Heckius critically examined a series of twelve *sententiae*, or "opinions," on the nova. In the context of these pronouncements by selected authorities, Heckius did indeed make a few important and stimulating claims. In the fourth *sententia*, for example, he insisted that since the nova was so plainly a natural phenomenon, it made no sense for anyone to argue for supernatural or divine causes.[59] Here he articulated a problem with which Galileo himself would soon have to grapple. In the second *sententia*, Heckius declared that the star seen by the Magi on their way to Jerusalem could not have been a nova, since that star, after all, was a divine miracle; and it was probably a comet anyway, belonging to the sublunary sphere.[60] Furthermore, in the course of discussing a number of other *sententiae*, Heckius efficiently dispatched the improbable views of a whole number of other misleading philosophical and theological authorities on the subject, from the

Pythagoreans to Agrippa von Nettesheim and even Paracelsus.[60] All this seemed up-to-date enough (though not excessively so).

But the most important part of Heckius's treatise was his opening section, preceding his discussion of these *sententiae*. Here he made a most careful and thorough use of the Tychonian method of parallax in order to establish that the new star was indeed supralunary and therefore formed part of the firmament—and as such belonged to the realm of the fixed stars. The *nova* was indeed a *fixa*.[62] This, of course, was exactly the conclusion of Kepler and the other *novatores*. Up to this point, then, Cesi could not but have been pleased.

But then the impetuous Heckius went and spoiled everything. On several occasions, and only with small variations, he asserted that Brahe was a stupid, reprehensible Calvinist—and an anti-Aristotelian to boot.[63] Against those who maintained that there always had been stars that could not be seen by human eyes (except very rarely), he took the thoroughly traditional position that there was "not a single star, however small, which can be seen but was unknown to ancient and modern astronomers."[64]

Most troubling of all were Heckius's heated assertions that the nova offered no evidence whatsoever against the Aristotelian doctrine of the incorruptibility of the spheres. Cesi had begun to adopt exactly the opposite position. Already then he was starting to formulate his own views on the unity and permeability of the heavenly spheres.[65] All in all, therefore, Heckius's piece was an upsetting mixture of the most modern views of the *novatores* and the most reactionary views of the Aristotelians.

What was going on in Heckius's mind, Cesi wondered, and what was he to do about it? To some extent Heckius's positions were understandable, though his tone was not. Since he hated the Protestantism of his native Holland, how could he now align himself with the anti-Aristotelian heretics, most of whom were Protestant, or more than a little tainted with heterodoxy? He could not. Moreover, to have adopted a strongly heterodox position would indeed have jeopardized the possibility of returning to Italy, at a time when he wanted nothing more than to do just that. And then there was the problem of his own emotional lability. He simply did not have the courage to renounce everything he had learned, or to bring himself to draw the conclusions that now stared him in the face. It was all too much for his fragile sense of self. He could not think of himself as a true *novator*, a revolutionary; and so he found himself in a quandary. The tension in all this was obviously too much for him, and it found its outlet in the series of intemperate outbursts against the anti-Aristotelians that so marred his book.

What, then, was Cesi to do? He knew that if he divulged his wide-ranging misgivings to Heckius himself, the latter would fall to pieces. From everything that had gone before, Cesi had good reason to fear a lapse into insanity. So he decided to correct Heckius's manuscript before delivering it to the printer, without letting Heckius know what he was doing. This must have caused Cesi much pain; but he took out his pen and made his corrections. By comparing the manuscript of the *Disputatio*[66] with the published book we may gain some idea of the extent of his interventions.

First, the tone had to be modified. The tirades had to go. Cesi would prove here and elsewhere to be a clever, tactful, and restrained editor. Accurately he gauged what could be allowed to stand and what had to be eliminated or changed. He struck out all the passages in which astronomers like Brahe were accused of being crude, irrational, audacious, and full of temerity.[67] Despite his assertion of the supralunary position of the nova, Heckius could not bring himself to go so far as to maintain, as the *novatores* had, that new stars could suddenly appear and shine brilliantly for a while in the firmament before finally degenerating and discharging further new stars. To refer to the *novatores* as "certain babbling new philosophers" was bad enough, but it was worse to reprimand them repeatedly for their stupidity, boastfulness, and ignorance.[68]

Such blanket dismissals were unacceptable to someone as courteous as Cesi. So he deleted them. Of Brahe Heckius had sarcastically remarked that "this good man even goes so far as to boast of his novel and unheard-of opinions, claiming that he has overthrown the whole of Aristotelian science (*disciplinam*), and that Aristotle himself barely deserves to be called a student of philosophy."[69] This, in Cesi's eyes, was also going too far, and it had to come out. A book that appeared under the Lincean auspices had to be tactful. It would have given altogether the wrong impression if it was so hostile to the positions that Cesi and his fellows in the Academy supported.

On March 19 1605 Cesi wrote to Heckius:

> I was so happy to receive your letter, since everything that comes from my brother *Illuminatus* is so wonderfully congruent with my own inclination, and since all the physicists and mathematicians are simply hallucinating on the subject. Only Maginus [Giovanni Antonio Magini, the famous professor of astronomy and mathematics at Padua, who had also written about the nova] has published his opinion. He has the absurd idea that its matter is made up of the conflation of vapors. The other Peripatetics seem to be mired in a great swamp of confusion, and are now beginning to deny the celestial doctrines of Aristotle. So that I can demonstrate the acuity of the Lincean eyes to these triflers, I handed your opinion over to the printer, which will bring you great honour, you the *Illuminatus* who illuminates the darkness of others.[70]

Such words may have encouraged Heckius, but the fact is that Cesi was not being honest. There was much in Heckius's treatise that was not so "wonderfully congruent" with Cesi's own inclination (Magini's views were not that different from Galileo's, for a start). He had purged the manuscript of its excesses and its areas of major disagreement, and sent it directly to the printer, without notifying Heckius of the corrections. Even Heckius must have wondered whether anything was amiss when Cesi said that the Peripatetics (followers of Aristotle) were "mired in a great swamp of confusion" and were even beginning to deny the celestial doctrines of their master.

But he soon found out. When the work was published shortly thereafter, Heckius did not fail to notice all the changes that had been made. And despite all Cesi's efforts to mollify him, he was upset. He wrote to Cesi that "that which was first sent to the press to demonstrate the acuity of Lincean eyes to the Roman public has now appeared; but Illuminatus is not at all pleased with it, since as a result of the interventions of certain jealous people, much in it has been changed into things which he himself never thought or believed."[71]

No wonder Heckius was displeased. In this, the first publication by a member of the Academy, and published not in Prague, Florence, or Padua (as the other works on the nova had been) but in Rome, Cesi had not only censored and muted his judgments. He had turned the treatise into a much more radical work than Heckius would ever have dreamed of producing. In his prefatory letter to Cesi, the printer, Zanetti, said that he had received the disputation, sent to him by Cesi, and that it was "praised by many, and sought after by scholars everywhere."[72] But this must have been cold comfort to Heckius. Not only did the book no longer express all his own opinions, even Zanetti regarded the book as partly the work of Cesi. Zanetti praised Cesi, not Heckius, as a man of sharp intelligence and singular wisdom, who was admired by the whole of Rome for these mature fruits of his youthful labors. All this must have been terribly galling to Heckius. His subsequent letters make it clear that he felt defeated.

Discouraged from further astronomical speculation, Heckius now returned with renewed vigor to his studies of botany, mycology, and entomology. If the hidden things of the heavens were still escaping his grasp, and if the envious people who surrounded him did not accept his views on them, he would investigate the obscure things of the earth, down to the last mushroom, lichen, butterfly, fly, and worm.[73] If further collaboration with Cesi on cosmological matters was impossible, he would devote himself to the other Lincean projects, such as the massive encyclopedia of fungi they often referred to as the *Icones fungorum*, or the huge compendium of plant life known as the *Syntaxis plantaria*. He sent seeds of "Indian" (i.e., American) plants, presumably obtained from Rudolf's gardens, to Cesi, who gratefully acknowledged them.[74] Once more he took up his studies of butterflies, insects, snakes, and monstrosities. He began writing again to scientific correspondents all over Europe, to men such as Clusius in Leiden and the aged Pisan authority on diet and exercise, Girolamo Mercuriale.[75]

At the same time Heckius continued his correspondence with Cesi and with the two other Linceans. He needed to. In the letters that ensued—tender, loving, and difficult as before—he wrote not only of his many vicissitudes as he wandered across Europe, but also of the development of the young society. These letters reveal both the ambitions and the promise of the Academy in these critical years. Recognizing their own significance and aware of their potential appeal to the more advanced thinkers of their time, the four young men planned every aspect of their society for the future: its organization, rules, and possible candidates for future membership. Back in Italy, Cesi, Stelluti, and De Filiis drew up a code of ethics (including, then, a ban on marriage and on joining the priesthood), made financial plans, and outlined their future projects, both antiquarian and natural historical.[76]

Through it all Cesi remained concerned with Heckius's fate. He arranged safer and more efficient ways of sending Heckius money, and he constantly encouraged him to expand his range of learned contacts and to continue buying rare books for the Linceans. Most significantly of all, he arranged for a secure salary for the young Flemish draftsman and engraver who accompanied Heckius everywhere on his botanical and zoological expeditions.[77]

There was little more important to Cesi at this stage than the need for the Linceans to draw everything they saw in the world of nature and to publish their results. In this way they would be able to disseminate their knowledge and provide the basis for the planned illustration of future books on the subjects that gripped them most—or that seemed to reveal aspects of nature unknown or unrecorded by the ancients and even by the most modern authorities.

## SCRIPTURE AND THE FLUIDITY OF THE HEAVENS

But ought not Cesi to have regarded himself among the troop of "babbling new philosophers," as Heckius described the innovators who believed that new stars were indeed capable of being generated in the heavens?[78] Perhaps Cesi deliberately allowed this phrase to stand so as not to alienate traditionalist readers of Heckius's treatise too much. Even before the appearance of the nova of 1604, Cesi had begun to work on a project that would occupy him for many years. However intensely he may have gathered information about plants, fossils, and fungi, he returned almost daily to the complex piece of research he called his *Coelispicium*.[79] This was the work in which he sought to demonstrate the fluidity and permeability of the heavens, a doctrine that was diametrically opposed to the Aristotelian one of the incorruptibility of the adamantine spheres.[80] There was scarcely a day in which he did not turn his head from the earth to the sky. Both occupied him equally.

In the course of preparing the *Coelispicium* Cesi found himself coming to conclusions that were directly in line with those of Brahe, and that were consistent with much of the most advanced thinking of the time on the subject of stars, comets, and the composition of the heavens. Just as he was beginning to doubt the authority of Aristotle and his followers in the field of botany (particularly with regard to issues of generation and reproduction), so too he grew profoundly skeptical, if not downright disdainful, of the ancient Greek's views on the generation of stars and the nature of the heavens. Every day he and his friends were discovering plants unknown to the ancients. Why should there not also be new things in the heavens, things not seen before and only now becoming known to their sharp Lincean eyes?

Cesi's cosmological views emerge not only in his notes and drafts for the *Coelispicium*, but also in his correspondence from this time forward. He reasoned that the heavens were fluid, unitary, and permeable, and capable of allowing the passage of innumerable hitherto unsuspected celestial bodies (as none other than Bruno himself had also maintained).[81] The Aristotelian doctrine of the adamantine spheres was thus no longer tenable. And in support of his positions, Cesi needed Brahe (as well, of course, as Galileo). After all, it was Brahe who had demonstrated that the heavens were of such thin matter that the stars, freed of the solid orbs imagined by Aristotle, did not necessarily face obstacles in their courses. This in turn allowed for the formation of new stars that could appear briefly in the sky and then disappear again. Brahe had also shown how a star, even though composed of the same matter as other stars, could burn briefly and then peter out, possibly disintegrating into new stars. In other words, a star was not necessarily so perfect that it could last forever—yet another nail in the coffin of the doctrine of the incorruptibility of the heavens. No wonder that Cesi could not allow Heckius's attacks on Brahe and the whole anti-Aristotelian school to stand. He had no choice but to eliminate them from Heckius's manuscript.

But there was another, perhaps still more complicated, difficulty. Not even Cesi, with all his fine skepticism about authority, could take the step of asserting that his views of the fluidity of the heavens and everything they entailed were incompatible with Holy Scripture (which, as we shall see, they seemed to be). So he strove with determination and passion to reconcile the irreconcilable. For all his realization that there were more things in the heavens (as well as on earth) than Aristotle ever dreamed of, still Cesi could not bring himself to admit that his cosmological views were at odds with the Bible.[82] This was the central problem—and it would take Galileo to resolve it.

Cesi now began to draw up a list of scriptural passages that he believed supported his theories. Yet his closest friend and colleague had repeatedly insisted in the very first work their society published that the theory of the fluid and unified heavens was in direct contradiction with Holy Scripture. It is easy to imagine Cesi's consternation. When Heckius wrote that the doctrine of the unity and the fluidity of the spheres was not only "bursting with many absurdities" but was "in no way to be accepted,"[82] Cesi balked. In this instance it was not just a matter of moderating Heckius's intemperateness. There was too much at stake to allow a Lincean to reject in print the very doctrine that Cesi was developing with such conviction and commitment. At this point, Cesi realized, he could not just take out the excess; there was nothing to do but to change the very sense of the passage. And so the manuscript's "nullo modo recipienda"—"in no way to be accepted"[84]—became "aliis probabilior"—"more probable to others"—in the printed version.[85]

Such a change may seem to be characteristically tactful, but because it so completely alters Heckius's dogmatic view of the subject, it is perhaps the most significant of all the changes Cesi introduced in the course of censoring Heckius's manuscript. It was certainly the one that made the differences between them plain, and would have the greatest reverberations in the near future.

It would be hard to overestimate the importance of Cesi's interventions in Heckius's draft for the *Disputatio* on the *nova* of 1604. They provide eloquent testimony to Cesi's growing awareness that his own views on the fluidity of the heavens actually aligned him with those who maintained that such views were incompatible with Sacred Scripture. Over the course of the next few years he developed these views further, often in conjunction with Galileo or as a result of his reading of Kepler. In June and again in July of 1612, just as he was putting the finishing touches to his *Coelispicium*, Cesi wrote to Galileo that Copernicus had never even mentioned the solidity of the spheres, that the work of Brahe and Kepler totally invalidated them, and that the Copernican system would be even more convincing if one accepted Kepler's discovery of the elliptical orbits of the planets.[86] In this respect Cesi went further than both Copernicus and Galileo himself. At this stage, Galileo was still cautious about wholly renouncing the basically Ptolemaic system of epicyclic and eccentric orbits for the planets. Cesi kept on pushing him to do so.

But however advanced and up-to-date the cosmological aspects of the *Coelispicium* may have been, Cesi could never bring himself to admit the incompatibility of the new views of the cosmos with the texts of the Bible. Whether he was just too aware of the dangers of doing so, or whether it simply ran against the grain, is difficult to tell. This is typical of the dilemma with which he never stopped struggling. For many years Cesi thought he would be able to reconcile the irreconcilable—and not just with regard to the relations between the new cosmological discoveries and the sacred old texts. Even as he saw Galileo heading closer to his collision with the Jesuits (and finally with the pope himself), he never ceased to hope that he could demonstrate that in the end their respective positions were not fundamentally at odds with each other. But of course they were.

FOUR

# *The Telescope*

## IMPERFECTION IN THE HEAVENS

### DELLA PORTA, GALILEO, AND THE TELESCOPE: EARLY CONFLICTS

For years Rome remained too dangerous. Cesi's father continued to do everything he could to block the progress of the subversive society his son had founded. Cesi himself still thought of setting up their "Lynceum" in the more propitious environment of Naples. If Heckius must return south, Cesi had written in June 1605, he should do so incognito, and only reveal himself when safely with De Filiis in Terni.[1] In 1606 Heckius did manage to return for a few meetings of the Academy in Rome, but had to leave again, still haunted by the elder Cesi; and he set out for Madrid, restless and paranoid as ever. Stelluti moved back and forth between Parma and his hometown of Fabriano in the Marches, with only fleeting visits to meet his friends in Rome. De Filiis stayed in Terni, for the time being a kind of anchor for the others.

In such ways were the Linceans constrained, right from the start. But they continued their natural historical researches, especially on fossils, fungi, and plants, observing and collecting as much as they could. Almost daily they discovered things never mentioned by Aristotle or any of his ancient or modern followers. Often they wrote in the special code they had devised to communicate their secrets and discoveries to one another. For the time being astronomy receded into the background.

But then news of the telescope and its possibilities began to reach them.[2] Sometime in early 1609 Galileo had turned his *cannochiale* to the night sky;[3] and from then on matters moved ahead with unanticipated speed.

For a start, Giovanni Battista della Porta, the sage of Naples and father figure to Cesi and the other Linceans, began snapping at Galileo's heels. Had he not been there first? "I have seen the secret use of the eyeglass and it's a load of balls [*coglionaria*]," he wrote to Cesi on August 28, 1609.[4] "In any case, it's taken from book 9 of my *De Refractione*," he pointed out, irascibly.[5] In the same letter he made a simple drawing of his own invention and explained that it consisted of one tube with a concave lens inserted into another with a convex lens.

> Looking just with the first one will see distant things nearby [*le cose lontane, vicine*, a phrase that would continue to reverberate in the next few years]; but since one cannot see within the tube [i.e., at the focal point], they appear obscure and indistinct. However, if one places the other tube with the concave lens inside, it has the contrary effect, and one sees things clearly and directly. It goes in and out like a trombone, and one adjusts it according to the eyesight of the beholder, since eyesight always varies from one beholder to the other.[6]

In other words, Della Porta had long since established the basic principle of the telescope. But by the time he wrote to Cesi it was too late. Galileo had beaten him to the post. He had made a far more refined instrument than Della Porta ever envisaged—and had actually begun to use it.

Night after night Galileo observed the heavens, and news of his discoveries began to leak out. At the beginning of 1610, Della Porta added a telling postscript to a note to Cesi (about Della Porta's recent works on atmospheric phenomena and on chironomy).

> I regret to say that the invention of the eyeglass in a tube was mine, but the Professor from Padua Galileo adapted it, and with it he has found four other planets in the sky, and thousands of new fixed stars, and just as many, never seen before, in the Milky Way, and great things on the orb of the moon, which fill the world with astonishment.[7]

Indeed. What Galileo saw in December 1609 and January 1610 so excited him that he wrote in white heat. By mid-March the book was out. In it he described the spectacular observations he had made with the aid of his telescope: the mountains and craters of the moon (fig. 4.1);[8] the stars of Orion and the Bull; the uncountable stars of the Milky Way and its nebulae; and the difference between the light points of the stars and the disks of the planets. Pride of place went to his discovery of the satellites of Jupiter, which he named the Medicean stars as a way of flattering Grand Duke Cosimo II de' Medici of Florence.[9]

The year before, Galileo had read Kepler's great *Astronomia Nova* proving that the orbits of the planets were not circular but elliptical. Now it became clear that Galileo's discoveries threatened to dismantle the whole basis of the old Ptolemaic cosmology still further. They subverted the Christian view of the divine perfection of the universe, and made it still more difficult to resist the claim of heliocentrism. And as soon as he demonstrated that the surface of the moon was pitted, creviced, cracked and riven, of the same matter as the earth, the uniqueness of our world must have seemed to be lost forever.

Once Galileo suggested that the earth moved round the sun, that the

Fig. 4.1. Engravings of the moon. Galileo, *Sidereus Nuncius* (1610), p. 10 verso.

moon moved round the earth, that Jupiter had satellites that moved round *it*, then it was clear that there were several centers of cosmic motion, not just one. In the next two months Galileo observed the strange appearance of Saturn and its rings (which he thought were two stationary satellites located near it), and by the end of the year he had verified the phases and variations in size of Venus. These showed that it circled the sun, and so too—by analogy—did the other planets. None had light of their own; they all received it from the sun.

Swiftly the debates erupted. Many applauded Galileo; others attacked

Fig. 4.2 Engraving of the moon. Galileo, *Sidereus Nuncius* (1610), p. 10 recto.

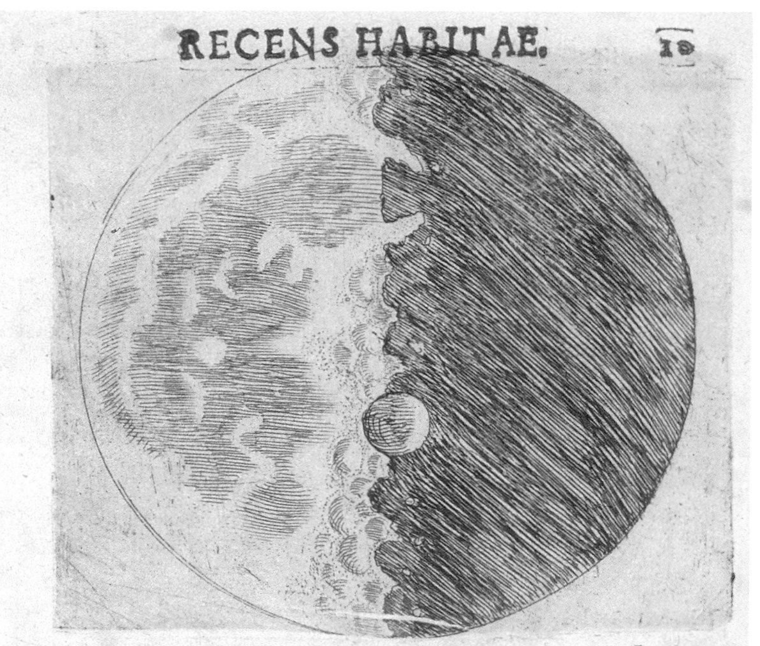
RECENS HABITAE. 10

Hæc eadem macula ante ſecundam quadraturam nigrioribus quibuſdam terminis circumuallata conſpicitur; qui tanquam altiſſima montium iuga ex parte Soli auerſa obſcuriores apparent, quà verò Solem reſpiciunt lucidiores extant; cuius oppoſitum in cauitatibus accidit, quarum pars Soli auerſa ſplendens apparet, obſcura verò, ac vmbroſa, quæ ex parte Solis ſita eſt. Imminuta deinde luminoſa ſuperficie, cum primum tota fermè dicta macula tenebris eſt obducta, clariora mõtium dorſa eminenter tenebras ſcandunt. Hanc duplicem apparentiam ſequentes figuræ commoſtrant.

him furiously. The story is well-known. In Prague Kepler read the *Starry Messenger* almost immediately, and while acknowledging Della Porta's role in the development of magnifying lenses, confirmed the importance and validity of the Galilean discoveries.[10] Already on April 19 he sent Galileo his own *Conversation with the Starry Messenger*, published shortly thereafter in Prague, Florence, and Frankfurt and bound together with a reprinted version of Galileo's work. Here Kepler developed some of the implications of the Galilean discoveries (he did not yet have a telescope himself) and set out some of the differences between them. But his support and approval were altogether clear.

So too was that of the Linceans. There could be no doubt of Galileo's genius. Already by then they felt that he was one of them. But having just elected Della Porta as their fifth member, Stelluti was in something of a quandary. Writing to his brother Giambattista in distant Fabriano in September 1610, he acknowledged the importance of Galileo's discoveries, but asked whether credit should not also be given where credit was due:

> by now I'm sure you've already seen Galileo, that is his *Starry Messenger*, and the great things he is saying. . . . but it's now thirty years since Giovanni Battista della Porta wrote about it in his *Natural Magic* and his *On Refraction*, so that poor Galileo will be shamed.[11]

On the other hand, though still fretting about the matter of priority, Della Porta soon began to realize that he could not really have achieved what Galileo had. Writing to Johannes Faber in Rome, he admitted that

> I really am delighted that my rather rude and slight invention has been elevated to such use by the talent and resourcefulness of the most learned mathematician Galileo Galilei, who has shown that many planets wander in the heavens, and that so many new stars shine again in the firmament, which had lain hidden for so many centuries. . . .
>
> I myself had previously observed the cavities and elevations of the moon, the galaxies, the Pleiades and the lesser stars, but the imperfection of my instrument and the defects of old age prevented me from seeing the stars wandering round Jupiter.[12]

It was too late to quibble, however. There were much more important matters at stake than the question of priority. It must have seemed to the Linceans and to the more forward-thinking astronomers of the time that only the deepest reactionaries—and there were many of them—could maintain their commitment to the solidity and incorruptibility of the spheres, and their hostility to the heliocentric doctrine of Copernicus. Even Jesuits such as Christopher Clavius, the famous professor of mathematics at the Collegio Romano, were beginning to waver.[13] If the appearance of the *nova* of 1604 had already called such positions into question, how much more so did Galileo's discovery of new planets and new moving stars! Serious difficulties were now faced by all those who attempted to cleanse the supposedly immutable spheres of the new celestial bodies, or suggested that the rugged surface of the moon was a result of atmospheric interference, or that the Medicean "stars" around Jupiter were actually only optical illusions produced by the lenses of the telescope. The appearances had still to be saved.

And so the forces of reaction gathered steam. The chorus of opposition grew. Much of it centered around a problem that had already emerged in the case of the nova controversy. This was the conflict between what Galileo observed with his telescope and what Holy Scripture had to say about the heavens and heavenly phenomena. How, for example, could the Medicean stars even exist, if the Bible made it clear that God had only created seven planets (the sun, the moon, and the five visible planets)? After all, every authority, whether Hebrew, Greek, or Latin, agreed that the planets were symbolized by the seven lamps of the golden candelabra referred to in Exodus 25:37 and Zechariah 4:2.[14] And so on and so forth. In such obscure and implausible appeals lay the seeds of the debates that soon followed, as well as the beginnings of Galileo's downfall.

The central conflict, however, was between Copernicanism and the Bible. Here Galileo's old adversary from 1605, Lodovico Delle Colombe, surfaced again. In his frankly titled *Contro il moto della terra* of 1611, he took up the arguments. Why was the Copernican notion of a moving earth untenable? Why was it irrefutably at rest in the center of the universe? Because the Bible said so. In support of this claim, Delle Colombe adduced a whole slew of passages from Scripture. Had not the psalmist said "you fixed the earth on its foundations, that it not be moved forever" (Psalm 104:5)? Had not the chronicler written "the world shall be stable, that it be not moved" (I Chronicles 16:30)? Had not Job said "He suspended the earth upon nothingness" (Job 26:7)? Nothing could have been clearer than these lines. Solomon himself had said in Proverbs 30:3 that heaven was up, and earth down, while the very first chapter of Ecclesiastes contained the crucial words "the sun rises, and the sun sets, and returns to its place, from which it arose" (Ecclesiastes 1:5). A barrage of further scriptural passages followed, all making the same point: Galileo's so-called discoveries could in no way be reconciled with the irrefragable truth of Scripture.

The situation was beginning to look menacing. In the wake of the discoveries announced by the *Sidereus Nuncius* there was still much work to do. But more and more everything Galileo saw and everything he concluded from his observations went against the grain. He realized that the only way to explain what he was seeing in the heavens with the telescope was by wholeheartedly embracing the heliocentric hypotheses of Copernicus. Since the pace of discovery was so great, and since the opposition was so rapidly gaining in vigor and strength, there was no time to lose. He had to go to Rome and stake his claims there.

In the first place he wanted to go to the Jesuit college, "the most eminent emporium of the arts and religion," as one contemporary described it.[15] It was responsible for the early education of the scions of some of the most powerful Roman families, and the Jesuits' own training was famously rigorous. The theological curriculum was Thomist, the scientific one strictly Aristotelian. Galileo decided to discuss matters with his old friend Clavius, author of the most influential geometrical textbook of the day, and with his talented pupil, Christof Grienberger, the German who had succeeded Clavius to the chair of mathematics at the Collegio Romano.

Clavius had begun as an unequivocal supporter of Ptolemaic geocentrism.[16] But by 1610 the open-minded professor had struck up a warm friendship with Galileo. Despite his initial skepticism about such discoveries as the new fixed stars and the wandering stars or satellites of Jupiter (he was unable to see the Medicean stars himself and may have shared the suspicion of many that they were either optical illusions, or no more than flaws in the lenses of the telescope),[17] he was beginning to come round. He is reported to have said that if such things did not actually exist, one would have had to put them inside the telescope (as a number of his colleagues actually went so far as to claim).[18]

Grienberger, on the other hand, was less certain. He wrote a letter setting out some of the mathematical problems with Galileo's proposals; and he shared the view that the discovery of seven new stars (that is, including the satellites of Jupiter), and the claim that they were moving and not fixed, was, as a Roman contemporary put it, "*contrary to the opinion of every ancient philosopher*."[19] These were all issues that Galileo now wanted to address openly. He also needed to confer with Father Giovanni Battista Lembo, well-known for his instrument-making skills, and with whose aid Galileo hoped to construct better versions of his telescopes.

Over everything, however, there still loomed the ominous figure of Robert Bellarmine. Once a Jesuit himself, no one was more influential at the Collegio Romano than he. Despite his initially friendly attitude toward Galileo (and he had always had cordial relations with Cesi), the cardinal had become more and more concerned about the heliocentric implications of the new observations by the rather too confident mathematician from Florence.

Bellarmine now set up a commission of four Jesuits—Clavius, Grienberger, Lembo, and van Maelcote—to judge the validity of Galileo's discoveries. Worried, Galileo had himself written a letter to the mathematicians at the Collegio Romano, asking them to resolve five questions pertaining to his astronomical discoveries. And so he decided to go to Rome himself to receive their verdict, in their heartland.

On March 30—Ash Wednesday—1611, barely a day after he arrived in the city, Galileo went straight to see Clavius. He had his telescope, or *cannochiale*, with him, so that together they might look at the heavens. Clavius was so convinced by what he saw that in reviewing Galileo's discoveries, he declared that "since these things are so, astronomers must reconsider how the celestial orbs are constituted in order that the phenomena can be saved."[20]

The roles of all these actors in the Galilean drama have long been well-defined; but there was one man who played a more crucial role than any of the others in bringing Galileo to Rome in 1611 and in arranging the events that attended upon it: Federico Cesi. The young prince accompanied Galileo at every step of the way, was present at almost every important occasion, and helped him devise the strategy, sometimes successful and sometimes a failure, by which he presented his discoveries. It was he, more than anyone else, who stood alongside

Galileo as together they fired the opening salvos of their pro-Copernican and anti-Aristotelian campaign. The evidence for his role is substantial.

## ROME, 1611

On the evening of April 14, 1611, Galileo took his telescope with him up to the garden of Monsignor Malvasia on the Janiculum in Rome, just behind Porta San Pancrazio. There he demonstrated the use of his new instrument to a group of eight friends and fellow investigators who had gathered specially for this purpose. They came at the instance of Cesi: Giulio Cesare Lagalla, medical doctor and professor of logic in Rome; Antonio Persio, learned philosopher and natural scientist; Giovanni Demisiani, the Greek, who was Cardinal Gonzaga's mathematician; Antonio Pifferi, the Camaldolite monk and lecturer in mathematics at Siena; Girolamo Sirtori, expert in optics from Milan; and the botanists Johannes Faber and Johannes Schreck, both Germans living in Rome.

All night long they looked at the stars, the moon, and the planets through the telescope, waiting for the clouds to clear and discussing each new sight with increasing enthusiasm.

As day broke, they trained the telescope on the three simple lines of Sixtus V's inscription above the door of the papal palace at San Giovanni in Laterano, way across Rome: SIXTUS V PONTIFEX MAXIMUS | ANNO PRIMO. They read it clearly, as if to verify the perfect efficacy of the instrument with which they had looked at the heavens.[21] But still, as they all well knew, this would not satisfy their many opponents who believed that while it was good for the magnification of things on earth, the telescope was deceptive and distorting when it came to celestial phenomena.

To the Aristotelians, everything in the heavens ought to be capable of being seen with the naked eye. Except as a result of optical illusion, or of flaws in the lenses themselves, the telescope itself was not capable of producing new and hitherto unseen bodies in the spheres beyond the moon—or so the strong Aristotelians maintained.

Not for nothing did Cesi and his friends call themselves lynx-eyed, for they knew, and were daily discovering, that there was indeed more to be seen on earth and in heaven than was immediately accessible to the ordinary human eye. As Galileo had already made clear in his satirical treatise on the nova, what could not be seen could indeed exist. After all, as he put it in a magnificent letter on the telescope and the powers and limitations of sight a few weeks later to Piero Dini, was it not commonly acknowledged that the eyes of eagles and lynxes could see better than those of humans?[22] And did not everyone know that lynxes and eagles could see things inaccessible to normal, unaided sight—just as that figure from classical mythology, also named Lynceus, who could see things clearly at a distance of one hundred and thirty miles and penetrate the trunks of trees with the eyes alone?[23]

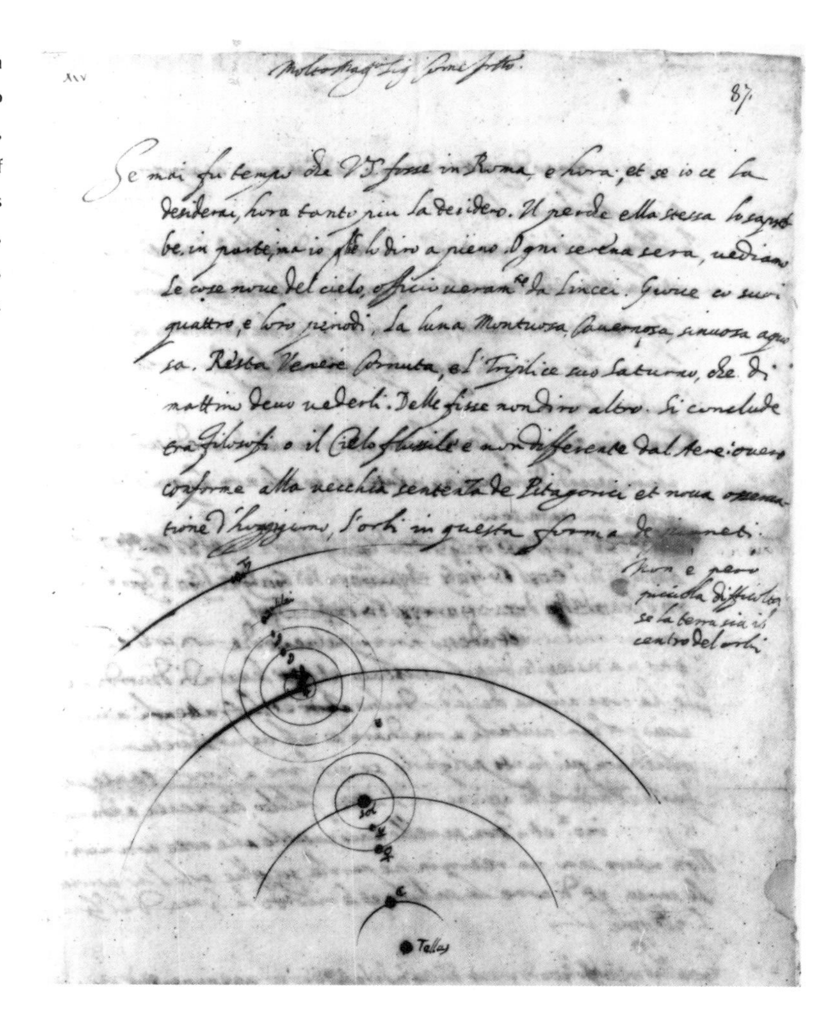
Molto Mag.co Sig.r Com.e Oss.mo

87

Se mai fu tempo che V.S. fosse in Roma, è hora; et se io ce la desiderai, hora tanto più la desidero. Il perde ella stessa lo saprà be, in parte, ma io glie lo dirò a pieno. Ogni serena sera vediamo le cose nuove del cielo, officio veram.te da Lincei. Giove co' suoi quattro, e loro periodi; la luna montuosa, cavernosa, sinuosa, acquosa. Resta Venere cornuta, et il triplice suo Saturno, che di mattina devo vederli. Delle fisse non dirò altro. Si conclude tra filosofi o il Cielo fluido e non differente dal aere, overo conforme alla vecchia sentenza de Pitagorici et nova openione d'huggiani, l'orbi in questa forma de pianeti.

Non è però piccola difficoltà se la terra sia il centro del orbi



Fig. 4.3. Letter from Federico Cesi to Francesco Stelluti, Rome, April 30, 1611, showing drawing of the orbits of the planets around the earth. Rome, BAV, MS Vat. Lat. 9684, fol. 87r.

At the end of the month Cesi wrote to Stelluti, away on family business in Fabriano:

> If ever there was a time for you to be Rome, it is now. . . . every clear evening we see new things in the heavens, truly the task for Linceans: Jupiter with his four [satellites] and their periods, the mountainous, cavernous, undulating, watery moon; horned Venus remains, and the triple aspect of Saturn, which I ought to see in the morning. Of the fixed stars I shall say no more.[24]

On the same sheet Cesi drew a simple diagram consisting of five large concentric circles (showing the successive orbits of the earth, the moon, the sun, Jupiter, and Saturn), intersected by the much smaller orbits of Venus and Mercury (circling the sun), and the satellites of Jupiter (fig. 4.3). As if to stress Galileo's achievement in discovering the Medicean stars, Cesi simply annotated the last of these with the word "Galileo." At the center of everything stands the earth; but at the bottom of this part of the letter, in the margin to the right of the diagram, Cesi added an afterthought: "Still, there is no small difficulty if the world is the center of the spheres"[25]—surely one of the most telling understatements in the history of astronomy.

Ten days after Cesi took Galileo up the Janiculum, Bellarmine's Jesuit commission delivered its verdict. On the face of things, it could not have been more favorable. The members of the commission declared that Galileo's observations were right. At the same time, however, they solidly refused to be drawn on the full implications of his discoveries. While they were prepared to allow the justness of several of his anti-Aristotelian conclusions, such as the fact that Venus waxes and wanes like the moon, his nascent Copernicanism was not something they could bring themselves to accept at all.[26]

Still, the occasion was one for celebration. Despite the murmurs of dissent, the Jesuits of the Collegio Romano wanted to acknowledge the distinction and fame of their visitor. On May 8, they held a glittering reception for Galileo in their new building. He made his entrance accompanied by the young prince Cesi. The buildings of the College were decked out; everyone was there—cardinals, princes, prelates, as well as the Jesuits themselves. Clavius and his junior colleagues attended: Grienberger, Lembo, and van Maelcote; and then their students, including Johann Adam Schall from Germany, who would later use the astronomical discoveries of Galileo to impress the imperial court of China; Paul Guldin from Vienna, who would go on to defend Galileo against the fiercest of his opponents; and the great French mathematician, so esteemed by Leibniz, Grégoire de Saint Vincent.[27]

Van Maelcote gave the introductory eulogy; passages from the *Starry Messenger* were read and discussed; and public demonstrations of Galileo's chief discoveries were held, along with experiments proving them. And Galileo showed that Venus revolved round the sun.[28]

It could hardly have been a more brilliant occasion, this triumph for Galileo. Or so it seemed. The professors of philosophy were skeptical, the theologians irritated—some, indeed, outraged. Galileo was asserting nothing less than the heliocentric hypotheses of Copernicus! At least Copernicus had left his heliocentric views just as hypotheses. But Galileo was bolder, and, not for the first time, was drawing the clear consequences of what he had plotted in the sky. For a brief while the skeptics were silent.

At the College itself they were waiting for the opportunity to ambush him. Many of the Jesuits were anxious about the threat to the doctrines on which all their teachings depended, and saw all too clearly the implications for the Christian view of the uniqueness and centrality of the earth. Bellarmine and half a dozen other cardinals had already written a cautionary note to the inquisitors in Padua, warning them to keep an eye on Galileo.[29] The Jesuits in Rome may have put on a good show for him, but beneath it all they knew perfectly well that if Galileo were to continue with his discoveries and keep drawing out their full implications, he would have no option but to become their implacable opponent—however covertly it might be necessary for him to occupy such a position.

Most of his friends were only too aware of the danger signals. In a justly celebrated letter written from Padua on May 6, two days before the reception at the Jesuit college, his old friend Paolo Gualdo did not mince his words:

> I have not yet found either a philosopher or an astrologer who is willing to support your opinion that the earth turns; even less the theologians. Therefore think very hard before you assertively publish this opinion of yours as true, because many things can be said for the sake of having an argument but which it is not good to assert as true—especially when the universal opinion, imbibed, so to speak from the very beginning of the world, has been exactly the opposite.[30]

He was right. In the wake of the honors accorded to Galileo by the Jesuits, the general of the order, Claudio Acquaviva, could stand it no longer. It was bad enough that Galileo should hold the views that he did, but that these should find a sympathetic audience among his members was intolerable. And so he decided to draw a line in the sand.

Just over two weeks after the celebrations in honor of Galileo, Acquaviva sent out a letter to the Jesuit houses, in which he emphasized the need for a return to what he described as "solid and uniform doctrine" in their teaching of theology and philosophy.[31] This he defined exactly. "Solid and uniform doctrine" henceforward would mean strict adherence to the teaching of Saint Thomas Aquinas in theology and Aristotle in philosophy. It was not enough just to avoid error; it was a matter of adhering to the basic principles of these two fundamental sources of Jesuit education.[32] The unswerving Jesuit principles of obedience first articulated by its founder, Saint Ignatius Loyola, were to be firmly applied. "I will believe that the white I see is black, if the hierarchical Church so defines it," Saint Ignatius had famously said in the *Spiritual Exercises*.[33] And Bellarmine himself had long before (in 1588) written a whole treatise on the subject: *On Obedience which is called Blind*.[34]

What problems these principles of obedience posed to those who were sympathetic to Galileo! There were many, like Grienberger, who were caught in the dilemma of choosing between their religious commitments and the new anti-Aristotelianism.[35] Even when they had the insight and good sense to see the validity of at least some of Galileo's opinions, such men did not dare indicate their agreement, except very guardedly. Galileo may indeed have retained some influential supporters—members of the Medici family in Florence, for example, and the occasional cardinal, such as Maffeo Barberini; but there was now only one group to whom he could turn in full and almost absolute confidence: the Linceans.

## BOTANY AND ASTRONOMY: THE ACADEMY EXPANDS

On April 25, 1611, Cesi formally invited Galileo to become the sixth member of the still young Academy. He agreed, and he signed his name in their album beneath that of the man who was almost his scientific antithesis, Giovanni Battista Della Porta. The Academy was on its way. This was just the spur it needed. From now on the task of the Linceans would be not only to dismantle the old forms of natural sciences but to rebuild them anew, on the basis of their sharp-eyed awareness of the latest discoveries, and by utilizing the most modern technologies of investigation available to them. For they did not stop at the simple collecting of material: the larger aim, from then on, would be to order the world of nature in the most efficient way possible, by analysis of the inner structures of outward form, and by a never-tiring search for the order and logic of everything on heaven and earth.

To this end, Cesi, often in consultation with Galileo, began to enlist the experts he needed. Eight days after Galileo signed his name in the album of the Academy, Johannes Schreck from Konstanz, who called himself by the latinized version of his name, Terrentius, joined their ranks. By the end of the year, Schreck had decided to become a Jesuit, and since the rules forbade any Lincean from joining a religious order, he was obliged to resign. But he continued to contribute substantially to the Lincean botanical projects; and when sent to China by his order, he wrote them remarkably detailed accounts of the substantial achievements of Chinese astronomy. Having been appointed by the last Ming emperor to reform the Chinese calendar and adopting a Chinese name himself, he became one of the very first conduits of astronomical (and also botanical) information between East and West.[36]

Next came Johannes Faber, the learned doctor from Bamberg, who, like Schreck, had also attended the meeting on the Janiculum on April 14. It was some years later that he penned the exultant lines about Galileo's telescopic observations: "O deed audacious! To have penetrated the adamantine walls of heaven with the help of so fragile a crystal!"[37] Faber was made a Lincean at the end of October.[38] For some time Faber had been a well-known figure in Roman intellectual and medical circles. He was an excellent plant pharmacologist, and Cesi had long been impressed with his expertise in the two fields that now seemed of most concern to the Linceans, astronomy and botany. Their abundant correspondence about the stars is interspersed with reflections on botanical problems; and Faber became one of the most diligent, devoted, sensible, and consistent of the Linceans.[39] Six weeks later the third German in this trio, Theophilus Müller, also a distinguished botanist and professor of the subject at the University of Ingolstadt, signed on.

All through 1611 astronomy and botany went hand in hand. In May Cesi invited Galileo to go to his Roman palace to look through some five hundred illustrations of American plants and to give his opinion on their identifications.[40] By the end of the year Schreck was writing about just these plants as well as engaging in "celestial speculations."[41] And in October Cesi wrote to Galileo from Tivoli about a botanical excursion that neither he nor the friends who accompanied him would ever forget. "In these last few lovely days I have visited and researched my Monte Gennaro nearby, with four very learned botanists [Faber, Schreck, Müller, and Henricus Corvinus, a Flemish pharmacologist living in Rome];[42] so I have not been able to reply to your very welcome letters until now."[43]

Preserved among Faber's voluminous papers is a brief account of this expedition, followed by a long list of the forty-one different plants they collected on that day, and another ninety-three collected in the course of various expeditions to Monte Gennaro during that intensely fruitful autumn.[44] The most notable aspect of this list is the fact that it contains a large proportion of binomial names, thus anticipating by more than a century the rules of nomenclature devised and codified by Linnaeus (a good number of the names are in fact the same as or very close to the modern Linnaean ones).[45]

The following year saw more inscriptions than ever before or after. Giovanni Della Porta's son, the eighteen-year-old Filesio, was allowed to join on January 20, 1612, perhaps only as a token of respect for his famous father; but from then on the majority of the new members were recruited for more compelling reasons. There were no fewer than eight more in that year. Before January was out, Nicolò Antonio Stelliola and Fabio Colonna, both of Naples, joined the society. Their membership was crucial. Condemned by the Sacred Office in 1595 for his Copernican views, Stelliola had been incarcerated in the same prison as the two great Dominican heretics, Bruno and Tommaso Campanella, who would soon be writing to Galileo himself in support of his Copernicanism. Stelliola emerged to devote himself not only to botany and pharmacology,[46] but also to mathematics and astronomy, and above all to the new instrument perfected by Galileo. His *Telescope or Heavenly Mirror* was posthumously published with the help of the Linceans and the financial support of Cardinal Francesco Barberini, while a mere sketch of his encyclopedic range of interests appears in the outline he published in Naples in 1616 of his proposed *Encyclopedia Pithagorea*. The title is itself significant, for the Pythagorean school had come to represent the antithesis of that of Aristotle, and Pythagorean cosmological doctrines were generally seen to be in line with Copernican astronomy. In a letter to Galileo about the two new members from Naples, Cesi described Stelliola as "philosopher, doctor, mathematician (and I believe a Copernican), excellent in literature, and especially Greek"; while Fabio Colonna was "a nobleman well-versed in Latin and Greek and an excellent naturalist, as one can tell from the two works of his that have been printed—and also a great expert in law."[47]

Colonna provided the link with the great Italian herbalists of the sixteenth century, chief of whom was Andrea Cesalpino, also Faber's teacher in botany. The two works referred to by Cesi in his letter to Galileo were the *Phytobasanos* of 1592, and its continuation, the *Ekphrasis* or "description of some lesser known plants," published in Rome in 1606 (of which a second part would be published in 1616). Colonna was much admired by the members of the growing Academy, not least because of the fact that in his *Phytobasanos* ("touchstone of plants"), Colonna had proven his skills at combining a deep knowledge of the traditional Dioscoridean material with a clear ability to revise that knowledge in terms not just of his own fieldwork but also of the most advanced writings about botany and the new specimens that were daily pouring in from home and abroad. It is a masterpiece both of critical observation and of pharmaceutical experiment. Indeed, Colonna's own commitment to botany had been inspired by his discovery after long excursions in the Campanian countryside that the plant Dioscorides called the *phu* was in fact valerian (fig. 9.1). Like the other Linceans, he believed in the importance not just of personal observation but in the critical role of accurate reproduction in communicating new knowledge and in resolving issues of nomenclature that could not be settled by purely verbal description. Both in the *Phytobasanos* and in the *Ekphrasis* (as also in his lesser works), Colonna pioneered the use of a new technique—etching—for the illustration of plants (e.g., figs. 9.1–2. Even more than copper engraving, and certainly more than wood engraving, etching offered the possibility of conveying more of the subtleties of the structure, morphology, and texture of plants.

The 1606 *Ekphrasis* also pioneered the use of a lens in the examination of cryptogams, those forms of life that seemed not, on the surface, to yield their reproductive secrets. Enlarging two forms of fungal growths—both of which he called "lichens"—with a lens,[48] Colonna anticipated the Linceans' own intense use of the microscope as a means to discovering the deep reproductive mysteries of fungi, bryophytes, and pteridophytes almost twenty years later.[49]

Five days after Colonna's enrollment on January 27, another Neapolitan joined the Academy. The Linceans needed an expert Arabist, at least partly in order to have him translate for them some of the classic works of Arabic science; and so they enrolled Diego de Urrea Conca, who knew Turkish and Persian too, thanks to his role as interpreter at the court of Fez and for the king of Spain.

In April Angelo De Filiis (the brother of the recently deceased Anastasio) was signed on and named librarian of the Academy.

In June yet another Neapolitan became a Lincean. This was Luca Valerio, whom Galileo later referred to as the "the great geometer and the new Archimedes of our age." Sixteen years older than Galileo, Valerio briefly played a central role in the Lincean dissemination of some of Galileo's most fundamental ideas. Known for his outstanding abilities in Greek as well as his important work in applied mathematics, he had been professor both of mathematics and of civil philosophy at the Sapienza in Rome. Galileo much admired his 1604 book *On the center of gravity of solids*, which occasioned the renewal of their friendship, begun in Pisa fourteen years earlier. It was Galileo who proposed Valerio to Cesi, and Valerio was the Lincean to whom the overall editorship of academic publications was assigned. He was given charge of the publication of one of the most critical documents for the history of the Linceans, the *Lynceographum*, the massive outline of the rules, regulations, aims, and ideals of their society,[50] in some ways akin to two even more Utopian projects: Tommaso Campanella's 1623 *City of the Sun* and Francis Bacon's *New Atlantis* of 1627.[51] It never appeared, except in summary form,[52] and Cesi continued working on it until he died. Valerio also played a crucial role, as we shall see, in the publication of Galileo's *Letters on the Sunspots* of the following year. Five years later, in 1618, Valerio died, after becoming their most serious apostate in the wake of the great controversy over the sunspots and at least partly as a result of a doomed affair with the pretentious but seductive poet Margherita Sarocchi.[53]

Two months later, in August, Giovanni Demisiani, the Greek from Zante who had also attended the meeting on the Janiculum in 1611, became a member. Obviously expert in Greek as well as in Latin, he had acquired considerable experience in both chemistry and mathematics, in addition to his literary and theological skills; but he also died too soon—in 1614—for his talents to be fully realized. It was Demisiani who offered Cesi his authoritative approval of the new term "telescope" for Galileo's *cannochiale*.

In September 1612 Cesi decided to enlist another German: Markus Welser, banker and *duumvir* of Augsburg and chronicler of its antiquities. Already in May of the previous year Welser had written to Gualdo about Galileo's demonstration of his telescope on the Janiculum, and had expressed his own reservations about the implications of Galileo's discoveries.[54] Born in 1558, Welser studied in Padua, Paris, and Rome before returning to Augsburg, where he became one of its most distinguished citizens. When admitted to the Academy he was described as endowed with "the greatest dignity, nobility, erudition and prudence."[55] An expert in the history and antiquities not only of Rome but of Germany too, Welser had published a number of important historical and archaeological works, including a compendium of the surviving antique inscriptions of Augsburg and the first edition of the *Tabula Peutingerana*, the great map of the military roads of the Western Roman Empire. He had also written a number of treatises on ecclesiastical history. In his learned correspondence with his friends in Italy, he wrote a polished and witty Italian, quite unlike Heckius's gauche use of the language, and expressed himself with genial wit and courtesy. When he died in 1614, so beset by family and economic troubles that some thought he had committed suicide, he was mourned by the entire Republic of Letters. By that time he had been honored not only with membership in the Academy of the Linceans but also in the *Accademia della Crusca*, the Florentine Academy more devoted to literary affairs. And in the last two years of his life he played a central role in the next great controversy that enveloped Galileo and his circle.

The final enrollment in the *annus mirabilis* of 1612 was that of Galileo's closest Florentine friend, Filippo Salviati. Salviati came from one of Florence's most famous families and was chiefly known for his mathematical work. Like Welser he had also been elected to the Accademia della Crusca. He had a comfortable villa at Signa, just outside Florence, where Galileo spent much of that year preparing his next two major works, the *Discourse on Bodies in and on Water* and the *Letters on the Sunspots*. It was Salviati's sister, Isabella, whom Cesi would marry in 1617, following the death of his beloved sixteen-year-old wife Artemisia Colonna in 1616. And it was Salviati himself who eventually served as the model for the Copernican interlocutor in Galileo's *Dialogue on the Two Great World Systems* of 1633. In every respect the fates of Cesi and Galileo were intertwined.

# FIVE *The Conflict of Truths*

## SUNSPOTS

*The History and Demonstrations Concerning Sunspots and their Phenomena, contained in three letters written to the illustrious Markus Welser, Duumvir of Augsburg and Counselor to His Imperial Majesty,* appeared in Rome in March 1613. On its title page Galileo was described as "Gentleman of Florence, Chief Philosopher and Mathematician of the Most Serene Cosimo II, Grand Duke of Tuscany." Beneath this came the lynx, the emblem of the Academy of the Linceans (fig. 5.1). For the next ten years, a substantial portion of their energy would be devoted to the support and encouragement of his ever more contested findings and to the publication of his most important works.

Long before 1612 was over, the group had decided to take on the responsibility of ensuring the publication of the *Letters on the Sunspots* (as *The History and Demonstrations* was always called). Recognizing that time was of the essence in yet another argument about priority, Cesi and the Linceans spurred Galileo on when he was unwell (which was often), arranged for the publication of the work, decided on printing policy, provided the prefatory material, counseled him on what to include and exclude (not that he always listened), and helped him negotiate the unexpected and ever-trickier demands of the censors.

In the *Letters on the Sunspots* Galileo openly declared his Copernicanism, made his most withering assaults yet on the Aristotelians, and dealt with the first of his two most intelligent opponents among the Jesuits. In every respect the history of this work presaged the future vicissitudes of both Galileo and his fellow Linceans. In the censors' responses to the work—and the Linceans' responses to them—are to be discerned the beginnings of the arguments that led to Galileo's downfall and that prevented the Lincean achievements from ever receiving the recognition they deserved. Above all, the arguments are about truth and the relations between Scripture and the evidence of nature. To examine these issues, and the strategies the Linceans adopted to deal with the censors and critics, is to be confronted with a set of fundamental problems about the relations between

Fig. 5.1. Title page, Galileo, *Istoria e Dimostrazioni intorno alle Macchie Solari* (1613).

ISTORIA
E DIMOSTRAZIONI
INTORNO ALLE MACCHIE SOLARI
E LORO ACCIDENTI
*COMPRESE IN TRE LETTERE SCRITTE*
ALL' ILLVSTRISSIMO SIGNOR
MARCO VELSERI LINCEO
DVVMVIRO D'AVGVSTA
CONSIGLIERO DI SVA MAESTA CESAREA
*DAL SIGNOR*
GALILEO GALILEI LINCEO
*Nobil Fiorentino, Filosofo, e Matematico Primario del Serenifs.*
*D. COSIMO II. GRAN DVCA DI TOSCANA.*

IN ROMA, Appreſſo Giacomo Maſcardi. MDCXIII.
*CON LICENZA DE' SVPERIORI.*

truth and interpretation. They also provide striking illustrations of the ways in which truths may sometimes be bent in the service of truth.

In April 1611, when he demonstrated his telescope to Cesi and his friends on the Janiculum, Galileo had pointed out the presence of a few spots on the face of the sun.[1] In a letter of October of that year to his friend the painter Ludovico Cigoli, he noted that these spots suggested the rotation of the sun on its own axis.[2] A series of letters throughout the following year made it clear that Galileo had consistently been observing sunspots from as early as January 1611 onward.[3] Still too involved, however, with his work on the behavior of bodies on and in water, he did not publish any of his sunspot material immediately. This was a mistake—for now someone else was claiming to have been the first to observe spots on the face of the sun. And the conclusions this other person was drawing were, for the most part, at radical and significant odds with those of Galileo himself.

When, following the enrollment of Schreck, Faber, and Müller into the Academy, Cesi wrote to Faber saying that his respect for German science was such that he regarded himself as a "germanofilo,"[4] little could he have known what trouble was brewing from just that quarter.

Already on November 11, 1611, Paolo Gualdo had written to Galileo saying that his Paduan friend, the antiquarian Lorenzo Pignoria, had heard from Markus Welser in Augsburg that sunspots had also been observed in Germany.[5] News traveled rapidly in the Republic of Letters. A week later Welser wrote to Faber in Rome saying that "some of my friends have observed with the optical tube certain spots appearing on the sun."[6] And a week after *that,* on November 25, 1611, Welser wrote to Gualdo in Padua reporting the same thing.[7] He insisted to both Faber and Gualdo that the spots observed by the unspecified German mathematician only *appeared* to be on the sun and that he was persuaded by the new evidence that showed they were stars.[8] The ever-solicitous Faber immediately communicated this news to Galileo.[9]

All this caused mild panic among the Linceans (even though there were others, including the Englishman Thomas Hariot,[10] who had also observed sunspots by this time). Galileo had told them about sunspots as early as April, but he had published nothing on the subject. Worse was to come. Apparently the German mathematician had published his own work on the topic! They did not yet know who he was; but who had sponsored the publication? None other than Welser himself.

Not that Welser was trying to keep anything from them. In the interests of scientific discussion, he sent a letter to Galileo on January 6, 1612, about the matter and enclosed with it a copy of the three letters on sunspots that the German had published the previous day. Their author had decided to suppress his own name and call himself "Apelles latens sub tabulam" (Apelles hiding behind the painting). Welser did not imagine that he was communicating anything especially new to Galileo; he simply wished to let Galileo know that the Germans too were keeping abreast of his discoveries.

Suffering from chest and kidney pains, insomnia, and blood loss, Galileo did not immediately respond to Welser's letter. In fact, he does not even seem to have received it until March. When he finally opened the unknown German's book, his state of mind can hardly have been improved by what he read. In the first place there was the provocation of the pseudonym itself. Galileo, who was as well-versed in literary as in scientific matters, would have understood the point immediately. He knew that Apelles was the famous painter of antiquity who, when in the service of Alexander the Great, had hidden behind one of his pictures in order to hear the comments of his critics without revealing himself. But Galileo was also aware of the sequel to the story, as recounted by Pliny the Elder in a passage in the *Natural History*—a passage to which Galileo had already alluded in his 1605 booklet on the new star. Unable to restrain himself in the face of the amateurish comments made on his painting by a shoemaker, the exasperated Apelles had finally revealed himself with the words: "Cobbler, stick to your last."[11]

The German's pseudonym was certainly a provocation; but even more infuriating and certainly more serious was the content of the letter on the sunspots. In it "Apelles" claimed that he had been the first to see the sunspots (in May 1611), and that rather than being actual spots on the surface of the sun, they were stars of one kind or another, either passing round the earth or around the sun. In this way he could avoid suggesting that there were defects and flaws (maculae) in a universe that was supposed to be spotless and immaculate. To top it all, he concluded that because he had been unable to see Venus at the time of its predicted transit on December 11, it revolved round the sun, in accord with the Tychonic system.

All this was too much for Galileo. In the first place he was sure that he had seen the sunspots before the German, who was probably lying about when he first observed them.[12] Much more to the point was the fact that "Apelles" was altogether wrong about the nature of sunspots—though at least he did not claim, as so many of Galileo's opponents had, that they were simply due to imperfections in the lenses of the telescope! Not only were they contiguous to the surface of the sun itself, they revolved with it, and thus proved the rotation of the sun on its own axis. And the fact that Venus revolved round the sun, as "Apelles" had admitted, was not just a matter of agreeing with the erroneous system of Tycho Brahe; for from his various observations Galileo could now say that it was "with absolute necessity that we conclude, in agreement with the theories of Pythagoras and Copernicus, that Venus revolves round the sun, just as do all the other planets."[13]

This statement comes from the first of three letters constituting Galileo's *Letters on the Sunspots*, which he immediately set about composing and which were dispatched to Welser on May 4, 1612. A week later he sent a copy to Cesi. In his covering letter he declared, in his usual robust way, that the news of his discoveries about the sunspots was "likely, I suspect, to be the funeral—or rather the Last and Final Judgement—of pseudophilosophy. . . . I now expect to hear the Peripatetics spouting big things to maintain the immutability of the heavens."[14] Cesi immediately began to worry about how best to illustrate the work.

The gloves were now off, and there was no retreating. As far as Galileo was concerned, the one thing the sunspots proved beyond any doubt (besides the truth of the Copernican hypotheses) was that the supposedly perfect heavens could not possibly be unalterable. In the next two letters, written in August and November, Galileo articulated the truth of the Copernican position ever more plainly, and in this too Cesi played a crucial role.

Still working on his *Coelispicium* and the problem of the fluidity of the heavens, Cesi wrote to Galileo on June 12 with an important question. Might not the theories of Copernicus (who had never said anything about the solidity of the spheres) be improved if one eliminated the Ptolemaic epicycles and eccentrics—as Kepler had also suggested?[15] Such a proposal was consistent with Cesi's ever-growing desire to reduce the fractiousness of multiplicity to order. At the end of the month Galileo responded that he was "delighted to hear that you are so occupied with the contemplation of the Copernican system, and that you're inclined to prefer it to the Ptolemaic one—especially if the eccentrics and epicycles could be totally removed." And then, articulating the central problem of all, the motto by which Cesi would try to live the rest of his life, he continued: "I only want to tell you what you know much better than I, namely that we should not wish nature to accomodate itself to what seems better ordered and disposed to us, but that we should rather accommodate our own intellect to what nature has made, in the certainty that this is the best and only way."[16]

A few weeks later Cesi wrote to Galileo:

> As for the qualities of the Copernican system, there is no doubt at all that one of the great satisfactions it gives us is the removal of the multiplicity of movements and spheres, and their very great and complex diversity; and it would be much better if in doing this, one did so totally. Not without reason does the human intellect, when it sees such a farrago of spheres and revolutions, with its lack of a stable or adequate point, . . . find it difficult to believe that this should be the work of nature.[17]

On August 14 Galileo fired off his second letter to Welser on the matter of the sunspots. Why make nature more complicated than necessary by suggesting that the sunspots were yet more stars circling in complex patterns round the sun? Now, with the aid of mathematical proofs and with a set of efficient diagrams showing his measurement of sunspot motions, he could take his next step: the sunspots either had to be extremely close to or actually on the face of the sun itself.

From then on things moved still more swiftly. In September Cesi wrote to Galileo telling him that while a certain Dominican father, in a debate held at the Collegio Romano, had upheld the Copernican system, most of the Jesuits were insisting that the sunspots were small stars, just as "Apelles" had claimed.[18] And then, on the thirteenth of that month, under the title of *A more accurate discourse on the sunspots and the wandering stars around Jupiter*, "Apelles" himself published three more letters on the subject, retrospectively dated to January, April, and July! In the first he said that when Galileo initially observed the phases of Venus others were doing so too, in the second he said he had discovered a fifth satellite of Jupiter, and in the third, describing the new instrument that he too had used as a *helioscope*, he implied that the diagrams that Galileo had just sent to Welser were not only the same as his own, but that Galileo had probably borrowed his methods of drawing them from him. Insisting that the sunspots were stars probably like the satellites of Jupiter, he repeated his assertion that they were not on the face of the sun.

The battle was truly joined. Even as Galileo worked on his third and final letter to Welser at Salviati's villa outside Florence, the Linceans decided that the letters must be published immediately. Already in September Cesi had begun to negotiate with Mascardi, the printer in Rome. But they had to be careful on several fronts. In the first place they did not wish to alienate Welser, whose name would appear on the title page as the addressee of Galileo's letters (just as he had been of those of "Apelles"). "I hope to expose the silliness with which this matter has been treated by the Jesuit," wrote Galileo to Cesi in an attempt to account for his delay, "but I want to make this resentment known, and the desire to do so without insulting Signor Welser is causing me no small difficulty, and is the cause of my being late."[19]

This was hard to negotiate. Cesi warned Galileo not to be too sharp with the Jesuits. On the one hand, a recent meeting with Cardinal Bellarmine had led him to believe that the Church would not be especially troubled by the attack on the doctrine of the inalterability of the heavens; on the other, when the Dominican lectured in favor of Copernicanism at the Jesuit college in Rome, he had encountered unexpected resistance there.

Galileo and Cesi were misreading the signals. They still thought they could allow Galileo's letters to voice strong declarations in support of Copernicus and openly attack Aristotelianism. In the third letter on the sunspots Galileo wrote, "I believe that there are not a few Peripatetics on this side of the Alps who go about philosophizing without any desire to learn the truth and the causes of these things, for they deny these new discoveries or jest about them . . . and they go around defending the inalterability of the sky, a view which Aristotle himself would probably abandon in our age."[20] He was so confident that when Cesi prepared a highly satirical series of letters purporting to be the Peripatetic response to Galileo, Galileo replied "I really have no need of this kind of artifice. It is sufficient for me only that the pure truth be known."[21]

Soon, however, Galileo had to give a great deal more thought to what he and others—especially others—intended by the idea of "pure truth." At a critical juncture in the final preparation of his manuscript, an unexpectedly malign figure appeared on the scene. This was the elderly Father Niccolò Lorini, also a Dominican, and favored by the Medicean court. In the course of an anti-Copernican conversation in Florence on All Saints Day, reported to Galileo, Lorini had said that "the opinion of Ipernicus, or whatever his name is, appears to be against Holy Scripture."[22] Lorini's intervention was a foretaste of things to come. Within less than a year, he and another Dominican, Tommaso Caccini, would turn out to be two of the most threatening of all Galileo's opponents.

Trouble loomed. The manuscript was already in the hands of the Vatican censors. Galileo and Cesi worked together—along with a little advice from Luca Valerio—in order to prepare it for them. They had hoped to avoid possible objections; but they had been lulled into a false sense of security. When they received the corrected manuscript back they must have been shocked. The implications of the censors' objections were much more serious than they had anticipated.

The Linceans had decided that it would be a good idea for the book to begin with a letter from Welser to Galileo soliciting his opinion on the matter of the sunspots; and they had this letter begin with what seemed to them a suitable verse from the Bible: "The Kingdom of Heaven suffereth violence, and the violent take it by force" (Matthew 11:12). A bad choice! From the very outset the censors made it plain that Galileo would not be allowed to exploit Holy Scripture for his own purposes. The line had to go. It was replaced with a vaguely apropos verse from Horace: "True worth, opening Heaven wide for those not deserving to die, sets out to try a way denied to others."[23] But this was followed by more fighting words: "Already the minds of men are making an attempt on the Heavens, and the bravest are now conquering them." Such changes were hardly calculated to placate the growing number of their enemies.

Galileo and Cesi might have thought they could risk declaring their Copernicanism openly, but in appealing to higher support they made another false move. When the censors came across a passage in the third letter stating that "Divine Goodness" ("Divina Bontà") had directed Galileo to set forth his Copernican theories to the world as a whole, they struck it out. Instead, Galileo was allowed to substitute "propitious winds" for that temerarious phrase.

The point could not have been clearer. God and the Bible had to be kept out of Galileo's daring speculations—for speculation, according to the theologians, was all that his so-called discoveries were. But Galileo repeatedly insisted on seeking theological justifications for his cosmological arguments. The two sides were heading for a clash. Galileo was not sticking to his last! On the contrary. The censors felt that he was meddling in dangerous and subversive affairs that were none of his business. In the manuscript he presented he had gone so far as to say the doctrine of the incorruptibility of the heavens was "not just false, but erroneous and repugnant to the undoubted truths of the Sacred Scripture, which in so many places openly and clearly refer to the unstable and failing nature of celestial matter."

This was intolerable. As early as November 10, Cesi wrote to Galileo saying that although the censors had finally approved of everything else, they did not want him to make such a claim about the irreconcilability of Sacred Scripture with the traditional view of the incorruptible heavens at all. The extent to which both Galileo and Cesi failed to appreciate what was really at stake is demonstrated by what happened next. Thinking to mitigate the force of the sentence that had offended the censors, Galileo wrote instead that his hypotheses "conformed to the undoubted truths of Sacred Scripture"; and then tried to get himself off the hook by praising the theologians, whose "sublime minds and subtle speculations know how to reconcile the apparent discordances between sacred dogma and the discourses of physics." But he also added that "I have no doubt that Aristotle himself would concede the truth of such sensible experiences [*sensate esperienze*]."

This was bound to irritate the censors yet again. Once more they objected. Expressing the conviction that every one of the sacred texts adduced by Galileo in favor of the corruptibility of the heavens could be "very well be interpreted by others in the manner of the Peripatetics," they told him to remove every allusion to Sacred Scripture.[24] This, of course, he had to do—at least if he wanted his work to be published; and it was already long overdue.

Even after the manuscript reached the hands of the printer, Galileo, Cesi, and Luca Valerio continued to work feverishly on it. Constantly they made changes they thought might lessen the potential opposition, especially from the Jesuits. Stelluti wrote an introductory poem in Galileo's honor, while Angelo De Filiis composed a fulsome dedicatory letter to Filippo Salviati and a laudatory preface to the book as a whole.

But this preface also presented difficulties. In the first place it was too long, and Galileo thought that De Filiis should be more moderate in his elaborate claims for priority in the discovery of the sunspots. To record the events of April 1611 was one thing; but to imply that "Apelles" had stolen Galileo's ideas was another. Why offend the Jesuits still further? Here was at least one case where Galileo could expediently mute what he knew to be true; which he did. Nor did he allow De Filiis to be too effusive about his superiority as an observer of the heavens.

But there was more to slim down (as Cesi himself put it to Galileo on February 22).[25] De Filiis had emphasized the role of the Linceans in supporting Galileo and in persuading him to write up and publish his discovery of sunspots. De Filiis also took the opportunity of alluding, right at the beginning of the preface, to the Lincean researches into the world of fossil remains and the ambiguous phenomena of the earth—a kind of lesser, terrestrial equivalent to the magnificence of Galileo's discoveries of new lights in the heavens.[26] The Linceans did not want to miss this opportunity to announce their own researches and aims in a more public domain than usual. But it was all too expansive and inflated, this preface, and Galileo and Cesi cut much of the fat.[27]

All this last-minute editorial activity must have been very frustrating to the publisher, Mascardi. Even while the book was in proof, Galileo and Cesi kept on making changes, either to prevent further difficulties from the censors or to satisfy the objections of their learned colleague Luca Valerio. They even worried about the printer's treatment of Galileo's excellent use of Tuscan, the preferred form of the Italian vernacular at the time. "Even if it is corrected two or three times, they still make mistakes," wrote Cesi to Galileo on December 28, 1612.[28]

This concern with the language of Galileo's text went beyond purely philological issues and Galileo's own affection for his native Tuscan. Now more than ever Galileo wanted his work to be as accessible as possible. It was not just a matter of public acknowledgment of his discoveries. The adoption of the vernacular went hand in hand with his belief that the mysteries of the universe should be neither the domain of the learned nor of ancient authority alone; they should be accessible to all. It did not speak only to professors; it was an open book. In the third letter on the sunspots Galileo wrote that the followers of Aristotle never seemed to wish to raise their eyes from his pages, "as if this great book of the universe had been written to be read by nobody but Aristotle, and as if his eyes had been destined to see everything for posterity."[29]

When Welser wrote to Galileo complimenting him on the charms of his Italian prose, he also expressed his concern that the drawings accompanying his work might be too large for a book.[30] But Cesi was thrilled with them, both because of the spectacle they presented and because of their accuracy; "it is now the turn of your adversaries to be vanquished by this sense experience, because in arguing with it they abuse reason," he wrote to Galileo on September 8.[31] Cesi told Galileo not to worry about the expense of having them engraved, and he and Cigoli immediately set about finding a suitably talented engraver.[32]

Soon—having eliminated two other possibilities—they decided upon Matthias Greuter, the German who would from then on be one of their favorite engravers. Together Cesi and Cigoli kept a close eye on Greuter's work, sending back his proofs when they were not accurate enough, or when the contrasts were not sufficiently bold to clearly show the movements of the sunspots. At every stage they consulted Galileo; and finally even Cesi was satisfied. In the end he had to give up the idea of having the illustrations printed in folio, but the reduction to quarto size meant not only that the book would be conveniently portable, but that every one of the thirty-eight engravings could be oriented in the same way, so that the motions of the spots over successive days could be instantly recognized. After looking at them, how could any reader possibly agree with "Apelles" that the spots were satellites of the sun, when their shapes so obviously changed?[33]

Despite the fact that the illustrations made at least one of the chief arguments of Galileo's work as plain as any words could, there was still much to explain and discuss. And this, as we have seen, had to be in the vernacular. Linguistic accessibility always went alongside the Lincean belief in the visual. The desire to write in a vernacular even overrode Galileo's awareness that if he wrote in Italian, "Apelles" might not be able to read him. But for Galileo this was a secondary consideration, as evidenced by his June 16, 1612, letter to Gualdo:

> I have received word from Sig. Welser that my letter has arrived . . . but that Apelles will not be able to read it right away because he does not understand the language. I wrote in the common language because I must have everyone able to read it, and for the same rea-

> son I wrote my last treatise in this language [the work on floating bodies]. I am induced to do this by seeing how many young men are sent through the universities indiscriminately, to be made physicians, philosophers, and so on . . . while other men who would be fitted for these are taken up by family cares and other occupations. . . . As Ruzzante would say, they are furnished with "horse sense," but because they are unable to read things that are "Greek to them" they become convinced that in those "big books there are great new things of logic and philosophy and still more that is way over their heads." Now I want them to see that just as nature has given them, as well as philosophers, eyes with which to see her works, so she has also given them brains capable of penetrating and understanding them. All the same, I hope that Apelles and other foreigners will be able to read this book too. . . . and so I should appreciate it if you and Sig. Sandelli would translate it into Latin as well.[34]

The sentiments were exactly those that motivated the language and form of the Paduan dialogue on the nova, itself written in a manner inspired by the faux-rustic poet Ruzzante. But now the issues were even more far-reaching.

The idea that the secrets of the universe could be revealed in the vernacular, in language plain to the common people, had always carried with it the overtones of the Protestant reformers, who had demystified the Bible by having it translated into language accessible to all. And now, in those very passages in the *Letters on the Sunspots* that the censors had objected to, Galileo was attempting to reconcile Scripture with the new discoveries in the heavens—and this he was doing in the common language. The threat was even greater, because the only way Galileo could reconcile the clear statements of the Bible with the circular motions of the earth as described by Copernicus was by maintaining that its interpretation was not the strict domain of the theologians, but that it spoke the language of the people.[35]

All this set every one of Galileo's opponents, even the favorably inclined ones, on edge. From now onward he would have to tread ever more carefully—and he was not always good at doing so. Even the Linceans, under such conditions, lost their footing.

Finally, on March 22, 1613, the *History and Demonstrations concerning the Sunspots* appeared. It was fifteen months after the publication of the first "Apelles" letters on the subject. Fourteen hundred copies were printed, of which seven hundred, for distribution in Italy, were accompanied by his opponent's letters. "Let us make it seem as if the printer himself wanted to add them," Cesi had written to Galileo on December 14. In order to achieve this, they asked Mascardi to state in his letter to the reader that he had decided to print Apelles's letters at the back, for the simple reason that only a few copies of the German's work had reached Italy, and that it was important to have it to compare with Galileo's. "And thus," Mascardi concluded, "through the labors and telescope of others, you will yourself be able to become a looker at the sun [*helioscopus*],[36] and know things hidden from all of antiquity itself." So much for the authority of the ancients! From now onward, everyone would be able to see for themselves.

Within a few weeks, the identity of "Apelles" emerged. On May 17, Giuseppe Biancani, a Jesuit who had already engaged with Galileo on the subject of the mountains of the moon,[37] wrote to Magini in Bologna saying that Father Christoph Scheiner of their order was the discoverer of the sunspots, and that it was he who had written under the pseudonym of Apelles.[38] At about the same time, there appeared in Antwerp a book on optics by another Jesuit, Franciscus Aguilonius,[39] in which he asserted that sunspots had first been seen by Christoph Scheiner, professor of mathematics at the Jesuit college of Ingolstadt in Germany, writing under the name of "Apelles behind the painting." "I certainly don't know to what purpose this Apelles has now come out into the open," wrote Cesi to Galileo when he learned of this.[40]

There were a few cardinals in Rome, and many others, who were impressed by both the science and the charm of Galileo's book. Among them was Cardinal Maffeo Barberini, who had already taken his side in the controversy over floating bodies the previous year. But there were many who did not like what they read at all. From then on Father Scheiner became one of Galileo's most implacable opponents; and soon he was joined by several other talented Jesuits. Moreover, Cardinal Bellarmine was losing patience. It was one thing to propose the new discoveries as hypotheses; but to suggest that they were truths that were capable of being reconciled with the Bible, as interpreted by Galileo, was something very different altogether.

## NATURE VERSUS SCRIPTURE

The full extent of the role of the Linceans in what happened next has not been fully acknowledged. Although the background is comparatively well-known, the core is not.

At the end of 1613, Benedetto Castelli, a Benedictine monk who had helped Galileo prepare the accurate drawings of the sunspots, breakfasted with the grand duchess of Tuscany at the Medici court at Pisa. Just as Castelli was leaving, the grand duchess called him back; while she believed in most of Galileo's discoveries in the heavens, she said, there was one problem with his claim that the earth moved: it had in it "something of the incredible, and could not occur, especially because Holy Scripture was contrary to that view."[41]

The grand duchess had put her finger on the crucial issue. It was left to others to fight it out. At the center of the controversy lay a series of critical passages in the Bible. In addition to the texts which Delle Colombe had already cited in his 1611 attack on the *Starry Messenger* (or rather on

Galileo's claim that the earth moved),[42] several more now occupied the concentrated attention of almost all of Galileo's opponents.

By far the most contentious were the lines in Joshua 10:12–13, describing Joshua's command to the sun before his victory over the Amorites. The grand duchess herself is said to have cited this text: "Sun, stand thou still upon Gibeon; and thou, Moon, in the valley of Ajalon. And the Sun stood still, and the moon stayed, until the people had avenged themselves upon their enemies." What could be clearer than this? If Joshua had to command the sun to stand still, then obviously it normally moved. This was the whole point of the miracle that enabled Joshua's victory.

Then there was Psalm 18:5–6 (Psalm 19 in the Authorized Version), which described the sun "as a bridegroom coming out of his chamber, and rejoiceth as a strong man to run a race. His going forth is from the end of the earth, and his circuit under the ends of it; and there is nothing hid from the heat thereof." Most modern readers would gasp at the notion of interpreting this beautiful simile in purely literal terms; but to the opponents of Galileo, it was to be read as yet another buttress in the massive argument in favor of the doctrine of the mobility of the sun; and, indeed, that of geocentrism.

The issue simmered for about a year. After all, the conflicts between Scripture and nature were not exactly unknown. Already in his *Astronomia Nova* of 1609, for example, Kepler had discussed the whole question of whether passages such as that in Joshua 10:12–13 should be taken literally; and in what contexts they might or might not be.[43] Galileo's opponents, like the always nettlesome Delle Colombe, kept on stirring things up; but Galileo, as always, was certain of himself. As far as he was concerned, it was all a matter of interpretation; there *could* be no conflict between scriptural truth and the truth of nature. This, of course, was also the claim of his opponents; the difference was that Galileo did not think that biblical interpretation should only be left to theologians. In replying to Castelli's account of how the grand duchess Christina had insisted that "Holy Scripture could never lie or err, but that its decrees are of absolute and inviolable truth," he was so sure of himself that he could hardly not inflame his opponents. In one of the most memorable passages he ever wrote, he insisted that

> Since Scripture is in many places not only accessible to but necessarily requires expositions that differ from the apparent meanings of the words, it seems to me that in physical disputes it should be reserved to the last place. . . . Moreover, it was necessary in Scripture, in order that it be accommodated to the general understanding, to say things quite diverse, in appearance and with regard to the [literal] meaning of the words, from absolute truth; yet on the other hand, since Nature is inexorable and immutable and cares nothing whether her hidden reasons and modes of operating are or

> are not revealed to the capacities of men, she never transgresses the bounds of the laws imposed on her. Thus it appears that physical effects placed before our eyes by sensible experience or concluded by necessary demonstrations should not in any circumstances be called in doubt by passages in Scripture that verbally have a different appearance. Not everything in Scripture is linked to such severe obligations as is every physical effect. . . . This being the case, and since, moreover, it is clear that two truths can never contradict each other, it is the office of wise expositors to work to find the true sense of passages in the Bible that accord with those physical conclusions of which we have first become sure and certain by manifest sense and necessary demonstrations.

The letter concludes with an extraordinarily pyrotechnical display of interpretive skill, in which he demonstrates that the Joshua text actually shows "the manifest impossibility and falsity of the Aristotelian and Ptolemaic world systems, and accords very well with the Copernican."[44]

The letter to Castelli stands as one of the noblest and most important of Galileo's statements about the priority of the divinely given laws of nature over the literal sense of the Bible. But it is not hard to see how such a claim would have caused offense, especially to the theologians; for were not *they* the ones who should be interpreting the Bible, rather than the mathematician from Tuscany?

The year 1614 was comparatively quiet. There was little in it to suggest the storm that would soon break. Galileo continued to work on his observations of the satellites of Jupiter and his forecasts of their positions. He consulted both Stelluti in Fabriano and Colonna in Naples, then chiefly and deeply involved in a variety of complicated botanical and zoological researches. Both offered welcome confirmations of his forecasts, with Colonna, typically enough, making some suggestions and corrections. Both kept him abreast of the lesser treatises contesting his discovery of one heavenly phenomenon or another. In Rome Father Grienberger remained sympathetic to Galileo's discoveries, though he admitted that he was restrained from expressing his full agreement by the recent gag-order on anti-Aristotelianism by General Acquaviva. He had made it perfectly clear that Jesuit professors were to adhere strictly to the teachings of Aristotle in science and St. Thomas in theology.[45]

On January 12, 1615, Cesi wrote to Galileo from Acquasparta. Encouraging Galileo to ignore "those enemies of knowledge who take it on themselves to interfere with heroic and useful discoveries and labors . . . who can never keep quiet," he attached an alarming note to this letter. In it he revealed that Bellarmine himself considered the earth's motion to be against Holy Scripture. Cesi warned Galileo of the possibility of censorship, cautioned him to lie low, and told him not to press his Copernicanism too far.[46]

Then the storm broke. It is not clear just how public Galileo had intended the views he outlined in the letter to Castelli to become (though he certainly wished them to be known), but several copies of the letter were circulating. At the beginning of 1615 a rather incomplete and inaccurate one fell into the hands of his most recent foe, the dogged Dominican monk Lorini. From then on the Dominicans would hound him with even greater tenacity than the Jesuits. The other Florentine Dominican, Caccini, had just preached against Galileo in Santa Maria Novella on the Sunday before Advent, warning that "no one is permitted to interpret the divine Scriptures contrary to the sense on which all the Holy Fathers agree, for this has been forbidden both by the Lateran Council under Leo X and by the Council of Trent."[47]

Fortified by the words of his colleague, Lorini sent a letter to the inquisitor-general in Rome, alerting him to the problems raised by Galileo's Copernicanism and by his high-handed approach to the interpretation of the Bible. With it Lorini enclosed his own copy of Galileo's letter to Castelli.

"I have written to Rome where, as St. Bernard says, the holy faith 'has the eyes of a lynx'," Lorini wrote.[48] There can be no doubt that this was intended as a glancing rhetorical blow at the claims of the very men who were then most committedly supporting Galileo.

The lines were drawn. It is true that old friends like Paolo Gualdo in Padua and Piero Dini in Rome were constantly telling Galileo to be more careful; but none did so with a more delicate sense of strategy than the Linceans. On the last day of February 1615, the twenty-four-year-old Giovanni Ciampoli, not yet a Lincean but soon to become one, wrote to his old teacher Galileo, reporting on a conversation he had just had with their friend and fellow Tuscan, Cardinal Maffeo Barberini. For all his sympathy for Galileo, Barberini had warned that

> greater caution is needed in dealing with the arguments of Copernicus and Ptolemy, and one should not exceed the limits of physics and mathematics, because the explication of the Scriptures is restricted to theologians who deal with such matters. . . . it is very necessary to emphasize frequently that one should submit to the authority of those who have jurisdiction over human reason in the interpretation of the Scriptures.[49]

In March the two troublesome Dominicans were brought to Rome to state their case to the Inquisition about Galileo's errors. They stopped at nothing. Caccini even asserted that though many claimed that Galileo was a good Catholic, "others hold him to be suspect in matters of faith, because they say that he is very intimate with the Servite Friar Paolo [Sarpi], who is so famous in Venice for his impiety, and they say even now letters still pass between them."[50] It began to seem that every step in Galileo's past, as well as in the present, could be held against him.

Suddenly support seemed to come from an unexpected quarter. On March 7 Cesi wrote to Galileo about

> a book which has just come out, namely a letter by a Carmelite Father, who defends the opinion of Copernicus while reconciling it with all the Scriptural passages. This work could not have appeared at a better time—unless it is harmful to increase the anger of our adversaries, which I doubt. The writer counts all Linceans as Copernicans, though that is not so; all we claim in common is freedom to philosophize in physical matters.[51]

But this was not the right call. The fact is that the Carmelite's book could hardly have appeared at a worse time.

In a letter to Galileo two weeks later, Ciampoli also referred to the unknown Carmelite. Ciampoli noted that the work was published in Naples and that it maintained that the motion of the earth and the stability of the sun was not contrary to Scripture. But he was much more attuned to the situation in Rome than Cesi. Warning that the new work ran "a great risk of being condemned by the Congregation of the Holy Office, which meets here in a month," he told Galileo not to say too much about Copernicanism. Not many people in Rome were worrying about it—at least not yet. And Ciampoli reiterated Bellarmine's view that in discussing the Copernican system there ought not to be any contradiction with Scripture, "the interpretation of which is to be reserved to the professors of theology who are approved by public authority." Finally he reminded Galileo that this had also been the opinion of their friend and supporter, Maffeo Barberini.[52]

The book in question turned out to have been written by Antonio Foscarini, an obscure Carmelite from Naples.[53] In it he eloquently set out the ways in which Scripture could actually be read in ways that supported both Kepler and Galileo; but instead of helping the Lincean cause, it simply precipitated matters.[54]

When Foscarini sent the work to Bellarmine, the cardinal realized just how much things were getting out of hand. On April 12 he wrote tartly to Foscarini, saying that he and Galileo should confine themselves to dealing with the Copernican system as a pure hypothesis, *ex suppositione*, as Bellarmine put it, and not absolutely.[55] He reminded them that it was forbidden to interpret Scripture contrary to the accepted sense in which it had been interpreted by the Holy Fathers of the Church.[56] Dini then sent Galileo a copy of Bellarmine's letter and told him that Cardinal Barberini had said that for the time being it would be better if he stuck to mathematics.[57]

The arguments went back and forth; Cesi began to worry about offending the Aristotelians;[58] and the Holy Office continued its damaging interviews. On February 25, 1616, it issued its famous conclusion. The propositions that "the sun is the center of the world and is completely immobile by local motion" and that "the earth is not the center of the world and is not immobile but moves as a whole and with a diurnal motion" were both censured and condemned as "foolish and absurd." The first proposition was also condemned as formally heretical, because of its explicit contradiction of many places in Sacred Scripture "according to the proper meaning of the words and according to the common interpretation and understanding of the Holy Fathers and learned theologians."[59]

No mention was made of Galileo in the decree; but he got the message. On March 5, the Congregation of the Index declared that "the Pythagorean doctrine of the mobility of the earth and the immobility of the sun," as taught by Copernicus and others, was "false and completely contrary to the divine Scriptures." Copernicanism was condemned, and the *De Revolutionibus* placed on the Index of Forbidden Books until corrected.[60] Foscarini's book was completely prohibited and condemned.[61] There could now be no question of publishing the *Letter to the Grand Duchess Christina* that Galileo had just completed. Besides, on the day after the Holy Office issued its decree, the pope ordered Bellarmine to meet with Galileo and to tell him to abandon his heliocentrism; and that if Galileo should refuse to obey, to "impose upon him an injunction to abstain completely from teaching or defending that doctrine and opinion or from discussing it; and if he should not agree, he is to be imprisoned."[62]

Galileo and his fellow Linceans had not expected that it would come to this. Clearly they could no longer count on the sympathies of either Cardinal Bellarmine or of the learned and generally favorable Maffeo Barberini.

## THE CASE OF LUCA VALERIO

In the extraordinarily difficult years between 1612, when he was elected to the Academy, and 1616, Valerio had worked as *censore* or reviser and editor of all the Lincean publications. He had been one of Galileo's most constant supporters and advisers. Galileo much admired him for his mathematical work. But as soon as Valerio heard of the Holy Office's February 25 decree, he declared that he could no longer accept Galileo's heliocentric view of the universe, and gave notice of his intention to resign from the Academy.[63]

Nothing was less expected than Valerio's abandonment of the views of his closest friend. The fact that the Jesuits of the Collegio Romano, who five years earlier had so lauded Galileo, were now silent was one thing; but from Valerio the Linceans expected greater consistency, to say the least. He had now aligned himself with the most blinkered and fanatical of Galileo's opponents. His astonished colleagues refused his resignation (for that was against their oath of loyalty); and in order to deal with this matter, as well as with the general crisis now facing them, they convoked what would turn out to be the most difficult meeting they ever had.

On March 24, 1616, the Linceans met at the Cesi Palace in Rome. They had one delicate matter after another to deal with. Before anything else there was the ever more worrying problem of Heckius (who in a sense had started it all). By this time they had come to the conclusion that there was nothing more to be done about his ever more irrational behavior and that he had clearly lost his mind. Finally they had to take the painful decision to exclude him forever from their deliberations. This is the last we ever hear of Heckius. But however difficult this decision was emotionally (if not for all of them, then certainly for Cesi and probably Stelluti too), at least it was clear.

The case of Luca Valerio was much less straightforward. But here too they really had no choice. They censured him for having betrayed his oath of loyalty and for having, by his conduct, offended both Galileo himself and the fundamental Lincean principle of mutual solidarity, or *Lyncealitas* as they called it. Reiterating their complete solidarity with Galileo, they deprived Valerio of his voting rights and forbade him from ever again participating in the sessions of their society. Bitterly they noted that by expressing the wish to resign, Valerio had implied that the Academy itself had committed a crime or a grave error in supporting the view of the movement of the earth; and that in accusing his old friend Galileo of this same "error," he had overlooked the fact that Galileo had only held the movement of the earth to be a hypothesis.[64]

This, of course, was crucial. In the month or so that followed, Galileo himself was concerned about rumors circulating in Rome that he had actually abjured his views and had been punished for them. For one last time he turned to Bellarmine for help, and Bellarmine obliged. He supplied Galileo with a letter stating that he had not actually abjured and that he had not been punished. But at the same time Bellarmine repeated the fact that Galileo had received the decree of the Sacred Congregation of the Holy Index, and that the doctrines of Copernicus could neither be defended nor even held.[65] This, to say the least, was a most ambiguous clarification.

At the meeting on March 24, the Linceans had once more insisted on their "freedom to philosophize in natural matters," their *libertas philosophandi in naturalibus*.[66] But more and more they came to realize that they were on their own. As Cesi himself would put it, "either we are alone—or let us be."[67] From this moment on the price of independence would be isolation. While they would succeed, by a variety of the most complex strategies, in getting many of Galileo's most important views published, the bulk of their own most important work, which they never for one moment neglected, would never see the light of public day—except in the most truncated and distorted of forms.

As for Luca Valerio, who had so profoundly infringed some of the most basic principles of their group—was it only the fear of persecution by the Holy Office that made him renounce the heliocentric views of Galileo so abruptly? It seems likely that he did so at least partly because of the pressure brought to bear on him by his possessive lover, Margherita Sarocchi, who had become increasingly jealous of his association with the Linceans; but that is another story.

## CESI VERSUS BELLARMINE

All through 1616 Galileo worked on the problems of longitude, on the satellites of Saturn, and on his planned treatise on the tides. But publication of the latter was out of the question, since it too was dependent on the Copernican thesis. In Galileo's view the tides were caused by the double motion of the earth around its axis; and even though he took the now usual precaution of emphasizing that this too was purely speculative, everything, for the moment, was clearly blocked.

No wonder Galileo was discouraged. His health began to decline as well. By 1616 he was already struggling with a variety of physical complaints; they grew worse the following year. Although he continued to enjoy a number of powerful and favorable friendships in Rome—even Bellarmine remained sympathetic to a point—he fell into a deep depression. It was on the support of Cesi and the other Linceans—Stelluti and Faber in particular—that he most depended. Joining them in their encouragement of Galileo, both practical and emotional, were two other young men, one of whom we have already encountered. Both Ciampoli and his close friend, the twenty-one-year-old Virginio Cesarini, had been proposed for membership in the Academy at the meeting of March 24, 1616.[68] Soon they became two of the most devoted Linceans (they were inscribed in the album of the Academy in 1618)—and two of the most tragic figures among them. The scion of an old and distinguished Roman family, Cesarini had been smitten with Galileo and his views since he first met him. Had it not been the continuing sustenance of these Linceans, it is unlikely that Galileo could have managed on his own, or achieved what he did.

In the meantime, Cesi returned to work on the project that had engaged him ever since the controversies around the nova of 1604, the *Coelispicium*. Already then he had begun to draw up a list of all the scriptural passages that could be interpreted in support of his theories of the fluidity of the heavens, in direct opposition to the views articulated by Heckius in the manuscript of his treatise on the nova and which Cesi himself had censored.[69] Although he had originally conceived of the *Coelispicium* as an ambitious cosmological treatise, Cesi realized that there was one particular aspect of his approach that was more pertinent than ever before. Galileo may henceforward have given up the increasingly futile attempt to reconcile Scripture with the newly discovered facts of nature; but Cesi certainly did not. That it was precisely this approach that formed the chief target of the 1616 decree did not daunt him. On the contrary: more keenly than ever he now felt the need to conclude the part of his project that had the greatest bearing on the argument brought to a head by the Holy Office in February and March.

Bellarmine himself seems to have solicited Cesi's opinion. Setting aside, for a moment, his other work on botany, mycology, and paleontology, Cesi decided to provide a summary of his views on the unity, permeability, and fluidity of the heavens—still a controversial issue.

The cardinal was torn. He could grasp the plausibility of the doctrine of celestial fluidity all too well, even though the new general of the Jesuits, Muzio Vitelleschi, was against it.[70] The problem was to reconcile it with Scripture, and this is what Cesi now attempted to do. He thought he could reply to Bellarmine's request and at the same time take some of the heat off Galileo. Finally, on August 14, 1618, he sent to Bellarmine his letter on the subject—which he had been working on for some time: "On the unity, tenuousness, fusion and permeability of the heavens, along with the nature of the movements of the stars, on the basis of Sacred Scripture."[71]

The title said it all. Cesi set out to refute, both on the basis of the new celestial observations and on theological grounds, the solidity, hardness, and multiplicity of the orbs and their movements. To his continued emphasis on the unity and permeability of the heavens, he added his Galilean belief that the "many and complex masses of large and small orbs" had to be eliminated "from the purity of nature."[72]

All this, Cesi asserted, was perfectly consistent with Scripture and with the writings of the early church fathers. Citing a large number of them, he insisted that they all believed the stars to be carried through an immobile (rather than a turning) heaven. Even more learnedly he discoursed on the meaning of crucial words for the heavens in the Hebrew Bible, especially *rakeaḥ* and *shamayim*, which he took to support his arguments in favor of a boundless, fluid space.

There was no shortage of erudition here, no gulf, it seemed, between scriptural and scientific knowledge. The alleged "firmness" of the firmament could not be derived from anything to be found in Scripture at all, Cesi asserted. Passages such as "He established the clouds [*aethera* in the Latin] above" of Proverbs 8 and "the tabernacle of the sun" of Psalm 18 [19] were references to the heavenly bodies themselves, and not to the movements of the spheres and the courses of the stars. Of these, he insisted, there was absolutely no mention in the Bible, nor any observable proof in physics.[73]

What was at stake was the importance of empirical observation (as exemplified by Galileo's great discoveries with the telescope) versus blind faith in the authority of the ancient philosophers. Cesi now went beyond Brahe's denial of the existence of solid spheres, in order to demonstrate—as some of the most radical thinkers of the time, including Bruno, had suggested—that the heavens were limitless and permeable, penetrated by countless stars and other heavenly bodies, and that by no means all of them were known to the ancients. One could not even *begin* to number these bodies, as Abraham himself had implied with his prophecy that the already numerous Israelites would become as the sand of the seashore.

The existence of so many observable stars only proved that there were many more hidden beyond the raw capacity of our naked eyes. "How could I not fail to deplore the weakness of so many of the philosophers of our own age, who not only abstain from experiment and observation, but often actually shy away from them?" asked Cesi in a critical passage. "Indeed," he continued,

> there are many who execrate not only the telescope, by which the sight of man is raised ever higher, but also Galileo himself, who has revealed to us so many things previously hidden in the heavens: new planets, new fixed stars, and the surfaces of new bodies in the heavens. Instead, men such as these do not even make use of simple observations of the naked eye, preferring rather to be blind, and to go around, as it were, in a some dark wood, as if so fascinated by the charm of the opinions of the ancient writers that they wish not to depart from them even a little, or, guided by both reason and the senses, to add or to change anything to their presumed rules and decrees.[74]

Cesi concluded that the heavens were unmoving, unified, fluid, tenuous, and bounded by no solid figure at all; that there was a multitude of stars both known to us and hidden; and that all this, in the end, was fully reconcilable with sacred Scripture—provided it was correctly interpreted. Whatever inconsistencies Galileo's opponents might find between his own views and the Bible, were, Cesi claimed, a largely hermeneutic matter, proceeding from the fact that there were a variety of possible interpretations and interpreters.[75]

It was quite a tour de force; but Bellarmine was unmoved—he would have none of it. Usually he managed to mask his differences with the Copernicans beneath a guise of benign tolerance; but in his response to Cesi eleven days later, he could not contain his exasperation:

> What I wanted from you is not to know that it can be asserted on the basis of Holy Scripture and the writings of the Holy Fathers that the sky is immovable and that the stars move, and that the sky is not hard and impenetrable like iron, but rather soft and very easily penetrable like air—*I knew all these things*. Rather I really wanted to learn from you how the movement of the sun and the fixed stars can be saved, which are and will always be together and make their circles larger or smaller according to whether they are further or closer to the pole.[76]

The great cardinal, trained Jesuit that he was, was not exactly an easy opponent. He could indeed say he knew all the things Cesi was claiming because, in a sense, he himself had once thought at length about the neo-Stoic possibility that the cosmos was indeed fluid and boundless.[77] Years later Cesi would write to Faber summarizing this part of their dispute, in which he explained the simple but fundamental difference between them: for Bellarmine the phenomena could not be saved unless the spheres were solid; for Cesi they could not be saved if they were.[78]

Much as he would have liked at least some of his work to be published now, Cesi's letter could hardly be published in the tense atmosphere of 1618. Not the least of the abundant ironies that attended the fate of his work is the fact that when it eventually appeared in 1630, just after his death, it was as an appendix to the chef d'oeuvre of the Linceans' first truly formidable opponent, Christoph Scheiner, "Apelles" himself. By that time Scheiner could reveal both himself and his work.

The *Rosa Ursina* was the massive and then definitive work on sunspots, replete with an abundance of diagrams based on Scheiner's own fine observations. Its first sixty-six pages contain a lengthy tirade against Galileo and a reiteration of his claims for priority in the discovery of the sunspots; its closing chapters are devoted to an exposition of the arguments in favor of the fluidity of the heavens and the scriptural authorities that could be marshaled in its favor. Cesi's letter was printed at the very end, as if to offer further support of Scheiner's more thorough investigations of the matter, along with Bellarmine's response to him. Fate had allowed one of Galileo's and Cesi's fiercest opponents—the man who arguably had prepared the way for the trial of 1633 ever since 1612—not only to have the last word on the sunspots, but to make it clear that in the matter of the fluidity of the heavens he had beaten them to the post as well. Finally Apelles could come out from behind his easel and deal with his critics. But there had been one more critical episode before then, one further celestial phenomenon that set the two sides apart.

## THE COMET OF 1618: LINCEANS AND JESUITS

However committed to Copernicanism he may have been, by 1618 Galileo was surely hoping that he could move on to other things. The battle over Copernicus must have seemed if not lost, then at least suspended. But this was not to be. Once more an object appeared in the heavens that revived the very issues that Galileo and his friends might have wished, at that point, to let go.

In October a small comet was sighted in the skies. It was followed by another in mid-November, and then a rather large one on or around November 27, so bright that it remained visible for more than a month. In Rome everyone was talking and writing about them. Soon a number of books and pamphlets appeared on the subject. Predictably enough, Galileo's friends wanted to know his opinions. On December 1 Cesarini wrote to him saying that he had been talking about the new sensation in the skies with the mathematicians at the Collegio Romano—Cesarini had always been close to the Jesuit order[79]—and that he was now sending Galileo their opinions on the comets.[80] Together with this letter he enclosed for comment a manuscript copy of a lecture by the brilliant young

Orazio Grassi, Giovanni Battista Ferrari's classmate and the favorite pupil of Grienberger. Just recently he had been made professor of mathematics at the College. In January Stelluti wrote to Galileo about the comet, mistakenly suggesting (as did Grassi) that since it did not seem to be magnified by the telescope, it must be very distant indeed. As with the nova of 1604, the controversy soon became unstoppable.

By January of 1619 Galileo was confined to bed with an even worse bout of arthritis than usual; he was unable to make the kinds of observations he felt he needed to. But opinions such as Stelluti's struck him as being wrongheaded, while several of the claims made by Grassi in his lecture on the three comets, which had just been published in Rome, particularly irked him. Galileo was anxious to set forth his own opinions on the matter; but now the complications were even greater than before.

When Grassi's *Disputatio de Tribus Cometis* first appeared, it was not attributed to him, but simply to "one of the Fathers of the Collegio Romano." Though reasonably good-natured and no more mistaken than most of the other views then circulating in Rome, it did contain several rebarbative moments, aimed directly at Galileo and his friends: "Only comets," this Jesuit tauntingly proclaimed, "have remained aloof from the Lincean eyes."[81] Moreover, at that very moment the Jesuits were spreading the rumor that their conclusions about the comet had finally succeeded (for reasons that are unclear) in overthrowing the basis of the Copernican system.[82] This, of course, was like a red rag to a bull.

But Galileo was forbidden from speaking on the whole subject of Copernicanism. It was a tight spot; and there was no clear way out. However much the Linceans may have realized the need for subtlety, Galileo had never been one for tact.

It is true that his response to the Jesuit father's *Disputatio* appeared not under his own name, but under that of his talented Florentine pupil and assistant, Mario Guiducci. But this seems only to have made him feel that he could speak as forcefully as he liked. It was a typical strategy. As Mario Biagioli has rightly noted, this was perfectly in keeping with a tactic advocated by both Cesi and Galileo's friend, the painter Cigoli, as early as 1612, when they gave him advice on how best to respond to the attacks on his work on floating bodies in water.[83] Cigoli then wrote that "such attacks ought to be answered by somebody young, or at least by someone under that semblance,"[84] while Cesi maintained that

> I have always been of the opinion that you should not answer such adversaries, but that they should be taught a lesson by having them answered by youths. Those in charge of the answers could be partially or totally guided [by us] or they may even be made to adopt already composed replies.[85]

Exactly the tactic Galileo now employed.

But it was a complex and fraught one. In those days when masks were sometimes real and sometimes not, it was often impossible to determine who the real author of a treatise was; and often, even if one *did* know, it was expedient to pretend that one did not. On other occasions one might write under someone else's name and yet do so in the knowledge that one's real identity was clear. In that way one could deny one's own authorship whenever it was necessary to do so. This was a climate of simulation and constant dissimulation, initiated to some extent by Galileo himself in his dialogue on the nova of 1604 and continued by Scheiner-Apelles in his own letters on the sunspots.

The manuscript of the greater part of the 1619 *Discourse on Comets*, though ascribed to Guiducci, is actually in Galileo's own hand; as are the comments on Guiducci's lectures of May that year (on which the book the book is based). There can be no doubt about the identity of the principal author of this *Discourse*. The main thrust of Galileo's attack was against Tycho Brahe, who had written the definitive book on the comet of 1577, and whose views on comets the Jesuit author of the *Disputatio* had followed. Once again the main issue was the measurement of distance and the use of parallax. But in this case Galileo maintained that parallax should be kept out of the discussion, since the comet was not an actual physical body in the heavens. In keeping with his view that comets were the refraction of sunlight on vapors or exhalations rising from the earth in the sky,[86] parallax was irrelevant to what was, after all, only an optical phenomenon. All the Jesuit's laborious calculations about its distance from the earth were thus not just wrong but completely misguided.

Galileo then criticized the Jesuit's claim that the small magnification of the comet by the telescope showed that it was very distant indeed, and certainly well above the moon; and he forcefully disposed of the Tychonic view, espoused by the author of the *Disputatio*, that comets had circular orbits (since it was clear that a comet could vary in its distance from the earth). For the rest, no more final view of the nature of comets was offered. Even in his old age Galileo remarked that he would never understand them, even if he lived a thousand years.

Despite the relatively moderate tone of the *Discourse on Comets*, Grassi was furious. Like many of his Jesuit colleagues, he took particular offense at the sarcastic remarks directed by Galileo at "Apelles." Although everybody knew by now that "Apelles" was in fact Scheiner, Galileo preferred not to acknowledge this. Ciampoli wrote to Galileo saying that although the *Discourse* pleased him and Cesarini, Galileo should not have picked an argument with the Jesuits, who had once received him with such honor but were now losing their respect.[87] The work was taken as a grave slight on Jesuit science; and Grassi set about composing his response.

Once again he did not do so under his own name. Instead, his *Libra Astronomica*, the *Astronomical Balance* of 1620 appeared under the name of an alleged pupil of Grassi's called "Lotario Sarsi," a slightly modified anagram of his own.[88] This was a genuinely hostile work, full of personal assaults on Galileo. It distorted his positions to the point of caricature. The Linceans immediately realized that its unpleasantness was such that it would be best not to respond in kind. A fuller and more vigorous response than Guiducci's *Discourse* was now needed. Together they began working on what would become one of Galileo's most famous works, the *Assayer* (deliberately named thus in response to Grassi's *Balance*) of 1623. The efficiency of the crude balance of the steelyard was to be measured against the much finer and more accurate instrument used in the weighing of pure gold.[89]

"Before I left Acquasparta for Fabriano, two days before Christmas, I had already received Father Grassi's *Libra Astronomica*, and he really seems to have gone much further in speaking against you, and against Guiducci, and against the Linceans," wrote Stelluti to Galileo in January 1620.[90] A few months later he suggested that "it would be better to address your reply to some friend, and not to that Lotario, or to Father Grassi—otherwise it will be never-ending."[91] Faber was worried too; Cesarini, suffering especially in this winter period from his pulmonary complaints, was in constant touch with Cesi about the latest developments; and Cesi was putting the finishing touches to the first proofs of the treatise on Mexican animals on which he and Faber had been working for so long.

As soon as he could, Cesi wrote to Galileo. "On not a few occasions, we [Cesi, Cesarini, and Ciampoli] have been thinking of you and your noble compositions, which we so much wish to be completed; and in particular we have unanimously made every possible consideration for the reply to the *Libra*. It really has seemed necessary to us that it appear quickly. But in every respect it's important that it doesn't come out in the form of a direct duel."[92]

On the very same day, May 18, 1620, Ciampoli wrote to Galileo as well:

> I've been in Acquasparta for fifteen days now, along with Signor Virginio and Prince Cesi. You would be envious of our conversation, with what cordial affection we speak of you, and how much we desire your company. . . . Only today we've been having long conversations about the desired reply to the *Libra Astronomica*. To all three of us, who live in affectionate jealousy of your reputation, a reply seems necessary—and as quickly as possible; but their opinion is that for the greater dignity of your person, either your name, which is so glorious, should not appear in the guise of some masked author, or that at least you should seem to be asking the opinion of some friend of yours, in the form of a letter rather than a book.[93]

And so it would be. From that day on the stops were opened. Discouraged and suffering from increasing arthritis, Galileo submitted his manuscript for the *Assayer* to his friends. Under Cesi's constant guidance, Ciampoli, Cesarini, and Faber worked unremittingly on the preparation of Galileo's definitive response to Grassi. Without them, this mature statement of Galileo's Copernican and anti-Aristotelian and anti-Ptolemaic astronomical views would probably never have been published.[94] To help them, they enlisted the aid of Cassiano dal Pozzo as a learned, well-connected editorial adviser and general enabler. By then Cassiano had grown very close to Francesco Barberini, Maffeo's intellectual nephew, whose scientific interests were deep and who looked on the Linceans with great favor.

As always the battle was fought on several fronts. There were skirmishes in Rome and the provinces. Guiducci worked on his own response to Grassi, which took the form of a letter to his old professor of rhetoric at the Jesuit college, Tarquinio Galluzzi.[95] Scheiner was preparing his own counterblast. Throughout it all, work on the *Assayer* continued.

Galileo grew even more discouraged. The three friends had their work cut out for them. They urged him to complete his manuscript and edited whatever they received from him. Stelluti was less present than usual, since family affairs often kept him in far-off Fabriano. But not for once did he neglect his commitment to his fellow Linceans. While they worked on the *Assayer*, he helped his brother Giovanni Battista write one of the few works in the whole Galileo affair that has not been analyzed to exhaustion. This was the *Sounding-out of the astronomical and philosophical balance by Lotario Sarsi in the controversy about the comets*, published in Terni in the first half of 1622. Purportedly written by Giovanni Battista Stelluti, the *Scandaglio*, as it is called in Italian (appropriately punning on "*scandalo*," or "scandal") raised exactly the kinds of questions that could be expected at this stage of the controversy. Who actually wrote it? And why was it published in provincial Terni, of all places?

A manuscript of the *Scandaglio* recently appeared on the New York antiquarian book market.[96] It contains annotations and corrections made by Cesi himself. This, in other words, was the text that one of the Stelluti brothers must have sent to Cesi for his comments, at the very time he was helping Galileo prepare the *Assayer*.

Cesi did not make many changes to the manuscript; but apart from a few minor stylistic improvements, those he did make were telling and significant. The book itself is a detailed response to Grassi's refutation in the *Libra Astronomica* of Guiducci's *Discourse*. In keeping with the Lincean strategy of pretending not to identify their opponent by name, every time Stelluti referred to "Father Grassi," or even "the Jesuit mathematician," Cesi struck it out, substituting "il Sarsi," or "Sarsi's Master" (both, of course, being Grassi himself). Stelluti repeatedly pointed out the many ways in which Grassi had either misunderstood Kepler and Brahe (with regard to such problems as the allegedly circular orbits of the comets), or willfully distorted their views in order to suit his own polemical purposes. But it was better not to accuse Grassi of ignorance or of tactical misquotation, and so Cesi set out to modify the strenuousness of these claims. He could see only too clearly where *that* might lead.

Better not to offend the other Roman Jesuits either. According to Stelluti, Grassi had used the dishonest strategy of attributing to Kepler views which were in fact those of his opponents or predecessors; but this rhetorical mode of exposition and argument, he continued, in which the views of the opposite side were set out only as a means of confuting them, was a particular favorite of the "Fathers of the Collegio Romano."[97] Cesi did not like the sound of this. It spelled trouble; and so he changed this phrase to read "of the Colleges"—something far more generic.

But the matter was complicated enough, especially since both Kepler and Brahe, whom Stelluti now seemed to be defending, were Protestant heretics. Cesi preferred the reader not to be reminded of this fact; and so, when Stelluti acknowledged that the movement of the earth was not admitted by Catholics, and that Grassi had said that the opinions of Kepler were repugnant to the Catholic faith and to be esteemed as nothing, Cesi eliminated all such references.[98] Truth was one thing, faith another.

All these were subtle but critical changes. As always, Cesi's sense of strategy was keen—though in the long run, as we now know, to no avail. He went through the manuscript moderating the references to Copernicanism; and when he found Stelluti arguing against the by-now rather tired claim that previously invisible stars were actually imperfections in the telescope itself, he deleted most of it. In place of this he substituted a notable passage about the nature of vision ("la potenza visiva," a strong and redolent phrase) as it applied to the relations between the eye and the telescope.[99] At that stage Cesi must have felt that explaining why the telescope failed to magnify the comets was a far more pressing issue. Galileo could not have wanted a more devoted supporter.

Soon the manuscript of the *Scandaglio* was ready to be sent to the publisher. An anonymous note records that once again Cesi arranged and paid for its publication, just as he had in the case of Heckius's treatise, which he had corrected so similarly seventeen years earlier.[100] At a time when it might have been difficult to find a publisher for so controversial a work in Rome, Terni was sufficiently provincial for it not too cause much of a stir. The town was also the chief center of the area around Cesi's estates in the neighborhood of Acquasparta and thus easily accessible to both Cesi and Stelluti, and there was much less likelihood of a delay in publication, as there might have been with firms in busier or more important centers. Furthermore, while the work might not immediately have become known to dangerous Romans, it would at least be available for circulation in areas where more propaganda might still be necessary.

But who was the real author of the *Scandaglio*—Stelluti or his brother Giovanni Battista? If Stelluti wrote it entirely himself, as has been suggested, it would have made a great deal of sense to have it attributed to his brother: blame for its contents could then be shifted to a provincial lawyer in Fabriano. But it is probably not entirely by Stelluti himself, although we know from his correspondence in these years that he guided his brother in its composition, suggested changes, and may even have been the initial inspiration. It is a learned, even vivacious work, and this, along with the knowledge of Semitic languages (used in the discussion of the telescope) does suggest at least the participation of Francesco.[101] In a letter written by Stelluti to Galileo on August 16, 1622, accompanying several copies of the *Scandaglio*, he gave not only a nice account of its genesis but also as good a reason as any for its composition:

> I can't deny that my brother has put himself at very great risk, and not only for having written about matters which are outside his profession. . . . He was not only moved by your own good arguments, as adopted by Signor Guiducci, but also stimulated by many discussions amongst his friends about Sarsi's *Libra astronomica;* and he was especially motivated by a number of country Jesuits, who held that it be impossible to respond to the arguments of the *Libra*. They kept on asking whether you yourself were going to reply to it. Then the printer in Terni wanted to print my brother's book when I was in Fabriano because of the near-fatal illness of my brother, and when I returned here I found the work already printed, and full of mistakes, which greatly displeases me. . . . Still I hope that you will look at the book when you have a moment, so that I can let my brother know where he has made mistakes or where he has weakly defended Signor Guiducci.[102]

But this was nothing. With the *Assayer* the situation was even more complicated and difficult. This time there could be no question of covering up Galileo's authorship. The Linceans discussed at length whether he should reply to "Sarsi" and to "Apelles" by these names—but to whom was the new work to be addressed? At first Cesi wanted to address it to Grienberger, one of the last of their friends at the Collegio Romano. But why put him in an awkward position with his colleagues? They gave up the idea; and then Ciampoli made the brilliant suggestion of addressing the *Assayer* to Cesarini himself. In this way the work would gain the protection of his influential friends, such as Cardinals Barberini, Bentivoglio, and Ludovisi (and perhaps Cardinal Bellarmine himself).

## THE PUBLICATION OF THE *ASSAYER*

Though Bellarmine died in 1621, Cardinal Ludovisi became Pope Gregory XV in the same year, and he took both Ciampoli and Cesarini onto his personal staff. Cesarini was especially close to both Bentivoglio and Barberini, and spent hours talking with them about the latest issues.[103] The more ill he became the closer he grew to the Jesuits, whose company had always pleased him. All in all, therefore, Cesarini was the right addressee for the "letter" into which the *Assayer* had been turned. His influential friends trusted him, and they would think twice before taking umbrage at the contents of the new work, however abrasive, by the mathematician from Florence.

The Linceans prodded Galileo, reminding him that the Jesuits were awaiting his answer to "Sarsi." Cesarini went through the final manuscript; then Cesi, Ciampoli, and Cassiano reviewed it as well. Finally it was ready. There was a brief hiccup, a moment of second thoughts, when in January of 1623 the Holy Office prohibited the heretical Dominican Campanella's *Apology for Galileo* (already written in 1616). But the Linceans covered all the bases they could. They cleared the final copy of the manuscript with the censors and arranged to have the task of deciding on the *imprimatur* given to a brilliant young Dominican theologian, Niccolò Riccardi, a man not especially moved by the Jesuits. Ten years later Riccardi would serve as master of the Holy Palace during Galileo's trial; but for the moment he was only too happy to approve the book. He did so in terms that went well beyond the norm: "I count myself lucky," he wrote in the *imprimatur* of February 3, 1623, "to have been born when the gold of truth is no longer weighed in bulk and with the steelyard, but is assayed with so fine a balance."[104]

The *Assayer* was printed in the course of the next few months. Despite the many errors it contained, it was rushed through the press and was an immediate success. Brilliantly argued and composed, it was direct, clear, robust, and often grippingly satirical. For once Galileo had found a way of defining exactly where he stood, not only philosophically but strategically as well. After attacking Sarsi for unmasking him as the author of Guiducci's *Discourse* ("instead of respecting my wish to remain incognito"), he wrote:

> I am aware that this name Lothario Sarsi, unheard of in the world, serves as the mask for someone who wants to remain unknown. It is not my place to make trouble for another man by tearing off his mask, after Sarsi's own fashion, for this seems to me neither a thing to be imitated nor one which could in any way assist my cause. On the contrary, I realize that to deal with him as an unknown person will leave me a clearer field when I come to make my reasoning clear and explain my notions freely.[105]

But the Jesuits were frustrated and furious. In writing of his pseudonymous adversary as "a lacerated and bruised snake having no vitality left except in the tip of its tail, but which nevertheless continues to wriggle, giving the impression of still being healthy and vigorous,"[106] Galileo must have known that he was letting himself in for more trouble. Grassi said he would reply in kind. Others among the Jesuits felt that however much Galileo's powerful friends may have prevented him from being sent straight to the Inquisition in 1616, the gloves were now off. In any case, as many of them must have realized, if left to his own devices, Galileo was capable of alienating even his closest friends in high places.

Then something happened for which the Linceans can hardly have dared to hope. Gregory XV died; and on August 6, 1623, Maffeo Barberini was elected Pope Urban VIII. Before the month was out, his nephew Francesco was made a cardinal and admitted to the Academy of Linceans. To the top of the frontispiece of the *Assayer*, with its description of Galileo as "Lincean, Florentine nobleman, Philosopher, and Chief Mathematician to the Grand Duke of Tuscany" and the crest of the Linceans below, they added the arms of the Barberini family (fig. 5.2). From now on the trigon of bees seemed to appear almost everywhere. Stelluti added panegyrics not only to Galileo but also to Francesco Barberini. The Linceans felt they could count on him for both support and protection. Urban himself is said to have enjoyed the *Assayer* so much that he had it read to him at table. Accomplished man of letters and poet that he was, he must have been delighted with its mordant witticisms. Not yet did he realize, as the Jesuits did, how little he would be able to restrain his irrepressible friend, and how much Galileo's work threatened the established order of things.

In the meantime Galileo and Cesi decided to regroup in Acquasparta. For months Cesi had been encouraging Galileo to come down from Florence, but bad weather and family commitments constantly got in the way. Finally, with the weather growing sweeter, Galileo made his way southward, arriving in Acquasparta on April 8, 1624, the day after Easter. They had another instrument besides the telescope to use and perfect.

But the joy of their reunion was swiftly broken, when they heard the devastating news of the death of the ever-ailing Cesarini on April 11, 1624, at the age of 28.[107] *Venit summa dies et ineluctabile tempus*, Faber wrote to Cesi from Rome. In Cesarini they lost one of their keenest allies, one of the steadiest hands at their helm, and their most direct conduit to the pope. Now, despite Galileo's fame and acknowledged brilliance, they were on their own.

There was much to discuss. The two Barberini, pope and cardinal-nephew, still seemed to be favorably disposed to Galileo. Had not Maffeo himself written, just four years earlier, a Latin ode in praise of Galileo, and had not Francesco just been elected to their company? But there was no doubt in either of their minds that the way would not be smooth. Already then, though ever mindful of the interdiction of 1616, Galileo was thinking of his magnum opus, the *Dialogue of the Two Great World Systems*, in which the Aristotelian cosmology would be set out against the Copernican one in its fullest form yet. We know from his subsequent letters how much Galileo valued the conversation and advice of Cesi, and Cesi must have done much to temper Galileo's natural impulsiveness by his own sense of the need for caution and prudence.

How intensely the two friends must have talked about their mutual interests and plans! The faithful Stelluti, now more often than not with Cesi in Acquasparta and Rome, was there too, and so was the German doctor, Johann-Baptist Winther. In a letter to his compatriot Faber in Rome, Winther recalls his arrival in Acquasparta:

> The elevated site of the place, along with the sheer loveliness of the green and fertile meadows lying beneath it, terminated by the most beautiful mountains, seemed, at first sight, to resemble nothing more

IL SAGGIATORE
Nel quale
Con bilancia eſquiſita e giuſta
ſi ponderano le coſe contenute
nella
LIBRA ASTRONOMICA E FILOSOFICA
DI LOTARIO SARSI SIGENSANO
Scritto in forma di lettera
All'Ill.mo et Reuer.mo Mons.re D.
VIRGINIO CESARINI
Acc.o Linceo M.o di Camera di N.S.
Dal Sig.r
GALILEO GALILEI
Acc.o Linceo Nobile Fiorentino
Filoſofo e Matematico Primario
del
Ser.mo Gran Duca di Toſcana.
FILOSOFIA NATVRALE
MATEMATICA
IN ROMA MDCXXIII
Appreſſo Giacomo Maſcardi

Fig. 5.2. Francesco Villamena, engraved title page, Galileo, *Il Saggiatore* (1623).

than those Elysian fields celebrated by the poets; or the celestial gardens whose beauty no mortal painter, with whatever skilled mixture of colors, however conscientiously applied, could ever attain. . . . I was overcome with awe at that scene, and still not sated by the wonders of Acquasparta, seemed to be wondering at an earthly paradise.

There was more:

> My astonishment only increased. It was no longer early when I descended from my horse on Sunday; and the good cheer and warmth of the Prince's court was such that it made my arrival seem most welcome. . . . I would immediately have had an audience with the Prince, if he had not already been engaged in conversation with Signor Galileo at that moment. . . . I handed the letter which you gave me to a servant of the house, an order was given to provide me with a meal, and after that I was given the loveliest room with an altogether delightful view. Finally, at two in the morning I was sum-

monsed to the company of the Prince, whom I found sitting in front of the fire in the company of Signor Galileo.[108]

It is easy to imagine how happy Galileo must have been in this respite between battles. Even Colonna in Naples heard of their meeting and wrote to Cesi, "I was so happy to hear of the pleasure you derived from the visit of Signor Galileo. You can be sure that I envy you; you must have been discussing wonderful things."[109] Cesi and Galileo walked in the countryside around Acquasparta and visited the local sites, such as the waterfalls of Velino near Terni. It is not hard to imagine Cesi presenting to Galileo his latest views on the origins of the lignites and other fossils he had excavated in the neighborhood, and Galileo raising some of the problems that had long occupied him on occasions such as these. From paleontology they would have moved on to the issue of spontaneous combustion that had always preoccupied Cesi (because of the underground fires and fumaroles of the region) and in which Galileo, because of his involvement in polemics about the nature of the sun and the light of the moon, had almost a vested interest. Galileo had always followed the botanical researches of the Linceans and, now that the great project on Mexican flora and fauna was being completed, was keen to hear of its progress. But above all they stayed up at nights, as was their wont, to survey the clear spring heavens—the April nights were still long enough—from the balcony of the palace at Acquasparta. No wonder that for years afterward Galileo would nostalgically recall his visit to that sweet spot, so far from the pressures of Florence and Rome.

He did not—he could not—stay long. He had to proceed to Rome, for the fourth time. Leaving Acquasparta on April 22, he arrived in Rome the next day. Though Bellarmine had died, there were other cardinals whom he had to see—Cobelluzzi, Boncompagni, and von Zollern—in a vain attempt to bring them round to his Copernican views. But if they remained skeptical of the conclusions Galileo drew from his use of the telescope, what can they have said about his demonstration of the use of the instrument he had even more recently perfected, the microscope?

Before leaving again on June 11, he managed to have no fewer than six audiences with his old friend Pope Urban VIII. Though assuring him of his friendship, Urban remained noncommittal on the subject of his astronomical opinions; and, perhaps with increased force, he impressed on Galileo the need for prudence, tact, and caution—never his strongest qualities.

The year 1624, therefore, was a pause between crises, the lull before the growing tempest. Although they continued to correspond with the same intensity, Galileo and Cesi never met again. For Cesi and his Lincean colleagues, it seemed that the work with the telescope was done; they had argued and written and propagandized endlessly about it. Now the time had come to turn their attention to the results they were beginning to obtain with that other instrument so closely related to it: the microscope.

PART THREE

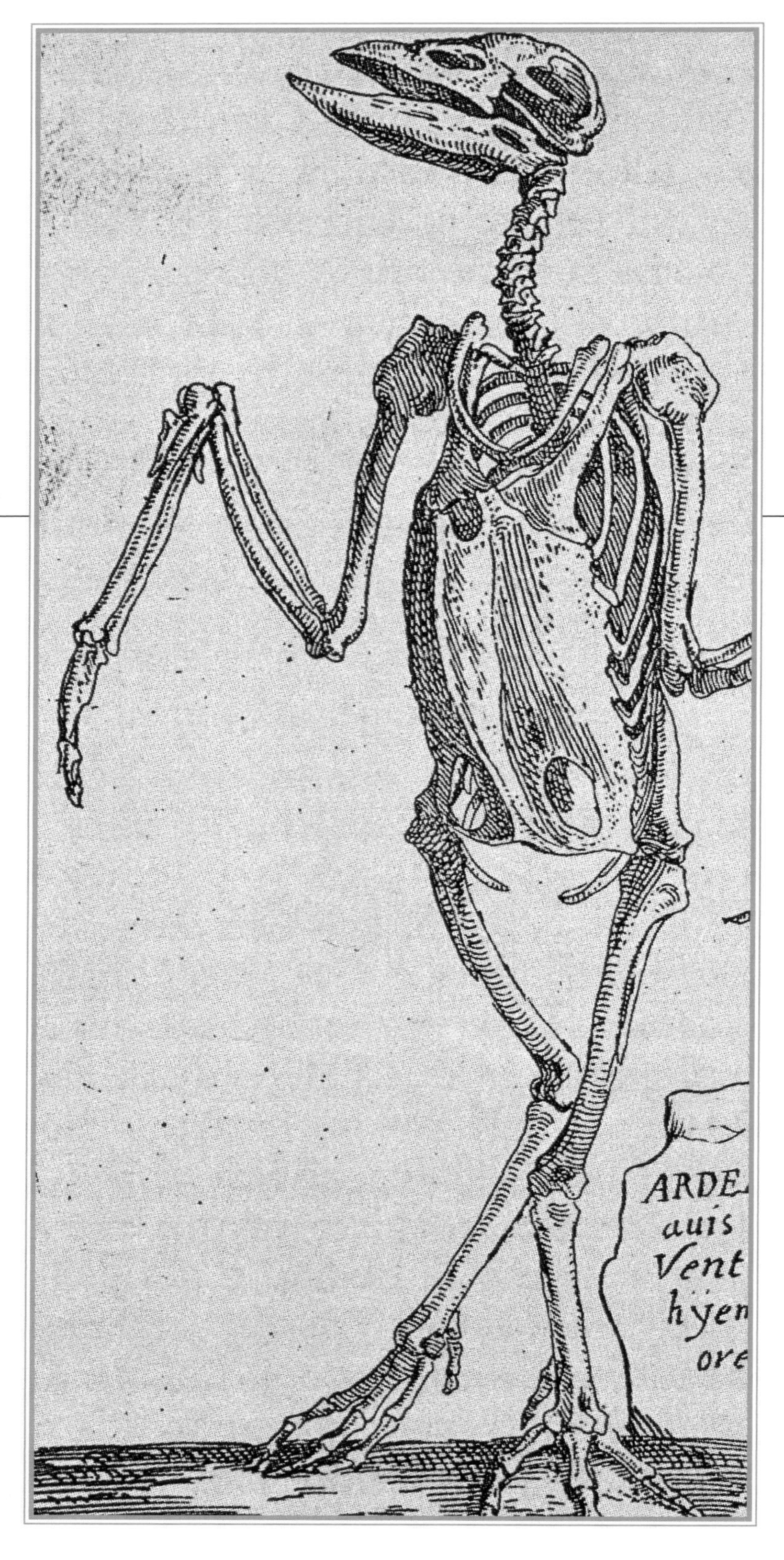

# *Natural History*

# SIX *The Chastity of Bees*

CESI ASKS FOR A MICROSCOPE

Although Galileo had surely discussed the microscope with Cesi on the occasion of his visit to Acquasparta in April 1624, it was only later in the year that Cesi asked for one of the new instruments for himself and for his wife. (Galileo had used the occasion of his trip to Rome to demonstrate the use of his microscope to Cardinals Cobelluzzi and von Zollern, and to confer with the latter about the two types of microscope then available.) On September 23, 1624, Galileo sent them his microscope, along with instructions for its use. The letter accompanying it is justly famous:

> I am sending your Excellency an *occhialino* to view the smallest things as if from nearby. I hope that you will derive no small pleasure and enjoyment from it, just as I did. I have been slow in sending it to you, because at first I was unable to perfect it, having had some difficulty in finding the correct way of cutting the crystals perfectly. . . . I have contemplated very many small little animals with infinite admiration: among which the flea is most horrid, the mosquito and the moth very beautiful; I have also seen with much pleasure how flies and other little animals walk on mirrors and are also seen from below.[1]

Then he concluded:

> But your Excellency will have a huge field in which to observe many thousands of specimens. I beg you to notify me of the most interesting things you observe. In sum, it [the *occhialino*] gives us the possibility of infinitely contemplating the grandeur of nature, how subtly she works, and with what indescribable diligence.[2]

Almost immediately Cesi saw the potential for his own researches of this adaptation of the principles of the telescope. A month later he acknowledged having received Galileo's recently devised instrument for seeing the smallest things.[3]

Here, of course, the Dutch were the experts; so Cesi wasted no time. Knowing that the eccentric Dutchman Cornelius Drebbel, son-in-law of the engraver Hendrick Goltzius, had made a better lens, Cesi immediately asked for one of the instruments belonging to Drebbel's brothers-in-law, the Kufflers from Cologne, recently arrived in Rome.[4] As soon as Cesi had one, he began his work on the problem of the reproductive system of the cryptogams, of the ferns and of the fungi they had all long been studying. At the same time, however, it occurred to Cesi and to his colleagues that there was another tiny specimen of nature that could be examined, for wholly different reasons. The study of cryptogams had no other motive but the purest scientific ones; but when it came to the examination of the bee, science, politics, and panegyric went hand in hand.

## MICROSCOPES AND MITES

For more than three years Cassiano's antiquarian friend from Aix, Peiresc, had been writing enthusiastically to his Roman correspondents about the microscopes of Drebbel and the Kufflers.[5] Peiresc had already sent one of these as a gift to Paolo Gualdo in Padua just before Gualdo died in October 1621.[6] In May of the following year he wrote a detailed note about a demonstration of Drebbel's "*lunette*" by Jacob Kuffler in the presence of Marie de Médicis in Paris. As usual, the chief objects that were scrutinized in those earliest days of the microscope were small insects: some cheese mites, several kinds of fleas, a louse, a vinegar fly that when magnified looked like a large cicada, a bedbug, and, as always, the eyes of a fly.[7]

Less than a month later, Peiresc wrote to Girolamo Aleandro asking him to introduce Kuffler to Cardinals Cobelluzzi and Barberini, whom he knew were interested in the possibilities of the new kind of *occhiale*. Kuffler, he said, would demonstrate his

> *periscope* or *occhiale*, a new invention different from that of Galileo, which shows a flea as large as a cricket and almost of the same shape . . . [as well as] the minute animals generated in cheese, called mites, so tiny that they are like dust-grains, but when seen with this instrument become as large as flies but without wings. They can be seen clearly as having very long legs, a pointed head, and every part of the body quite distinct.[8]

While Galileo's first microscope had used a convex objective lens and a concave ocular lens (as Della Porta had in fact proposed using many years previously),[9] the "German" one was a true compound instrument with two convex lenses (as already described by Kepler in his *Dioptrice* of 1611). Its advantage was that it offered better magnification than the earlier forms. Galileo was very much taken with it. From then on he probably only used a compound microscope, following the "German" invention, but perhaps modified it to use three biconvex lenses, two ocular and one objective. It is not clear to which microscope Johannes Faber was referring when he wrote to Cesi on May 11, 1624, about Galileo's meeting with von Zollern, but he could barely contain his excitement:

> I spent yesterday evening with our Signor Galileo, who is staying near the Maddalena. He has given a very beautiful *occhialino* to Cardinal von Zollern for the Duke of Bavaria. I examined a fly which Galileo himself showed me; and I, remaining astonished, said to Galileo that this was another Creator, given that it makes things appear that until now one wouldn't know that they had been created.[10]

"Occhialino" (or "occhiale") was then a standard word for any instrument containing lenses; but in this case it clearly referred to a microscope, which had still not received its modern name.[11]

But this was soon to change. Already in the *Assayer* Galileo had mentioned the use of the inverted telescope "adapted to see things from very near";[12] and by mid-April of 1625, Faber had discovered the right word for the new instrument: "I should also mention that I am calling the new *occhiale* for looking at minute things a *microscope*."[13] The term seems immediately to have taken hold—though it would be a few months before it became current among all the Linceans. On June 6 Fabio Colonna wrote to Cesi that he too had used the new instrument:

> I have called it—as I wrote to Signor Stelluti—an *Enghiscope*, which means an *occhiale* for seeing from nearby, as opposed to the other instrument which sees from far and from near, but not so closely, because of the distance between the lenses and the way [in which they are placed]. This is rightly called the *Telescope* because it is perfect, but it could also be called *Ponoscope*, which sees distant things minutely.[14]

At the beginning of this letter, Colonna noted that although the *occhiale* was capable of enlarging small things seen from nearby, the clarity of the image remained a problem. This, he suggested, was probably in consequence of the difficulty of polishing the lenses adequately. Describing some of his own observations with the new instrument, Colonna also tells of the further difficulties he had with what was evidently still the primitive form of the instrument:

> I have seen those little animals in the grains of cheese, in truth a stupendous thing; while the structure of the eye of the fly with those eyelids around them, larger than our own, amazed me. But the golden reticulation could not be seen with the telescope, because however small it is—and I have a very small one—it has to see at a distance of at least two palms; and with the same small lenses composed in the manner of the telescope, I still could not see too well, not just because it does not enlarge very much but also because when it sees from a certain distance it doesn't distinguish minute things so well.[15]

Examination of the eyes of insects was one of the chief subjects of early microscopic investigation; and here, once again, it was Galileo who was the real pioneer. Already in 1610 a Scottish student at Padua, John Wodderburn (or Wedderburn), in the course of defending the *Starry Messenger*, had commented on Galileo's use of his *perspicillum* to distinguish

> the organs of motion and of sense in the smallest animal, notably in a certain insect in which each eye is covered by a thick membrane, which, however, is perforated with holes like the visor of a warrior, thus affording a passage to the images of visible things. Here you have new proof that the *perspicillum*, concentrating its rays, enlarges the object.[16]

The eye was examined because of all the senses it was the chief instrument of observation. But it posed a problem that continued to plague the Linceans for many a long year: what were the limits of vision, and how far could it be aided? Even by 1625, when the microscope was clearly ready for use, and the Linceans were ready to tell the world—or at least a part of it—of its possibilities, they remained all too aware of its limitations.

## BARBERINI BEES

At the tensest moment of the conclave that would soon elect Maffeo Barberini to the papacy, a fateful and prophetic event occurred. A swarm of bees entered the Vatican palace from the meadows facing Tuscany and settled on the wall of the cell where he waited. It seemed that Divine Providence had sent this portent to announce the imminent accession to the papacy of a member of that Tuscan family whose coat of arms had long since been transformed from three wasps into an emblem of three bees.

Within a few years it was impossible to go anywhere in Rome without

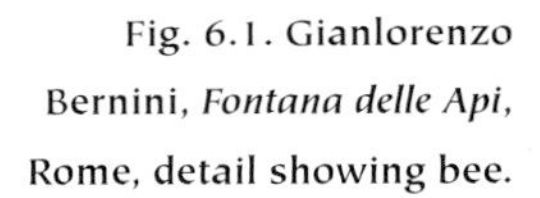

**Fig. 6.1. Gianlorenzo Bernini, *Fontana delle Api*, Rome, detail showing bee.**

Fig. 6.2. Gianlorenzo Bernini, Baldacchino, St. Peter's, Rome, detail showing trigon of bees.

encountering the Barberini bees. Even today it is hard to avoid them. Almost immediately upon his elevation in August 1623, Urban VIII, his brothers, and his nephews embarked on an ambitious program of architectural and artistic patronage. They built and rebuilt, and restored the remains of ancient Rome as well as the monuments of the early Christians. As if to ensure that their munificence would never be forgotten, the trigon of bees, their triple emblem, was attached to every example of their enterprise and intervention.

Bernini's fountains display them (fig. 6.1), and so do the freshly restored walls of the ancient city itself. From ceiling to floor, from cornice to pavement, from triumphal entryways to modest sacristy doors, on tombs and every imaginable piece of church furniture, one may still discover the bees that bear lasting witness to the patronage of the Barberini (fig. 6.2). Giant bees fly above Divine Providence in the center of the vast ceiling that Pietro da Cortona painted for them in their newly rebuilt family palace on the Quirinal (fig. 6.3). Bees crawl up the twisted Solomonic columns that support the mighty bronze and gold baldachin fashioned by Bernini over the main altar of Saint Peter's; while on Urban's tomb in the apse minute bees climb toward the lid of the sarcophagus (fig. 6.4). They do so not to smell the decay announced by the hooded skeleton who inscribes the name of the deceased pope on the black page of death, but rather to rise toward the sweet odor of sanctity, the *odor sanctitatis*, that issues from the tomb of the Barberini pope.

But there are two larger bees on Urban's tomb as well (fig. 6.5). These

Fig. 6.3. Pietro da Cortona, *The Triumph of the Barberini*, Palazzo Barberini, Rome, detail showing trigon of bees.

give the impression of having flown free of the sarcophagus in order to settle—but only briefly, it seems—above the grim reminder of mortality itself. Since they appear so clearly to be on an upward course, they must be the ancient symbol of the immortality of the soul; for in antiquity the bees that allegedly emerged from the corpses of animals were taken to stand for the spirit's ability to rise from the bodies of the dead.

As well as these great works in bronze and stone, endless poems were written in order to ensure the immortality of the Barberini. Their title pages were all illustrated with the trigon of bees. Hardly a medical, technical, scientific, or geographical work published in Rome between 1623 and 1644 lacked that symbol of sweetness, industry, and power. When Galileo's *Assayer* was published just two and a half months after Urban's accession, the papal bees, as we have seen, appeared at the top of the title page (fig. 5.2), as if to encourage the support of the same Maffeo with whom Galileo had discoursed in friendship in Florence many years earlier.

Fig. 6.4. *Opposite:* Gianlorenzo Bernini, *Tomb of Urban VIII.*

The great Flemish painter Rubens designed a title page for the splendid

Fig. 6.5. Gianlorenzo Bernini, *Tomb of Urban VIII*, detail showing bees.

Antwerp edition of Maffeo's own Latin poems (fig. 6.6). In it he illustrated the scene from Judges 14 in which Samson tears open the jaws of the lion to bring forth the sweetness of honey. Bees pour from the mouth of the lion, but in a lightened space just behind the vivid scene, a trio arrange themselves in the formation of the Barberini trigon. "Out of the strong came forth sweetness" is the central paradox of the riddle posed by Samson after his defeat of the young lion; and so the allusion is not just to the antique notion of the immortality of the soul, or to the honeyed beauty of Urban's youthful poetry. It is to the benevolence and magnanimity of the Latin-loving poet who had attained the rank of supreme pontiff.

But bees, when irritated, also sting; and under the pressures of doctrine, politics, and a cosmos that was being transformed, the liberal be-

MAPHÆI
S.R.E.CARD.
BARBERINI
NVNC
VRBANI PP. VIII.
POEMATA.
Pet. Paul. Rubens pinxit.
ANTVERPIÆ, EX OFFICINA PLANTINIANA
BALTHASARIS MORETI. M. DC.XXXIV.

Fig. 6.6. Cornelis Galle after Peter Paul Rubens, *Samson and the Lion*. Engraved frontispiece to Maffeo Barberini, *Poemata* (Antwerp, 1634).

nignity of Maffeo Barberini turned to sternness; and then into something chilling and unyielding. Before the first decade of Urban's pontificate was out, Galileo was sent to the Inquisition by the man who had once been his friend. Maffeo had supported his anti-Aristotelian views of the suspension of bodies in water, had enthused over the *Letters on the Sunspots*, and had written an ode in his honor. When Bellarmine issued his injunction to Galileo in 1616 to renounce his view that the sun was the center of the universe and that the earth revolved round it, Maffeo was instrumental in preventing him from actually being condemned for heresy. Even in 1624, during the course of his six-week stay in Rome, Galileo was warmly received by him. But within a few years Urban realized that he could no longer protect his stubborn and headstrong friend.

The reasons for Galileo's progressive estrangement from the Barberini and for their sudden abandonment of him have been much discussed. But there is one crucial document in this complex personal and scientific drama that has not: the Linceans' panegyric on the family of bees.

## THE *MELISSOGRAPHIA*

The Jubilee of 1625 could not have come at a better time for the newly elected pope. This was the moment to celebrate the power of the papacy, the triumph of the Catholic Church over the German heresies, and the glory and prosperity of Rome under the Barberini. It was a year full of celebrations, processions, feasts, and showy displays. The poets prepared epigrams, odes, and panegyrics in their honor. Even the mathematicians and philosophers made it clear that their discoveries could not have occurred under any auspices other than those of the Barberini.

Toward the end of that festive year, the Linceans were working feverishly to complete an elaborate celebration of the chief Barberini emblem, the bee. This celebration was not just antiquarian, philological, and archaeological—it was profoundly natural historical as well. Just as the year turned they published three separate works, in which archaeology, philology, panegyric, and scientific investigation were combined in a wholly unprecedented way. Appearing in swift succession, these works bore the titles *Melissographia*, *Apes Dianiae*, and *Apiarium* (figs. 6.7–11).

The first of these three works was an unusually large engraved sheet, measuring 41.6 by 30.7 cm (fig. 6.7). Copper plates, in those days, did not come much larger. Dedicated to Urban VIII, the sheet carried a title written in Greek capitals, *Melissographia*. Signed in the lower right hand corner by Matthias Greuter, it carries the date of 1625.[17] "The sheet included herewith," wrote Cesi to Galileo on September 26, "has been made all the more to show our devotion to our Patrons, and to exercise our particular commitment to the observation of nature."[18] How could the supreme pontiff not have been impressed? In the center is the trigon of bees, framed by a flourishing pair of bay branches. Two putti, holding aloft the papal tiara and the keys of Saint Peter, sport above it.

But this is no ordinary trigon. These are not bees whose spiky forms have been reduced and mollified by the usual strategies of art. On the contrary. They are bees that have been magnified many times larger than life-size, each one represented with extraordinary attention to anatomical detail. The Linceans surely knew a long passage in Giovanni Rucellai's famous poem on bees, *Le Api*, written just about a century earlier and published several times since its first edition of 1539.[19] In it, Rucellai had described how he had made an anatomy of the bee, and how he had magnified its parts with the aid of a concave mirror. Thus he was able to see things "almost invisible to our eyes"—the legs, the back, the wings, the sting, the antenna, the proboscis ("appearing like the trunk of an Indian elephant"), and the complex structure of tongue and palps ("sheaths") within the mouth of the bee.

Fig. 6.7. Johann Friedrich Greuter, *Melissographia*. Engraved broadsheet showing trigon of bees under magnification, with details of microscopic examination of the bee.

But the Linceans went much further. They had their magnified anatomy of the bee engraved and its details reproduced with undreamed-of precision. Until then, no one could have predicted the visual possibilities of such a combination of magnification and skilled engraving. It is not just the ease and clarity with which one can make out the structure of the head and proboscis of the bee, as well as the details of the thorax, the abdomen, the all-important legs, the antennae, and the sting that impress; it is also the precision of every one of the mouth parts, including glossa, galeae, and labial palps—to say nothing of the success with which the engraver managed to convey the diaphanous and flimsy quality of much of the bee's body, especially its wings and the delicate extremities of its legs.

The *Melissographia*, in fact, is the first printed illustration in the history of natural history to have been made with the aid of a microscope.

"Observed by Francesco Stelluti, Lincean of Fabriano, by means of a microscope" runs the proud inscription. This is no simple illustration; it is an examination, a close observation from the life. Whereas other representations of the Barberini emblem simply repeated the same view of the bee disposed at appropriate angles to form the trigon, Stelluti did something quite different. In order to examine the bee, Stelluti viewed it from above, from below, and from the side. The idea was to show the bee just as it appeared in life—or rather, as it appeared under the microscope.

Then, as if yielding to more purely aesthetic considerations, a number of other parts of the bee are prettily displayed across the scroll that unfolds with its texts across the bottom of the page: on the left there are profile and then frontal views of the head with its eyes, antennae, and the proboscis with its undivided elements separated; on the right, an antenna, a single eye, the sting, and remarkable details of the mouthparts forming the proboscis; and finally, in emulation of some graceful printer's vignette, a pair of hairy posterior legs extending neatly across much of the width of the sheet.[20]

"To Urban VIII, Supreme Pontiff, When this more accurate description of a bee was offered to him by the Academy of Linceans as a symbol of their perpetual devotion." Thus the inscription at the top of the sheet. On the scroll below, with its illustrations of the parts of the bee, runs a much longer text.

> O great Parent of Things, to whom Nature willingly submits itself, behold the BEE in the BARBERINI escutcheon. Nature has nothing more remarkable than this. Surveying it with a keener gaze, the work of the Linceans has set it forth in these pictures, and explained it. The genius of the Cesi family has stimulated this sacred labor; the art of Pallas has aided these willing men. Great miracles have emerged as a result of their work with the polished glass, and the eye has learned to have greater faith. Had it not been for the divine discoveries of the new art, who would have known that there are five tongues[21] on the Hyblean body [i.e., the bee's], that the neck is similar to a lion's mane, that the eyes are hirsute, and that there are two sheaths on each mandible? Thus it is fitting that while the world looks up to you in wonder, your BEE shows itself even more worthy of wonder.

The Linceans' pride and their desire to promulgate the results of their achievement had turned the *Melissographia* into propaganda for themselves as much as for the pope. Even though fewer than half a dozen prints survive—suggesting that it was distributed to the pope and his family alone—this was the most visible and public statement they had made so far of their commitment to empirical investigation and experiment, and their belief in the power of sight to penetrate the mysteries of nature.

The eulogistic text of the *Melissographia*, in eight distichs and in more or less immaculate classical form, was written by Justus Riquius, "the

Belgian Lincean," as he regularly signed himself.[22] A great friend of Faber's, he had been elected to the Academy only a few months previously, the second Netherlander to be admitted as a Lincean after the ill-starred Heckius. His official role in the Academy was that of panegyrist, because of his command of ancient rhetoric (which won him the appointment of professor of rhetoric at the University of Bologna in the very year he wrote the text of the *Melissographia*). In 1617 he had published a learned treatise on the archaeology of the Roman Capitol, and this was followed by the publication not only of his own elegant letters and poems but a number of panegyrics, funeral orations, and antiquarian tracts.[23] He can scarcely have better suited to work on the second of the Linceans' apiarian offerings in the autumn of 1625.

## DIANA'S BEES

*The Bees of Diana recently observed on ancient monuments* is a ninety-line elegiac poem in honor of Urban VIII. Dated November 1625, it was printed hard on the heels of the *Melissographia*.[24] But the two works could hardly be more different. Where the *Melissographia* points to the future, the *Apes Dianiae* remains suspended in the past. While the *Melissographia* uses a wholly new technology—that of magnifying lenses—to examine a real bee, the *Apes Dianiae* is a poetic consideration, followed by deeply recondite explanatory notes, on the representation of bees on ancient coins. The engraved broadsheet underscores the Linceans' commitment to firsthand observation as the basis for natural history; the long poem, in every line and erudite note, reflects their immersion in the classical past. Everything in the *Melissographia* that is seen through new lenses is here only reflected in the dim mirror of mythology and ancient religion.

The *Apes Dianiae* begins with a paean to Urban and continues with a short poem on a bee carved in an ancient gem owned by Francesco Gualdo (a member of Urban's household who already then was building up an important museum of his own). Then comes the main poem. It is dense, difficult, and highly allusive. Without the learned notes that follow it, its rich archaeological and philological references would have been largely incomprehensible—even to someone as skilled in Latin and as familiar with the byways of classical mythology as Urban himself. They also identify the three coins illustrated on the frontispiece, which are central to the meaning of the work as a whole.

> O muses of Helicon, who cherish the Cecropian bees on the summit of Hymettus with liquid dew, if you were once able, in those patriarchal caverns, to pour the Gnossian honey on Jove, and to favor all the swarms of Ariston. . . . now come to my aid, summon the laurel from Mount Pindus. . . and let me do homage to URBAN.

This is how the poem begins, and it is typical. The Barberini may have known that bees came under the tutelage of the muses, and that "Gnoss-

Fig. 6.8. Three coins from the collection of Cardinal Antonio Barberini. Engraved frontispiece to Justus Riquius, *Apes Dianiae* (1625).

ian honey" (i.e., the honey of Crete) refers to the nurture of Jupiter (a nice parallel for Urban) by the nymph *Melissa* in a Cretan cave, but without Riquius's dense note to this passage, they certainly would not have known to find confirmation of such matters in Aldrovandi's *De insectis*, Lilius Gregorius Giraldus's *Syntagma de musis*, Antonius Liberalis's commentary on Ovid's *Metamorphoses*, Servius's commentary on the *Aeneid*, or Didymus the Grammarian citing Lactantius. So it goes throughout the work. The Barberini were probably amused, if not actually flattered, by these many further proofs of the richness of the tradition in which their chief family emblem was embedded.

The notes also explain that the three coins illustrated on the frontispiece (fig. 6.8) came from the collection of Urban's ascetic elder brother, Cardinal Antonio Barberini. They are central to the meaning of the work as a whole. Here Riquius could show off his numismatic expertise and explain that one of the coins (which Cesi also owned) was from Ephesus, another from Naples, and the third from Megara Hyblaea in Sicily.[25] But no one, according to note D, had yet commented on the fact that each of the coins showed bees that fell under the tutelage of the Goddess Diana.

What exactly was all this about? That Diana was the goddess of the threefold intersection known as the trivium—"*Diana in Trivio*"—only enhanced the connection with the Barberini trigon. But there was much more to it than that. It was not just a matter of alluding to the relationship between the honeyed poetry of the pope and that of ancient poets such as Pindar (to whom Urban liked being compared) or Theocritus. The point, according to Riquius, was that bees had dripped honey on Pindar's mouth as he lay sleeping, and that Theocritus' hometown was Syracuse, which was in turn very close to Megara Hyblaea, where bees flourished and which produced the sweetest honey. But there was a much more important theme to be worked out here.

As everyone then knew, Diana was the stern goddess of chastity. As goddess of the hunt as well, it was natural enough that on coins she should often be accompanied by a stag. Even the famous many-breasted statue of Diana at Ephesus, according to Riquius, did not mean that she was in any way unchaste. Her abundant breasts were not for any sexual purpose, but for nurturing and nourishment.[26] Despite the *horned* stag that so often accompanied her representation on coins, she was also shown with bees, which were themselves the model of chastity.

Why? Because, as the ancients maintained, bees were autogenetic: they reproduced without any kind of sexual congress,[27] and were therefore particularly pleasing to Diana.[28]

Already in his introductory "Epigram" to the pope, Riquius observed that the world was all the purer because of the chaste and virginal model of the supreme pontiff.[29] The leitmotif of the poem would thus be the parallel between the chastity of bees and that of Urban himself.[30]

But there is more. When Riquius writes in his elegy that bees are dear to Diana because of their chastity, because they do not engage in sexual congress, and that they are in fact autogenetic—*ex sese genita*—he adds a note to the following effect:

> As Pliny noted, the way in which bees are generated is a great and subtle dispute among scholars. But it is certainly agreed that they produce a foetus without coitus and that they lack either sex. Therefore, since they are virgins, they are consecrated to the Virgin Diana.[31]

There he leaves it.

If there was a central natural historical problem—let us not yet call it biological[32]—that occupied Cesi and his colleagues at the time, it was precisely the problem of generation and reproduction. They were concerned with both the general issue and particularly difficult and enigmatic cases. Bees were a special crux, for exactly the reason that Riquius alluded to in his note. The matter demanded further investigation.

Little better exemplifies the confluence of classical learning and natural historical exploration than the ways in which the Lincean panegyrics to the chaste morals of the pope were used to allude to the problem of the reproductive system of bees. But could the strategies of panegyric and of scientific investigation still come together, or were the two forms inevitably divided?

## THE *APIARIUM*

The third and final document published by the Linceans in celebration of Urban's first Jubilee year was the most important of all: a huge broadsheet titled the *Apiarium* (or *Apiary*) that Cesi began to prepare almost immediately upon Urban's accession in 1623 (figs. 6.9–11). At every stage it depends on wordplay, allusion, and insinuation to refer, often obliquely, to the object of its praise. Ordinary readers would not have been able to catch the full range of its veiled, abstruse, and arcane references. But deeply buried within this *Apiarium*, and hidden beneath a surface that glitters with classical learning, is a plea for tolerance, benignity, and restraint, and a foreshadowing of the implications of Galileo's discoveries for the sciences of life.

The work was certainly the most thoroughgoing and imaginative examination of the archaeology, history, literature, and science of bees ever. Its publication was announced by Cesi in a letter of September 26, 1625, to Galileo. In it he declared that the *Apiarium* would serve not only as an expression of Lincean devotion to the Barberini but also as an instance of their special and distinctive commitment to the observation of nature, *il nostro particolar studio delle naturali osservazioni*.[33]

Printed on four separate sheets joined together, the *Apiarium* measures 107 x 69.5 cm, much larger than most other broadsheets published until then. At the top of the sheet is the papal stemma with its trigon of bees, flanked by four ancient bee coins shown both in obverse and reverse.[34]

Beneath this comes the title: the *Apiarium*, "taken from the Frontispieces of the Theatre of Natural History by Prince Federico Cesi, Lincean, first Prince of Sant' Angelo and San Polo, second Marquis of Monticelli, Baron of Rome, in which the entire family of honey-making bees is derived from its first origins, distributed into its species and differences, and presented *in physicum conspectum*."

The frontispieces referred to here were the introductions in tabular form to the individual sections of Cesi's "Theatre of Natural History." This was his great attempt, left uncompleted at his death, to survey all of nature.

**Fig. 6.9. *Opposite:* Federico Cesi and others, *Apiarium*, broadsheet with engraving and letterpress.**

A more daunting and forbidding-looking sheet than the *Apiarium* could hardly be imagined. It is composed of immense paragraphs, printed in tiny letters, little relieved by indentation; a complex system of brackets

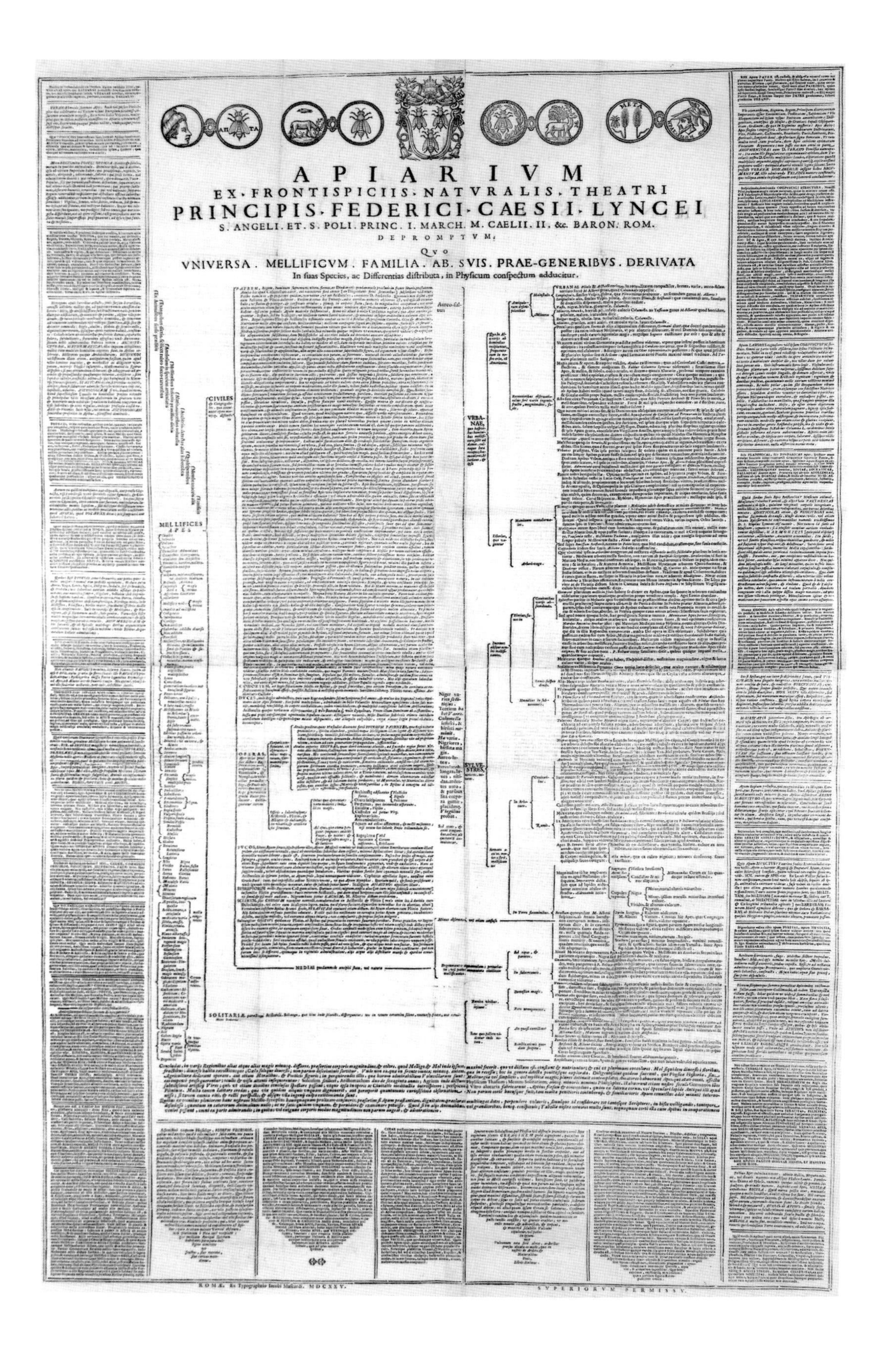
APIARIVM
EX FRONTISPICIIS NATVRALIS THEATRI
PRINCIPIS FEDERICI CAESII LYNCEI
S. ANGELI ET S. POLI PRINC. I. MARCH. M. CAELII II. &c. BARON. ROM.
DEPROMPTVM;
QVO
VNIVERSA. MELLIFICVM. FAMILIA. AB. SVIS. PRAE-GENERIBVS. DERIVATA
In suas Species, ac Differentias distributa, in Physicum conspectum adducitur.
ROMAE, Ex Typographio Iacobi Mascardi. MDCXXV.
SVPERIORVM PERMISSV

Mellea te forsan dulcedo ex Physico Vatem reddidit Plini, qui VRBANAS Apes, qui BARBARAS protulisti: hoc siquidem nectare, hoc cælesti lymphatus latice, VRBANAS moribus, virtute, dignitate præsensisti: agmine, puritate, concentu, BARBARAS.

*VRBANAS recole summas Apes. Audi vel priscos* Philosophos *has celebrantes ad* Vatum *vsque Energiam peruenisse: ipsummet oraculum attigisse. Ecce verè habes* Vrbanas, *multiplici quæ mellificio, nectarea plenissimo ex Apiario virtutum effusi one, beare vndequaque probos velint, vndequaque probis doctisque fauere.*

Quæ CONVICTVS bene instituti sunt, pete ab Apibus luculenter. Prouocant siquidem ad vniuersam MORALEM PHILOSOPHIAM. Ab hisce Volucellis, præter ipsam morum puritatem, probitatem, discenda, quæ ad domum & familiam: quæ ad Ciuitatem: quæ ad opera, labores, munimenta, defensionem ipsam, spectent. quæ denique ad omnem rationem vitæ ciuilis.

*Mira REGIS, mira POPVLI OFFICIA: parētis & sobolis, morum in quolibet animalculo. Dominus ipsis, qui à dotibus, ipsa ab virtute Imperium habet. qui præcipiendo, regendo, fouendo, dirigendo, adhortando, labores: qui proli initia, qui educationem dederit. qui instruxerit, opere ditauerit,* Pater. *Populus, filij qui pareant, custodiant, defendant, sequantur; nec vel alienis oculis* Dominū *lædi permittant: qui proprijs subleuent, deferant humeris, ceruicibus: qui vel nascentes suppares Regulos venerabūdi respiciant: qui distributis assiduè laboribus officijsq. addicantur, certam in optimè instituto cōuictu normam seruantes: Vigilias, somnos, cibos, horas, ordinem, sed & mores ipsos, vt ait* Plinius, *ritè rectèque habentes. Ita, vt nec tolerent domi suæ ignauos, nec prodigos: sed eos omnes, qui aut congesta disperdunt, aut ab opere cessant, castigent; probatos aut in socios mutuis semper officijs prosequantur: ad ipsa vsque funera & mæstitias.*

Si validis viribus, si prudenti doctoquè consilio, si locorum apta constitutione constat MILITIA, quæ aut inuadit, aut defendit; Regem, Ducem, Militem, Apem adi; Apem consule. Metus abest: ars, agmen, ordo adest. Concordia pariter inter se, erga Maiores obedientia. Expeditiones miræ, copiarum motus, & illa agminatim procedendi ratio, BARBARVM etiam nomen Cimbris induxerit. Sunt castra & vigiles, & qui defendant. Est quouis in milite ea fortitudo, ea vis; vt absque vlla ferri ope, quo inermes nos insolescimus, hostem obtruncare prosternereque possint ingenitis armis; quod & ipsi vidimus, imò & quæ impetum in inuadentes facerent, quamuis exaculeatæ. Porrò nec durioribus parcunt; dum vires suas in ipsis pertundendis, diffindendisque ligneis corporibus exercent. nec ad ingentes moles maiorum animalium tremiscunt; dum homines & equos nocentes adoriuntur & conficiunt. Nam aduersus innoxios nunquā iram conceperint; nunquā extra pugnam læserint quemquam: ea vel in bello Apibus iustitia est. Nota Hermonactis causa. E prædonibus Dominum inuadentibus, in illos missas, tres occidisse, reliquos prorsus abegisse, domum Dominumq. liberasse scimus. Scaliger & equum, & tyrunculum equitem occisos notat. Virtus certè & vis, in Apibus militaris.

*Hexagona æquis lateribus cellula, senis forsan à cruribus, amussis adhibita nullius ope, ad quamuis tamen amussim elaborata: Mira, & artificiosissima substructionum ratio, pendentiumque ædificiorum, & connexio & paritas, absque vllo ponderum, & ruinæ periculo: constructa domicilia, & domorum ad viuentem continendam ritè rectèque familiam, disposita & decenter commoda: Regia, ciuiles, plebeia & seruiles cellæ, dignitatis, meritorum, ipsiusque operis ratione habitâ, constituta. Collabentium ab exterioribus præsertim damnis, reparatio, fulcra, subductiones, statumina apposita; vnde Antonomasticum nobis admirantibus Fabricæ nomen; ARCHITECTVRAM, MATHEMATICAS, illas inquam disciplinas; quibus tam multi ex hominibus expertes sunt, in Apibus celebrant. Militarem quoque Architecturam, MVNIENDI vtilissimam illam artem, antiquiorem forsitan, quàm apud vllos homines inuenias, & methodicè ac diligenter vsurpatam, materijs* Physicè *insumptis, Mathematicè in figuras dispositis; si quæ circumcirca obducunt, & adaggerant ad arcendas & propulsandas quascunque iniurias, consideres; si in lapideos tubulos, constrictos aditus, & quæ huiusmodi in Aluearijs plura, respexeris. STATICAM ipsam, si præcisius exquiris, arreptis, dum se saburrant aduersus furiosos ventorum impetus, lapillulis, aduerte. ASTROLOGIAM si vis, haud inanem, haud superstitiosam [ qualis in hominibus persæpe ridenda*

A P I A

EX · FRONTISPICIIS ·

PRINCIPIS · FEDER

S. ANGELI . ET . S . POLI . PRINC . I.

D E P R

VNIVERSA . MELLIFICVM . FAMILIA

In suas Species, ac Differentias dist

{Ex Animalibus sensu præditis

{Exanguibus dictis, scilicet rubro succo carentibus

{Amlosam in speciem artic

{Mollioribus Insec

{A

*PATREM, Regem, Dominum supremum; nitore, forma, ac* Diadematis *quodammodo præclaris in fronte lituris; præ*
*Aculeo hic apud* Columellam *caret, cui recentiores ferè omnes [ ex* Diogenianis *fortè sermonibus ] subscribere vi*
*Potiùs tamen eo, & præpotenti quidem, non vtitur; vt ait* Aristoteles, *accedente* D. Ambrosio. *Quamuis non de*
*cum* Aeliano *&* Plinio *dubitent. Verùm decorus hic* Princeps, *adeò omnibus numeris absolutus est, vel ipsis ab*
*bulis; vt statim & penniger, & conspicuis artubus, plenoq. in corporis statu, suaq. in magnitudine compareat;*
*aculeo expertem illum facilè quisquam existimauerit; à quo omnes aculeatæ Apes producuntur, cæteriq. pariter ortu*
*habent: ipsi enim genitalis Apum materia substernitur. Hanc nil aliud à melle* Cardanus *cogitat; quo Apes om*
*gnantur. Respuit, sed ita* Scaliger, *vt Melleam tamen fateri videatur; & concinnè insuper architectum calorem*
*dere. Atqui legitimo* Plinij *&* Aristotelis *testimonio compar ea melli è floribus est; præsertim Cerinthe, Olea & Aru*
*Nutriuntur melle Apes, quo constant. Mellea exordia: melleam compagem sagaci præsensione, appetitu, facie ipsa, re*
*Mel autē sedata iam natura liquor; illa fermenti quadam virtute quæ in motus ire debeat, materia est, ex ipsis florum*
*selectaq. visceribus, proprijs scilicet mellis latibulis. Sed minutiùs quoque inspicere vt conemur, operæpretiū videtur; &*
*hic aliquantulum subsistamus, dum* Melissophilos *ab exteriori tabula ad interiores auocare discupimus.*

*Succo materia prægnans haud æquè disposita, indigestis prioribus magisque simplicibus figuris, particulas in rudiusculum*
*geneum concrementum interclusas ita habet; vt madore ab ipso, dum euaporatio impeditur, primùm calores concipia*
*meditullijs pulsent & cieant, intestinis in feruorē suffocationibus. Ebulliente verò mole multiplici remixtione variī co*
*varijs & solutionibus & appositionibus, alia atque alia contexuntur superinducta figura, ad organorum vsque claustra*
*incipientibus concitationes replicantur, ea concutiunt non parum, ac secernunt, mutui ab internis colluctationibus*
*plexus, siue abscessus fiunt, perturbatæ ac confligentes particulæ, aptis intus extraque finitionibus, eam, quæ comprehen*
*compescat, connexionem paulatim subeunt. Quæ prouidè stabilitur; dum, fluentes ductibus, spirantes meatibus, factis; p*
*statuminibus labiles parietibus appositis; & omnibus rectè distributis, construuntur partes: donec placida coagmentatio*
*peraccōmodo, corpus constiterint in vnum. In quo, reducta dispositio spatia reliquerit peruadentibus, frenū furentibus ini*
*& in fluorem alijs, alijs quidem in duritiem actis, quàrarius, quà spissius iungendo; sicciora, solidiora quæ contin*
*contentis ipsis mollioribus dispertierit; & ne quid prorumpere, aut concidere vel posset, vel deberet, motiones quietè, lib*
*blandis obligationibus temperauerit. Ita vt expiratio ad vitalis caloris opera citra febrim prohibita; citra resolutio*
*missa sit, mediaque in furoris ac torporis moderatione, calor ille suauis, veluti factis radicibus, insideat; quem cælest*
*à stellis dicunt mysticè magis respondentem elemento Stellarum; quamuis origine & domicilio terrestris sit. Hoc, Ari*
*ctiones spirituum [ tenuiorum scilicet & assiduè concitatarum productarumq. partium, interiori pulsatione ad anima*

Fig. 6.10. *Apiarium*, broadsheet with engraving and letterpress, detail.

whose logic is far from apparent; and a kind of "frame" made up of forty-three compartments containing further paragraphs, equally dense and unrelieved, around the main body of the text. Not many people—if any at all—would have had the stamina, let alone the capacity and patience, to penetrate so great a mass of words in tiny letters. When Fabio Colonna received a proof copy of the *Apiarium* at the beginning of 1626, he expressed his pleasure in its many minute observations, but confessed that "in truth the reading is really difficult and troublesome."[35]

The sequence in which one was supposed to read these long paragraphs is wholly unclear. Whatever order there may be in this mass of words seems beyond comprehension. Instead of focusing attention, the array of brackets only distracts it, while the many typographical strategies still further confuse the gaze. In every respect, the *Apiarium* resists rather than encourages scrutiny. It buzzes with words. It induces vertigo rather than invites attention. But it opens up a world that is as essential to the

RIVM

TVRALIS. THEATRI

CI. CAESII. LYNCEI

CH. M. CAELII. II. &c. BARON. ROM.

P T V M:

. SVIS. PRAE-GENERIBVS. DERIVATA

in Physicum conspectum adducitur.

history of science as the great discoveries of Galileo, in the very years that Cesi and Galileo were the best and most mutually supportive of friends.

Where, however, to begin, and in what order are the paragraphs to be read? How are the bracketed groups and various compartments related to one another?[36] The difficulty is understandable, since such principles of organization are by no means apparent from the sheet itself. The problem is complicated by the fact that in the *Apiarium*, poetry, mythology, literary history, agriculture, and even anthropology are jumbled together, in much the same way as the Aldrovandian treatises on animals. The archaeology and philology of bees are mixed with genuinely new observations, sometimes made with the aid of the microscope, about the anatomy and reproduction of bees. Empirical facts seem to be randomly embedded in extravagant allusions to Urban, in the complex allegories on his family and papacy, and even in the constant wordplays that form so integral a part of the work.[37]

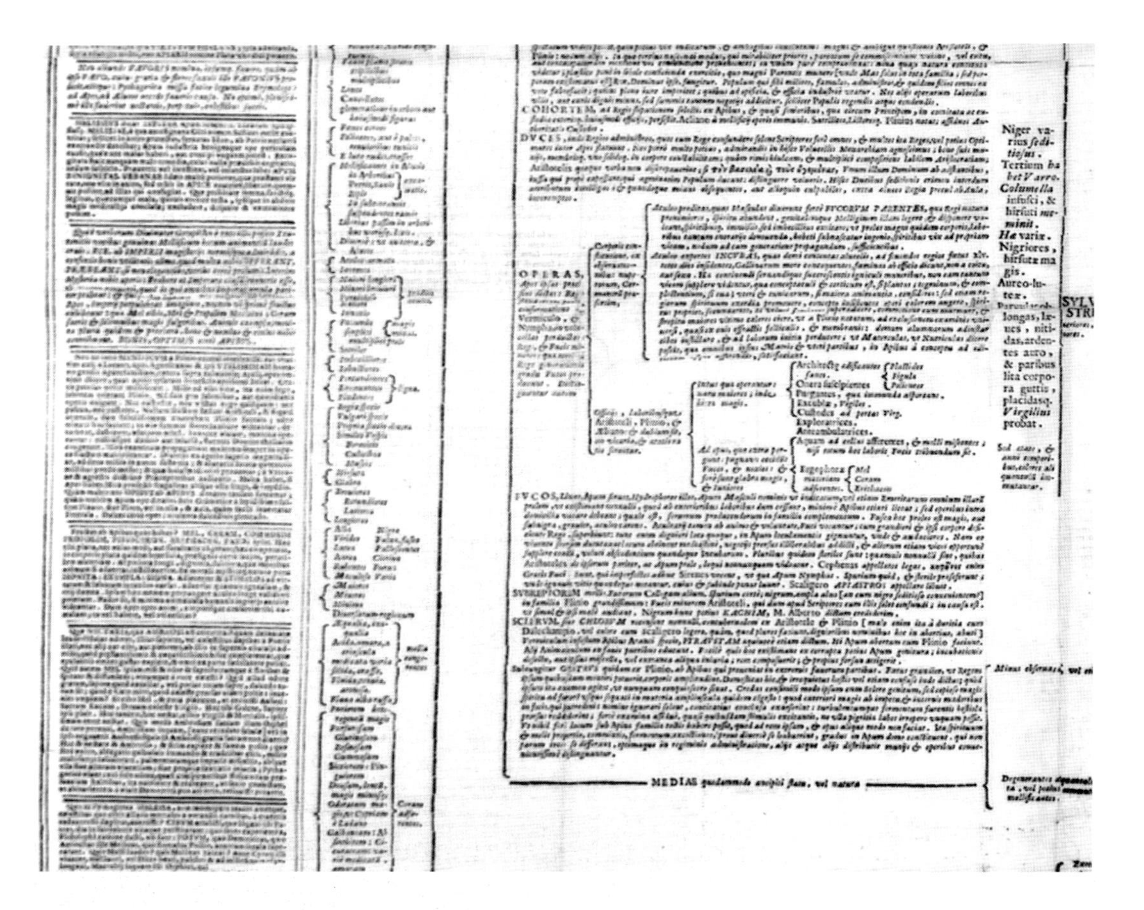

Fig. 6.11. *Apiarium*, broadsheet, detail.

No wonder that on two occasions the *Apiarium* refers to itself as a *Melissosynopsis*. Like the other "frontispieces" to his "Theatre of Nature," it was intended to classify all the information Cesi and his friends had managed to assemble on the subject of the bee. But in this particular synopsis Cesi's habitual desire for order and classification was even more thwarted than usual. Whatever order of thought there may be in it, it is endlessly subverted by the puns and the etymological games, often alleged to have much deeper significance than one may at first detect.[38]

The multifold and tedious digressions in the work ought not to throw one off track, however. Significant clues to its order and structure are provided by the fourth of the thirty-two manuscripts that form the core of the surviving archives of the original Academy of Linceans. It contains more than 250 pages of notes for the *Apiarium*. These notes are written in a variety of hands, including those of Cesi himself, Colonna, Faber, and Riquius—and possibly some others as well. The very first page contains a terrifyingly complex draft for the central section of the work (fig. 6.12), and this is followed by several diagrammatized versions of the whole sheet (one in Cesi's own hand), indicating the relationship between the central section of the *Apiarium* and the surrounding paragraphs (fig. 6.13).[39]

The fifth volume in the Lincean archive contains proofs of the printed

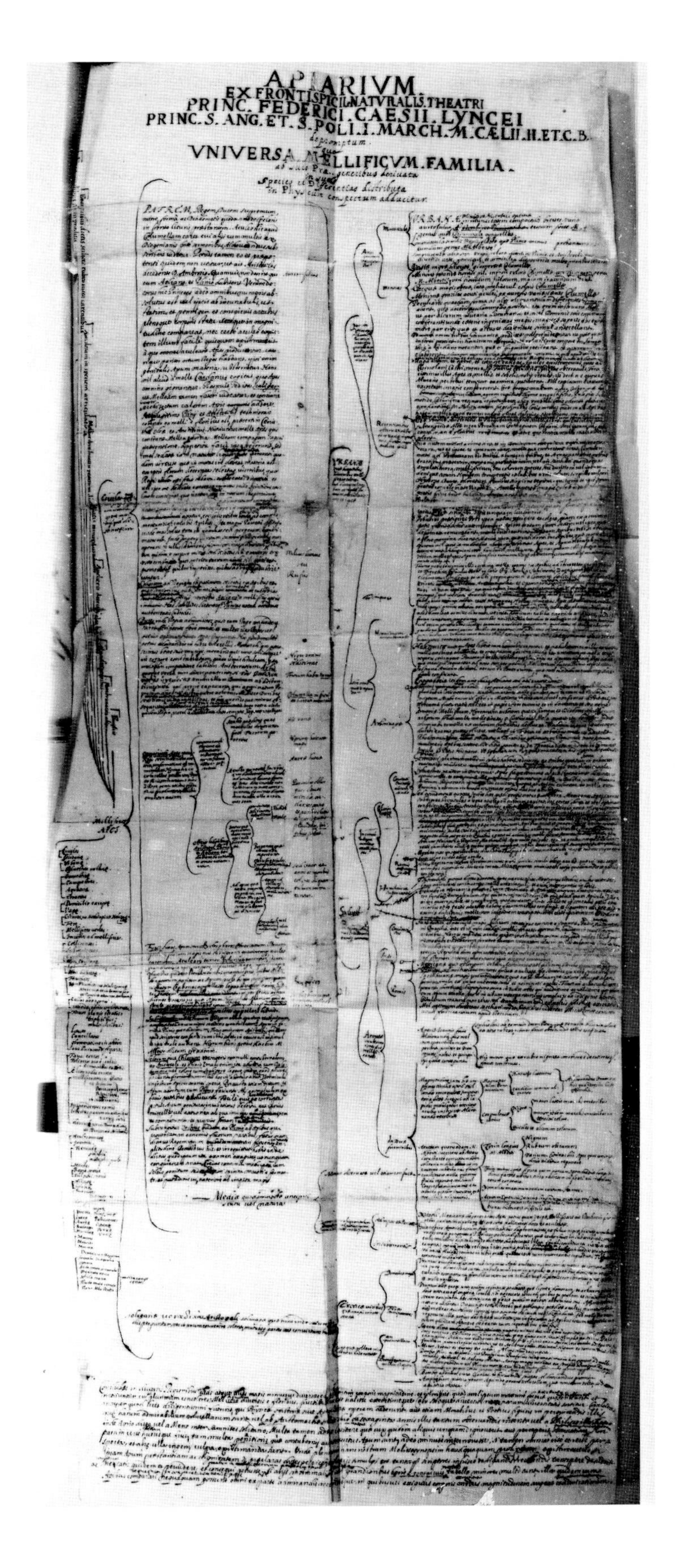


Fig. 6.12. Federico Cesi, draft diagram for the *Apiarium*. Rome, BANL, Archivio Linceo IV, fol. 415.

Fig. 6.13. Cesi, draft diagram for the *Apiarium*. Rome, BANL, Archivio Linceo IV, fol. 466.

version of the *Apiarium* cut up into sections, with brief annotations and insertions by Cesi himself. Together the two manuscripts make it clear that the central section, which seems to be all about the classification of bees, was to be read first, with the forty-three paragraphs that enframe it serving as a kind of commentary.[40] These "emblems," as both manuscripts call them, were intended to serve a more purely laudatory role on the pope and his beelike virtues, but it did not quite turn out that way, since science and panegyric were so ceaselessly interwoven.

Much of the material for this central section is first brought together as a separate part of the fourth Lincean manuscript titled *On the Generation and Families of Bees*. This, in fact, is what the main body of the *Apiarium* itself is about. It presents the world of bees and attempts to classify it into the groups and subgroups indicated by the brackets; it outlines, in considerable detail, the social organization and hierarchy of bees;[41] and it offers a complex view of the ways in which reproduction takes place.

The emblem titled *GIGNENDI purissima ars*, "the purest art of *GENERATION*," encapsulates just this issue. It is the central topic of this huge corpus of information, and the one that seems to receive the most obsessive attention. In turn the subject is framed within a discourse of "stinglessness" that is critical to the wider understanding of the purpose of the work as a whole.

The *Apiarium* opens with a flurry of classical citations. They serve not only to praise Urban but to testify that for all his exceptional power and beauty, the father, king, and supreme lord of bees does not sting.[42] Throughout the work, ironically enough, the panegyric turns on the parallel between the king bee and Urban VIII. This was before the days that it was known that the leader of the hive was actually the queen.[43]

## THE KING'S STING

But why this emphasis on the sting—or rather, the lack of the sting—of the king bee? Because it was precisely the appeal to Urban's benignity and goodwill that motivated this panegyric. Certainly the desire to give evidence of the Lincean researches and the use of a new scientific technique was present here too. But the need to panegyrize Urban was not simply a matter of making one more contribution to the celebrations of the Jubilee year. Everyone else who eulogized Urban in that year hoped thereby to win his favor; but in the case of Cesi and his fellow Linceans the need was especially urgent. After all, it was they—or rather, their core members—who had encouraged and facilitated the publication of Galileo's *Assayer* less than two years before. That work had been printed and presented to Urban only two months after his accession to the papacy. It was instantly in demand, and Urban, for all his broad-mindedness, must soon have wondered about the wisdom of having his papacy associated with theories that threatened the foundations of the Church itself.

"If you should irritate his sting, flee," begins one of the final emblems in the *Apiarium*. But he only stung the wicked and unjust. Cesi and his friends had to ensure that the Barberini bees would not turn against them. They had to hope that even if Urban had a sting, he would not use it against them or against Galileo. They had to insist on his benignity. "The BENIGNITY of BEES," the *Apiarium* anxiously proclaimed, "heads off both innate and acquired faults." "Much more powerful than the sweet bees of Mount Ida are the URBAN bees, who by their outstanding virtue are able to nourish and to liberate, not only in some hidden place, but at the apex of the world,[44] all those who are able to flee to them for refuge." Here the Linceans remembered the assertion of the ancient agricultural writer Columella to the effect that although the king bee had a sting he was magnanimous and capable of subduing even the inclination to anger.

Every conceivable quality of bees is here related to the pope; and vice versa. Urban is king and father of bees; we are his offspring, his family and his workers. Like the king bee, he protects us, makes laws for us, organizes us, and sees to our education. He has no self-interest and thinks only of the well-being of his subjects. Everything in the king bee's domain is well-ordered and efficient; everything he does is in harmony with nature. The description of the social life of bees in the *Apiarium* is so careful and so detailed that in this respect at least Faber may be said to precede Fabre.

As the penultimate emblem puts it, bees provide proof of all the highest virtues and stand as arguments for the virtues of Urban himself. These are all spelled out at length. Classical parallels are drawn at every step.[45] Like Urban, bees are studious, inventive, watchful, ingenious, wise, solicitous, diligent, guileless, pious, thrifty, and chaste. As Pliny notes, they have regular hours for sleep and for work. They are prudent, temperate, fearless, pure, clean-living, honest, and above all kind. *Multa habes, si apes habes*, Cesi declares at one point—if you have a bee you have a lot (punning, of course, on "habes" and "apes"). In the context of the *Apiarium* this is some understatement.

Aristotle may have claimed that bees were deaf, but in fact they could not have been more musical; they were irrevocably associated with Apollo, and with the muses on Mount Helicon. Even Apollo's name had the same root, *ap-*, as that of bees. *Melicus*—derived from *mel*, or honey—also meant tuneful; and *melica* was the name for lyric poetry. How could Urban not have been gratified by the invocation of his own poetic talents when Cesi named that distinguished foster child of bees, the poet Pindar? It was he who had sung so well of the bees as ministers of Ceres and associated their qualities with those of the gods. Inimitable by any of the ancient poets, Cesi continued, Pindar had at last been imitated, in our own days, by the URBAN BEES, the true bees of lyric poetry, *vere* MELICAE *apes*. Had he read it, the association would have been most pleasing to the man who often longed to return to the lyrical poetry that had given him such pleasure when he was younger. It would again be invoked by the great title page that Rubens devised for the collected edition of his poems in 1634 (fig. 6.6).

## CHASTITY AND INSEMINATION

But it is the insistence on the chastity of the father or the king bee that sounds like a drumbeat throughout.[46] He procreates, he even inseminates, without sexual desire. He is not even remotely libidinous. You have to admire him: he knows none of the soft pleasures of sex, none of its mad irritations.[47] He is entirely chaste, yet immensely fecund.

What is all this really about, other than the need to panegyrize yet another quality—and an important one—of the pope? It arises, of course, from Cesi's central concern with the problem of the generation of bees, and the paradox already noted by Riquius: that whatever their differences about the way in which bees are generated, most authorities, both ancient and modern, agreed that bees produce their offspring without coitus.[48]

For Cesi the question of generation was one of the central mysteries of nature: "Behold the admirable work of making offspring—the purest ART OF GENERATION, far beyond the gates of desire, the most singular and mysterious spectacle in all the Theatre of Nature." These are also poignant words. Could it be that at least part of the intensity with which he pursued the subject had to do with the persistent loss of his own children during childbirth? Presumably;[49] yet from the very beginning Cesi believed that insight into the order of nature had to begin with the problem of generation and reproduction. Other than the reduction of multiplicity to order, nothing, not even the microscopic examination of the anatomy of the bee, seems to have occupied his attention more than this.

In the case of the bees the processes were indeed singular and remarkable. For the most part, these were not things that could be directly observed, certainly not with the naked eye. Cesi was aware of this. The theory he presented had as much to do with his interest in the behavior of liquids and solutions at different temperatures (in this case honey), as with his view of the social organization of bees (the king gives rise to all his workers and subjects, who are protective of, and protected by, the king), and his need for order in the midst of an immensely elusive phenomenon (which he pointedly called "this unique and multifarious work of nature").

For Cesi, like many before and several after him, the process of generation started with the honey itself. This was the basic formative material out of which the bee emerged. Once one grasps this, one can begin to comprehend why so much of the rest of the *Apiarium* is devoted to honey and its varieties—a matter that to us would seem much less of a scientific than a nutritional, or even gastronomic, problem.

Much of what Cesi has to say in his theory of the generation of bees depends on Galileo's own corpuscular view of matter. Freshly collected from flowers, the honey—or rather the honeylike liquor—is still in liquid form. It is in a constantly active state—"bubbling" is the term Cesi uses. This heterogeneous substance contains a mass of disorderly particles, which nevertheless show relations that suggest rudimentary but still unformed "figures." Gradually, by means of mutual "strugglings and embraces" these agitated and conflicting particles come together (not for the last time does Cesi transfer the language of sexual congress to describe the behavior of the primal substance itself). As these particles form connections one with the other, the material is progressively stabilized. Order is established within it, and the figures—Cesi's word—of the future parts of the bee slowly begin to take shape. Boundaries are firmly established, parts are formed. The bubbling activity of the earlier material is thus sedated and calmed. But within it is still some gentle but vital heat, remaining from the source from which it was drawn. This keeps the liquid from settling altogether, so that gradually the actual form of the animal begins to take shape. The parts and figures of its body are thus formed by a process in which the basic particles both move and are moved, dissolve and are dissolved, until the necessary connections for the formation of the animal are coordinated, ordered, and established.

None of this, however, is sufficient to explain how the figures and parts are animated. Cesi declares that the white substance from which the bee will eventually emerge is spermlike and foamy, and is thicker than when first collected. It has to undergo a kind of fermenting process from which the living form will appear. At this point the work of the father, that is, the king bee, is crucial. It is he who provides the necessary additional warmth for the production of what Cesi calls the "seminal spirits," which he then introduces into the primordial substance. By this means it is set into still faster motion, and the various figures are then combined in order to complete the parts and organs of the bee.

You have to admire the activity of the father, Cesi declares, since he does the work of creating offspring without engaging in any overtly sexual act. One cannot even speak of the introduction of a seed, or semen, since this would have a substance that the seminal spirit lacks. Rather, the king bee procreates by means of the introduction of the vital spirits of the king himself, which themselves contain the shapes and figures of that which is impressed on the genital material. It is from the outside, therefore, that the necessary connections for the formation of the bee are created.

These generative spirits, Cesi repeatedly insists, cannot be seen, or at least are not available to the investigator's eye. How, then, does one know of them? The issue is especially crucial to someone who puts such stock in direct observation. Cesi's answer is strangely evasive: he draws a parallel with perception by means of the other senses, notably smell. Then, maintaining that we perceive as a result of the blows of small particles striking the sense organs, he claims that these particles contain the *maternal* characteristics, or rather, that they reflect the maternal composition of the figures that make up the constitution of the bee. What these maternal characteristics actually are Cesi never explains, but it is clear that he feels the need to acknowledge some kind of maternal presence in the genital constitution of the bee.

Having received its similitude, as Cesi puts it, from the outside (via the vital spirit of the father), the embryo—so to speak—progresses from a nymph into the beginnings of an egg. Then, as the egg begins to form into its round shape, the nymph it contains takes strength from the egg. It puts forth its legs, arms, and wings, and becomes a bee. Here the account is at its most straightforward; but then Cesi introduces a complication. It is at exactly this moment, he says, that the "little mothers" stand by. They incubate the kingly fetuses until they are complete. Although it is great Mother Nature herself who works on these little bees, declares Cesi, he nevertheless concludes: "thus the father excites, promotes, and confirms the work; finally come those foundling nurses," as he refers to the mothers. This account is rounded off with a brief summary, in which Cesi acknowledges the role of both parents in the formation of bees and admits the similarity between the movements of the material in which they receive their primal "figures" and those of sexual congress itself. But still he emphasizes that the actual actions of the father in producing his manifold progeny are pure and free from any possibility of lust. Once again science and panegyric are inextricably intertwined.

Though not, perhaps, what we would now call science. In so many respects Cesi's account is simply wrong. It was not for him to discover, as Charles Butler did and reported in his *Feminin' Monarchi'* of 1634, that the leader of the hive was in fact the queen bee; or that it was the male drones who fertilized her, as was only proved much later.[50]

But there are plenty of other mistakes too, such as the statement that bees carry small stones in their feet to serve as ballast when they fly in high winds, or the long discussion of different types of honey that rain from the sky (a phenomenon Cesi also intended to include in his planned treatise on celestial rains, the *Taumatombria*). Claims such as these arise from an excessive trust in the literature, whether ancient or modern. Ought one therefore to conclude, as his sole modern English editor has, that the *Apiarium* is "unscientific"?[51]

True, Cesi places great store in the old authorities, but he is hardly uncritical. New knowledge and new facts in vast abundance demanded a critical rethinking of authority. The evidence of the old writers as well as the conclusions drawn from the occult sciences could now be tested—*had* now to be tested—by means of direct observation. New empirical techniques in turn compelled a new theoretical style that focused on the inner operations of things not seen by the eyes. This was a mode of thinking that was wholly divorced from the old theories—such as the many and varied physiognomic ones—that postulated equivalences and correlations between outer appearance and inner mechanisms and functions.

True, the abundant digressions and excursuses of the *Apiarium* rarely seem to transcend the level of the anecdotal. But one ought not to overlook those aspects that are indeed valuable from an empirical point of view, such as the long and detailed listing of American bees, based largely on firsthand sources.[52] Passages such as the description of the behavior of the stimulated mass of recently collected honey are also unprecedented in their striving toward internally consistent explanations and in their sense that knowledge of the behavior of a whole—even a liquid—cannot be achieved without a close investigation of the behavior of its parts.

Although Cesi's meticulous account of the behavior of the particles of an effervescent liquid cannot have been based on direct observation, it must have been derived from his observation of seemingly comparable phenomena that formed part of other treatises he was planning. These include the *Taumatombria* and the 1618 letter to Cardinal Bellarmine, the *De tenuitate caeli* published in Scheiner's *Rosa Ursina*, on the behavior of bodies in the atmosphere, where celestial bodies are supposed to pass through the essentially fluid substance of the skies.

Equally important in Cesi's lengthy discussion of the formation of the "figures" of bees within the thinnish liquid of freshly collected honey is his sense that, in order for such figures to take shape, the particles of the liquid need to establish particular connections with one another.[53] In this way the spaces between them are clarified, and the disorder reduced to order. Cesi's awareness of the necessity of boundary in the determination of order plays a critical role in his thinking about classification; his discussion of the way in which figures of future bees somehow emerge out of a restless and inchoate mass is proof of his continuing inability to renounce the terms of visual description altogether—whatever the claims and merits of abstraction. His clear anticipation of the preformation theory adopted by the great bee anatomist Jan Swammerdam and by many others throughout the next century is emblematic of the acute tension in his work between pictures and the structure of order.

Throughout the *Apiarium* Cesi seems constantly to be struggling to arrive at internally consistent hypotheses that might, at least eventually, be proven or falsified by empirical and experimental data.

The commitment to direct evidence emerges many times in the work. Cesi refers to his own close observation of the behavior of bees in the walls of his house and around Acquasparta. He accurately describes a large variety of flowers and plants, just as is to be expected from someone who studied botany. The observation of the calyx where honey is received is minute. He notes in detail how pollen adheres to the legs of bees, as reported by Colonna. Where he can—particularly in the case of Colonna's researches and those of that favorite Lincean source on Mexican matters, Hernández—he cites the reports of those able to provide firsthand information.

For the close examination of honey and wax in the hives Cesi coins the word "oposcopy," *oposcopia*—only one of many compound words used by Cesi with the crucial suffix indicating the science of examination by direct looking. But the desire actually to see and observe is most apparent on those occasions when he expresses his frustration at what cannot be seen with the naked eye: the particles within the liquid honey, the seminal substance of the bees, and the place in which the nymphs are formed (and here, in referring to the *gynaeceum apum*, Cesi seems to show some sense of the possibility that female bees may play even more of a role than the rest of his discussion suggests). For this purpose only the brand-new instrument, the microscope, could serve—and even it would leave some of the desire to see unassuaged.

# SEVEN *The Microscope and the Vernacular*

## THE PARTS OF THE BEE

When it came to the microscopic examination of the bee, the *Apiarium* was heroic in its struggle to convey even the smallest detail in words (figs. 6.9–11) In its opening passage, Cesi wrote that Pliny himself "knew that nature is never more complete than in its smallest parts." Only imagine, he continued, if Pliny had had the microscope at his disposal![1] What would he then have said of all the parts of the bee that particularly interested him, especially its mouth, lips, and the multiple "tongues" it used in the production of honey?

At this point Cesi described these parts of the bee with infinite care and attention to minutiae. Although he acknowledged the help he had received from Fabio Colonna's descriptions, what really fascinated him, just as in the case of Robert Hooke in his *Micrographia* of forty years later, were the precisely defined compartments and fine reticulation of the bees' eyes:

> [They] appear to be distinctly divided into very beautiful small golden dice-boxes, in a kind of reticulation made up of very fine hairs marking the dividing lines.[2] These are also to be found at the extremities of the wings, but there they are sparser and still smaller.

Thus he continued, always insisting on the necessity of seeing the clear articulation and structure of each part. Never before had anyone offered such meticulous proof of the potential of the eye.[3]

The closeness of observation in the whole of this passage is spectacular; but through all the densely packed details, the struggle to order remains. The passage begins with structure and ends with structure. And this, for Cesi, was the real beauty of the work of God in the bee.

After the initial exaltation of the complexity of the mouth and "tongues," Cesi emphasized over and over again the articulation and relationship of the parts to one another. The instruments of the new science had enabled him and his colleagues to see things never seen before. In their use of these instruments, they could not have made their challenge to the old authorities more explicit. Describing exactly how bees scrape up pollen with their hind legs, Cesi continued:

> Just as Aristotle wrote, so the most learned Doctor Fabio Colonna saw and observed that they carried these things by means of their legs, with the yellow globules clearly sticking to their posterior limbs. For bees work with all their rough legs and the whole of their hirsute body, and carry away their pollen with these hairs.

"Just as Aristotle wrote, so Fabio Colonna saw and observed"—it was not just a matter of seeing; one had to observe as well. Obviously the Linceans were influenced by Aristotle in what they saw; but what a phrase such as this implies is that whatever the influence of authority, one had always to go *beyond* the simple processes of seeing. And this in turn entailed the impossibility of ever seeing enough. It was not just a matter of using the microscope, for however much one might enlarge the object under examination, there always remained something that was there but was forever beyond the reach of the organ of sight. Sight was essential for good science, but sight alone could never be sufficient.

All this was already implicit in Cesi's account of the generation of bees; but it was made explicit in the final passage on the microscope in the *Apiarium*. Attacking Aristotle for having said that bees were deaf, Cesi insisted that this was not so—and not just on the evidence of the ancient texts:

> You say that the apertures which take the place of the vestibules for hearing look very little like ears. But not to our microscope. For nature makes bodies much much smaller than can be perceived by our senses, and many can be perceived if you apply the microscope. If you can discern these many subtly constructed things with it, you will conclude that there are still other much smaller things yet, which escape and elude even the sharpest of instruments constructed by us. This applies to our telescope as well. Though it draws farther things closer to our eyes, you can also judge that there remain other things even further away, which it could never reach. Therefore get used to the fact that there is an immense number of very small and very distant things that cannot be seen.

Nothing could be clearer than this acknowledgment of the limitations of the microscope and the telescope. But they are limitations that have to do with those of sight itself, and the impossibility of ever seeing everything in the constitution of natural bodies. The microscope was fundamentally inefficient because it was fundamentally imperfectible—however much it might be refined.

The implicit conclusion is always that there remain further structures to be discovered, further clearly articulated relationships to be established. You might find some, but within these lay still others.

What all this implies is that the use of the senses alone can never reveal the essence of things. This is what Cesi and his colleagues had learned from their readings and their knowledge of Galileo's work, above all from the *Assayer* of 1623. As Galileo so clearly insisted there, the evidence of the senses was superficial and too subjective to reveal what was constant and essential about the things of nature.[4] The use of the microscope, then, was only the beginning of the real work.

## ORDER IN THE *APIARIUM*

At first glance the *Apiarium* is a huge mass of undigested and disorderly anecdote, history, and observation.[5] But it is much less promiscuous than it seems. In the end, if one reads it carefully, what emerges is the courage of Cesi's attempt to order this great and seemingly disjunct mass of information. Not for nothing do the passages on the microscope begin and end with reflections on structure. Even amid the apparent randomness of so much information, the obsession with structure and the relationship of part to part is apparent throughout the work.

For example: Cesi observes—just like Pliny—that the beehive is made up of hexagonal cells. He also notes, however, that the sides of these cells are equal, that the number of sides may well be related to the number of legs, that they are as regular as if a carpenter's rule had been used. The order (*ratio*) of the substructures is just as worthy of admiration: they are connected in such a way that there is never any danger of collapse. In the end, says Cesi, one has to pay homage to the way in which the bee uses the disciplines of both architecture and mathematics.

In passages like this, along with others such as those that deal with the clear reticulation of the bee's eyes, one senses a movement away from the raggedness of ordinary visual description. What is at stake, finally, is the ordered, mathematically determinable structure of all things. Ironically enough, it was the microscope that provided further corroboration of this fundamental perception of the world. On the face of it the microscope might have been expected to yield only more in the way of descriptive density; but it too had begun to reveal the kinds of geometrical patterns underlying natural forms that Kepler discovered with the naked eye on the bridge in Prague. By now, of course, Cesi had begun to realize that such geometrically structured forms could provide basic clues not just to the secrets of the heavens but to those of the earth as well.

Such themes haunted the history of microscopy for the rest of the century. Copernicus had begun to reduce the complexity of the Ptolemaic epicycles to simple diagrammatic form; Galileo would continue the work; but it was Cesi and his fellow Linceans who embarked on the task of reducing the apparently impenetrable density and complexity of the terrestrial world to structure, order, and number.

It is true that they never came close to completion, and that they did not succeed in reducing the complexity of pictures to the mathematically determinable simplicity of Euclidean diagrams (in any case, Cesi would die too soon for that); but this was the road on which they now set out.

Cesi's efforts to organize and structure the dense mass of information in the *Apiarium* are not, to be sure, of the order of the geometric diagram. Instead, he uses the system of hierarchical bracketing that found its most tireless exponents in the work of Peter Ramus and his school.[6] As its title proclaims, the *Apiarium* attempts to present the entire (or "universal") family of honey-making bees in such a way as to distribute it into its species and differences. It is a synopsis—a *Melisso-synopsis*, as Cesi calls it—of a family of animals that had been studied in greater detail than any other. The synopsis had to be reduced to its logical component parts. Otherwise it would have been even *more* prolix and unwieldy than it already was. The struggle for order could not be more apparent than in the way in which Cesi sets out to embrace as much as can possibly be grouped together within single brackets, and then to break down these groups into more closely definable species, so that even the smallest division may bear a logical relation to the whole.

But the desire to determine classes and relations remains poignant in its failure. No modern biologist or taxonomist would have much patience with the Cesian classes, or the differences by which they were distinguished, since both differences and classes were established by means of what appear to be inconsistent and labile criteria. But there is one small aspect of Cesi's mighty effort to classify every kind of bee according to its appearance, habits, and habitat that provides a further clue to the full significance of what he and his friends attempted to achieve. Like so much else in the *Apiarium* it is easy to overlook.

The first and major division in the *Apiarium* is between the social (or urban) bees and the solitary (or rural) ones. The social organization of the urban bees is discussed at length; that of the rural or solitary ones hardly at all. Then comes a subdivision that seems very slight: the "middle [bees], of a doubtful status or nature." These few words are of great significance—though they are not entirely easy to understand. A critical component of this "middle" group are those that are "somewhat degenerate" [*degenerantes*, i.e., defective, or somehow lacking in the distinctive characteristics of the family bees]. But the point of this group, "of a doubtful status or nature," is really that it is composed of all those bees (or beelike animals) that are impossible to place in the existing classificatory categories. Such animals cannot be placed in particular groups (or boxes), either because they lack a defining characteristic of honey-making bees (e.g., the sting, the habit of making combs with hexagonal cells, and so on) or because they share some—or too many—characteristics with other insects, or even with bees in other groups. That is, they partake of the defining characteristics of more than one category and thus seem to overlap already defined groups.[7]

Almost from the beginning Cesi realized that the problem of order and classification depended on the resolution of the problem of what he repeatedly called things of a middle nature. In the overarchingly ambitious outline that survives in the Biblioteca Nazionale in Naples for his great and never completed *Mirror of Reason*—in which he hoped to classify all of human knowledge—he had a heading for

> things of doubtful nature, or doubtful species, or ambiguous things. On [the nature of] ambiguous Nature. Two different natures joined in a single species . . . species participating in two natures.[8]

It was essential to resolve the problem of intermediate and overlapping classes. Long before Linnaeus, Cesi realized that the key to a solution lay in the possibility of finding a sign—or a *characteristic* as it would later be called—that provided the sole and unique marker of each class, in such a way that there could never be classes that overlapped or were intermediate. Like Linnaeus, Cesi concluded that the key to a stable classificatory system lay in a better understanding of systems of reproduction. The elements of reproduction, rather than any more superficial feature, would provide the distinctive signs of classes. But such signs were hidden ones. For the most part, the signs Cesi sought lay beyond the reach of direct observation with the naked eye. But they did lie within reach of the microscope.

## COLONNA, FABER, AND THE MICROSCOPE

The *Melissographia* and the *Apiarium* were barely off the press when Stelluti, Colonna, and Faber began writing to one another and to Cesi about new and better observations. At last they could work on their results less encumbered by panegyric. Even before the works of the Jubilee year appeared, Faber had already inserted a notable passage about the microscope in the commentary he was preparing on the animals of Central America.[9] It follows an exalted account of Galileo's discoveries with the telescope. Along with a much better-known passage by Francis Bacon in the *Novum Organum*,[10] Faber's text is one of the most fundamental in the early history of the microscope:

> A few days ago I looked through an optical tube of marvelous clarity, and was astonished by what I saw. It was made with great skill and craft by two Germans who brought it to my house and presented it to me. Since it was made for the observation of very small things, I decided to call it a *microscope*, by analogy with the telescope. I examined a louse, that dirty little animal, and not infrequent companion of man, and saw not only its mouth, but its eyes, beard and two little horns on its forehead. I examined its three very long and articulated feet on either side of its body; each had two curved claws, one long and one short, which took the place of the thumb. With these it grasps the skin, and then crawls by fixing its foothold on it.

How much care and perfect diligence Nature devoted to this tiny digit, and to every similar detail of these most abject little animals.[11]

Proud as they were of such observations, the Linceans were by no means done yet. Three hundred pages later in his work Faber returns to the subject of the microscope and praises both Cesi's work in discovering the sporangia of ferns (cf. figs. 8.24–25) and Stelluti's observations of the bee.[12] In fact, neither Stelluti nor Colonna were satisfied with what they had achieved so far; and so they pursued their studies of the bee for several years more.

The two friends worried greatly about the differences between their observations. Sometimes they became quite competitive with each other. Was it possible that bees in Naples differed from those in Rome? On January 9, 1626, Colonna wrote to Cesi that he had observed and drawn the mandible and the sting of the Neapolitan bee,

> which is probably different from the Roman one, as I have indicated to Signor Stelluti, who has better observed it; but he hasn't replied to me at all. Now that I have seen the copper engraving [the *Melissographia*], which has several differences from mine, I have decided to send my drawing to you. And since you are doing a history of the bee and its differences, I suggest that you take another look at the Roman one, and see if its tongues are similarly articulated, and if the internal tongue is similarly turned back and divided at its extremities. I do not believe this to be the case, so I suggest that you carefully open the tongues with the point of a pin and dilate them, and then observe the central tongue by the light of the sun. This, I believe, will appear more clearly with the aid of your microscope, since it is considerably better than mine.[13]

In closing this letter, Colonna laments the deficiencies of his own drawing, unfortunately now lost. Once more we sense the importance the Linceans attached to the role of drawing and visual reproduction in their work, as well as their belief in the necessity of internal examination and dissection.

By February 2 Colonna had received the *Apiarium*. Ten days later he sent Cesi his enthusiastic response to it (though he admitted he had found it a little long-winded).[14] The *Apiarium* had provoked him to pursue his own bee investigations with renewed vigor. Promptly, on February 20, he sent Cesi the drawing of a head of a bee

> so that you can see that the mandibles of our bees are different from those of Rome. In the drawing I am sending you, you can see that there are two thick internal tongues, longer than the external ones, broad and curved like a ship's keel, and with the points turned back and prickly, and the tongue very long, articulated, and prickly too, with the *acetabulum* at the end.[15]

An extraordinarily detailed description![16] Exactly a month later Colonna invited Stelluti as well to examine these differences, which he had "observed and seen so clearly" himself.[17]

For the rest of 1626, Colonna, Cesi, and Stelluti continued their exchanges on the use of drawing and engraving and on ways of improving the microscope. "I'm sending you the image of the foot of the bee, very different from the one already published," wrote Colonna to Cesi on July 17. "The plan," he said,

> is to make the whole of the image on large paper, in such a way that one bee will take up a royal folio so that it will show everything that you see now. I believe that the more you observe the more you will discover, so skilled is the master who made it.

He had a new microscope; and with it, he wrote, he could see still more things than before, right down to the feathered hairs of the tongue (here he is obviously referring to the glossa). This new *occhiale* he would soon be sending to Cesi. With it, he said,

> you will be able to observe a whole day long without tiring your eyes, and the image will be upright and clear. It is the invention of a friend and I am still helping him to publish it.[18]

But who was this friend who had found a better way of making a microscope than the Kufflers from Cologne? It was Francesco Fontana, also from Naples, who would later claim to have made a microscope as early as 1618;[19] but this, as Colonna's own letters make clear, was probably not true at all—another squabble about priority in the matter of the new instruments for looking! In any case, Fontana continued to help Colonna greatly in his investigations and it is not surprising that from then Colonna simply referred to him as "the friend of the bee."

"The friend of the bee has not been well lately," wrote Colonna to Cesi on September 19. He had, nevertheless, been working on a microscope that did not dazzle the eye with the kinds of chromatic aberrations that had marred the Kufflers' instruments. That the issue of visual clarity was a constant and critical one emerges from the discussion of a small telescope that now follows:

> This friend of mine has also invented another small *occhiale* one palm long, which inverts the image, but enlarges it enormously. What is more important is the fact that it brings the object so close that even if it is as far as a musket shot away it seems close to the eyes. So far, however, I haven't worked out why it brings things so close, since two convex lenses make the object seem distant and small, but nevertheless clarify it. . . . I will see if I can procure one for you.[20]

These letters of Colonna's of 1626 are of immense importance for the early history of the microscope, and yet they have never been studied. They also make clear that whatever initial rivalries about precedence there may have been, a mood of collaboration had again become the order of the day, in the true spirit of *lincealità*.

## STELLUTI'S PERSIUS: SCIENCE AND THE VERNACULAR

By the middle of 1629 Stelluti was ready to publish a new and improved engraving of a bee, along with a better, clearer, and much fuller description of its anatomy than that provided either by the *Melissographia* or the *Apiarium*. But where would he do so? The answer could hardly be more surprising—at least to modern eyes. He published it in his translation of that most difficult of Latin poets, Persius.

"I have received your letter of last week," Colonna wrote Stelluti on November 15,

> in which you tell me that you want to complete your *Persius*. This will be a good thing, especially since our Prince [Cesi], distracted by his domestic affairs, is taking too long and is postponing the printing of his Tables. But with the publication of your *Persius*, it will be clear that the Academy is not sleeping.[21]

Colonna was right to be fearful. Cesi died in August of the following year. But what of the other publication– Stelluti's *Persius*–about which Colonna was so much more sanguine?

It was a work whose likes had never been seen before. For many years Stelluti, like Galileo, had taken an interest in the literary possibilities of the vernacular Italian of his day, particularly Tuscan. He had written poems on the marriages of Cesi and his brother; and where men like Riquius wrote dedicatory poems in Latin, Stelluti contributed direct vernacular poems in praise of Galileo to the preliminary pages of both the *Letters on the Sunspots* of 1613 and the *Assayer* of 1623. He had barely finished collaborating with Cesi, Faber, Colonna, and Riquius on the *Melissographia* and the *Apiarium* when he turned to his next project. This was his translation into Tuscan blank verse of Persius, whose six elegies contain some of the most allusive and obscure lines ever written in all of Latin literature. A less promising project for translation would be difficult to imagine.

The *Persius translated* was published in 1630. It is an extraordinary performance, not only in its blunt use of the Tuscan dialect but also in its immensely digressive notes—digressive, but not irrelevant to the Linceans themselves. The language of the translation, deliberately meant to echo Persius's own Etruscan heritage, is in that vigorously free form of verse known as *verso sciolto*. This is all very accomplished, but it is the notes that really command attention. They are not just literary, or text-critical, or archaeological, as one might have expected. In any case, Stelluti's attempts at textual criticism are especially feeble and haphazard, and variant readings are only mentioned when they provide the opportunity for discourse on some other, more interesting subject. The point is that whenever he possibly could, Stelluti took a word or phrase in Persius—almost any word or phrase—and used it as an excuse to refer to one or another aspect of the natural historical researches of the Linceans. The most insignificant reference in the elegies sparked long and short excursuses on the Linceans' work. There are hundreds of such references, sometimes critical and learned, sometimes not. A vigorous literary exercise thus turns out to offer some of the most crucial information we have on the Lincean projects in the very year in which Cesi's life and work were so abruptly cut short.

The references are above all to the unfinished projects of Cesi, to the achievements of Galileo, and to the writings of Faber and Colonna.[22] Persius scoffs at the soft luxury of poets who scribble on couches made of the wood of the citron (I.53); Stelluti cannot resist commenting on the gleam of that same wood when polished, and then uses this as a pretext for a long description of the fossilized wood that Cesi found in the hills around Acquasparta, and which he called *metallophytes*, "of a middle nature between plants and minerals" (and from which a polished table was made for Francesco Barberini).[23]

Persius derides the effeminacy of a line such as "The maenad who, ready to guide the lynx with reins of ivy, cries Euhoe" (I.101); and the merely secondary reference to his favorite animal gives Stelluti the chance to expatiate on the characteristics and zoological status of the lynx, to eulogize its superior sense of sight, and to explain why it was chosen as an emblem of Cesi's Academy (fig. 7.1, cf. fig. 5.1). After all, did not the lynx symbolize not so much corporeal sight, but intellectual insight? This, Stelluti continued, was "necessary for the contemplation of natural things, where one has all the more to penetrate to the interior of things, in order to understand their causes and the operations of nature, and how it works from within—just as it is well said of the lynx that it sees not only that which is on the outside, but that which is hidden inside."[24]

Persius jokes about poets who write as smoothly as if they were drawing a line with one eye shut (I.66); Stelluti adds a long discourse about the advantages of looking with one eye only, especially in carpentry and in the use of the crossbow while hunting birds. But this is merely a prelude to a detailed discussion of Della Porta's views on binocular and monocular vision in his *De Refractione*, which in turn is followed by a long passage praising Della Porta's work as a whole, insisting on his primacy over the Dutch in the invention of the telescope, and hailing the way in which Galileo, *nostro Accademico Linceo*, perfected it (pp. 26–28).

By far the most important notes to this satire are those in which Galileo appears. They include notes on the astrological doctrine of the influence of the planet Jupiter on our lives and character (V.50), and on the incompetent sailor who cannot find the morning star (V.102). In the first of these notes, Stelluti deals swiftly and efficiently with astrology; and

Fig. 7.1. Lynx. Colored drawing from BANL, Archivio Linceo IV, fol. 244.

then moves on to astronomy, summarizing Galileo's major discoveries and recalling Galileo's meetings with Cesi and the other Linceans to scan the heavens. Finally Stelluti alludes to the forthcoming appearance of the *Dialogue on the Two Chief World Systems*, the work that would finally send Galileo to the Inquisition (pp. 148–49).

In the second of these notes, there is an extended account of Galileo's observations about the planet Venus: after them, Stelluti affirms, one can no longer say, as the ancients did, that Venus stands in the third firmament. But for the composition and constitution of the heavens Stelluti refers his readers to yet another of Cesi's never-published works, the *De Coelo*, or *Coelispicium* (pp. 161–62).[25]

Let us return, however, to the bee. Stelluti had not yet finished with it.

In the course of the first satire (I.29), Persius has an obscure reference to pompous citizens in Arezzo. Noting the many ancient writers who mention Arezzo, and commenting on the fact that some critics have read not "Arreti" for "at Arezzo," but rather "Ereti" for "at Eretum," Stelluti points out that the ancient town of Eretum in the Sabine countryside was in fact the present-day Monterotondo, seat of the Barberini country estate.

Stelluti has his pretext. He can return to panegyric through science.

The Barberini were endowed with the greatest of gifts and virtues. Under the patronage of the bee, the arts and sciences had flourished as never before—for example, Cesi's *Apiarium*. This Stelluti describes as "full of erudition and original and novel concepts, restricted to one large folio, it is true, but so full of data and theories that it could just as well form a large volume."

Then he recalls his own observation of the bee under the microscope, as published in the *Melissographia*. He had discovered things not known to Aristotle or any other ancient naturalist or philosopher. Making his investigations with the aid of Faber, he had had them drawn by Fontana in Naples and engraved in Rome (referring to the *Melissographia*);[26] but now he had examined the bee with still greater diligence; and the results of this examination, he says, he has placed at the end of the satire. He does not, after all, wish to interrupt his reader with too long a digression (pp. 46–48).

The full-page illustration of a bee and its parts that appears at the end of the satire is perhaps the most unexpected ever to appear in the edition or translation of a Latin poet (fig. 7.2). It still has the capacity to arouse the wonder of modern experts. The clarity of its detail is unprecedented. At first sight the illustration in the Persius seems less ambitious than its predecessor (fig. 6.7). It is smaller and completely unadorned, and seems not to offer much beyond what was already available in the earlier—and much less circulated—print.

But here one must be careful. The method of presentation is strikingly different from the *Melissographia*. In the suppression of the decorative bay leaves and putti surrounding the bees of the *Melissographia*, and in the absence of elegant cartouche and beautifully calligraphic typology, it is—to put it simply—much plainer and more direct. Here one may comfortably study every detail of the antennae, legs, sting and head.[27] The structure of the proboscis (the "tongues," as Cesi put it) is precisely shown.[28] For the first time the hairs of the bee are presented as clearly plumose (this is the feature that instantly distinguishes bees from wasps). Only the venation of the wings is inaccurate.

However artfully the parts of the bee may still seem to be arranged on the page, they are less contrivedly disposed. The facts of investigation are not embellished. There is a simple frame. The elements of the illustration are numbered, correlated with the unadorned listing below. And the listing of parts is not embedded in a poem consisting of fine Latin distichs; it is written in the plainest and most direct vernacular. The facts are presented for what they are, and not for their antiquarian or panegyric implications.

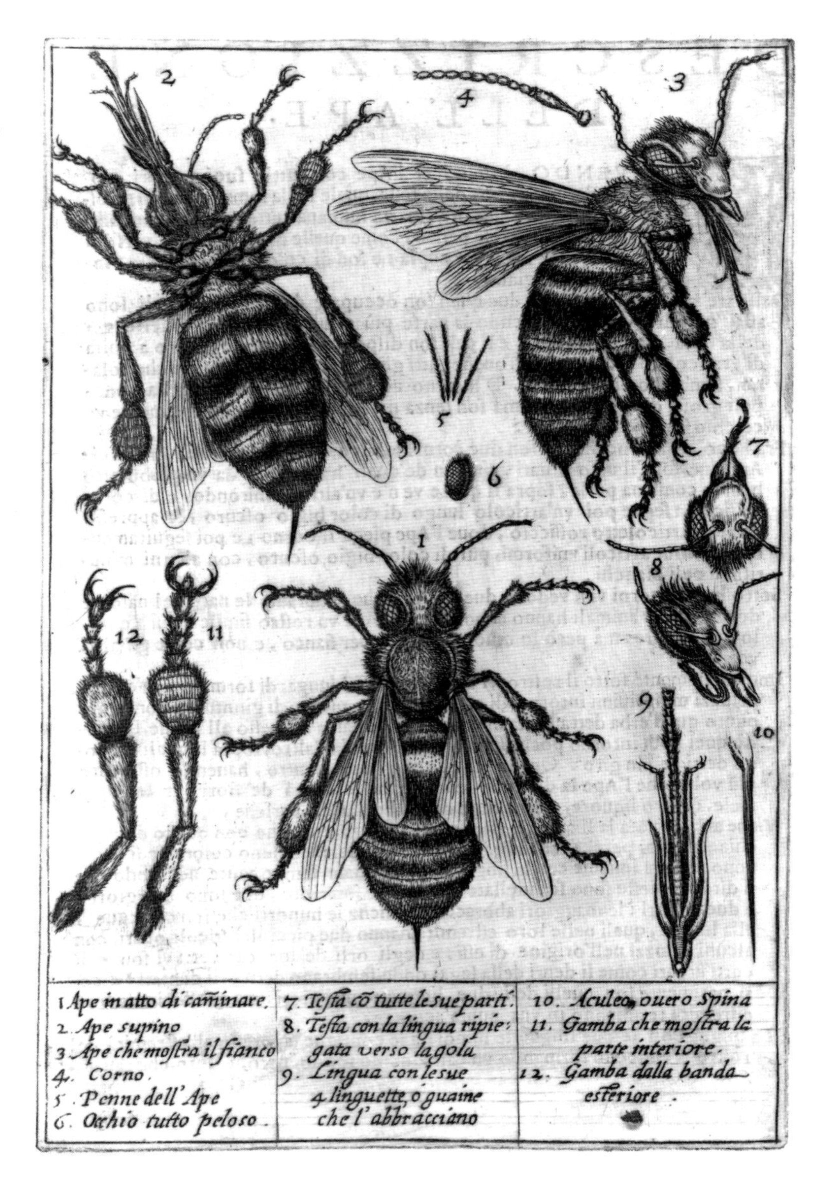

1 Ape in atto di caminare.
2. Ape supino
3 Ape che mostra il fianco
4. Corno.
5. Penne dell'Ape
6. Occhio tutto peloso.
7. Testa cō tutte le sue parti.
8. Testa con la lingua ripiegata verso la gola
9. Lingua con le sue 4 linguette, ò guaine che l'abbracciano
10. Aculeo, ouero Spina
11. Gamba che mostra la parte interiore.
12. Gamba dalla banda esteriore.


Fig. 7.2. Anatomy of a bee. Engraving and letterpress, Stelluti, *Persio tradotto* (1630), p. 52.

All this applies just as well to the long prose description opposite the illustration. Here are the full fruits of Stelluti's observations. There can be no question that the verbal description marks the real breakthrough, not the illustration of the bee. In the end the latter is not that much different from the *Melissographia*. When Colonna first received the *Apiarium* he commented on the difficulty of reading it;[29] but when he received his copy of the *Persius* he complimented its author on the fact that while the illustration of the bee was "excellently and clearly made," it was even better in the description—by which he meant (lest there be any misunderstanding) the *verbal* description.[30]

It is true that Cesi's account of the microscopic examination of the bee in the *Apiarium* is written in a rather more direct way than the rest of that forbiddingly dense and convoluted work; but still it is embedded in panegyric and written in a difficult Latin, with the struggle for dispassionate description painfully obvious, and the lure of rhetorical devices ever present. Stelluti, on the other hand, separates out his description from his divagations. He assigns it an appendix of its own. It is clearly detached from the rhetorical, panegyrical, and digressive agendas in the rest of the book; and is written in a straightforward vernacular, clear and easy to understand. It is also methodically presented, beginning with the head of the bee and working its way systematically down to the sting. Having described each part clearly, Stelluti notes the colors of the bee with great precision. This attentiveness to color is almost wholly absent from the description in the *Apiarium*. But at the same time it is nevertheless *secondary*, and always comes firmly at the end of the description of each part. Stelluti fully realizes the importance of describing the structure and articulation of the parts first.

Whereas the *Apiarium* devotes disproportionate attention to the "tongues" of the bee, and the means by which it draws out honey from flowers (the description of the other parts seems almost cursory in comparison), Stelluti neglects none of the parts he has examined; and the space he devotes to each part and its relation to the others seems just and proportionate. The *Apiarium*, for example, has little about the legs and almost nothing about the structure of the wings or sting (although there is much on the metaphorical implications of the latter). "There are four wings," writes Stelluti,

> two are large and two are small. They are always open, as in the case of the fly. They originate at the extremity of the shoulders, adjoining their sides. They are veined and sclerotized. The veins are very hard, similar to those of the wings of the bat. On the exterior ridges on the circumference of the wings are very fine hairs.[31]

This is typical of Stelluti's mode of description. It is brief and to the point. To us this kind of description still sounds amateurish and too attentive to superficial appearance. But one cannot fail to be struck by the directness of Stelluti's language. In the case of his account of the legs, verbal description does become very intense, as if he really believed in the potential of words to convey what one might otherwise have assumed visual description could better do. But he did not lament the impossibility of ever conveying biological complexity in words, as Cesi had. Once method was imported into description, everything seemed possible.

For all the density of Stelluti's description of the bee's legs, his account of the parts and connecting joints of the bee, right down to the extremities of the claws, is clear and well-ordered. It proceeds methodically from part to part, from the part closest to the body down to the last member of each claw. In the tortuous description of the abdomen to be found in the *Apiarium* one easily loses one's way; but Stelluti's description of both abdomen and thorax, as well as of their articulation, could not be clearer.

It is impossible to exaggerate the difference between the complex and insistent density of the *Apiarium* and the lucidity of the description of the bee in Stelluti's *Persius*. In contrast to the high allusiveness of the notes elsewhere in this book, not one ancient writer is mentioned in the account of the microscopic examination of the bee. It is free of the rhetorical devices that so encumber the *Apiarium*, even in those parts of the *Apiarium* where panegyric loses its accustomed grip on exposition. There are light-years between the classical Latin of the *Apiarium* and the suppleness and clarity of Stelluti's vernacular.

This distinction, broad though it may be, is crucial. For the Linceans, the vernacular—and by no means the Italian vernaculars alone—played a fundamental role in the new forms and subjects of science.[32] Unlike the older and often occult science, the secrets of the new were plain and accessible, at least in principle. Just as one could now appeal to the clear evidence of the eyes (at least at the outset), rather than to the testimony of ancient books and writers, so too one could rely on the quotidian directness of the vernacular. This could appeal to all. The vernacular was robust and unfussy, flexible and witty, as Galileo showed time after time. It was capable of all the clarity the new science demanded. It did not cloak obscurity in the lengthy periods of Ciceronian Latin. It would be deliberately unrhetorical. Colloquial and invitingly dialogic, it did not appeal to tradition and traditional authority in the way the older languages of science did. No wonder that the major works of Galileo's maturity, the *Assayer* and the *Dialogue on the Two Chief World Systems* were written in the same robust vernacular as Stelluti's *Persius*.

For Galileo as well as for Cesi, tradition and dogma stood in the way of the discovery of scientific truth. Indeed, unless viewed with appropriate skepticism, they actively hampered it. Aristotle, Pliny, and the others could go; it was imperative to rely on observation and hypothesis. Physics was separated from metaphysics; methodology took its place. And this methodology was not arcane. Even geometry could be learned and grasped without the study of Greek and Latin.

But in all this there is paradox. The Linceans were swift to appreciate the significance of the vernacular as a vigorous means of communication. Its adoption was a clear sign of the ways in which the field of science was now expanded to include the everyday and the most lowly. But at the same time the Linceans often seemed to hold tightly to their proud and sometimes hermetic antiquarianism. Indeed, their passionate involvement and immersion in classical culture was always evident. However skeptical they may have been of ancient authority, they often seemed slow to renounce it.

But they did force it into a new mold. It could no longer be regarded as exclusive. Its fruits had to be adapted—and if necessary rejected—by the

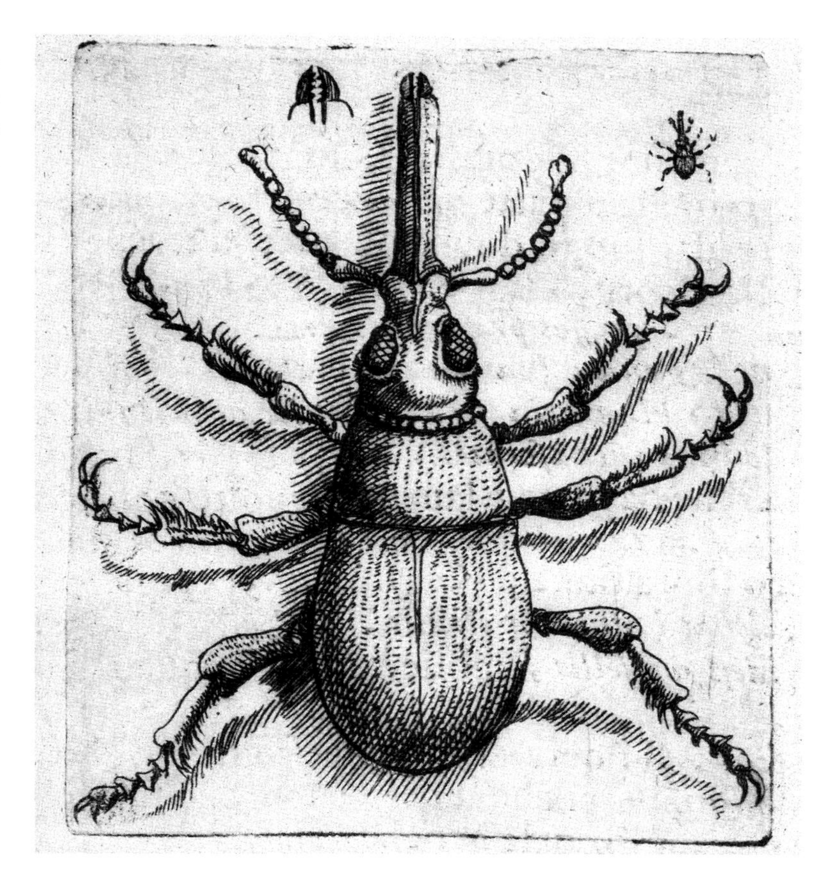

Fig. 7.3. Weevil (fam. *Curculionidae*). Engraving, Stelluti, *Persio tradotto* (1630), p. 127.

pressure of new knowledge. Latin—and the other classical languages, Greek and Arabic—were the languages of the Old World; but now the evidence of the New was emerging abundantly every day. One of the major projects of the Linceans in the very years they were editing and supporting Galileo was the great work on the fauna and flora of Mexico. This work is astonishing in its insistence on the importance of vernacular names, and vernacular knowledge. Hernández and his successors had found and recorded plants whose equivalents could never have been found in Aristotle or Dioscorides, and the Linceans knew that it would be vain to seek them. The new natural history, like the new science, had burst beyond the bounds of the old books.

Of course the earlier botanists, from Brunfels, Bock, and Fuchs on, knew this too; but they were much less systematic in the adoption of vernacular terminology. Just as the evidence of the New World pressured old authority, so too did the evidence of the local. For almost as soon as old authority was shown to be incomplete, it became clear that New Worlds could be found on your doorstep too. Even more than in the case of the great sixteenth-century botanists, the plants, fungi, and fossils that Cesi and his friends found in the hills and valleys round his home presented problems that could not be resolved by the ancient writers.

The language of science was thus liberated from its old bondage to authority and tradition; indeed it could only find adequate exposition in the vernacular, whether the Aztec vernacular of Central America or the regional languages that alone provide the terms for the local and the unexceptional. By its very nature, the microscope was especially suited to the examination of the lowly, the everyday, and the seemingly insignificant. No wonder that from the very beginning Galileo, Colonna, and Stelluti constantly insisted on the microscope's revelations of the scientific interest of the tiniest insects, from the louse to the lowly flea.

It comes as no surprise, then, that another highlight of Stelluti's work was his examination of the weevil that infested the grain harvests of Tuscany (fig. 7.3). His description, as we might expect, is neatly terse and robust, and of course he expands on its etymological origins.[33]

One of the most noteworthy aspects of the work of the Linceans, one soon comes to realize, is its ethnographic range. They expanded the region of select knowledge—the region of classical languages and the archaeology, literature, and general culture of the ancient world—to embrace the wider and more intimate regions of local cultures. When the Linceans and their friends wrote about ancient sports they knew that the field of modern games had been even less studied, and so they turned their attention to such matters too.[34] Their work was thus deeply concerned with folkways and with the acceptance of popular cultures as much as with the pondering of ancient ones. If ever there was an unnoticed aspect of their work it is this. It offers testimony not to the courtly and exclusive nature of their researches, as has recently been claimed and insisted upon.[35] Science could no longer be the exclusive domain it once was; its secrets were essentially accessible.

But still the situation is a complicated one. Like Galileo, Cesi often railed against the stranglehold of tradition, against received authority and dogma. But it was only Galileo who could freely loosen himself from their bonds. It was he who firmly separated physics from metaphysics, and insisted that the best road to the modes of science still broadly called philosophy was via physical experiment and mathematical demonstration.[36] The Linceans knew this too, but they had too many other agendas to fulfill. When they assembled the material in the *Apiarium*, they had still to ensure that the new approaches it so discreetly announced were veiled in classical allusion and wrapped in the shining cloak of panegyric. Whatever threatened to be subversive about it would never be discerned. To cursory readers of works such as this (as, lacking the patience to penetrate to its full implications, almost every reader inevitably was) it must still have seemed that the problems of science as philosophy could be resolved in terms of rhetoric and poetry. But this could only lead to a dead end. The Linceans did realize that there was a way out, but they did not, as we shall see, go far enough.

# EIGHT *Plants and Reproduction*

Nothing offers more eloquent testimony to the energy and versatility of the Linceans than the ways in which they turned their eyes to the earth at the very time Galileo was discovering new things in the heavens. The more he looked at the skies with the telescope, the more he discovered; the more they looked down at the earth, the more they found there too. Just as he came to the conclusion that the heavens and the bodies therein were less than perfect, so too they had to reckon with the imperfect things of the earth. In particular, they worked on plants, fungi, fossils, and the flora and fauna of Central America. In each case they were undaunted by the staggering accumulation of new data, and by the need to search for ways of ordering it. They had as much as possible drawn and reproduced in watercolor. But soon they came to realize that surface and the superficial, subjective appearances of things did not always provide the most satisfactory guide to the mysteries of nature. It was necessary to penetrate the innermost recesses of both heaven and earth, and find the underlying structures of things.

The Linceans planned to publish the results of their botanical, mycological, and paleontological work, but only in the case of the project on the New World did anything substantial ever see the light of day—and this, for the most part, long after all but Cassiano dal Pozzo had died. Cesi's thoughtful and innovative work on fossils was palely reflected in a timid little treatise published by Stelluti seven years after the prince's death; and nothing on the fungi, lichens, and other cryptogams ever appeared at all. What happened?

Galileo's dramatic and often confrontational way of presenting his work helped create a situation of high risk and jeopardized everything the Linceans did. But at the same time, he also knew that he had to be careful, and the other Linceans too were obliged to pretend that their discoveries were not as threatening to the fabric of authority as they really were. In this they were all too successful; and the result was precisely the oblivion from which at least something may now finally be recovered.

The fact is that the natural historical work of the Linceans was a casualty of the very situation they had helped to create in the years between the publication of the *Starry Messenger* in 1610 and the *Assayer* in 1623. In the decade that led up to the trial of 1633, Galileo withdrew to write his *Dialogue on the Two Great World Systems*, and the Linceans were obliged to stand aside while the once enthusiastic support of the Barberini withered and died on the vine.

But their terrestrial investigations, begun even before their involvement with Galileo's discoveries, continued for several years after Cesi's death in 1630. In each of the four natural historical spheres to which they devoted their energy, there were two pairs of related preoccupations, which we have already encountered: on the one hand, the problems of classification and generation; on the other, the concerns with the vernacular and the possibilities of visual reproduction.

Right from the outset, the aim of Cesi and his three friends was to make direct observations of everything in nature, and to collect and to draw as many specimens as possible of everything they saw. Eventually they wished to make their many discoveries accessible to others, and to record every one of them by means of drawing and engraving. In this manner they began to confirm, as they had already intuitively realized, that there were things on this earth, just as there were in the heavens, that were not only wholly unknown to the ancients, but also to every modern authority as well.

## HECKIUS AND HIS SIMPLES

Strictly speaking, a "simple" was a single plant—rather than a compound of plants—from which a medical remedy could be prepared; but by the beginning of the seventeenth century, "simples" (or *semplici* in Italian) was the term used for any herbal remedy prepared from plants. Many botanical gardens were little more than gardens of simples, and a doctor such as Faber, expert in botanical cures and in charge of the Vatican gardens, was known as the *semplicista* of the pope.

Heckius too was a *semplicista*, though in a much more modest way than Faber. His beginnings presaged the future, in more ways than one. Even before he met Cesi, he took a passionate interest in simples and in the remedies they offered. Shortly after he arrived in Italy in 1597 (for the second time),[1] the young Dutchman went to live with the Gelosi family in Spoleto. When not dispensing remedies to the local population, he relaxed by studying literature, natural history, and astrology in the small museum the Gelosi had established in their villa. Having decided to pursue his medical studies at the University of Perugia, he graduated in August 1601; whereupon he moved yet again, and entered the employ of the Caetani family in Maenza. Once more he ministered to the sick and the poor, and applied himself with considerable intensity to the study of *i nostri semplici*. Ceaselessly he pounded and distilled the remedies that he derived from them, and recorded the results in his notebooks.[2]

Throughout Heckius's life, but especially in his early years in Italy, he turned his knowledge of the pharmaceutical use of plants to benefit the sick and dying around him. He made it a point to know about the "*vires et effectus*," the "powers and effects" of plants.[3] The phrase was a pregnant and critical one, as we shall see. Already at the age of seventeen Heckius had composed a *regimen sanitatis*, a diet and program of healthy living for scholars, and from then on he wrote incessantly. His early writings are full of the most detailed accounts of people's illnesses and handicaps, and of the various remedies he applied to salve their ailments. They contain page after page of recipes for pills, antidotes, opiates, and remedies of every kind. He wrote about outbursts of apoplexy and occurrences of the plague, and noted down whatever he could about dermatological problems and a whole variety of internal pains.[4] Since his concern with pharmacology was so intimately bound up with botany, he recorded plants of every conceivable description, both local and exotic.

From Maenza, Heckius moved on to the only slightly larger town (or rather village) of Scandriglia, where he had another powerful patron, Giovanni Antonio Orsini. Here he conducted a whole series of what he called experiments (by which he meant firsthand medical investigations, rather than experiments in the sense we now understand them): on the internal pains of a fifty-year-old woman, on the intestinal worms of a young boy, on the ringworm—or some such infestation—of a four-year-old child, on the quotidian ague of a twenty-year-old, on a woman with subcutaneous fluid, on a woman of sixty with pains in her ribcage, on the dysentery of an obese patient. Heckius notes the effectiveness or otherwise of his remedies, such as the one he concocted for a man whose handsome face was disfigured by a red blemish or scar on his cheeks and large pustules around his nose.[5]

In his earliest letter to Aldrovandi of November 14, 1602, Heckius wrote that while in Maenza he had been so successful in his study and use of simples that within a short time, he had exposed two charlatans (*truffatori*, or swindlers, is the word Heckius uses) who had been selling Indian medicines, which the Italians called "specials" (more specifically herbal remedies for particular diseases). The charlatans had left town in exasperation. Heckius had proven to his patrons that local plants and minerals could serve just as well as the more exotic plants of the Indies. Indeed, as he proudly recorded, for the whole six months in which he had stayed in the Caetani castle in Maenza, he had never ceased to practice "speciaria," or the science of specials, and with the help of God, no one over three and below sixty had died during the whole period he was there. Then he launched a tirade against all those who pretended to be experts on the basis of their knowledge of the plants of the New World, "as if all the things of Europe were already well-known to them."[6]

Little could Heckius then have imagined that the greatest of the Lincean works to be published would be the huge treatise on the plants and animals of Mexico. But in the meantime it was no wonder that he should have gotten into trouble with his competitors among the apothecaries. Vigorously accusing the apothecary in Scandriglia—who was already jealous of Heckius's success among the locals—of shortchanging his customers by not mixing his prescriptions properly, he spoke as plainly as ever. The two got into a fight, ending with Heckius stabbing the apothecary to death. Thrown into a Roman prison, he was only freed as a result of the intervention of his new friend, Cesi.[7] From then on their bond was sealed.

The Academy was soon founded, and Heckius banished from Italy. But he took his small pocketbooks of drawings and notes wherever he went.[8] In Prague, though distracted by the excitement of seeing unprecedented objects in the heavens, he continued to write about medical problems, often with an astrological slant. He had not yet freed himself from the old bonds of astrological and Paracelsan medicine, whereby the influence of the stars was taken into account in determining the etiology and diagnosis of diseases, and the remedies for them. Given the overheated, hermetic court of Rudolf II, this was only to be expected.

Even in his misery, both in Prague and later in Madrid, Heckius went on ministering to the sick. The problem of contagion, especially with regard to an epidemic in the Low Countries, continued to worry him. In October 1605 his treatise *On the Plague and Why particularly it has spread throughout the Netherlands for many years now* was published in Deventer by his devoted brother Willem. On its title page, its author was proudly announced as "Doctor Joannes Heckius of Deventer, Lincean Knight." Once again the work was dedicated to Cesi. Only two copies of this short treatise are known.[9] The subtitle declares that the work contains a description of the "Electuarii Lyncei," the special nostrum that Heckius used as an effective antidote to the plague.

Discouraged by the reception of his treatise on the nova of 1604, Heckius decided to return to his botanical and pharmacological concerns. Cesi had emasculated his manuscript to such a degree that his original ideas were barely recognizable in the published version. The public seemed wholly uninterested. The first Lincean publication sank without a trace, leaving not the slightest influence on the burning astronomical issues of the day. Undaunted, Heckius returned with renewed vigor to his old studies of plants—and of insects as well. If he could not convince his friends of his insights into the phenomena of the heavens, at least he would show them what he could achieve with the lowly things of the earth.

On the calends of August 1605 Heckius sent a new manuscript from Prague to the Linceans in Rome. It is the earliest of a group of four manuscripts sent by Heckius in the fall of that year. All are extensively illustrated, in his own rather crude hand. They contain over six hundred drawings of the plants, fungi, insects, and butterflies he had encountered in the course of his travels all over northern Europe.[10] Titled *Fruits of my Journey to the Countries of the North,* they are preserved in the library of the old medical school in Montpellier along with a quantity of other manuscripts originally from the collection of Cassiano dal Pozzo.[11] The drawings are generally accompanied by Heckius's untidy inscriptions in Italian, Dutch, Latin, and sometimes Arabic, testifying to his intense preoccupation with the names of things—a preoccupation he shared with many others at the time.

The first of the manuscripts contains more than one hundred drawings of mushrooms, lichens, mosses, flowers, roots, and other parts of plants. Several pages have just their names, suggesting that the drawings were yet to come. Often the local and Latin names are given, while there are frequent notes in Dutch on how each specimen looked. References are provided to their illustration (or the illustration of similar species) in the most popular herbal of the time, that of Pierandrea Mattioli.[12] One had to be sure that the plant one drew (or planned to draw) was the same as its equivalent in Mattioli—whatever the difference in their names. On the first page of this manuscript, Heckius referred to his subject not as *plantae,* as one might have expected, but as *vegetales.* In this way he subtly made clear the Linceans' interest in the growth of plants, and in their generative and reproductive aspects.

Two of the manuscripts in Montpellier contain well over three hundred pages of drawings of butterflies, moths, mosquitoes, wasps, flies, other insects, mollusks, and snakes large and small. The first of these manuscripts—sent from Parma "in haste" in October 1605—is devoted to species found in Pomerania, Poland, Bohemia, Franconia, Austria, Bavaria, Saxony, and northern Italy; while the second covers species from the British Isles, Denmark, Norway, Sweden, and France. Most of the drawings were colored, a few are in pen and ink, and about twenty-five pages have no drawings at all, just the Latin names of particular species, as if Heckius was in such a hurry to send off his manuscripts that he had no time to complete this one. But no one could doubt Heckius's taxonomic obsession: most pages contain at least two and sometimes three names for each species—not just their Latin names, but also their vernacular ones, whether in Italian or in Dutch.

The rudimentary drawings in the first of these two manuscripts are mostly of butterflies, with a notable interest in showing the details of chrysalises and fetuses.[13] Another manuscript contains a number of curiosities, ranging from a series of drawings of fleas to a malaria mosquito and finally to a strange beast described as a "basilisk" kept by "Hegmundus Princeps."[14] Indeed, alongside the precise descriptions and drawings of the flora and fauna Heckius saw on his travels are those of a number of monsters and prodigies, both fabulous and real. His concern with such matters is most strikingly revealed in a notebook treatise (now in the Bibliotheca Vallicelliana in Rome) devoted wholly to monstrosity.[15]

"Not only has my pen sent you new stars never seen in the heavens, it also sends you new living forms born on earth," wrote Heckius in the letter accompanying the first of the Montpellier manuscripts. "In looking more acutely at the viscera of the earth, I have found new plants neither described nor known by anyone before."[16] Always, from the very beginning, the Linceans knew to look to the innards, beneath the surface of things; no wonder that Heckius drew roots and tubers as well as leaves and flowers. But more than that: "I send them to you so that you may speculate about them, and thoroughly investigate their powers and their effects."[17]

In the prefatory letter to the first manuscript on butterflies, insects, and snakes, however, another set of preoccupations emerged as well. "These things," Heckius began, "are the games of nature and marvels of mixtures." Like so many other naturalists of the time (and earlier) the Linceans were fascinated by the so-called games of nature, the *lusus naturae*. They included animals and plants (such as Heckius was now encountering) that played with the rules and transgressed the boundaries. Sometimes these "jokes of nature" seemed to belong to neither one class of things nor another; sometimes they belonged to more than one (just like fossils, for example). How was one to make sense of them, and how were they to be ordered if they were miracles of mixedness, if they combined in one species the characteristics of several? This was the problem that Cesi would soon confront head-on. For all their interest in the *lusus naturae*, the Linceans soon realized that the very notion of playfulness stood at odds with the order of nature.

In Heckius's all too labile character, however, there was little to suggest a capacity for order or for ordering things. With him the priorities were different. "Even if you cannot know these things in their essence, you can still make conjectures about their mode of acting, on the basis of their particularities," he insisted. Heckius was less concerned with order than with operation and effectiveness. As usual, however, he wanted to have it both ways, and asserted that "modes of acting" were certain and invariable—even in the case of what he called the imperfect plants and animals. All such things, he said, were "the arcane things of nature and neglected."[18]

When Heckius returned to Rome he continued to write not only to Kepler and the aging expert on medicine and exercise Hieronymus Mercuriale in Pisa, but also to the great botanists of the day, like Jean Robin in Paris and Clusius in Leiden. He asked them to send him information about new plants as well as seeds, so that he could investigate their effects.[19] And in a remarkable letter of March 19, 1606, unaddressed but almost certainly written to Clusius, Heckius broached the problem of one of the more difficult aspects of the "arcane no less than unknown nature" of plants that have the "innate power to kill," *insita vis necandi*.

Not, said Heckius, that he had anything to add to Clusius's great knowledge (Clusius had just published a treatise on the larger fungi of Pannonia, the first ever printed book to be devoted exclusively to the subject);[20] but he did want to know the great expert's opinion on a number of the imperfect plants he had found in the course of his travels. He also enclosed some drawings of them along with his letter. "I have no doubt," he wrote, "that they are to be counted among the poisonous species, not so much on the grounds that they are fungi, but because they seem to be accompanied by the external signs of virulence." For Heckius, as for many of the exponents of the traditional forms of natural history, it was precisely the external signs that offered the clues to inner nature, and, even more, to the inner forces—the *vis* as well as the *virtus*—of things.

By then Cesi and Heckius had already begun working on the problem of fungi, and on the ambitious illustrative project later referred to as the *icones fungorum*. In the passage that follows, Heckius offers not only a close description of the "external signs of virulence" but also a dramatic account of the effects of two such plants. The first, he says,

> is found on the rocky mountains of Norway growing on rotting tree trunks, about the size of a large abscess. It looks like what others call the dog-nut. . . . It has a very thin, short, and strong stalk, so that it deceives the eyes and seems to appear without any support. . . . I brought back some dried specimens and [simply] showed them to a dog. Although death did not follow, by its many contortions, the animal seemed to give evidence of great tension and pain.

The second fungus was even more poisonous. Frequently to be seen in the woods of central Germany,

> it has a horrible red color, just like blood. It varies in size. . . but is generally a few inches high, with the color of the lower part now a blue out of yellow, now on the contrary, a yellow out of blue, and spattered with black spots. I thought it was of the kind which kills by smell. . . . Attracted by its colors I descended from my horse and collected it, but it caused me pain even when I held it at some distance from my nostrils with my hands outstretched, rather in the manner of a magician. But this hardly helped, because I was so affected by its horrible fetidness that I immediately began to have a serious headache. Even with my horse proceeding very slowly I would barely have been able to stay in the saddle for the space of an hour, had I not brought with me a powerful antidote, which I always kept handy in case of plague. Otherwise I would have been overtaken by fever or my illness would have worsened. My own servant, whom I was using for the drawing of plants, was aston-

> ished, and thought that I had contracted the disease for some other reason. He was himself stricken with an even stronger pain because of the odor of the plant, but he got better as a result of spontaneous vomiting and by taking the drug, though the headache remained with him for several days following.[21]

If you were someone like Heckius, concerned with both the beneficent and the poisonous effects of plants, then precise identification on the basis of external characteristics such as color and odor was indeed important. You had to be able to identify what might be dangerous; and you ought to be capable of predicting how things worked on the basis of how they looked.

Soon the Linceans discovered just how illusory this was, or could be. But for the time being they remained in the grip of the old paradigm, believing that the secrets of things—both their essence and their mysterious effects—could be read from the logic of their external signs. They knew the influential books on physiognomy and phytognomy by their old mentor, Giovanni Battista Della Porta, all too well. In these early days of their researches, long before they had the microscope at their disposal, works such as Della Porta's reinforced their view that the indicators of inner powers and inner character lay on the surface of things—provided that one knew how to interpret them correctly. It was not yet time to be skeptical about the possibilities of description.

Many of the Linceans' letters are striking for their close descriptions of what later would be called the secondary qualities—qualities perceived with the senses and whose assessment depends on subjective perception: color above all, but also texture and smell. This was especially the case in the early correspondence. For the moment, briefly, the Linceans thought that superficial indicators could tell them about the differences between things. In his first surviving letter to Cesi, Heckius wrote that he was sending him some seeds of a simple, not at all dissimilar to the common melon, which he had found in a deep cave in the mountains near Spoleto. The only difference was that its leaves were more serrated, while the flowers were

> sometimes yellow and sometimes white; it has some long fruits, some round, and some squashed. At first the color of some of them is dark green, then they become even golder than gold, and give so pleasing and sweet an odor that their fragrance can fill any large room. They are so beautiful to the eyes, so pleasing to the nose, and so satisfying to the mind that it is an incredible thing.[22]

The appeal to the senses could hardly be clearer. Heckius here takes a kind of pure delight in the aesthetic appeal of the objects. Even the pleasure it affords the mind is turned into something sensual. But it is the meticulous attention to the nuances of color that is most striking. This is the kind of description that for good reason is called pictorial; it seems to cry out for representation by picture. When it comes to this kind of attentiveness, words never seem quite good enough.

Of course, all the sixteenth-century botanists and most of the other students of natural history realized this perfectly well too. When Fuchs wrote his introduction to the epoch-making *Historia Stirpium* of 1542, he referred, as others did too, to the old claims for the superiority of pictures over words. With pictures you were able to know *exactly* what plant, or plant specimen, was being referred to. You could avoid the inherent vagueness of words in natural history. The colors were there before your eyes, and the forms too; you did not have to interpret pictures in the way you always had to with words. When it came to representation, words were essentially ambiguous (exactly the opposite position from the one held by naturalists from Galen through Linnaeus).

But there were two fundamental problems with pictures: the hand of the artist was always liable to import a subjective element into representation, and the unique ability of pictures to represent particularity militated against the differentiation of the things they represented into classes.

Concerned as he was with the "faculties" of things, and not particularly inclined to order, Heckius did not see these problems as especially critical. For Cesi, however, the tension was great. On the one hand he recognized the need for pictures; on the other, he discerned the necessity for order in multiplicity—not just the multiplicity of drawings, but the multiplicity of nature itself.

All along, however, the Linceans continued to make their drawings—or have their drawings made. They did so in abundance and with unparalleled intensity and attention to detail. With black-and-white drawings, they used words to supplement what could not be shown in color. Even then (as with the great bulk of the fossil drawings in Windsor), one senses their frustration at the basic imprecisions of the language of sensual perceptions, and of color and texture in particular. The denser the language, the more one wanders in the dark, the more frustrating it all seems. The color drawings in the fossil corpus, on the other hand, offer a sense of relief: finally one can *see* exactly what Cesi's finds looked like. But immediately the question arises: what is the use of so close a pictorial description of mere color and surface—the subjective appearance of objects?

In the early years of their explorations and investigations, the Linceans—and Heckius in particular—were not too daunted by such problems. Cesi realized that they had simply to press ahead with the recording of nature. Soon they found themselves with a growing corpus of illustrations, one that expanded so rapidly that there was hardly any time to order it at all.

## THE *ERBARIO MINIATO*

Only a few pages of Heckius's illustrated notebooks in Montpellier and Rome have ever been reproduced, let alone commented upon. But they are almost as nothing beside the rest of the Lincean corpus of drawings; they are mere prologues, so to speak, to the massive and more careful visual projects that follow them.

Probably the first and certainly the most rudimentary of these is represented by the so-called *Erbario miniato*[23] in Windsor Castle. Like the manuscripts in Montpellier, it passed from Cesi's collection to Cassiano's in 1631, and then from the Dal Pozzo family to the Albani in 1714.[24] It contains over two hundred pages of drawings of plants and fungi, often more than one to a page.[25] They chiefly show Italian species, though some, like the Crown Imperial (*Fritillaria imperialis*), the Indian heliotrope (*Helianthus annuus*), the aubergine (*Solanum melongena*), and the tomato (*Lycopersicon esculentum*) are of imported specimens (figs. 8.1–2). Their range is very wide, showing everything from ordinary flowers to the tobacco plant (*Nicotinia tabacum*) and that rare phytogeographical curiosity, the *Periploca graeca* L. Close attention is often paid to bulbs, leaves, and the details of flowers, buds, and roots. Though all the drawings are considerably higher in quality than Heckius's in Montpellier (in terms of

Fig. 8.1. Aubergine (*Solanum melongena* L.). Watercolor, bodycolor, and touches of black ink and wash over traces of black chalk, 423 x 275 mm. *Erbario miniato*, fol. 21. Windsor, RL 27711.

Fig. 8.2. *Opposite:* Tomato (*Lycopersicon esculentum* Miller). Watercolor and bodycolor over black chalk, 423 x 275 mm. *Erbario miniato*, fol. 23. Windsor, RL 27713. 

521

Pomi doro· somno buoni da Mangiare como le meli insanie.
1136. 56 se ritrouano di doi sorte· luna rossa è laltra
gialli che rassembrano oro·

La Balsamina. hanno le sue frondi virtù
di consolidare tutte le ferite. e masime de
nervi. el olio, che per infusione si fa del suo frutto. conferisce a tutte le fe=
rite, alle posteme. e ulcere delle mammelle, levandone il dolore: ~~e parimente~~
~~all'ulcere, posteme, e dolori della madrice.~~ quando vi si getta dentro con la siringa.
vale ai dolori del parto. e a quelli dell' HEMORRHOIDI mirabilmente. il perche
si fa egli particolarmente. infondendo i suoi frutti nell' olio di mandorle dolci; e
mettendo per ogni libra de olio una oncia di vernice liquida, spegne il fervore
delle cotture del fuoco. e di tutte le calide posteme, vale alle ponture de i,
~~ne~~ nervi, e asotiglia le cicatrice, Referiscono alcuni de i moderni, che se le
donne sterili entrano prima in un bagno fatto con herbe matricali, e posci
si ungano la bocca della madrice con questo olio, e si congiungano poscia
con il marito. facilmente si ingravidano: e salutiferissima per l'ulcere del
la madrice: Giova benissimo alle crepature INTESTINALI ungendone spesso
il luoco con esso caldo. La polvere dell' herba data alla quantità d'un cucchiaro
con acqua di piantagine. consolida le ferite de li interiora. ancora che la
ferita passasse da una banda e l'altra. e altri dicano che la medesima polvere
vale a i dolori colici, e delli budella. del che opera con mirabile prestezza.

1349

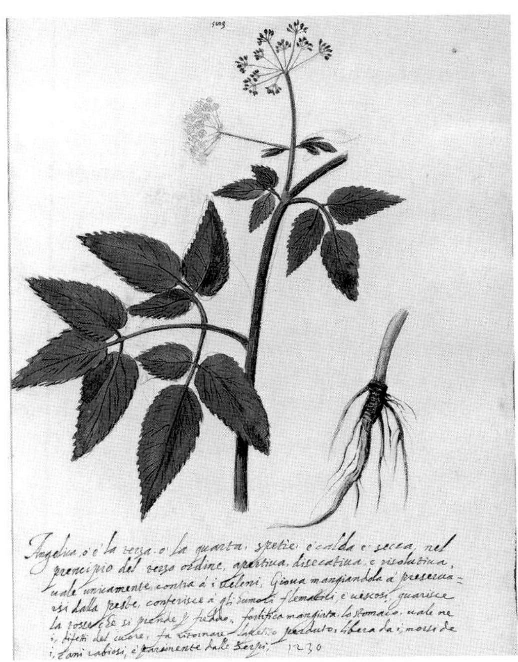

Fig. 8.3. *Opposite: Sechium edule* (?). Watercolor and bodycolor over traces of black chalk, 423 x 275 mm. *Erbario miniato,* fol. 140. Windsor, RL 27830.

Fig. 8.4. *Verbascum cf. sinuatum* L. Watercolor and bodycolor over black chalk, 423 x 275 mm. *Erbario miniato,* fol. 71. Windsor, RL 27761.

Fig. 8.5. Angelica (*Aegopodium podagraria* L.). Watercolor and touches of white bodycolor over black chalk, 423 x 275 mm. *Erbario miniato,* fol. 86. Windsor, RL 27776. All The Royal Collection © 2000, Her Majesty Queen Elizabeth II.

color, execution, and attentiveness to general morphology), they do not begin to attain the same level of analysis or meticulous description as most of the other drawings in the Cesian corpus.

The inscriptions on these sheets are in two hands. One is probably that of the young Cesi, the other (appearing in far fewer instances) is perhaps the more distinctively foreign hand of Heckius.[26] Many of the inscriptions contain cross-references to the same plants in the 1585 Venice edition of Mattioli's famous herbal;[27] a few also have additional references to other herbals, such as Tabernaemontanus's 1588–91 *Eicones Plantarum*. Occasionally other authorities are cited as well—Monardes and Dodonaeus among the "moderns," and Pliny, Dioscorides, Mesue among the ancients (or near ancients).

At first sight, the aim seems to be to correlate the drawings in the *Erbario* with the relevant pages in Mattioli—and in the same order. The inscriptions below the illustrations are copied, cut, or closely adapted from Mattioli's text too. But then—especially in the final pages of the manuscript—come a number of inscriptions recording that the plant illustrated is *not* to be found in Mattioli (or in any other herbal either, for that matter). "Species of ginestra not described by Mattioli," "Species of alsine not described by Mattioli," "Tobacco plant—Mattioli did not know it" are all typical remarks of this kind. They are made in the pleasure of discovery and in the full awareness that the old authorities were incomplete and often inadequate.

Then, too, are the many drawings of plants mentioned but not illustrated in the edition of Mattioli. Many are of New World plants, such as

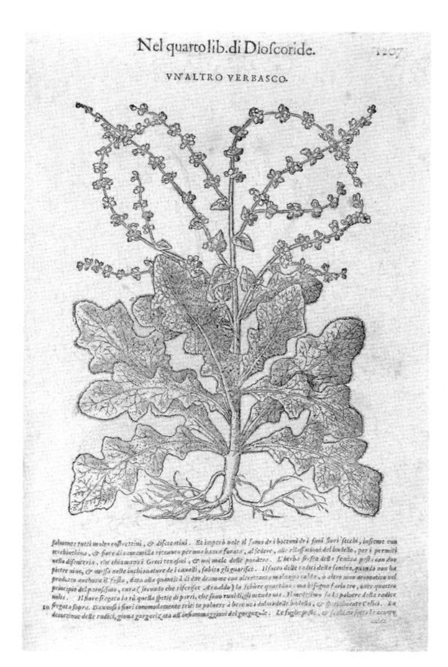


Fig. 8.6. *Verbascum* species. Woodcut. Mattioli (1585), p. 1207.

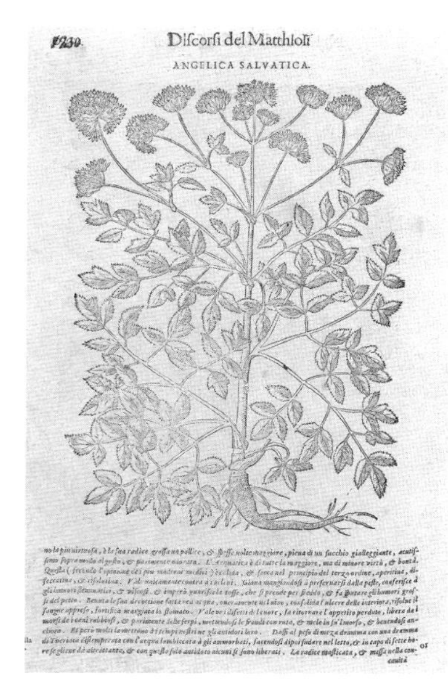


Fig. 8.7. *Aegopodium* species ("Angelica Salvatica"). Woodcut, Mattioli (1585), p. 1230.

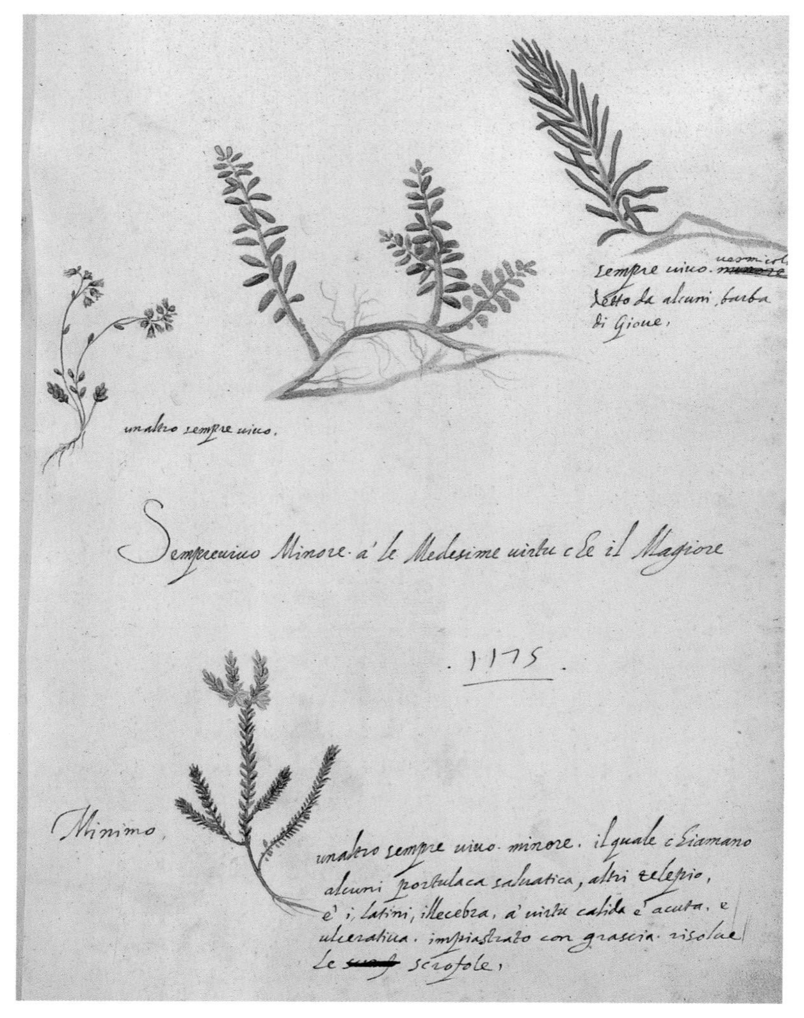


Fig. 8.8. *Semprevivo minore* (*Sedum* species) (*Sedum dusyphyllum* L.; *Sedum album* L.; *Sedum* species; *Sedum aere* L.). Watercolor and touches of bodycolor, 423 x 275 mm. *Erbario miniato*, fol. 59. Windsor, RL 27749. The Royal Collection © 2000, Her Majesty Queen Elizabeth II.

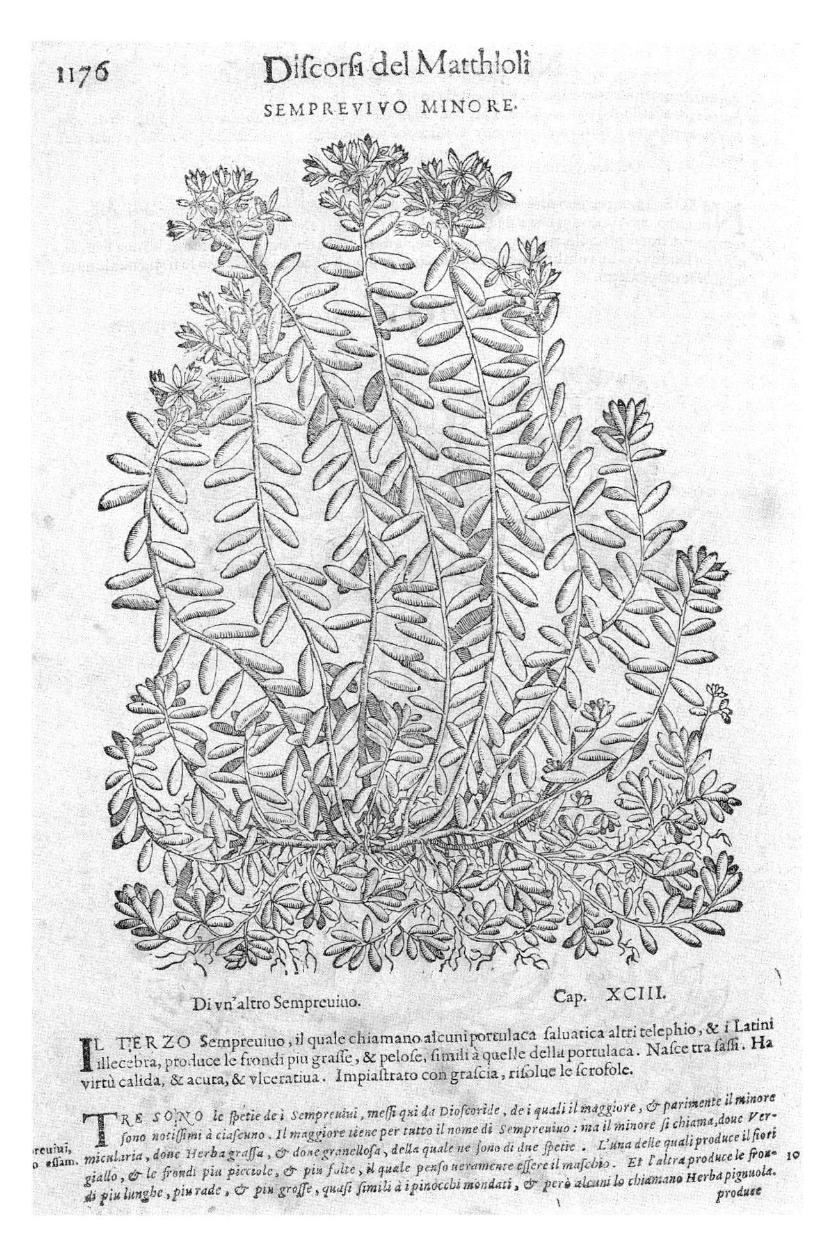
1176 Diſcorſi del Matthioli

SEMPREVIVO MINORE.

Di vn'altro Sempreuiuo. Cap. XCIII.

IL TERZO Sempreuiuo, il quale chiamano alcuni portulaca ſaluatica altri telephio, & i Latini illecebra, produce le frondi piu graſſe, & peloſe, ſimili à quelle della portulaca. Naſce tra ſaſſi. Ha virtù calida, & acuta, & vlceratiua. Impiaſtrato con graſcia, riſolue le ſcrofole.

*TRE SONO le ſpetie de i Sempreuiui, meſſi qui da Dioſcoride, de i quali il maggiore, & parimente il minore ſono notiſſimi à ciaſcuno. Il maggiore tiene per tutto il nome di Sempreuiuo: ma il minore ſi chiama, doue Vermicularia, doue Herba graſſa, & doue granelloſa, della quale ne ſono di due ſpetie. L'una delle quali produce il fiori giallo, & le frondi piu picciole, & piu ſolte, il quale penſo ueramente eſſere il maſchio. Et l'altra produce le frondi piu lunghe, piu rade, & piu groſſe, quaſi ſimili à i pinocchi mondati, & però alcuni lo chiamano Herba pignuola.*

produce

Fig. 8.9. *Sedum* species ("Semprevivo"). Woodcut. Mattioli (1585), p.1176.

the tomato and the various kinds of American balsam (e.g., figs. 8.2–3). But even where the same plants are illustrated, the differences in the way they are represented could hardly be greater. The plants of the *Erbario* are shown either still growing in the field, or as if they had just been taken out of the ground for examination by some rhizotome or another. Mattioli's woodcuts, on the other hand, seem artificial and often improbably schematic (compare figs. 8.4–5, for example, with figs. 8.6–7).

In a way, it is this alone that makes the taxonomic knowledge and classificatory diligence of the *Erbario* so remarkable. Both qualities emerge from the correlations that the compiler of the *Erbario* establishes

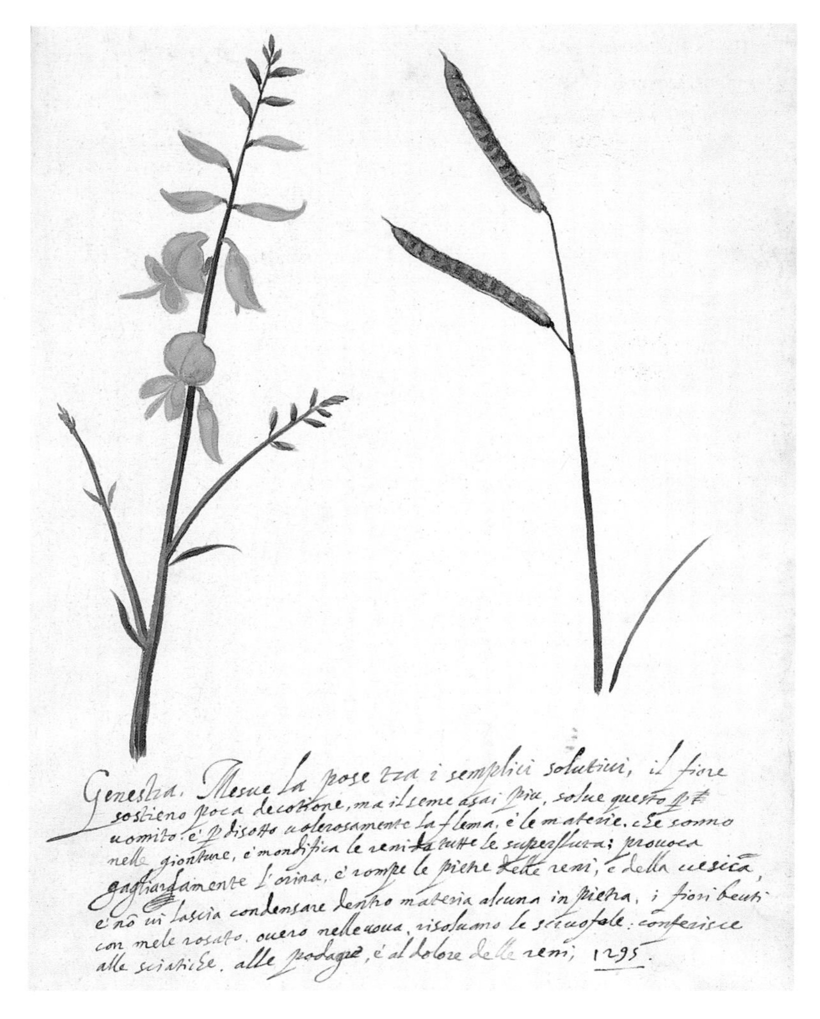


Fig. 8.10. Spanish broom (*Spartium junceum* L.). Watercolor and bodycolor, 423 x 275 mm. *Erbario miniato*, fol. 119. Windsor, RL 27809. The Royal Collection © 2000, Her Majesty Queen Elizabeth II.

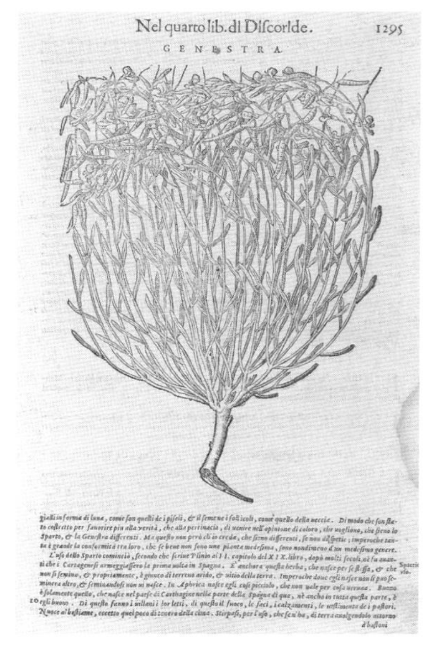


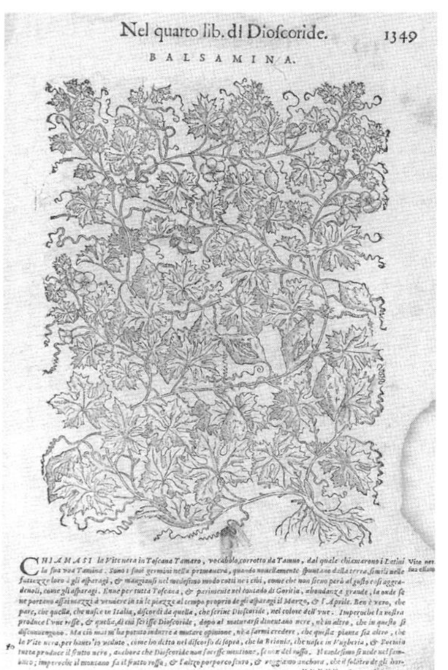


Fig. 8.11. *Spartium* species ("ginestra"). Woodcut, Mattioli (1585), p. 1295.

Fig. 8.12. *Sechium* species (?). Woodcut, Mattioli (1585), p. 1349.

Fig. 8.13. *Nuphar lutea.* Woodcut, Mattioli (1585), p. 945.

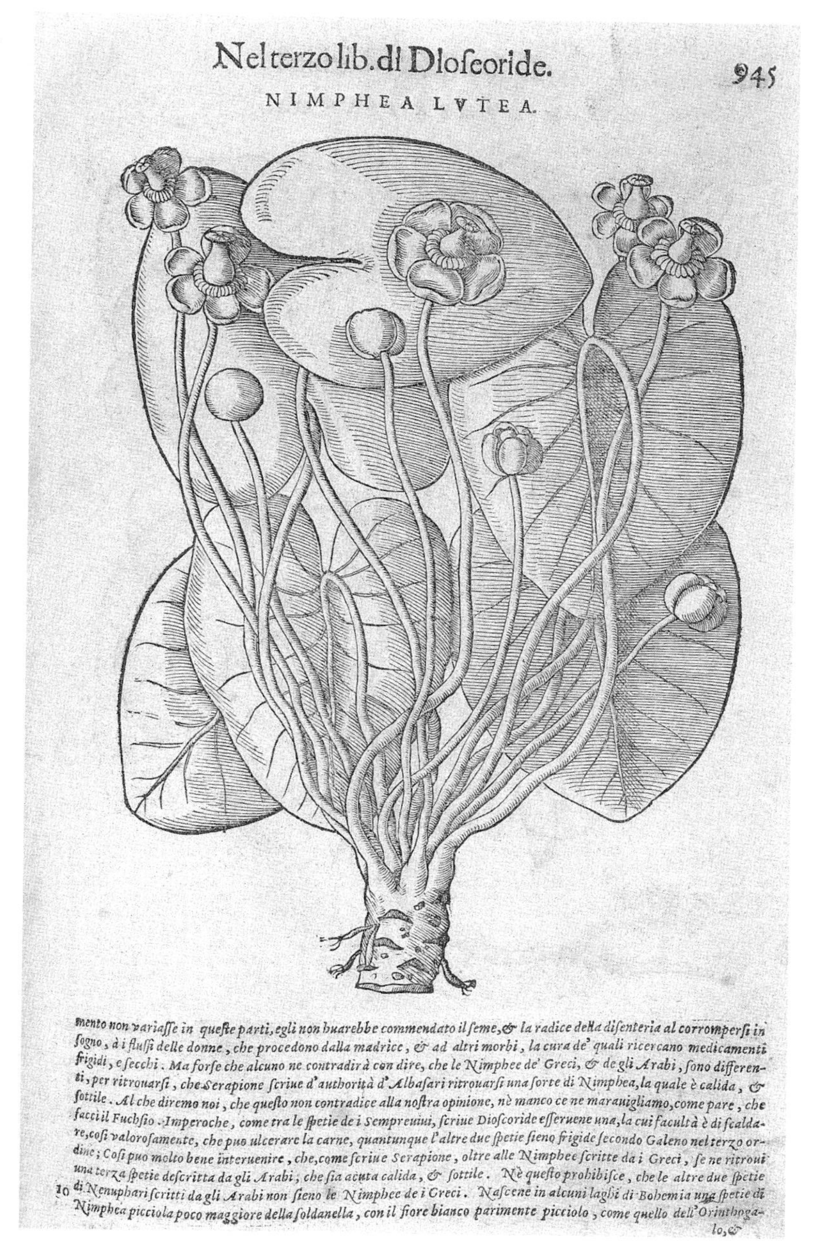

Nel terzo lib. di Dioscoride. 945

NIMPHEA LVTEA.

*mento non variasse in queste parti, egli non huarebbe commendato il seme, & la radice della disenteria al corrompersi in sogno, à i flussi delle donne, che procedono dalla madrice, & ad altri morbi, la cura de' quali ricercano medicamenti frigidi, e secchi. Ma forse che alcuno ne contradirà con dire, che le Nimphee de' Greci, & de gli Arabi, sono differenti, per ritrouarsi, che Serapione scriue d'authorità d'Albasari ritrouarsi una sorte di Nimphea, la quale è calida, & sottile. Al che diremo noi, che questo non contradice alla nostra opinione, nè manco ce ne marauigliamo, come pare, che facci il Fuchsio. Imperoche, come tra le spetie de i Sempreuiui, scriue Dioscoride esseruene una, la cui facultà è di scaldare, così valorosamente, che puo ulcerare la carne, quantunque l'altre due spetie sieno frigide secondo Galeno nel terzo ordine; Così puo molto bene interuenire, che, come scriue Serapione, oltre alle Nimphee scritte da i Greci, se ne ritroui una terza spetie descritta da gli Arabi; che sia acuta calida, & sottile. Nè questo prohibisce, che le altre due spetie* 10 *di Nenuphari scritti da gli Arabi non sieno le Nimphee de i Greci. Nascene in alcuni laghi di Bohemia una spetie di Nimphea picciola poco maggiore della soldanella, con il fiore bianco parimente picciolo, come quello dell'Orinthogalo, &*

between the illustrations here and in Mattioli.[28] Mattioli's plants seem to be wholly constrained by their medium. They are forced into highly decorative patterns, often dense with repetitive and not very significant detail. Each seemingly replete woodcut betrays a high degree of schematization, with little attention, if any, to individual or idiosyncratic detail, or to the plain untidiness of how things grow in nature. It is precisely this which the Lincean drawings—genuine field drawings—manage to show so effectively. One has only to compare the various illustrations of Cesi's and Mattioli's versions of the slightly different forms of the evergreen

Fig. 8.14. Aubergine (*Solanum* species). Woodcut, Mattioli (1585), p. 1135.

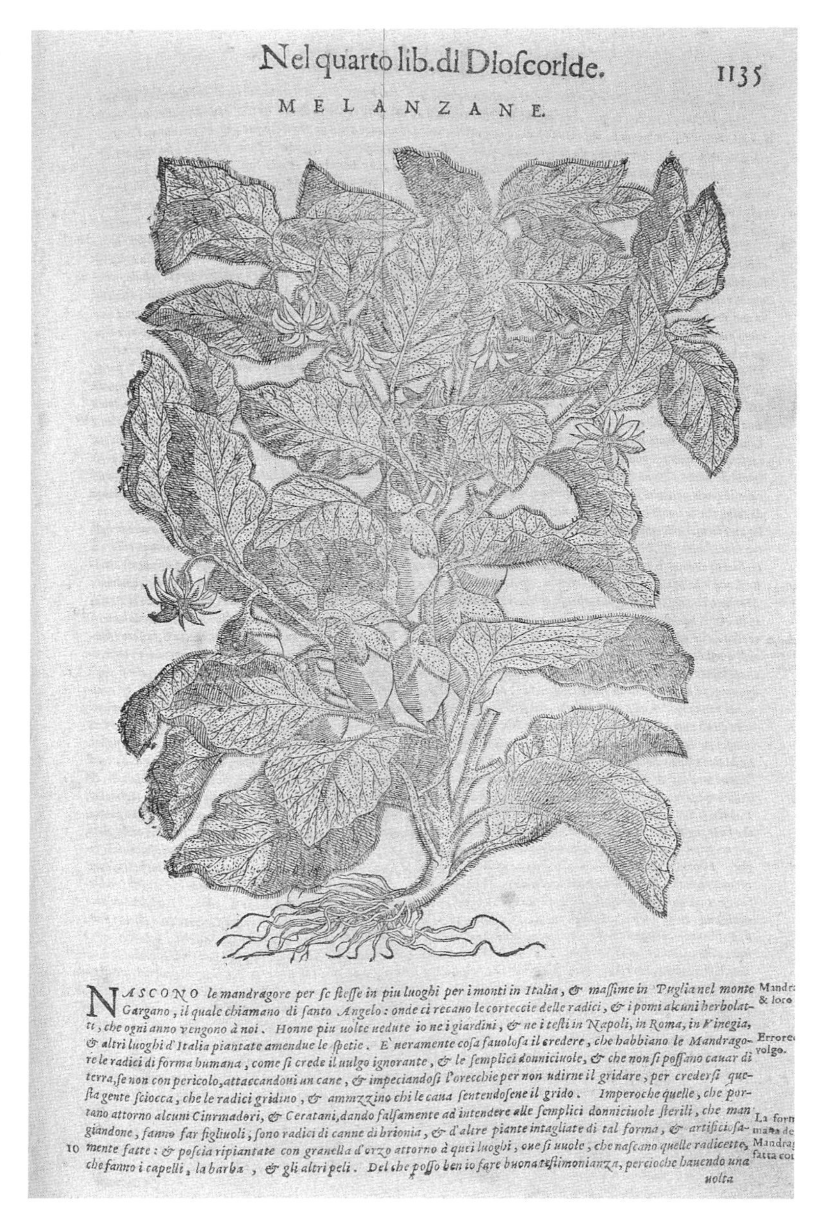

Nel quarto lib. di Dioscoride. 1135

MELANZANE.

NASCONO le mandragore per se stesse in piu luoghi per i monti in Italia, & massime in Puglia nel monte Gargano, il quale chiamano di santo Angelo: onde ci recano le corteccie delle radici, & i pomi alcuni herbolatti, che ogni anno vengono à noi. Honne piu uolte uedute io ne i giardini, & ne i testi in Napoli, in Roma, in Vinegia, & altri luoghi d'Italia piantate amendue le spetie. E' ueramente cosa fauolosa il credere, che habbiano le Mandragore le radici di forma humana, come si crede il uulgo ignorante, & le semplici donniciuole, & che non si possano cauar di terra, se non con pericolo, attaccandoui un cane, & impeciandosi l'orecchie per non udirne il gridare, per credersi questa gente sciocca, che le radici gridino, & ammzzino chi le caua sentendosene il grido. Imperoche quelle, che portano attorno alcuni Ciurmadori, & Ceratani, dando falsamente ad intendere alle semplici donniciuole sterili, che mangiandone, fanno far figliuoli, sono radici di canne di brionia, & d'altre piante intagliate di tal forma, & artificiosamente fatte: & poscia ripiantate con granella d'orzo attorno à quei luoghi, oue si uuole, che nascano quelle radicette, che fanno i capelli, la barba, & gli altri peli. Del che posso ben io fare buona testimonianza, perciòche hauendo una uolta

Mandr… & loro

Errore… volgo.

La form… mata de… Mandra… fatta co…

*Sedum* species, for example,[29] to get some sense of the differences (figs. 8.8–9).[30]

Here lies one of the real achievements of these illustrations. The plants in the *Erbario* are revealed as the products of natural growth, often in their own habitats, never as constrained by the medium of expression as the woodcuts are. The illustrations in the *Erbario* show flowers, seeds, seed pods, and fruits much more clearly than in Mattioli, where particular features of the plant are often sacrificed to overall decorative needs and impulses. This contrast appears very strikingly when one compares drawings such as those of the cucurbitaceous *balsamina* (fig. 8.3),[31]

Fig. 8.15. Teasel (*Dipsacus pilosus* L.). Watercolor and bodycolor over black chalk, 377 x 260 mm. Whereabouts unknown.

antirrhinum,[32] Spanish broom (fig. 8.10),[33] water lilies (fig. 1.14),[34] and—perhaps most beautifully of all—the aubergines (fig. 8.2), with their equivalents in Mattioli (figs. 8.11–14). In Mattioli's illustrations they are all but buried amid so much graphic sinuosity. They are also boxed in by the square or rectangular forms of the woodcut, in exactly the opposite way to that in which Cesi's seemingly informal and highly "natural" drawings are unconstrained by the limits and format of the page.

At first sight it may seem as if the old tendencies to anthropomorphism and "explanation" by way of similitude are as strong as ever, but these tendencies are just a reflection of what Cesi found in his Mattioli. When one turns to illustrations such as those of the *labbro di Venere* or "lips of Venus" (actually a teasel, fig. 8.15), and the *sopraciglie di Venere* or "eyebrows of Venus" (actually an achillea),[35] and—most strikingly of all—the

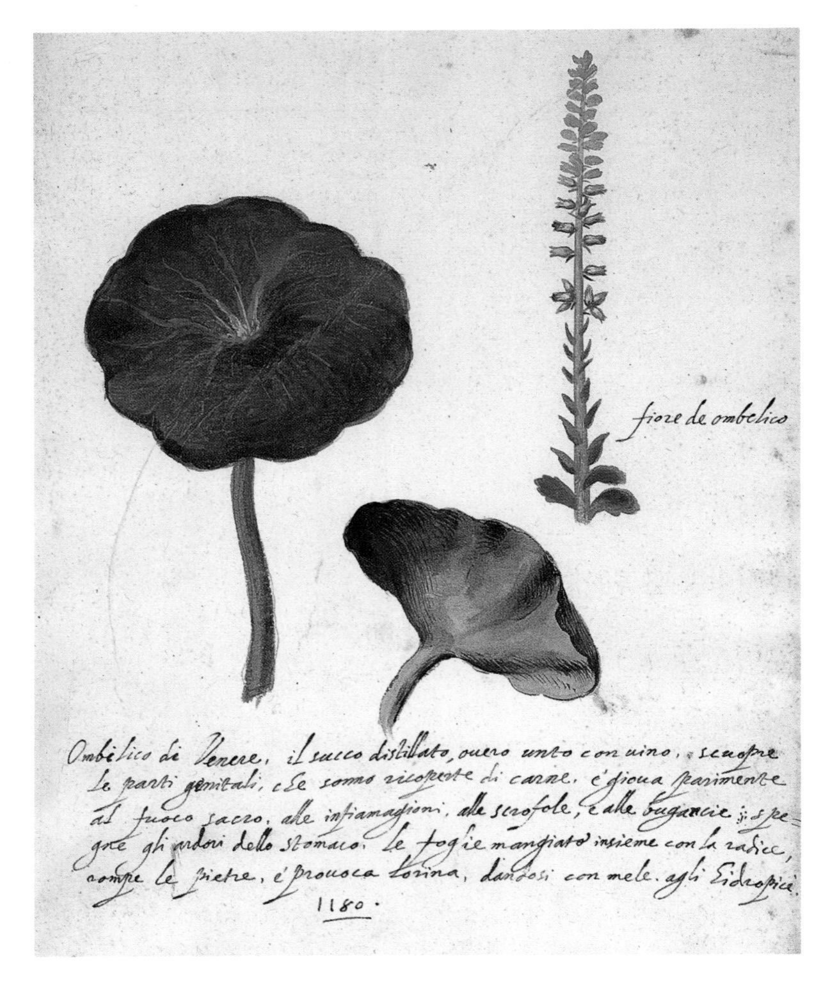


Fig. 8.16. *Umbilicus rupestris* (Salisbury) Dandy. Watercolor and bodycolor over traces of black chalk, 423 x 275 mm. *Erbario miniato*, fol. 61. Windsor, RL 27751. The Royal Collection © 2000, Her Majesty Queen Elizabeth II.

*ombilico di Venere* or "Venus's navel" (actually a foxglove *Umbilicus rupestris*, fig. 8.16), it is easy to see just how closely the artist of the *Erbario* hewed to the path of natural accuracy instead of succumbing to the old temptation to similitude, so artificially sought after by Mattioli's woodcutter (fig. 8.17).[36] The same goes for the much more accurate illustrations of the aconite (with its scorpionlike roots that so distracted Mattioli's illustrator, fig. 8.18),[37] and the wonderfully named "dog's testicles" by which the purple *orchis pratense* was also known (fig. 8.19).

Many other traces of the older paradigms also survive in these drawings. For example, Cesi lifts from Mattioli some of the traditional explanations of names in terms of their mythological etiologies.[38] In the drawing of a blue hyacinth, for instance, all the inscription says is "A Hyacinth, which the poets say was born from the blood of Ajax, where one still finds the letters of his name,"[39] while the vernacular names of "dog's testicles," and "lips" or "navel" or "eyebrows" of Venus require nothing by way of further explanation: all one has to do is look.

But there is one area in which the *Erbario* reveals its roots in the past,

Fig. 8.17. *Umbilicus* species. Woodcut, Mattioli (1585), p. 1180.

Fig. 8.18. So-called aconite. Woodcut, Mattioli (1585), p. 1139.

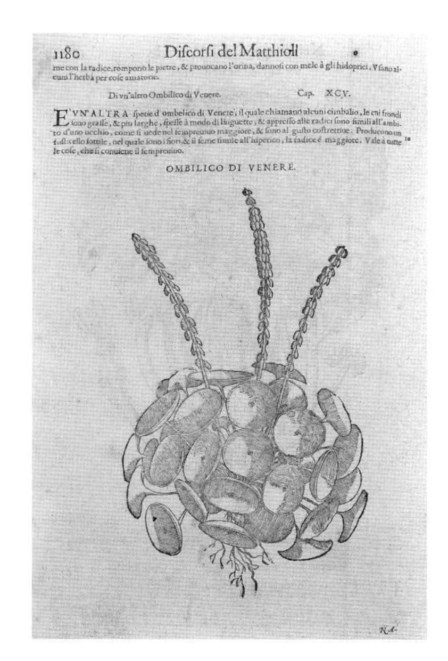


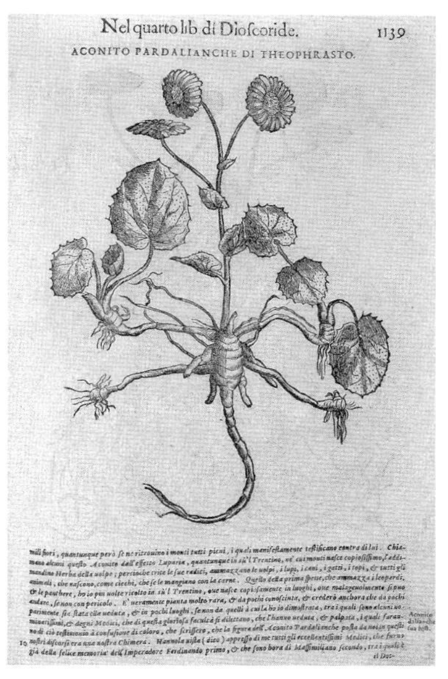


as well as one of the central botanical preoccupations of the early Linceans. Perhaps the most notable and consistent feature of its inscriptions is their resolute emphasis on the medicinal properties of plants. They give brief accounts of what each plant is good (or bad) for, and of how to make the relevant preparations from them. If a plant is particularly useful, or especially to be avoided, this too is carefully noted. The inscriptions are largely more or less verbatim extracts from Mattioli; but whereas Mattioli offered a comparatively large amount of information, ranging from the taxonomic to the morphological and the medicinal, the inscriptions in the *Erbario* confine themselves almost exclusively to the medicinal parts of Mattioli's original text (with a smattering of taxonomic material). Sometimes lengthy excerpts are provided; but for the most part they are relatively short. In any case they provide a conspectus of the medicinal properties of the plants that is both charming and valuable at once.

In these pages one may find any number of cures for headaches, stomach pains, sties in the eye, coughs, rabies, children's diseases, skin problems, various forms of "scrofula," hemorrhoids, intestinal ruptures, and ulcers more generally (the comfrey, of course). There are cures for insomnia, depression, and a variety of gynecological ailments.[40] Plants such as Our Lady's bedstraw (*Galium verum* L.) serve to stanch blood flow;[41] others serve as diuretics; still others help with the opposite problem, that of too-ready urination. Often advice is given on how to make simple mixtures, or to apply the remedies.

The root of the teasel or "Venus's lips" (from its appearance of course), "cooked in wine and then pounded until it is almost like wax, heals anal fissures and fistulas; and it is said that if one binds the worms that one finds in the heads of the thistle in leather, and then attaches these to neck or the arm, they can actually heal the quartan ague."[42]

The houseleek (*Sempervivum tectorum* L.), according to its long inscription, has the capacity to reduce the temperature and to restrict the blood flow, and its

> leaves help heal holy fire, malignant, persistent and creeping sores, and are of assistance for inflammation of the eyes, for burns, and for gout, either applied on their own or mixed with polenta. The juice mixed with polenta or with rose-water is useful in the case of headaches, and when drunk works against the bites of those spiders called *phalanghi*, against dysentery and other bodily flows; when drunk with wine it expels worms from the body, and when applied directly with wool, stanches menstrual flows and helps with eye defects.

To this lengthy list of uses, all in Mattioli as well, Heckius (if it is he) cannot resist adding, "and is also really perfect for hemorrhoids."[43]

Special attention is paid, in these excerpts, to the effects of poisonous plants and fungi. Altogether typically, Mattioli's pleasant account of different ways of cooking aubergines is omitted in favor of a kind of pharmacological elaboration of its etymology. As both Cesi and Mattioli note, *melanzane*—that is, *mele insane*—are a species of mandrake, which (as their name suggests) can cause one to go seriously off color or have long fevers, and generate almost everything from melancholic humors to suppurations, cankers, leprosy, headaches, sadness, and blockages of the liver and the spleen.

Often these remarks about negative potency, poison, and venom, are made in specific domestic and agricultural contexts. The leaves and the flowers of the oleander (here still called a "rhododaphne or rhododendron") are "lethal poison for mules, dogs, asses and many other quadrupeds, but for men they are useful against the bites of snakes; sheep and goats die if they drink water in which the leaves of this plant have been infused."[44] The root of the colchicum "kills by strangulation, as in the case of certain mushrooms; cow's milk is an excellent remedy for those who happen to have eaten this root."[45] The belladonna, if copiously eaten, makes one sleep long, and kills children who eat it.[46] Aconite may sometimes be useful in that it can be "used to kill wolves";[47] the *apocino* (*Periploca graeca* L.), when ground up with fat and pasta, can be made into a kind of bread that if fed to dogs, wolves, foxes, and even panthers, can kill them.[48] In short, the interest in dangerous plants such as the deadly nightshade, the hellebore, and the monkshood is almost as notable as the interest in the pharmacological qualities of more beneficent species.

Although the medicinal component of the inscriptions is stronger than any other, one other interest also emerges in these pages: that of classi-

Fig. 8.19. Purple orchid (*Orchis pratense*). Watercolor and bodycolor over black chalk, dimensions unknown; whereabouts unknown.

fication and taxonomy. This may be judged not only from the intensity and efficiency with which the Mattiolan equivalents for the very differently depicted plants are discovered, but also from the concentration on and interest in names. Above all, however, it is signaled by the attempt to separate one type of a species from another. This occurs even when plants carry identical names, as in the case of the large range of "poppies" at the beginning of the Windsor volume (some mentioned by Mattioli, others not), the different kinds of "solanum" or "solatrum" that follow, and the many kinds of aconite, and grapevines.

The preoccupation with names, both in Latin and the vernacular, is even greater than one might have expected—certainly more so than in most sixteenth-century herbals. Sometimes the references to the relevant name in Mattioli are scratched out and changed, revealing the Lincean preoccupation with getting the names (and obviously the identification) right. Names are deleted and better or more likely ones substituted. When doubt remains, the inscription records that a plant is called one thing by one authority, another by another. And throughout the vernacular name is noted, especially when it seems to offer a clue to classification by similitude. With a well-known kind of plant such as the heliotrope, the delight in giving an expanded list indicating the variety of names applied to it is unmistakable and infectious.[49]

The pleasure in having new plants illustrated, beyond those recorded in Mattioli and elsewhere, is high. There is a clear concern with color and getting the colors right, much more so than in any other preponderantly taxonomic work of the time. But for the rest the *Erbario miniato* does not come across as especially revolutionary—except in its effort at comprehensive coverage in the form of colored drawings. Certainly there is no parallel for the range of illustrations here. By later standards (and in comparison with the botanical drawings by artists such as Joris Hoefnagel and Jacopo Ligozzi), these drawings may be comparatively rudimentary, but the faith in pictures signaled by the *Erbario* announces one of the key aspects of the Linceans' approach to the recording and analysis of nature.

## THE DRAWINGS IN PARIS

In the early 1980s, news began to surface about another group of Lincean drawings. They were in the Institut de France in Paris, and they turned out to be very different and indisputably more important than those of the *Erbario miniato*. They too were said to have belonged to Cassiano dal Pozzo, and before that to Federico Cesi.[50]

There were no fewer than eight volumes of drawings. Three had *Fungorum Genera et Species* inscribed on their spines;[51] five had *Plantae et Flores* written there.[52] Each contained several pages of an alphabetized index written in the same seventeenth-century hand—probably Cassiano's scribe—as the index of the *Erbario miniato*.[53] Beyond these indices come almost fourteen hundred pages, often with many drawings per page.

Fig. 8.20. Unidentified specimen. Paris, Bibliothèque de l'Institut de France, MS 978, fol. 176. ©Photo RMN—Gérard Blot.

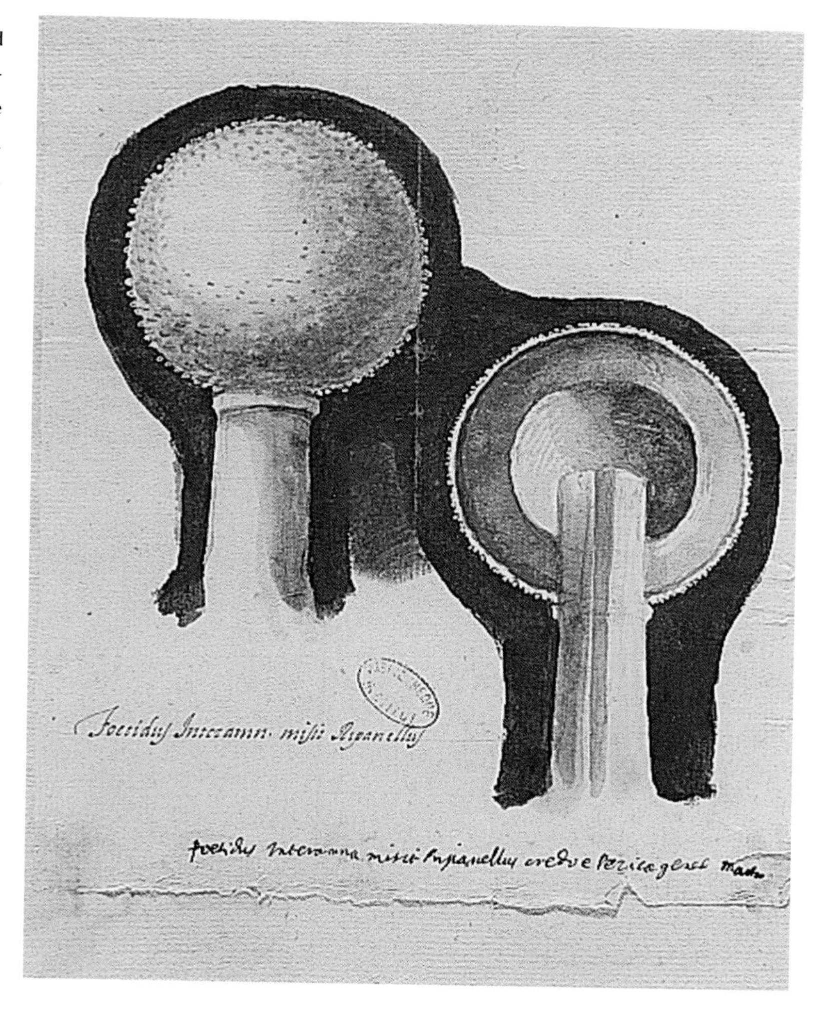

Most are colored, of a quality and range far beyond anything in the *Erbario*. The fungi, lichens, mosses, and local and exotic plants in the Paris volumes are mostly represented in situ, from a variety of angles, at different stages of their life cycles, often sectioned, often anatomized, and frequently, as clearly stated in the inscriptions, seen with the aid of a microscope or lens (e.g., figs. 8.21, 8.25, 8.28, 8.35). Sometimes these details are shown highlighted against a striking dark blue background, as if they were special parts to be observed and studied (e.g., figs. 8.21, 8.27, 8.32–33, and the still wholly unidentifiable fig. 8.20).

In comparison with the manuscripts in Montpellier and Windsor, the Paris drawings reveal an art that is essentially one of depth, however meticulously attentive to surface it may be. Instead of always focusing on exteriors, it concentrates on interiors. It is as if the artist of the Paris codices were constantly instructed to look deeper, and to penetrate beneath the surfaces of things.

The range of graphic techniques used to emphasize details of these

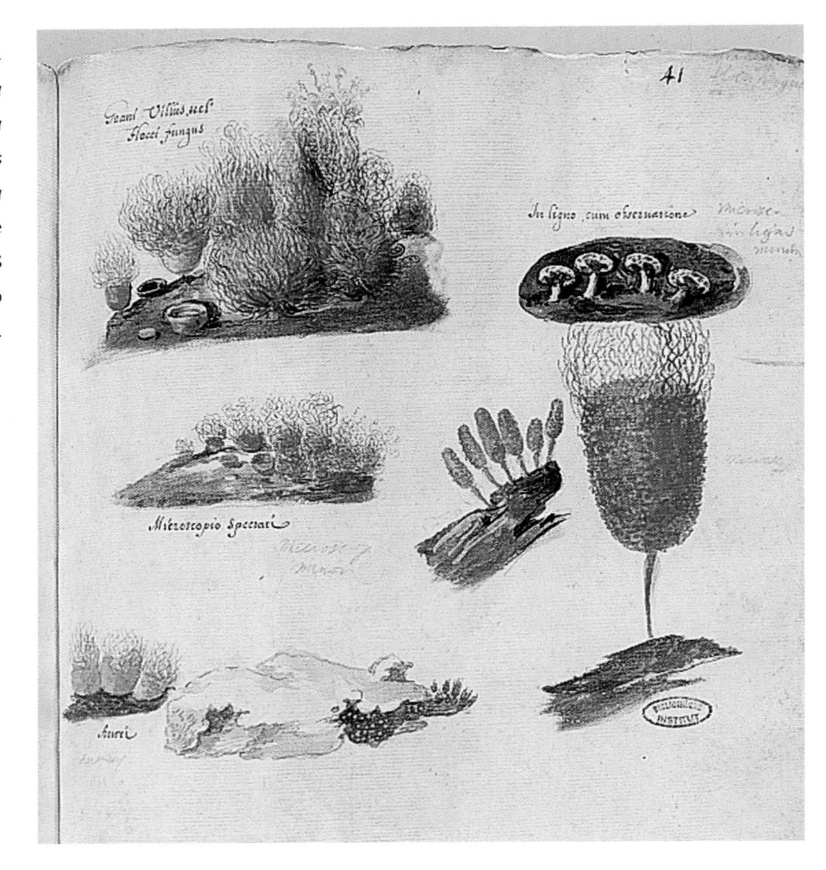

Fig. 8.21. Slime molds. (*Myxomycetes*: *Arcyria* species, *Arcyria denudata* Wettst., *Physarum nutans* Pers., and *Trichia favoginea* Pers.). Paris, Bibliothèque de l'Institut de France, MS 968, fol. 41. © Photo RMN—Gérard Blot.

specimens vastly exceeds in number and sophistication anything else in the corpus. While the volumes titled *Fungorum Genera et Species* are indeed devoted just to fungi, lichens, mosses, and the like—every imaginable form of fungal growth it sometimes seems—those inscribed *Plantae et Flores* contain not only local and exotic plants and flowers, but also large numbers of ferns, polypods, and several extraordinary drawings, often just in pen and ink, of insects, animals, and a few other curiosities as well. There are some beautiful watercolors of individual plants and flowers—sunflowers, daffodils, lilies, a few oranges too—but these are not the most significant ones, by any means.

Each of the volumes in Paris carries a brief inscription recording that it had belonged to "Orteil Directeur des Postes Militaires a Rome, 25 Novembre 1798." They have the distinctive stamp of the Albani family on their spines. In other words, these volumes were also among those confiscated from the remnants of the great Albani library by the French occupying troops in Rome in 1798.[54] Orteil seems to have kept these volumes for himself, before they were sold to Benjamin Delessert (1773–1847), the famous financier and amateur botanist, the first reader of Rousseau's *Letters on Botany* (commissioned by Delessert's mother) and a great collector of botanical books, whose library was bequeathed to the

Institut by his son in 1874.[55] The catalog of the Institut notes simply that both sets of these manuscripts were "executed at Rome in the first third of the seventeenth century."[56]

But it is certainly possible to go further than this. Since a number of sheets give the exact location where the specimens were found, as well as the actual date on which they were drawn, Delessert himself concluded on the flysheet of the first of these volumes (MS 968) that they were done between 1623 and 1628, and that "they were valuable because of the localities indicated for many of the plants." Indeed. "The work is composed," he continued,

> of a large number of illustrations, sometimes mediocrely rendered, but all by a practised hand. The colored drawings are the most remarkable. The large volume of the *Plantae et Flores*, which is probably the earliest (1623), contains drawings by different hands, mostly rather poor, though several are interesting . . . ; it also contains several old engravings of plants.[57]

Although these remarks barely give more than the slightest indication of the scientific interest of all these volumes, they offer a reasonable (though somewhat cool) assessment. Beneath Delessert's inscription is another, more ignorant one: "There is nothing," it reads, "to indicate the author of this collection. It is likely to be the work of some Roman monk, and it comes from the sale of some monastic library."[58]

Some misnomer for the great Albani collection, into which these volumes had passed in 1703, like all the Cesi volumes in Cassiano's museum on paper![59] And while we may still not know who actually did the drawings—he is, of course, much more likely to have been a professional rhizotome or artist than "some Roman monk"—there is in fact a definitive set of clues to their origins. In the top right hand corner of hundreds of these pages (and occasionally elsewhere) are annotations both in pen and ink, by a hand that is instantly recognizable—that of Federico Cesi.[60] His precise, slightly spidery handwriting is unmistakable. These inscriptions, generally quite brief, are mostly transcribed into a more flowing script by another now familiar hand—that of Cassiano's (or possibly Cesi's) favorite scribe. And from them we learn not only when many of these drawings were made, but where.

The earliest specific date is December 23, 1623;[61] the latest October 15, 1628.[62] When specific days or months are not given, the relevant season is noted. "Found at Acquasparta in autumn" says one; "An autumnal *setivillus carnosus* [a fairy-club mushroom] found in the woods at Acquasparta" says another; and so on and so forth.[63] The localities recorded are either in Rome and the Sabine countryside around it, or Acquasparta and its Umbrian environs. Sometimes the specimens are from Cesi's own gardens, sometimes from those of his friends. This attentiveness to the pre-

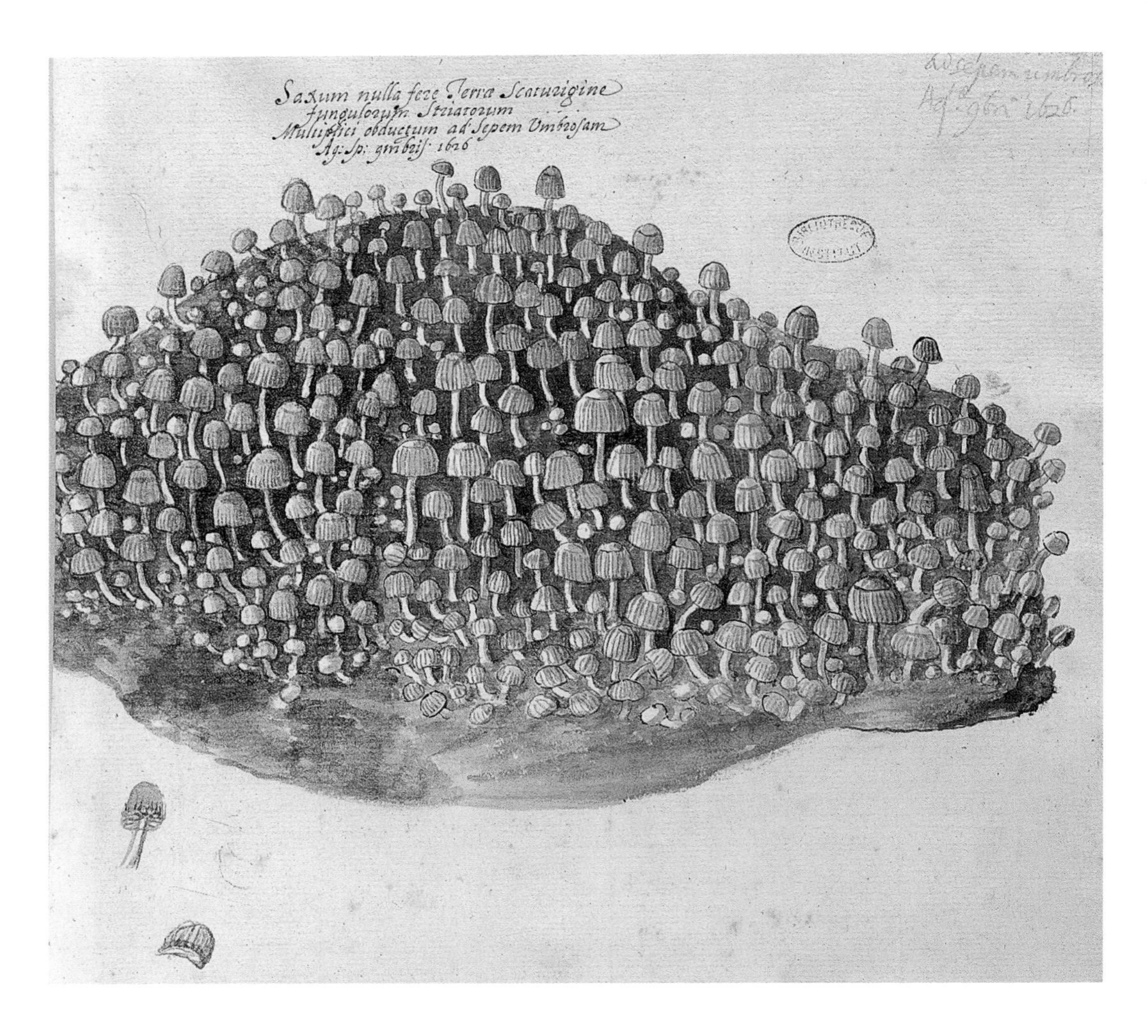
Saxum nulla fere Terra Scaturigine
Fungulorum Striatorum
Multiplici obductum ad Sepem Umbrosam
Aq: Sp: 9mbris 1626

Fig. 8.22. Mature trooping crumble caps (*Coprinus disseminatus* [Pers.: Fr.] Gray), found at Acquasparta in November 1626. Paris, Bibliothèque de l'Institut de France, MS 970, fol. 54. © Photo RMN—Bulloz.

cise recording of habitat and season is in itself remarkable; but of even greater interest is Cesi's use of the microscope.

Time and again he records that specific parts of the plants illustrated in these volumes were observed with the aid of a microscope or a lens: "lente observatus," "ex microscopio," "microscopio observatus," and "microscopio spectatus" are frequently inscribed on these pages (e.g., figs. 8.21, 8.25, 8.28, 8.35); sometimes just an "m." occurs to indicate the same. In these manuscripts observation often becomes synonymous with observation with the microscope: thus, when Cesi writes "microscopio" in his notes on the edge of the page, the scribe, in copying, frequently substitutes "observatio" or "microscopio observatus."

The order of magnification is generally within the range of twenty to one hundred times. Some of the enlargements may have been achieved with a handheld lens, but Cesi surely used one of the primitive compound microscopes so recently given to him by Galileo.[64] The enlargements of details of fungi thus precede by some forty years the much better-known ones in Robert Hooke's *Micrographia* of 1665.

The first three volumes of the Paris series (numbered 968–970) are devoted almost entirely to fungi, slime-molds, and some mosses and liverworts. The first, MS 968, contains the best of the drawings and was assembled with the greatest care. Other than the drawings that Clusius had made in Pannonia in preparation for his own book on larger fungi of 1601,[65] there is no antecedent for such a concentration on field drawings of fungi, or as clear a distinction between the plant and fungal kingdoms. But the drawings Clusius commissioned were nothing like as nuanced or as attentive to detail, and there were far fewer of them. In comparison with Clusius's drawings, moreover, Cesi's are much clearer in their definition of the parts of each fungus: they even depict, for example, the lamellalike attachment on the underside of the pileus in the agaric drawings, "a character now recognized as being of the greatest taxonomic importance," as one distinguished modern mycologist has noted.[66] Many of the drawings, such as the marvelous page of trooping crumble caps (fig. 8.22), are by far the earliest illustrations of the fungus shown. And then there are the many remarkable drawings of fungal fruit bodies, generally accompanied by annotations referring to the use of the microscope. The plain fact is that it is with the researches illustrated by the volumes in the Institut de France (themselves copied in a second set of drawings now preserved in the Library of the Royal Botanical Gardens at Kew),[67] that modern mycology begins—not with Clusius's observations on Hungarian fungi, or with much later treatises such as Lancisi's and Marsigli's in the eighteenth century, or Persoon's and Fries's in the nineteenth.[68]

On the face of it, the *Plantae et Flores* manuscripts (974–978) are much less focused (at least in terms of subject matter). MS 974, for example, contains many local plants, while MS 975 concentrates on plants with traditional pharmacological functions, such as belladonna, valerian, and cannabis. But the finest of this group—and certainly the most significant for the history and science of botany—is MS 976. It contains a quantity of exotic plants (though most of them seem to have been grown in Rome and its immediate environs).[69] They range from aloes, yuccas, and prickly pears to ornithogala, gladioli, and jasmines; from an extraordinarily bold series of orchids to the ferns and mosses that were so important to Cesi himself. These are almost all shown with close attention to their seed-bearing (or spore-bearing) parts; and many of these parts, in turn, are illustrated under magnification by the microscope (cf. figs. 8.24–25, 8.28–29, 8.35).

MS 977 is a strangely beautiful volume. It is a veritable treatise on algae, certainly the first of its kind. But it also contains a variety of other forms of marine life, from mollusks to seaweeds, sea grass, and other mysterious underwater phenomena. Often the artist depicts rocks, stones, and pieces of flotsam covered with just the kind of apparently inexplicable fungal and mosslike growths in which Cesi, ever searching for the principles of biological reproduction, was especially interested. There

Fig. 8.23. Marine algae. All almost impossible to identify, though the red algae in the center may be *Gracilaria* and the brown on the left below perhaps *Cystoseira*. Paris, Bibliothèque de l'Institut de France, MS 977, fol. 106. © Photo RMN—Gérard Blot.

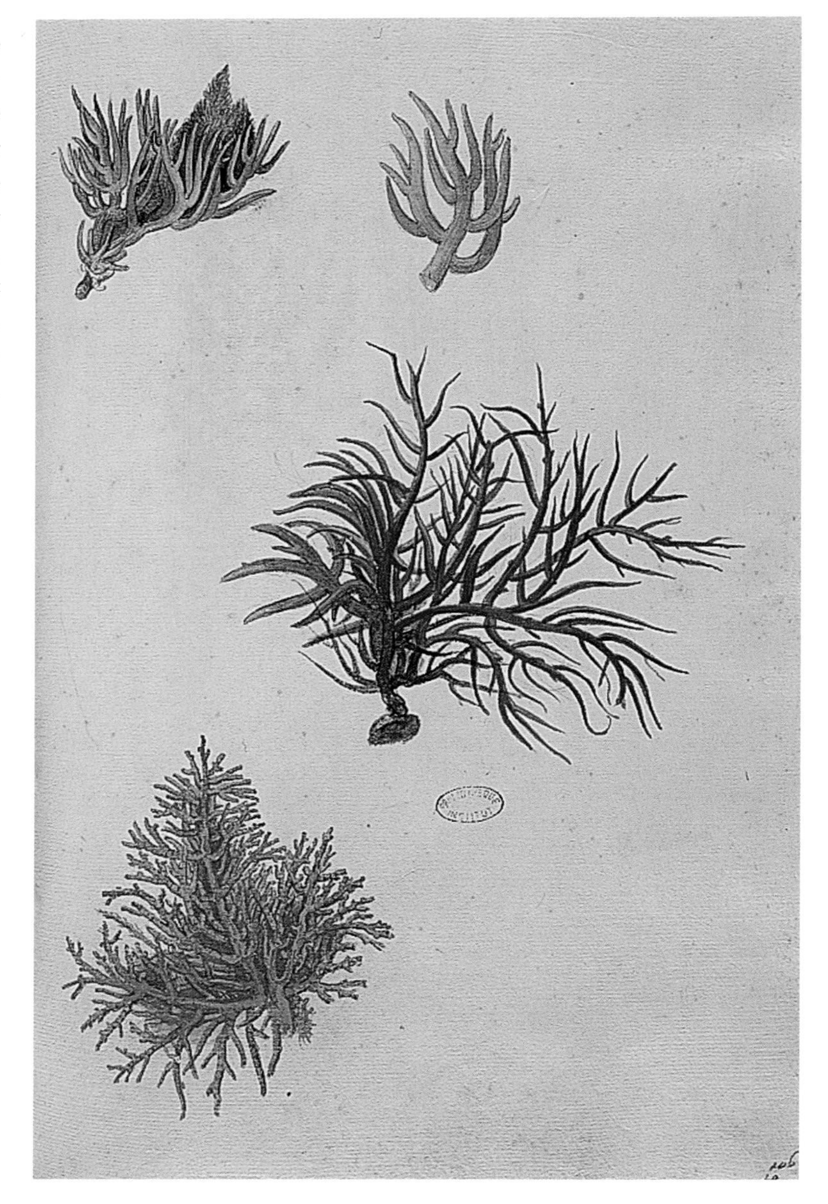

are crabs almost totally hidden by the algae that sprout from them, and an amphora, clearly retrieved from under the sea, covered with mollusks. Goose barnacles sprout from a piece of flotsam, and a sea hare is shown both whole and in dissection.[70] The colors used in this volume are sometimes of a brilliance and an unearthly splendor scarcely matched elsewhere in the corpus (e.g., fig. 8.23).

The last of the volumes, MS 978, is a miscellany, perhaps the oddest of all the volumes left behind by Cesi. Here too are a large number of algae, but it also contains many hybrid and hermaphroditic forms, often displaying a strangely old-fashioned penchant for anthropomorphization

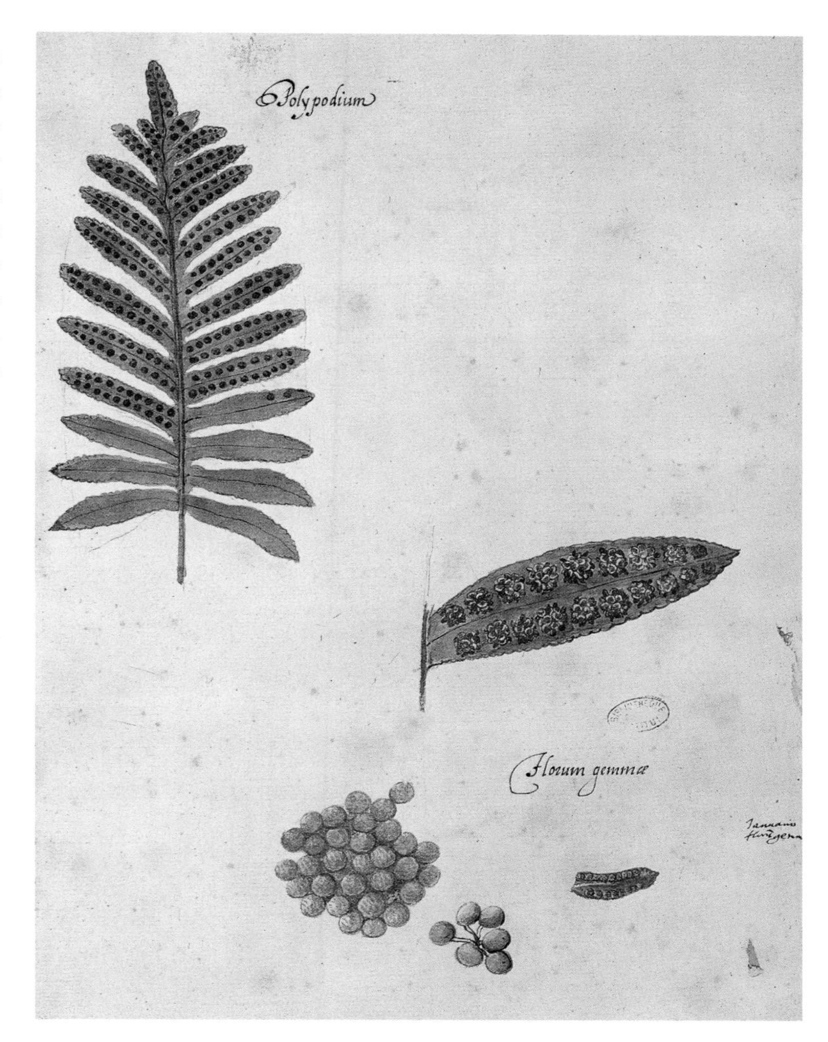


Fig. 8.24. Common polypodium (*Polypodium vulgare* L.). Along with figure 8.25, the first microscopic depiction of the sori and sporangia of a pteridophyte. Paris, Bibliothèque de l'Institut de France, MS 976, fol. 63. © Photo RMN—Bulloz.

(such as is to be found elsewhere in the corpus, but nowhere as frequently as here). It includes material from the earliest days of the Lincean researches as well as from their latest. Alongside some of the drawings, especially the early ones, are pasted some of their engraver's earliest attempts at printmaking (e.g., figs. 8.36–37).

Once they had their microscope, Cesi and his friends—especially Faber, Stelluti, Colonna, and the young German doctor, Winther—could not resist using it. They did this not just out of random curiosity, or because they were fascinated by the miraculous enlargements of things, or because they were amazed at the order to be discerned in the details. They used the microscope for the most specific and systematic of reasons.

For Cesi, fungi, ferns, and algae—and mosses, lichens, and all the tuberous things that grew underground, like truffles—posed a central problem. How did they and all the other cryptogams reproduce? What if

Fig. 8.25. Sori and sporangia on the underside of various ferns, seen under a microscope. Paris, Bibliothèque de l'Institut de France, MS 976, fol. 65. © Photo RMN—Bulloz.

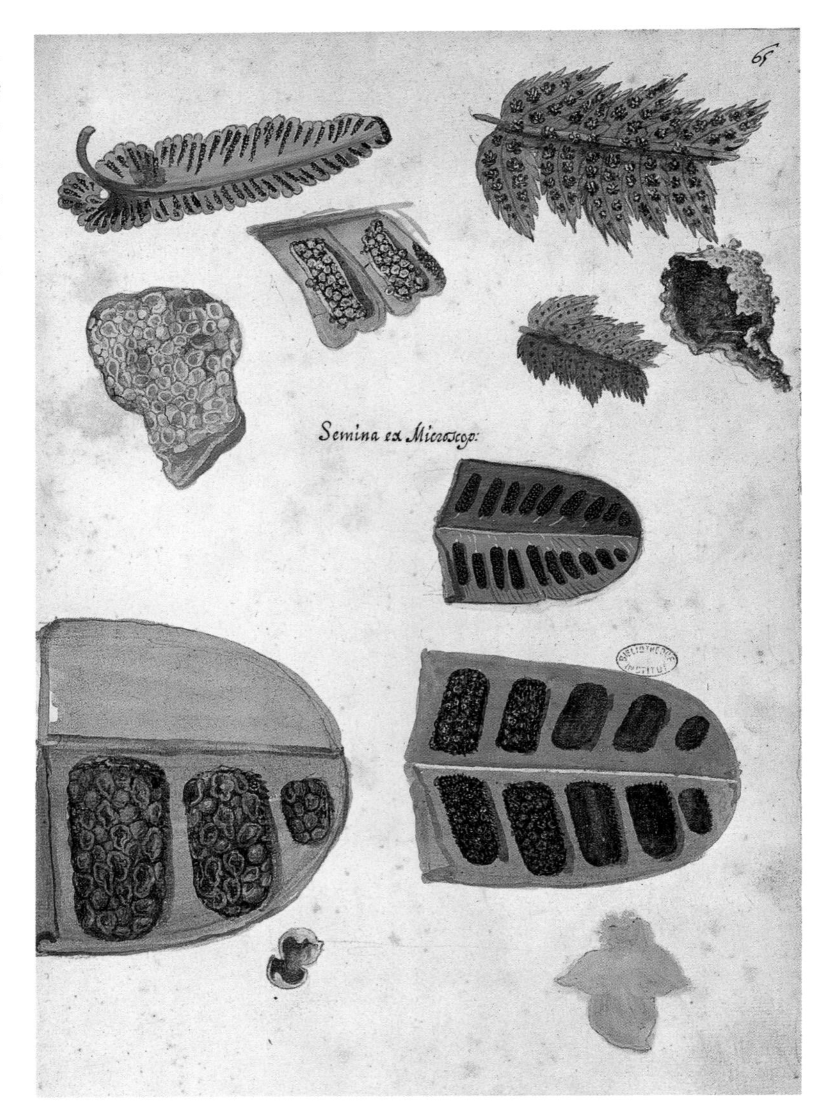


one could not actually find their seeds, or anything that even looked like seeds—to say nothing of the seed-bearing organs?

"The seeds [sic] of the ferns were discovered by Bobart," wrote Linnaeus in his *Philosophia Botanica* of 1751, "and those of the mosses by me."[71] In each case Linnaeus was wrong. Cesi had already discovered the "seeds" of both. Not just the history of mycology but also that of a number of other critical botanical issues has to be rewritten in the light of the microscopic observations in the Paris volumes.[72]

It was Cesi who first illustrated (or had illustrated) the spores of bryophytes and the sporangia of pteridophytes, as may be seen on folio after folio in Paris. With lens and microscope he observed and had illustrations made of the spores, sporangia, and sori of ferns (figs. 8.24–25; cf. fig. 8.26). Few plants interested him and his friends more than the polypod,

Fig. 8.26. Common polypodium (*Polypodium vulgare* L.). *Erbario miniato*, fol. 143. Windsor, RL 27833. The Royal Collecton © 2000, Her Majesty Queen Elizabeth II.

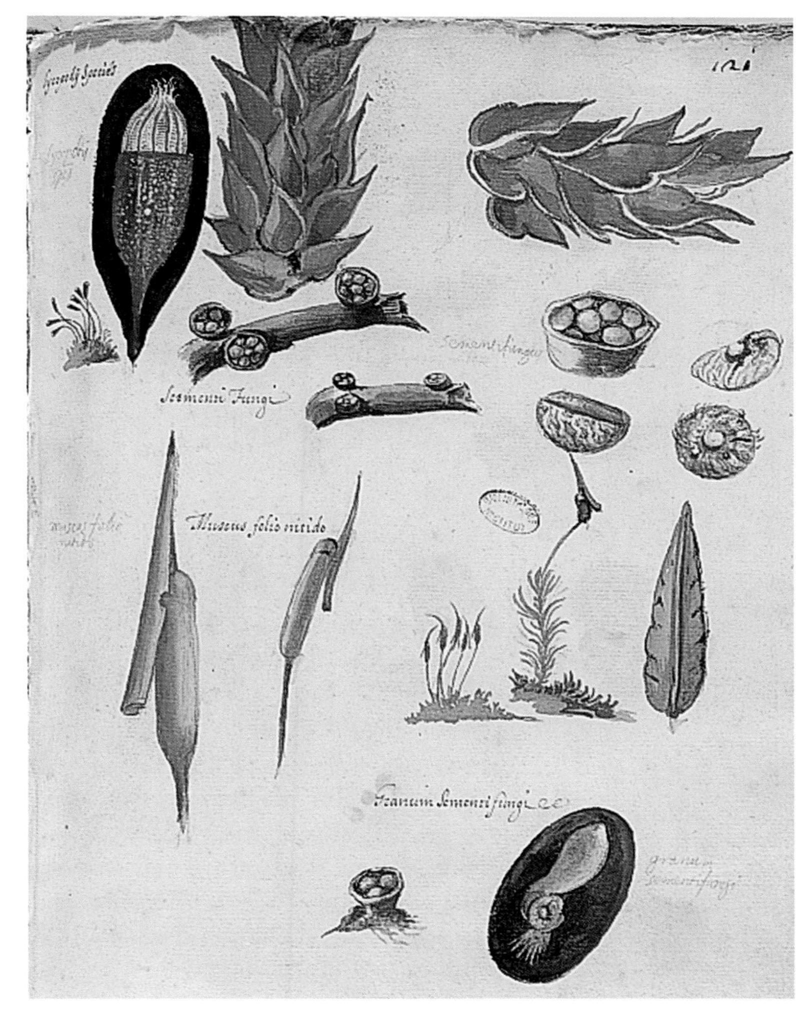

Fig. 8.27. Bird's-nest fungi and mosses (*Nidulariales* and *Bryophytes*: *Crucibulum laeve* [Huds. ex. Rhelan] Kambly; *Cyathus olla* Batsch: Pers.). Paris, Bibliothèque de l'Institut de France, MS 968, fol. 121. © Photo RMN—Bulloz.

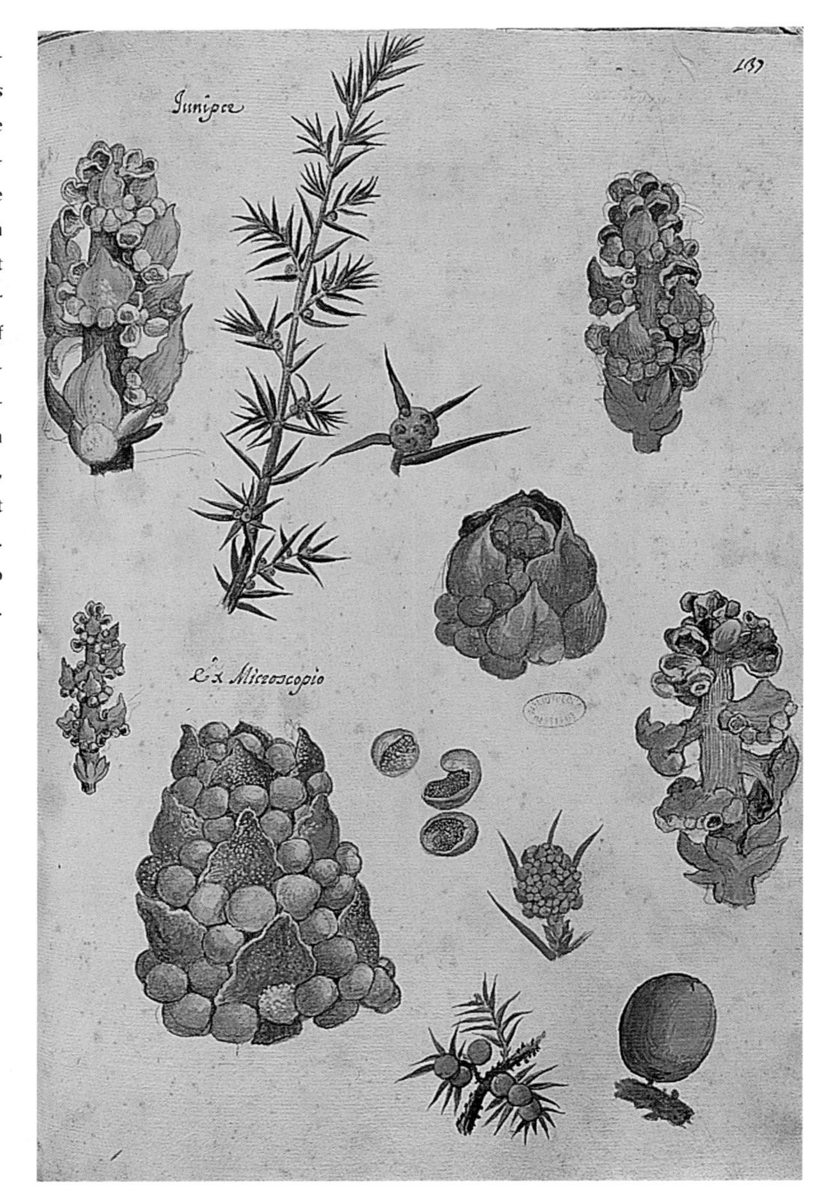


Fig. 8.28. Common juniper (*Juniperus communis* L.), fertile twig of male plant with various microscopic enlargements of the flower. Note the depiction of the individual dehiscent pollen sacs. At the lower right is a fertile branch of female plant with immature galbuli, the first microscopic representation of a gymnosperm. Paris, Bibliothèque de l'Institut de France, MS 976, fol. 137. © Photo RMN—Bulloz.

which Colonna had named the *Planta caesia* in honor of him.[73] Often the spores were enlarged as much as thirty times. The sporophytes of mosses in genera such as the *Polytrichum* and the *Dicranum* were magnified so that one could see both the setae and the sporangia (fig. 8.27). These were shown in section as well.[74] The first of the three volumes of fungi contain several enlargements of the small gastromycete *Cyatus olla* of the family of bird's-nest fungi (*Nidulariaceae*), including some remarkable details of their peridia (e.g., fig. 8.28).[75] In the course of these intense examinations, Cesi came very close to discovering the fungal spores, which in this species could never have been seen by the naked eye alone.[76]

And all this with regard to the cryptogams alone! The same interest in the visual recording of seeds and organs of reproduction runs through the illustrations of the gymnosperms and angiosperms as well. A juniper branch is broken off and its flowers carefully examined with a microscope in order to show its microsporophylls and its dehiscent pollen sacs; a female plant is shown with its immature galbuli, in the first microscopic examination ever of the flower of a gymnosperm (fig. 8.28);[77] and so on and so forth, for hundreds of pages.

## FERNS, SPORES, AND REPRODUCTION

The "Additions and Annotations" that Fabio Colonna added to the Linceans' great work on the flora and fauna of Central America offer a series of clues both to the intensity and meaning of Cesi's researches and to the provenance and significance of the volumes in Paris. One of the most important of these clues occurs in the course of Colonna's discussion of the *Planta caesia*, the polypodium that the Aztecs called *Tuzpatli*.[78]

Having begun with a description of the smell and colors of the *Planta caesia* (including bluish-gray, or *caesius*—a typically convenient double meaning),[79] Colonna turned to the more serious issue of its generation and reproduction. Did this puzzling plant really bear anything that could be described as flowers and fruit (in the way Hernández's text had)? Colonna informed his readers that Cesi had eloquently treated this problem in a work titled *De Plantis imperfectis*, or *On Imperfect Plants*. This is the first reference in the *Tesoro Messicano* to one of the most frequently mentioned of all Cesi's projects.

According to Colonna, Cesi made the important discovery that the "uterus" and the "fetus" of a plant could be found in other than the usual places, even, for example, on the undersides of leaves. "These things have not only long been observed by our illustrious Prince," he wrote,

> they have now been clearly and openly demonstrated with the aid of that little telescope which, in a great and wonderful way, much enlarges even the smallest objects and makes things so tiny that they are like atoms [Colonna's own word] appear so large to the eyes that one can only be amazed.[80]

Colonna could barely find adequate words to describe this new and fundamental increment to the natural powers of sight. But he persisted. He described how the "seeds" (as they were still called) of the polypodium and its "congeners" could now be submitted—*subjected*—to visual examination. Instead of looking like grains of sand, they looked like peas or chickpeas. With the aid of the "little telescope," wrote Colonna, Cesi was the first to examine the hirsute swelling on similar examples and to see the "seeds," as well as the umbilicated part above them that stuck to the fronds and provided them with food and growth. Without the "telescope" these "seeds" would have seemed little more than glistening black sand. One could now see so much and so distinctly! The hirsute yellow swelling on the back of the leaf, for example, was composed of clusters arranged in two rows with about fifty flowers or fruits with seeds in each. To Colonna it seemed that these plants were not really to be thought of as "imperfect," as Cesi described them, since they possessed both "flowers" and "fruits." In fact, since they were "dorsiparous"—giving birth from the back (of the fronds)—they were actually perfect.[81] How great the dangers of analogy! Cesi was on the better track.

All these references may clearly be applied to the volumes in the Institut de France, with their hundreds of drawings of ferns, polypodia, and the microscopic observations of the sori on the undersides of their leaves. They certainly cannot refer to the meager drawing of a polypodium in the *Erbario miniato* in Windsor (fig. 8.26).[82] But there are several other illuminating references in the *Tesoro Messicano* to Cesi's work on "imperfect plants" that secure the connection with the microscopic observations of the Paris manuscripts.

These references occur in the course of the much longer—and more entertaining and vastly more digressive—commentary on Hernández's text by Faber. After an overview of some of the problems relating to the generation of truffles, tubers, and fungoid stones (all to be found in abundance in these manuscripts), Faber suddenly stopped himself:

> I am not going to say anything more about fungi: Clusius has described them with sufficient precision in his book on Rare and Exotic Plants.[83] Yet we are expecting a much more exact treatment from Prince Cesi, not only historically but also theoretically and physiologically, in his books *On Imperfect Plants*, which will shortly see the light of day.[84]

They never actually did. It is worth remembering that the first edition of the *Tesoro* appeared in 1628, and Cesi died only two years later. But at least we now have the manuscripts that document the achievement and the industry of Cesi's indefatigable work in this domain, as he continued to search incessantly for a solution to the problem of reproduction.

Two hundred and twenty pages later Faber returned to the instrument that Colonna still called a "little telescope" but that Faber himself had recently named "microscope," as he proudly informed his readers. He declared that it could magnify up to thirty thousand times, and that

> with this as an aid to the eyes, our Prince Cesi saw to it that many plants, previously believed by botanists to be lacking in seeds, were drawn on paper by the painter whom he had designated for this work, in such a way that they were shown swelling clearly with the most distinct seeds. In the polypodium you could see those extremely minute particles sticking to the backs of leaves, looking about the size of a grain of peppers, which until now were believed given by Nature simply as an ornament to these plants. . . . and to

these, even before the use of the microscope, he rightly gave the name of *plantae tergifoetae*.

Time and again the Paris volumes offer examples of microscopic observations of just this kind. From them one gains a sense of what might have been, had Cesi lived to set out his views on the reproduction of cryptogams. To read Faber's next sentence is to understand how cruelly and how soon the Linceans' hopes would be dashed, how optimistic and somehow presumptuous they were: "But soon we shall have innumerable other observations similar to these, as well as the new natures (one could almost say) detected by Cesi, when his books and considerations on imperfect plants are published." How could Faber then have known that within a year or two this promise would be so abruptly cut short?

At least he could still boast of the one microscopic observation that was in fact published. "With the same microscope," he recalled, "Francesco Stelluti showed us the marvelous anatomy of all the external parts of that tiny animal, the bee—eyes, tongue, proboscis, antennae, hair, sting, feet and digits, and he has recently commissioned an engraving of these."[85] The passages on the microscope in the *Tesoro Messicano* were thus ready by 1625, when the *Melissographia* finally appeared.

As if to summarize the whole point of this section of the *Tesoro*, a marginal note reads: "Prince Cesi demonstrates that plants not believed to have seeds actually abound in them."[86] But neither the woodcut of the *Tuzpatli* on page 147, nor the drawing in the *Erbario Miniato* in Windsor (fig. 8.26),[87] show anything at all on the underside of the fronds of the polypodium. The drawings in Paris, on the other hand, most certainly do—and abundantly so. They are the precious record of the intensity of Cesi's researches in this area. No wonder, then, that he should have added manuscript references to the "Polypodium tuberosum" and to the "Polipodii species" alongside the illustration of the *Tuzpatli* in his personal copy of the earliest printing of the *Tesoro*,[88] as well as making several adjustments to the proof plate, surviving in the Vatican Library,[89] for the very plant named after him.

In the end, what unites all the researches witnessed by the Paris volumes is the concern with problems of biological reproduction. On the gross level, mushrooms are shown growing off stones or in dung; lichens cover every imaginable piece of rock or bark; mosses grow in dark crannies; phallic growths are endlessly recorded; algae, grasses, seaweeds, and every kind of rampant growth are shown in as much detail as possible. On the microscopic level, the concentration on the seeds and the seed-bearing (or spore-bearing) parts of plants borders on the obsessive. Over and over again Cesi examines the organs of reproduction and lays bare their structures, as best he (or his artist) can. Ceaselessly he inquires into the principles of reproduction and growth. Where do the generative powers of plants come from? How is one to explain not just the fertility but also the multiplicity of things? How is one to account for the generation of plants that not only seem to produce no seeds but also to have no obvious source of nourishment? Nothing could illustrate such questions more dramatically than the many drawings, both in the manuscripts in the Institut de France and elsewhere in the Lincean corpus, of fungi that seem to grow directly out of rocks and stones (figs. 8.36–37; cf. also fig. 8.22). The telescope had helped Galileo discover not just how new stars were born, but also stars that could otherwise not be seen; now the microscope was helping Cesi and his friends discover things hitherto unseen on earth and resolve the problems of how they were generated.

## HABITAT AND ABUNDANCE

It is in precisely this context that one can begin to disentangle what is new and what is old in the Paris codices. Of course, there had been a few herbals with more or less rudimentary indications of the habitat of plants—a good example is Hieronymus Bock's *Kreutter-Buch* of 1551—but there was never anything like the concentration on immediate habitat to be found in Cesi's manuscripts. When Clusius had his draftsman record the larger fungi of Hungary (in the codices that served as the basis for his treatiselet on the subject),[90] there was only the slightest visual indication of where particular specimens grew. But Cesi had his draftsman show specimen after specimen in situ, in exactly the mound of earth, the stone, the twig on which his fungi grow. And of course there was a reason. He wanted to plumb—was *compelled* to plumb—the mystery of what it was about such sites, often puzzling ones, that allowed such specimens to grow and reproduce. Never before had there been such faith in visual reproduction.

Cesi also investigated why some things failed to grow, and why others grew in such fine abundance. How was one to account for the ways in which fungi, mosses, and lichens so easily proliferated, in such seemingly uncontrollable and inexplicably rampant a fashion? Abundance and multiplicity are the hallmarks of the Paris codices. Its folios, like the fields themselves, are covered with every conceivable form of growth. Fungi pullulate, on the page and in the ground, on branches, leaves, stones, mollusks, and in every conceivable cranny and excrescence of nature. And Cesi *counts*, a perfect exemplar of what Foucault regarded as a new epistemological order; and when he does not do so explicitly, he gives himself and his readers the opportunity, through clear visual reproduction, to do so.[91] This concentration on counting, and on number, especially with regard to the elements of biological reproduction, is surely modern—certainly in the realm of mycology, and probably the rest of botany too.

What Cesi was trying to do was to subject abundance to measure, to see if multiplicity, variety, and spontaneity were somehow beholden to rule and to order. On the one hand, there was proliferation and abundance; on the other, seeming sterility. Was there any way for him to ex-

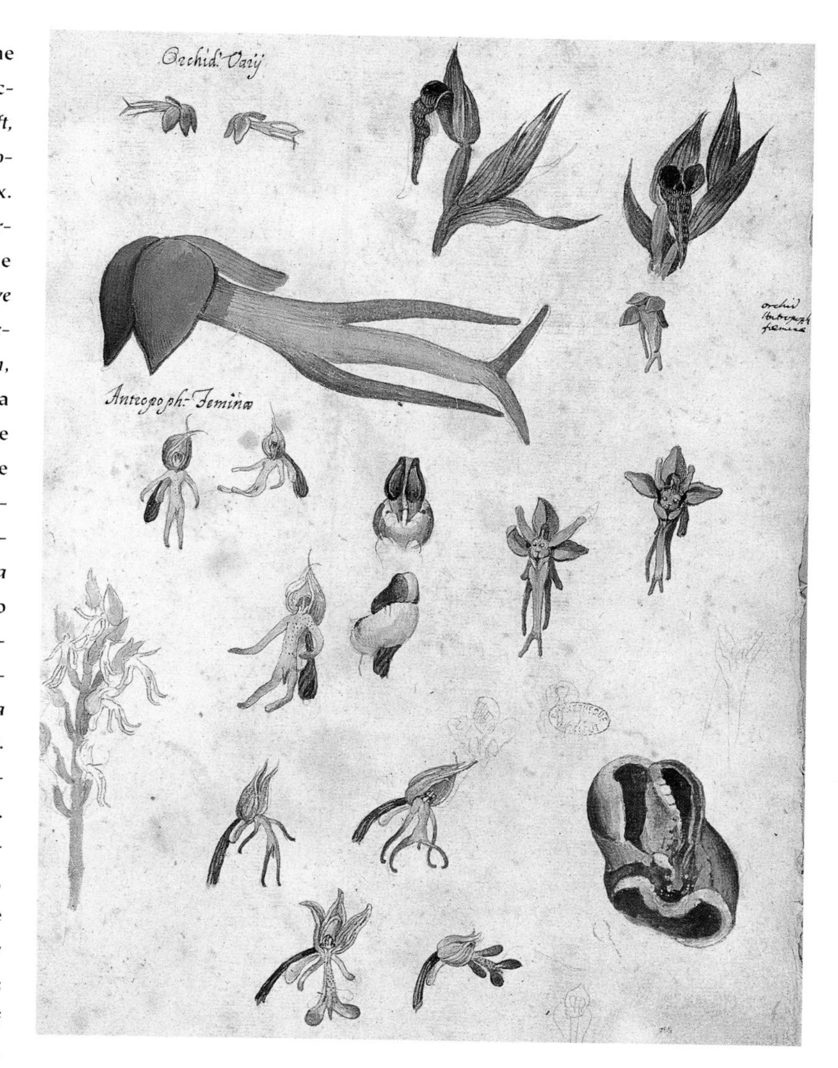


Fig. 8.29. Details of the flowers and floral structures of orchids. *Above left,* flowers of *Aceras anthropophorum* (L.) R. Br. Ex. Ait. Fil; *above center, Serapias* (cfr. *lingua* L.); the same *above right*). *Above center,* large flower of *Aceras antrhopophorum,* which on the right has a small flower of the same species. Beneath the large flower inscribed "*Anthropof. Femina*" are two flowers of the *Orchis italica* Poir. in Lam., then two pollinodia of an indeterminate species. *Right center,* two flowers of *Listera cordata* (L.) R. Br. Ex Ait. Fil. *Left below,* an inflorescence of *Orchis italica* Poir. in Lam.; beside it, two details of the same. *Below,* two more flowers of the *Orchis italica,* and below that two of *Orchis militaris* L. *Right below,* a piece of the gynostemium of an indeterminate species. Paris, Bibliothèque de l'Institut de France, MS 976, fol. 47. © Photo RMN—Bulloz.

plain how living things could grow out of dead stones, old pieces of wood, dried up and broken branches? He would never be satisfied with an explanation in terms of spontaneous generation. No; he needed to discover where fungi came from, and what nourished their growth. The last issue was the simplest and could be illustrated easily enough; the first was more complicated and took him back to the problem that underlay all his studies and researches.

Once more Cesi's approach was terribly divided. The modern would eventually win out over the traditional; but there are very old-fashioned, even reactionary elements in these pages—in particular the signs of the old drive to anthropomorphism and similitude. Both with resignation and surprise we come across the all too predictable mandrake roots; the homunculi in flowers, leaves, and roots; and the various forms that seem to be *forced* in the direction of resemblance to an animal.[92]

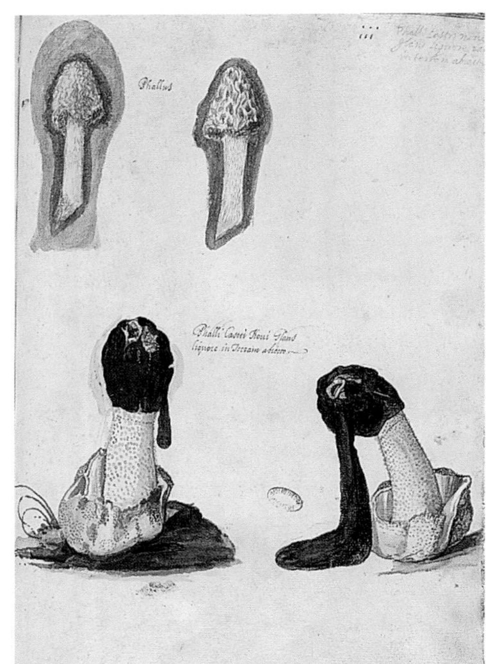


Fig. 8.30. Dune stinkhorn (*Phallus hadriani* Ventenat: Pers.). Paris, Bibliothèque de l'Institut de France, MS 968, fol. 110.

Fig. 8.31. Common stinkhorn (*Phallus impudicus* L.: Fr.). Paris, Bibliothèque de l'Institut de France, MS 968, fol. 111.

Fig. 8.32. Common stinkhorn (*Phallus impudicus* L.: Fr.). Paris, Bibliothèque de l'Institut de France, MS 968, fol. 114. All © Photo RMN—Gérard Blot.

Fig. 8.33. *Left,* Two unidentified flowers ("Christoforiane flos" and "Sandalida Plinii"); *right, Arisarum proboscideum* (L.) Savi (Araceae; *below right,* perhaps a withered male inflorescence of a walnut (*Juglans regia*). Paris, Bibliothèque de l'Institut de France, MS 974, fol. 26. © Photo RMN—Bulloz.

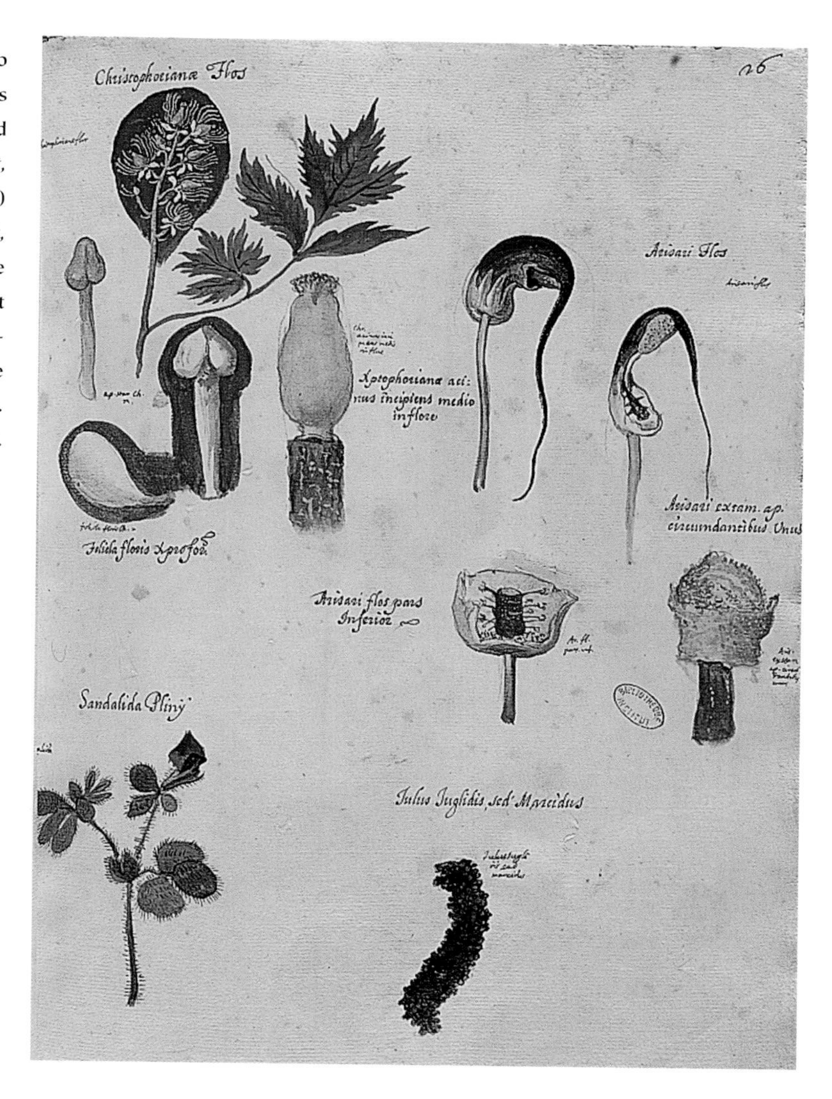


## SIMILITUDE: ANTHROPOMORPHIC AND GENITAL

Among these, the most striking are perhaps the "man-eating orchids" (with its flowers made to look bizarrely like naked little men and women),[93] and the many strange homunculi in the great manuscript of exotic plants, MS 976 (fig. 8.29). Then too there are phenomena such as the oval agate in MS 977, which seems like a complete throwback to an older tradition. It could easily have come from an old-fashioned cabinet of curiosity, with its inscription telling us that it shows a bee, a bull, and the portrait of Pope Urban VIII.[94] We may not recognize Urban VIII, but the bee is clear enough, and so too are the many charming little heads that the artist has managed to discern—and depict—within the agate. Typical enough, especially in precious stones; but in the context of these manuscripts, disappointing. Emblem and panegyric here seem to take over once again from anything we might like to call science.

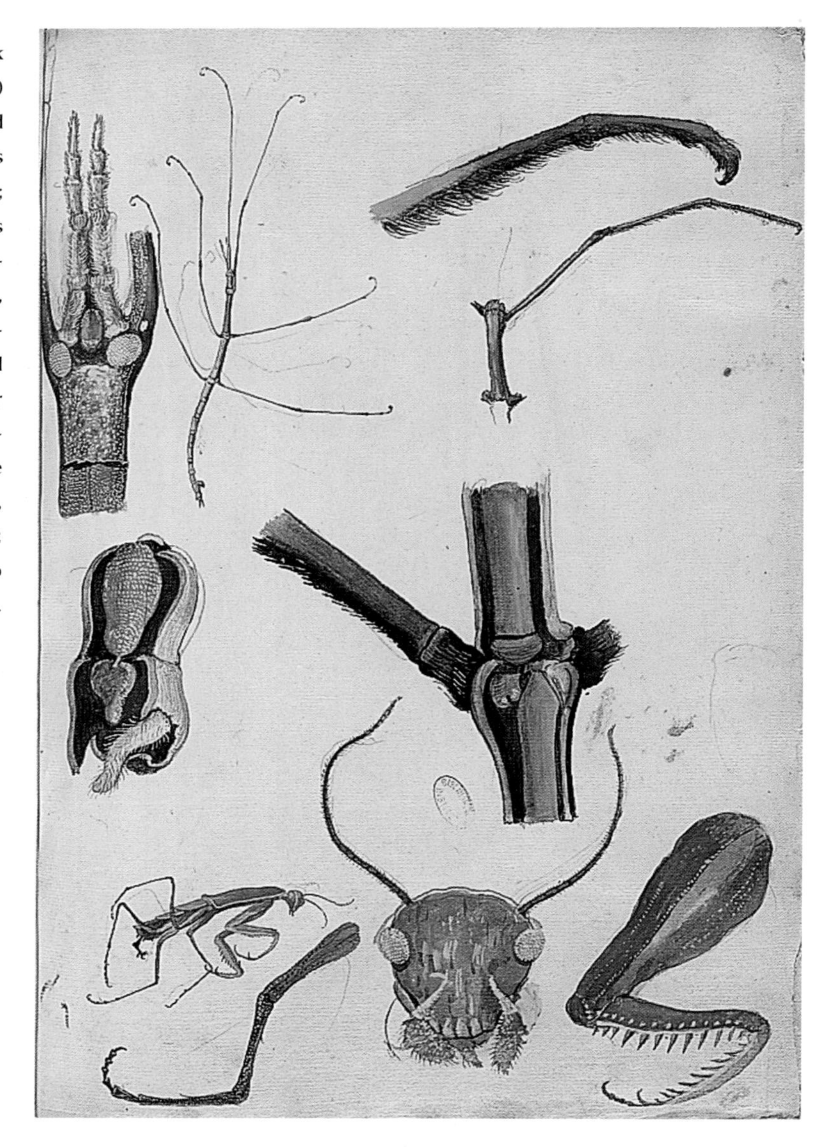

Fig. 8.34. *Above,* stick insect (fam. *Phasmatidae*) with details of head and legs; *center,* further details of parts of a stick insect; *below,* praying mantis (fam. *Mantidae*) with details of head and of legs, including, *right below,* details of tarsus, femur, and elongate coxa (trocaster not shown). Paris, Bibliothèque de l'Institut de France, MS 974, fol. between 112 and 113. © Photo RMN—Gérard Blot.

None of this anthropomorphism, however—though mild in comparison with much else at the time—can match the drive to genital similitude in the Paris codices. Given the overall reproductive concerns of these volumes, it comes as no surprise to find that the sexual organs of plants and a strong parallelism with human genitalia should be one of their chief themes. The *Phallus impudicus,* the stinkhorn, had long been a staple of mushroom description,[95] but there is little in the earlier illustrations to equal the sheer phallicity of drawings such as figs. 8.30–32, with their erect, drooping, ejaculating, and dripping phalli.[96] The fruit bodies burst open to reveal a fast-growing columnar receptacle, which in turn spill their spore-bearing slime onto the earth (fig. 8.31). These are only a few of the more blatant instances, but there are many others where there seems to be a kind of unrepentant *jouissance* in drawing the phallic or testicular parallel, or making specimens seem to be as genital as they possibly could (e.g., figs. 8.33, 8.35).[97] If it took the human genital parallel to make the anatomical-reproductive point, then so be it.

Over and over again the morphology of plants—and to a lesser extent their terminology—is sexualized in human terms. Counterbalancing all this is the careful—one could almost say objective and clinical—examination, with the aid of the microscope, of the organs of reproduction and of the seeds and whatever contains them. Cesi repeatedly uses his new aid to the eyes to examine what he calls the *vascula* and the *vagina seminalis*. He investigates and illustrates the surfaces of flowers and florets as well as their pistils and stamens; and then he goes deeper, in order to examine seedpods (often hirsute), nuts, and finally pollen, seeds, and anything that could be eggs or spores. Indeed, the concentration on eggs in almost all the volumes of *Plantae et Flores* is one of their most striking features. But it emerges with startling clarity in a number of the beautiful drawings in MS 974, probably the most egg-obsessed of all these manuscripts. Though in many respects the most rudimentary manuscript in the group, it contains several meticulous pen-and-ink drawings of animals and insects (for how do such things reproduce?) and even the eggs of the latter (fig. 8.34).[98]

Most of these manuscripts display a clear bifurcation between similitude and identity (though perhaps it is more of a transition than a bifurcation). The interest in reproduction continues unabated in MSS 975 and 976. But in MS 976, which contains many imported plants such as yuccas and aloes, the drive to genital similitude is again extraordinary. Despite the more generalized anthropomorphism of many of its illustrations, what is especially notable is the abundance not just of phallic and testicular shapes, but of vaginalike and uterine forms as well. There is also a series of illustrations that belongs to an utterly different visual regime altogether—and one that is certainly prophetic. This is best exemplified by the illustration (and subsequent ones) of the *orchidi testiculati* (fig. 8.35), previously drawn in the Windsor *Erbario miniato* (cf. fig. 8.19). But where the illustration of the *testicoli di cane* (or "orchis pratense") in the latter manuscript contained an inscription that gave an account both of its genital similitude (clear enough from the drawing) and of its medicinal function, here there is no inscription other than the precisely descriptive *vascul: cum semine* alongside the name of the plant. This sheet does indeed show the usual testicular shapes of the roots, but they are not exaggerated at all—and they are insignificant beside the microscopic study of the interior of the flower. The examination is then enhanced and enlarged in the details on the pages that follow.[99]

This, then, is one of the chief tensions with which Cesi struggled. Surface representation never seemed to yield what it promised. It allowed too

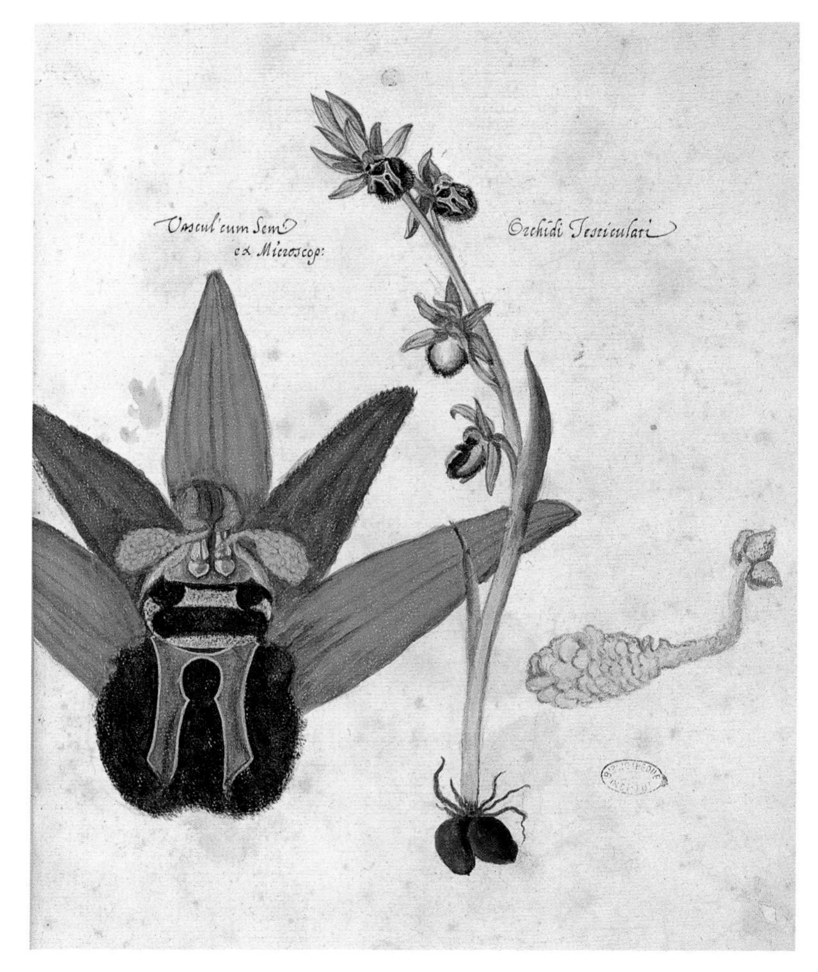


Fig. 8.35. *Ophrys incubacea* bianca ex. Tod. *Center,* the whole plant; *left,* a flower with two pollinia projecting from the gynostemium; *right,* a pollinium with tail and retinaculum. Paris, Bibliothèque de l'Institut de France, MS 976, fol. 46. © Photo RMN—Gérard Blot.

much to the subjective imagination. In offering the possibility of an alternative scopic regime to the old physiognomic and phytognomic one, the microscope promised to unlock the deepest secrets of life on earth.

There remained the problem of rampant growth—of how things grew out of nothing, or in the most inhospitable of places. MS 977 concentrates on this problem with exceptional intensity. It contains hundreds of drawings of local lichens and mosses, growing from every conceivable place, on stones and mollusks as well as on twigs and leaves. But in comparison with the other manuscripts, there are fewer enlargements, fewer microscopic examinations, and far less in the way of similitude too. What is left is the impression of abundance: abundance of growth and abundance of information—especially visual information—about the lowliest and scruffiest things.

In some ways MS 978 is the most puzzling, the most random, and the most crucial of all. It is a veritable scrapbook of information and of different representational media. Justly described by its indexer (and perhaps its originator) as being a miscellany,[100] it contains not only the

Fig. 8.36. Fungi, sponge or seabiscuit, and blastoids (?). Paris, Bibliothèque de l'Institut de France, MS 978, fol. 150. © Photo RMN—Gérard Blot.

usual watercolors and pen-and-ink drawings (some meticulous, others not), but also a large number of rather primitive etchings and engravings (e.g., figs. 8.36–37). Most of the drawings are directly on the page, but many are pasted in, as are all the prints. Some are rather precise and rigid, but a number are rather primitive. The illustrations are chiefly of fungi, but some show flowers (including a few exotic ones from America), grasses, leaves, algae, and, grouped together toward the end, a small section of insects and other animals. Many of the fungal drawings concentrate on growths out of rocks and on bark, on what Cesi and the scribe describe as "tuberi fungiferi"[101] and "tubifungi,"[102] and on hairlike mosses and algae.[103] Drawings of commoner species include charming ones of chanterelles and morels, while a special section, bound in toward the end, is devoted to the phosphorescent sea pen, the *Pennatula phosophorica* (what Cesi seems to have known as the *Mentula pennata*).[104] Giovanni Battista Della Porta's name occurs here, not in

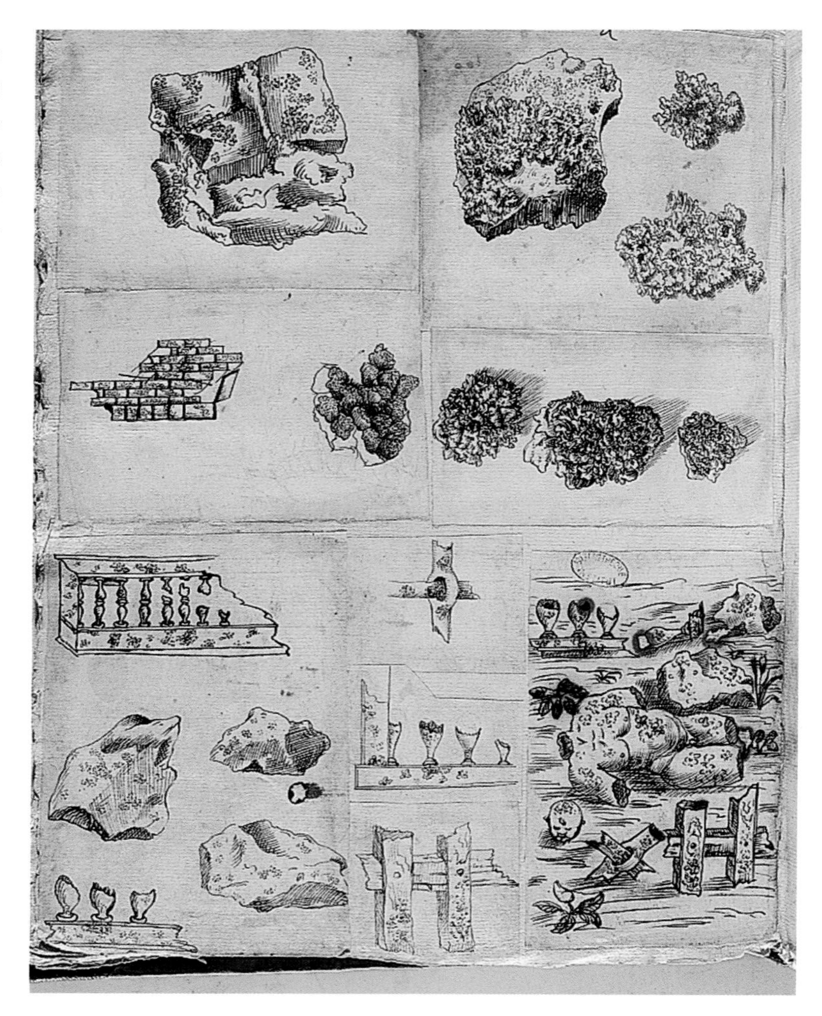

Fig. 8.37. Lichens on natural and shaped stones. Paris, Bibliothèque de l'Institut de France, MS 978, fol. 148. © Photo RMN—Gérard Blot.

connection with any physiognomic notion but rather as having given Cesi an example of strange green algae, perhaps a species of Codium.[105] Right at the end of the volume, for reasons that remain to be discovered, are four engravings showing different views of a hydrocephalic child (fig. 8.38).

MS 978 was clearly made over a number of years with illustrations by many different hands. It contains some of the very earliest drawings in this whole corpus, as well as some of the latest, dating from between 1623 and 1626.[106] Opening with a comparatively innocent—and typical—group of illustrations of growths on oak leaves and other pieces of wood, it continues with a section of small drawings and prints, largely pasted in, of a quality that is very much cruder than anything that we have seen so far. As far as one can make out, they show pieces of stones and growths out of stones (figs. 8.36–37),[107] as well as some exotic plants and oddities, such as the *Coccus maldivus* later used in the *Tesoro Messicano*.[108] But one of these very primitive etchings is inscribed "Mercurius

Fig. 8.38. Hydrocephalic child. Paris, Bibliothèque de l'Institut de France, MS 978, fol. 361. © Photo RMN—Gérard Blot.

Vegetativus Heckij."[109] It is thus yet another instance of the borderline phenomena that so absorbed Cesi from the very earliest days of his researches.

Suddenly the place of this crucial manuscript in the Linceans' endeavors becomes much clearer. The rough-and-ready etched illustrations, and probably a number of the other drawings as well, are by the Flemish engraver who is recorded as having accompanied Heckius on his travels, paid for by Cesi.[110] Of course they were never published—they were too imprecise ever to have been. Nevertheless, they are critical early *jeux d'essai*. At this early stage of the Linceans' endeavors, the drive to anthropomorphism and genital similitude was very strong indeed. Typical are the strange pieces of stone (or are they wood? or tubers?) shaped like a phallus and testicles on folio 133; the stones covered with growths, stones that on perusal we realize to be fragments of architecture and a statue (itself a "fossile," dug up from the earth; fig.

Fig. 8.39. Bat. Paris, Bibliothèque de l'Institut de France, MS 978, fol. 355. © Photo RMN—Gérard Blot.

8.36); and the truly rudimentary drawings of soft coral, the *Alcyonium palmatum*, serendipitously shaped like hands with twisted fingers and titled "manus marinae."[111] These surely point forward to the marked concern with digitated specimens in Ferrari's book on citrus fruit of about twenty years later.[112]

But what of Heckius's "vegetative mercury"? "Whether metallophytes [or not]" reads the inscription on the etching next to it, and in Cesi's own hand above it. This shows a strangely inconsequential tuber—or is it a piece of wood? Or stone? Suddenly we become aware of the basic issue that underlies not only the puzzles on these pages, but perhaps of every one of the many diverse specimens reproduced throughout the manuscript. It is the problem of the borderline case, of species that stand on the boundaries between things or that share in the characteristics of more than one category, class, or group. Now at last we begin to realize the theme that unites so many of these illustrations. We begin to understand

Fig. 8.40. Partial dissection of a hermaphroditic rat. Paris, Bibliothèque de l'Institut de France, MS 978, fol. 357. © Photo RMN—Gérard Blot.

their motivation, and we comprehend the fundamental issues that lie behind the illustrations of pieces of wood that look like stone (presumably fossilized wood), of fossilized fungi and corals (animal, vegetable, or mineral?), of the stick insects and the caterpillar on folios 318–328 (cf. fig. 8.34),[113] of the wonderful bat (fig. 8.39) from the crypt of a church at Narni (bird or rodent?), and of the dissection of a hermaphroditic rat (fig. 8.40)—presumably made to illustrate the dissection of just such an animal by Müller at Faber's house in Rome in 1611.[114]

## IMPERFECTION AND CONSTANCY

From the beginning Cesi gave the term "imperfect species" to what I have here called borderline cases. In doing so he intended to suggest that all such species were somehow deficient, or lacking the essential (or necessary) characteristics of a class.[114] And in coming to the intuition that in order to determine a class one had to find what was essential, or constant, or unchanging about it, he reached out for the clues to constancy that could be provided by the sexual organs of things.

This, then, is the trajectory followed by the Linceans' botanical manuscripts. They begin with the joy of discovery, and with a kind of high pleasure in the plenitude of nature and the new and unexpected things within it. They convey the Linceans' studies of the names and virtues of plants, of their positive uses and their dangerous, even lethal, effects. But in the end what the manuscripts forced the Linceans to try to come to terms with was the unmanageability of abundance and the chaos of nature itself. More than any of the others, Cesi felt the need to find the order of nature, and to devise a system for conveying it. In this way he and his fellow Linceans were motivated by the desire to understand fertility as well as by the drive to classify; and they set out to uncover the basic, unchanging principles of reproduction in nature—as if the secret to order lay, deeply but ironically enough, in the fundamental promiscuity of reproduction.

# NINE *The Mexican Treasury*

## TAXONOMY AND ILLUSTRATION

NAPLES, 1604

When, seeking to escape from the hostility and machinations of his father, Cesi went to Naples in April 1604, he was not entirely sure of what to expect. He knew that he could count on the support of at least two of the leading scientific figures there, Giovanni Battista Della Porta and Ferrante Imperato. Della Porta, as we have seen, was one of the most famous all-around scientists of the previous generation, while Imperato was renowned not only for his studies of plants and fossils (just as his son Francesco would soon become too) but also for his gardens and his exceptional collection of natural historical specimens.[1] Cesi struck up instant friendships with both men, but especially with Della Porta, the greatest master of the sciences of signature and surface.[2] "However little I am worth," he wrote to Cesi and his fellow Linceans, "I shall devote myself to you"; but he also warned them, as if in full awareness of the perils of not looking beneath the surface of things, "do not be deceived in me: my forehead promises more than there is inside."[3]

> Naples, paradise of delights, full of amenities and pleasures, relaxed in its loveliness, most beautiful, gentle, and charming; the abode of fertile Ceres, abundant Neptune, and most pleasing and courteous Venus . . . there is too much to tell you, but know how much I would have liked, if I had the Linceans with me, to establish my abode here.[4]

Cesi could hardly have been more enthusiastic when he described Naples to Stelluti upon his return to Rome in July 1604. Everywhere in the city he seemed welcome. From then on he and his three friends remained in close touch with many of the Neapolitan scientists and philosophers of the day. For years afterward Imperato sent them packets of seeds from his rich garden, full of exotic specimens,[5] and Della Porta dedicated one work after another to the young prince he so admired.[6] No wonder that Cesi wanted to base his new society—or at least to establish a branch of it, a "Lynceum"—there.

It was probably at the time of this first visit to Naples that Cesi caught a glimpse of a manuscript and a group of illustrations that would change his life forever. Perhaps he only *heard* of them then, and only saw them at the time of his second trip to Naples in 1610. In any case, the story of the material he found in Naples is a complicated one. It concerns one of the fundamental documents in the history of American natural history, and is about the central project of Cesi's life, as well as that of his fellow Linceans.

## HERNÁNDEZ AND HIS MANUSCRIPTS

In January 1570, Francisco Hernández was appointed personal physician to Philip II, king of Spain and all its dominions. His letter of appointment specifically nominated him as "*protomédico general* of all the Indies."[7] In the next year Hernández set out for New Spain in order to investigate and record its plants and their medicinal uses, as well as to do research into other aspects of the natural history of the region. He remained there for seven years. Using the huge sum of eighty thousand ducats allocated to him by the king, Hernández began to make as complete as possible a record of Mexican and other Central American flora and fauna.

Soon Philip was expecting some return on his investment. In February 1575, Hernández wrote excusing himself for not having sent the material he had promised on several occasions earlier. Philip, intensely interested in the pharmacological possibilities of the plants of New Spain, was much irritated by the delay: "Write to the Viceroy, and tell him that this doctor has many times promised to send the books of this work, and that he has never complied; and that he should put them together and send them on the first fleet that goes out."[8] Twice more Hernández sent his excuses for not delivering; but at last the sixteen "cuerpos de libros grandes de la Historia Natural de esta tierra" arrived, in the last week of March 1576.[9]

It is still not entirely clear what exactly these sixteen "cuerpos de libros grandes" were. But we do know that Hernández sent back three volumes of twenty-four books on the plants of New Spain, one volume of six treatises on the animals, and eleven volumes of colored illustrations—almost four thousand of them—as well as at least one further volume of dried specimens of Aztec plants (there may, in fact, have been more).[10] The pictures were either acquired from native artists or commissioned by Hernández. Many were made on the basis of plants he actually saw, but others portrayed specimens from the last of the great Aztec gardens at the hospital of Huaxtepec, as well as a few from the collection put together a century before by the Aztec king Nezahualcoyotl of Texcoco.

All this material first passed into the possession of the king himself, then to the Council of the Indies, and then finally deposited in the Escorial.[11] In 1605, the librarian of the Escorial, Fray Jose de Siguenza, described the contents of Hernández's manuscripts:

> [The library] has a curiosity of great value, worthy of the mind and the greatness of its founder. This is the history of all the animals and plants that could be seen in the West Indies, painted in their native colors. It shows the actual colors of the trees and the grasses, the roots, trunks, branches, leaves, flowers and fruits. Also the colors of the caiman, the spider, the snake, the serpent, the rabbit, the dog and the fish with it scales; the very beautiful plumage of so many different birds, the feet and the beak. . . . It is something that offers great delight and variety to those who look at it; and no small profit to those whose task it is to consider nature, and that which God created as medicine for man.[12]

The emphasis on the lively and accurate color of the illustrations and the acknowledgment of the medicinal scope of Hernández's *History of the Plants and Animals of New Spain* would be leitmotifs of every subsequent description of the work.[13] But from the outset the trouble was that the books were in a mess, and constituted a much larger body of material than anyone could have imagined.[14] Hernández himself acknowledged the problem a month after the books were shipped off from Mexico: "they are not so clean, so polished or so well-ordered that they don't still require some final touches before they are printed, especially since there are many illustrations mixed up in them that were painted just as they were encountered."[15] Everyone who subsequently saw this material referred to its disorder and confusion.[16] Philip decided to do something about all this, and called upon his next personal physician to help with the problem.

## RECCHI TAKES OVER

Nardo Antonio Recchi was a learned doctor from Montecorvino, near Naples.[17] In 1580 he succeeded Hernández as Philip II's personal physician, and in his letter of appointment was given two specific commissions: firstly, "to exercise the office of *simplicista*, taking special care to cultivate medicinal plants in our gardens and other convenient places"; and secondly, "to look at the writings that Dr. Francisco Hernández brought back from New Spain and to organize them and put them in order."[18] As the king's personal *simplicista* Recchi was in charge of cultivating medicinal plants in the Spanish Royal Gardens, especially those at Aranjuez, as well as of making a manageable selection from the immense body of material, both written and pictorial, sent from the New World by his predecessor. In just over two years he completed this difficult task.

On March 24, 1582, Juan de Herrera, the architect of the Escorial, wrote to the king's secretary that Recchi was done, and that the work, "de mucha calidad y grandeça," should be published as soon as possible. Someone should immediately be sought out to make copper engravings of the illustrations in Hernández's manuscripts to accompany it.[19] Seven weeks later, on May 5, Herrera wrote again, confirming that Recchi had completed his "recopilación" of Hernández, extracting from it, just as the king had commanded, the most useful simples. And since Recchi would soon be returning to his native Naples as the king's *protomédico* there, he would himself assume responsibility for publishing the work. "In order for it to appear, I've tried to find some people to engrave and to make the prints of the simples and plants in the form and size as they appear in the said books,"[20] he wrote to secretary Vazquéz. If the work were not published, he insisted, His Majesty's whole intention would come to naught; and it had to be illustrated with adequate and appropriate engravings. The sum of fifteen hundred ducats, which Herrera estimated the project would cost, could hardly be better spent.[21]

Other than a few engravings of Hernández's illustrations, nothing ever came of Herrera's efforts.[22] But Recchi's labors were better than fortuitous, since in 1671 the great Library of the Escorial burned down in a fire that raged for five days, and Hernández's original manuscripts were lost forever. Had it not been for the Linceans, for Cesi, Colonna, Terrentius and Faber at the beginning, and Stelluti and Cassiano at the end, we would know very much less than we do—indeed almost nothing—about this extraordinary material, whether Hernández's original or Recchi's remarkable though sometimes misleading redaction.

True, there were others who recognized the significance and utility of Hernández's and Recchi's efforts.[23] But none of the extracts, excerpts, copies, and compilations made from them can even remotely match the interest of what the Linceans produced on the basis of Recchi's manuscripts. Some of what they had ready they had printed in 1628–29; but the full fruits of their labors would only be published some twenty years later, between 1649 and 1651.

## AMERICAN PICTURES IN NAPLES

When Recchi retired to his native Naples in 1589, he brought back with him his redaction of Hernández's manuscripts, as well as approximately six hundred illustrations copied from among the several thousand in Hernández. Their fame spread rapidly. Many had heard of Hernández's work, but now everyone seemed to be eager to see what Recchi had done with it. By the end of 1589 Aldrovandi was already writing to Della Porta inquiring about Recchi's manuscripts.[24] Della Porta replied by recalling how Hernández (whom he mistakenly called "Cortes") had died of sadness when Philip II was told by his council that the descriptions and illustrations of four thousand plants and animals were of little use "since they were of Indian plants that could not be used in Spain; and besides, the book had no order to it." Recchi, Della Porta continued, had selected more than six hundred plants and animals out of the mass of information in Hernández,

> and the King is now having them engraved, and we will have the book soon. The said doctor has an income of 400 ducats and has come to Naples with the originals. These he shows courteously to all. And they are certainly beautiful, rare, useful and most extravagant things. Since he is my close friend, I asked him intensely [to let me have them]; but he said that if he published them before the King did, he would be in danger, not just because the King would take away his income, but also his life.[25]

Such were the perils of stepping out of line when it came to the publication of new knowledge.

Indeed, as we shall shortly discover, it was by no means easy to see everything that Recchi had brought back with him; but it is clear that by now natural history had become almost inconceivable without picturing. In the same letter to Aldrovandi, for example, Della Porta touches on a number of other natural historical subjects of interest to them both, including that of the true remora, which some Neapolitan fisherman had found, but which they had had great difficulty in detaching. Though it greatly clarified the relevant passages on it in both Aelian and Pliny, Della Porta went on, "if you don't have a picture of it, I will send you one" (since the fisherman didn't want to sell the specimen itself).[26]

This just shows how intense and wide-ranging were the interests of the Neapolitan natural historians in those days. Della Porta was chief among them—or at least among the older generation, for Fabio Colonna was already a rising star. As we have seen, Colonna's *Phytobasanos* of 1592 had already made a pioneering contribution to Old World botany.[27] But what of the plants of the New World, about which so much information was now beginning to pour in from the Indies—not least in the form of Recchi's manuscripts? That Colonna knew about these emerges from a passage in the *Phytobasanos* where he refers to the *Solanum manicum* (*Datura stramonium* L., fig. 9.2) and says: "the first to give me a sample of this exotic plant was Leonardo Antonio Recchi, most skilled doctor of Philip III . . . through whose dedication and labor there has emerged a description in Latin of innumerable exotic plants, which will be printed as soon as possible thanks to the generosity of the same monarch, at his command and his costs."[28]

Colonna now became the main conduit of information about the Recchian manuscripts to the growing number of botanical experts interested in Hernández's achievement. Aldrovandi wrote to him at the end of 1595 asking for more information (he was probably dissatisfied with what he had been able to glean from Della Porta); and his name features in a remarkable exchange of letters between the Nuremberg natural historian Joachim Camerarius and the greatest of all European botanists, Clusius, in Leiden. The context is noteworthy, to say the least.

Having discussed a number of urgent topics, including the problem of

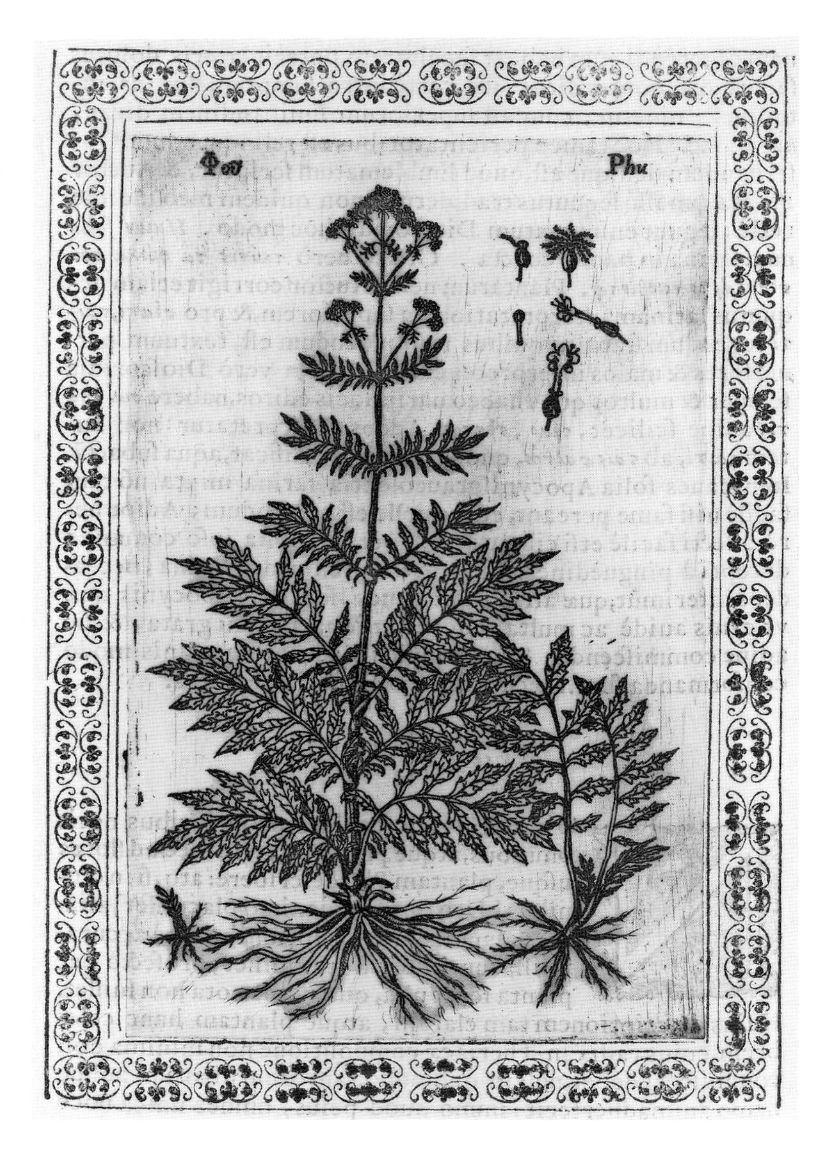


Fig. 9.1. Fabio Colonna, etching of valerian (*phu*). Colonna, *Phytobasanos* (1592), p. 114.

illustrations in herbals, the correct classification of Solanaceae, the use of comparative evidence from the Americas, and botanical books in the hands of the Leiden branch of the Plantin press, Clusius suddenly reflects:

> In twenty years I've received just one letter from Benito Arias Montano, which he sent to me last year. He lives near Seville in a place far from all disturbance, where he can dedicate himself to his studies. He seems to retain the memory of his old friends. Whether he knows anything certain about the great volume of plants, which Fabio Colonna recalls, I do not know.

What is the meaning of this reference to Montano, one of the great dissimulators of the sixteenth century, the man sent by Philip II to be his censor in the Netherlands, and who yet associated himself with the same

Fig. 9.2. Fabio Colonna, etching of Solanum manicum (*Datura stramonium* L.), Colonna *Phytobasanos*, p. 47.

heretical and irenical group to which none other than the Royal Printer, Christopher Plantin himself, belonged?[29] What could he, the editor of the great Plantinian project of the Polyglot Bible, possibly have to do with a compendium of American plants? Was it just because he was close to both the king himself and to Plantin, whose chief publishing domains—outside that of the Greek and Latin classics—were precisely those of religion and botany? Certainly he was particularly attuned to problems relating to the publication of controversial texts. But the real reason for the reference to Montano was surely Clusius's awareness of the fact that Montano had always been Hernández's friend, and his protector at court. Indeed, Hernández himself seems to have hoped—in vain—that Montano would take care of the publication of his work.[30]

We have gained a sense of how much difficulty Della Porta had in getting a good glimpse of Recchi's selections from the Hernández material; and even now there was something about it that seemed to make its owners want to keep it hidden from sight. Why?

Clusius continued:

> I remember that when I was living in Frankfurt, some people returning there from Naples told me that they had seen a huge number of plants, birds and quadrupeds depicted in their own colors, at the home of a certain doctor of the King of Spain, to whom they had been introduced by Giovanni Battista della Porta. But this doctor showed them the pictures only cursorily, and did not permit them to look any more attentively, saying that he had been restrained by the King from showing them to anyone, and that he ought not even to divulge their names; for the King had it in mind to publish the natural history of these plants and animals. But those who saw them did not think that this would ever happen; because they said that it was all in a state of disorder, and that he who had been entrusted with those pictures did not seem to them to be capable of writing such a natural history.[31]

The letter not only confirms what we already know about the disorder of the manuscripts, it also casts some light on the mystery of the splendid color illustrations they contained, on why they remained hidden, and on the strange aura of sensitivity that surrounded them. At the same time it tantalizes us with a sense that we may never get to the whole truth about their early history.

For some reason the king of Spain remained determined to have the work published under his auspices, and to publish it before anyone else. Perhaps he simply did not want to diminish the dramatic effect of novelty that he hoped the account of the expedition he had sponsored would have. But still, why this extraordinary secrecy, this half-whispering of the names of the plants in the illustrations? What secrets were at stake?

Despite his best efforts, Clusius himself never got any further—even though he was probably the greatest European expert on exotic plants and was at that very moment preparing his *History of Rare Plants* (1601). Ferrante Imperato did not fare much better. In January 1598, he wrote to Clusius saying that he had indeed known Recchi a little, but that the good doctor had never done a thing with his Indian material during his lifetime.[32] Indeed, said Imperato, he had had great difficulty in persuading Recchi to show him the drawings of Indian animals and plants, but that

> when I saw them, they were indeed new things. The difficulty—as it seemed to me—was that while the colors were particularly beautiful and vivid, they were rather coarsely done pictures, about 100 [sic!] of them. At that time, when he showed them to me, I tried to

> find out whether there was any intention of publishing them, either those or others, in Naples or elsewhere. In the end I managed to get him to admit that at the court of his Majesty some Spanish doctor or another had been commissioned to make a book of those simples of the Indies, but that some other doctors did him an ill deed, with the result that those of the Council blocked the whole business; and he never spoke of it again.[33]

For many years an ill star hung over these "pictures," whose importance everyone recognized. No one took up the challenge of publishing the material from New Spain. If Cesi did see some of it on his first visit to Naples in 1604, he still cannot have seen much. Every report confirms the continuing reluctance to show it. But Cesi's appetite was whetted by a tantalizing reference in one of Heckius's characteristically miserable letters, this time written when he was staying in Madrid in 1608.

Even amid his degrading work as a doctor to the poor, reduced to being a "pulse-feeler and urine-checker,"[34] Heckius never ceased "to investigate the nature and differences of things." In fact he had "about 100 Iberian plants which have never been published" in his house and he was about to go to the Royal Library in the Escorial, "where they say that the King keeps all the Indian plants fixed with glue."[35] Almost certainly this refers to the herbarium of dried plants that Hernández had sent from Mexico or brought back with him to Spain. But when Heckius went to see it, did he see the rest of Hernández's material as well? Or at least some of it? Quite probably.

In early spring of the same year that Heckius paid his visit to the Escorial, those two promising young experts in botany, Johannes Faber and Johannes Schreck, made a trip from Rome to Naples. Their ostensible aims were to gather plants for the Vatican gardens, meet some of the city's famous natural historical authorities, and examine the Recchian manuscripts. Under cover of their botanical trip, however, Faber also went to negotiate with the viceroy of Naples on behalf of the imprisoned Tommaso Campanella.[36] This was a dangerous mission, since Campanella continued to write ever more heterodox works than before, and was experiencing the manic and ecstatic episodes that continued to subvert the efforts of his friends and supporters.[37] Faber was repeatedly cautioned by the sponsors of his mission, chiefly Gaspar Scioppius, and perhaps the latter's then friend, the philosopher Antonio Persio, to circumspection and silence.[38] The result is that all that remains of the evidence for this trip are a few fragmentary indications[39]—and the outlines of a new botanical project.

Whatever Faber and Schreck actually did on behalf of Campanella remains unknown, but they did meet several of the distinguished scientists of Naples, including Della Porta, Imperato, and Colonna. Through them they were able to look at Recchi's manuscripts and illustrations. Imperato started sending packets of seeds to the two Germans, in order to aid them in the production of their book on the subject of Mexican plants.[40]

## COLONNA AND THE NAPLES LYNCEUM

Cesi kept on hearing more about the American material in Naples. Faber, Schreck, Heckius, and a number of their Neapolitan contacts were all talking about it. Suddenly it seems to have become more available for scrutiny than ever before.

In 1610, the manuscripts belonged to a poverty-stricken and obsessive bibliophile, Marco Antonio Petilio,[41] who had inherited them from his uncle, Recchi. Cesi was now determined to break the issue open by getting hold of Recchi's manuscripts and publishing them, along with reproductions of their beautiful color illustrations. In any case, he wanted to return to Naples, stay with Robert Bellarmine on the way, and investigate the possibility of setting up a branch of his Academy there. There were several other famous men whom he had yet to meet. Some, he thought, could be elected to it—after all it had not yet expanded beyond its original membership. What better project to engage such a body than the publication of Recchi's material?[42]

It was at this stage, during Cesi's second visit to Naples, that the problem of reproducing the illustrations of the Recchian manuscripts became crucial. Little could Cesi then have known how far inferior a standard he and his friends would eventually have to settle for!

Even at this point Cesi must have known that he would want Colonna—another friend of Campanella's—to head the Naples "Lynceum." But who now accompanied Cesi to see the material which Petilio had inherited from his uncle—Della Porta, Colonna, or one of the many other natural historians and doctors of Naples? We do not know. But what Cesi saw changed his life, at least as much as his encounter with Galileo shortly thereafter.[43] Petilio's material offered a larger number of illustrations of the natural world than he had yet seen—and not even of a world he knew. Nothing of what it illustrated could be found—or at least, easily found—in Pliny, Theophrastus, or Dioscorides. Suddenly there were more things on earth than he or his friends had ever imagined; and most of these had yet to be descried by the eye of the lynx.

## A NEW BOTANY: THE CLASSICAL AND THE VERNACULAR

It is hard now to imagine the impact that Hernández's material must have had on those trained in botany based on the traditional classical sources. Here were not just a few plants but several thousand, unlike anything ever described by the ancients. No wonder, then, that Hernández and then the Linceans should have placed such an emphasis on the Nahuatl and other Amerindian names! There were simply no others available.[44]

The strikingly local taxonomy of the plants in the *Tesoro Messicano* has as much to do with this necessity as with any other ethnographic concern.

The Linceans' turn to the vernacular should be seen in the context of its usefulness not only as means of promulgating scientific data, but also as a critical element in the actual understanding of that data. On the one hand, the vernacular was familiar and therefore accessible; on the other, it could be astringently unfamiliar—certainly when it came to the languages of other cultures, languages vastly more remote than Latin. But such languages, the Linceans came to realize, were by no means too arcane to use. On the contrary, they had much to reveal, especially where the old sources were silent. One *had* to attend to the locals, however difficult it might be.

Petilio gave Cesi full access to the actual text of the manuscripts he inherited, either selling him the original pages or allowing him to have copies made. But he always seems to have been reluctant to let Cesi have the illustrations, or even to see them. For a very long time Petilio only allowed Cesi and his colleagues to see the illustrations one by one, thus displaying the same nervous reticence as his uncle Recchi.[45] But by May 1611 the illustrations were in Rome.[46] Years later Faber would recall the beauty and precision of their colors.[47] Immediately Cesi embarked on his project, still not clearly worked out, of publishing all the material he could lay his hands on, whether Recchi's or whatever of Hernández's originals he could find.

## FIRST DELAYS

*Habent sua fata libelli*. The story of Hernández's manuscript trove was already a halting and troubled one. Practically none of it appeared in the first fifty years after it had been put together.[48] None of the famous illustrations in color had appeared, not even in black and white.[49] Few books can ever have been quite so long in the making.

Originally conceived by Cesi in 1604, he and his fellow Linceans worked on it until the day he died in 1630—and even then it was not complete.[50] Although a significant portion of the *Tesoro Messicano* came off the press in late 1628, Cesi never saw the work in its final form of 1649–51. The project occupied the most stalwart of the Linceans until almost the very last one of them, the *ultimus superstes*, as Stelluti described himself on one of the final pages of the book, was dead. How prophetic were the words of Markus Welser in August 1611: "From my experience of the delays and tardiness of Rome, I'm afraid that this book will make one lose many years before it appears."[51]

Exactly forty, in fact. True, some nine hundred pages and almost all of its eight hundred illustrations had been printed in 1628, and a few copies were actually in circulation by 1630; but it was only in the years between 1648 and 1651 that anything like a definitive form of the *Treasury of Med-*

*ical Matters of New Spain or the History of Mexican Plants, Animals and Minerals* appeared.[52]

The title alone gives some sense of the density of the contents of this *Thesaurus*.[53] No wonder that it swiftly became known as the "Mexican Treasury." Its story reveals many of the vicissitudes of the Linceans' lives, both personal and scientific; and, like the book itself, offers testimony to some of their greatest achievements and saddest failures. If we survey the distance from Schreck's initial contributions on the plants to the conclusion of the portions by Faber on the animals, we pass, at a stroke, from the days of Galileo's greatest triumphs, through the controversies over the sunspots and the comets, and on to the days in which Faber and Colonna used the microscope to unveil some of the secrets of the things of this earth.

Barely a month after Galileo demonstrated his telescope to Cesi and his friends on the Janiculum on April 14, 1611, the Linceans called him in to help resolve some more down-to-earth problems: "In the last few days," he wrote to Piero Dini on May 21, 1611, "when I was in the house of His Excellency the Marquis Cesi, I saw the pictures of 500 Indian plants, and I was expected to affirm either that this or that one was a fiction (denying that such a plants were to be found in the world) or that if they existed, they were deceptions or supposititious [*frustratorie et superflue*], since neither I nor anyone else present knew their qualities, virtues and effects."[54]

How fitting—and how poignant—that the Linceans should have turned to Galileo, whom they inscribed as one of them four days later, for advice about the project that from then on ran alongside every one of their labors on his behalf, almost until the very year of his trial. And how fitting, too, that this letter of Galileo's should be the very first mention in the Lincean correspondence of the great trove of illustrations that proved to be one of the major obstacles to the timely publication of their own most ambitious project.

Just over two months after Galileo's letter to Dini, Welser wrote from Augsburg to Faber in Rome (the German connection was working again). Soon news would reach Welser of Scheiner's sightings of sunspots in Germany and his receipt of the Ingolstadt Jesuit's book on the subject. From *then* on Welser's energies would be absorbed by the sunspot controversy; but in the meantime he was still able to devote himself to the antiquarian matters he loved, and to problems in natural history.

> That book on the West Indies . . . certainly deserves to be called a *Treasure*, rather than just a book, and certainly Marquis Cesi could hardly spend his money better than having it published; but the figures need to be done with greater refinement than that which is apparent from the one you sent me. After all, the beauty of the publication should correspond to the singularity of the work.[55]

The team was swiftly put together. Schreck was elected to the Academy just nine days after Galileo, on May 9, 1611; Faber followed on October 29 ("on Saturday I think I'll make Dr. Faber a Lincean," wrote Cesi to Galileo the day before); and Colonna on January 27 of the New Year (just three days after his Neapolitan compatriot Stelliola). Sometimes they turned for assistance to the third German in this group, elected on December 13, Theophilus Müller. Together the newly elected Linceans provided a fresh injection of botanical and other natural historical expertise. Could it have been the project of publishing the Recchian material that determined these early choices, and that so revivified the Academy? Almost certainly.

There can be little doubt that it was his sense of the importance of the Mexican book that refreshed Cesi's enthusiasm for his Academy. Both the German and the Neapolitan enrollments of 1611–12 were motivated by his belief that they now had a major and worthy task on which to base their scientific lives. Schreck, Faber, Colonna, and Cesi worked with attack and energy on the book. Soon there were plans to go to press. On September 15, 1611, Faber announced to a friend that Schreck was printing his book on the plants of Mexico.[56] Two days later, Cesi wrote to Galileo that "once again I must tell you that I have begun to have the book of Indian plants printed, and Sig. Schreck will make us a little bit of commentary on it."[57] At this stage he can hardly have suspected just how slow the Roman printers could be. In October he reports that the book was being prepared for the printers;[58] at the beginning of December, the same: Schreck has been working as best he can on his commentary on the Indian simples "which you saw," and the book is well on its way to the press.[59] All this urgency perhaps, because by then Schreck was on his way out. He had decided to become a Jesuit and was about to leave the Academy he had only just joined.[60]

On December 10, Welser sent Faber a proof engraving from Basilius Besler's sumptuous *Hortus Eystettensis*, noting that that work would still not stand up to comparison with Cesi's Mexican book—except, he declared, with regard to the beauty of its copper engravings.[61]

How true this was! The woodcuts of the published *Tesoro Messicano* look positively schematic and crude beside the exceptionally accomplished engravings in the *Hortus Eystettensis*. Cesi never even aspired to illustrations of such quality—but then he had more than eight hundred to produce, and the costs would far have outweighed the means he had at his disposal. In any case—to polarize the issue perhaps more than it was at the time—the main purpose of the *Tesoro* was informational rather than aesthetic. Nevertheless, the reproduction of Hernández's apparently fine originals remained Cesi's constant preoccupation and became one of the most critical stumbling blocks on the road to publication.

There was another stumbling block, too. This was the ongoing difficulty of getting the relevant permissions and privileges. In a long letter to Faber on June 20, 1612, Cesi wrote about the urgent need for the appropriate Roman privilege. Outlining some of the technical difficulties involved in obtaining it, he observed that they and Schreck should immediately begin working on the problem. Of the two thousand scudi they had for the project, they had already spent three hundred. Everything now lay in the hands of Mgr. Cobelluzzi.[62] Since Cobelluzzi was a friend of both Schreck and Faber, the latter should approach him and tell him about the beauty and importance of the work, its enormous expense, and Schreck's intense labors in editing it. Cesi concluded with a remark that takes one by surprise (though perhaps it ought not to): "in speaking of the Academy, speak of it as an ordinary Academy of Philosophy, and be sure not to say where we got this book from."[63]

All this, we should remember, took place a few years before the outbreak of the great Copernican debates, before the breakfast of December 14, 1613, at the court of the grand duchess Christina, before the controversy over the sunspots. What had the Academy to hide? Even then, however, Galileo's enemies must have suspected that the Linceans were up to more than most other academies at the time (there were many more than the twenty-five referred to by Cesi in this same letter). Such people may well have begun to feel that the Linceans were not sticking to their last, and that they were upending too much of the Aristotelian knowledge that lay at the basis of traditional science. And could they still have been afraid that King Philip III was waiting to be the first to publish the book based on the fruits of an expedition sponsored by his father some forty years earlier? Or was there some other more mysterious reason to cover up the details of their acquisition of Recchi's manuscript?

For all the initial energy invested in the project, it soon seems to have lost some of its steam. At the Academy's meeting of October 15, 1612, the printing was reported to be proceeding well;[64] but six weeks later Cesi wrote to Galileo that "the Indian book is going slowly, there's nothing to do about it." In any case "everyone's awaiting the *Sunspots* with great impatience and eagerness." Three thousand copies should probably be printed.[65]

All attention was now focused on the sunspot controversy. This was one compelling reason for the slowing down of operations on the Mexican Treasury. The other was the sheer scale of the project. To get a sense of just what it entailed, and how much work was involved, let us look more closely at how the *Tesoro Messicano* was actually put together, from this time on until that far-off day when Stelluti signed his afterword as the "last survivor" of the Linceans (but even this was not entirely true, since the great facilitator himself, Cassiano dal Pozzo, was still alive).

## SCHRECK AND FABER AT WORK

It was Schreck, about to become a Jesuit and then depart for China, who worked most actively and enthusiastically on Recchi's manuscripts in the early years of the project. At least *his* attention was not deflected by the sunspots controversy. How could it be? By the time the controversy broke out, he had already gone over to the Jesuits. And they stood on the opposite side of the matter from Galileo and his friends. Was it a relief for Schreck to lay down his pen, and not to have to side himself openly with the enemies of his Roman superiors and his German Jesuit colleagues, like Scheiner? To some extent it must have been.

For a while Schreck continued to work for the Linceans, if sometimes covertly.[66] There are scattered references to the Mexican project throughout the correspondence of the next few years, but by August of 1615 it was clear that Schreck had completed the bulk of his contribution and Cesi was already correcting page proofs.[67] Schreck had edited the largest part of Recchi's compendium and added a brief commentary and some notes to most of the chapters in each of Recchi's ten books on plants, animals, and minerals.[68] Each book was extensively illustrated with woodcuts, except for the last on minerals. This subject would have to wait until later.

To each of these books, Schreck appended a brief introduction, made some corrections, and added a very few notes, chiefly on classificatory problems and on colors.[69] He also added valuable references to the mentions of similar or related plants in the other great sixteenth-century herbals of American plants.[70] But only in the case of the *Guyacan* and the related *Hoaxacan*,[71] the tobacco plant (*Nicotinia rustica* L.),[72] and the plant that would become their symbol, the *Lincis flos* (*Stanhopea tigrina* Batem.)[73] were the commentaries longer than a few lines.

Following these books came a 110-page section titled "The Images and Names of the Other Plants of New Spain of Nardo Antonio Recchi with a Commentary by Johannes Terrentius, Lincean."[74] How often pictures and taxonomy went hand in hand—and how well this fitted with Lincean preoccupations! This section was based on a part of Recchi's manuscript that comprised only illustrations and names, with no other text at all; but it contained some of the most interesting plants in the corpus. All of them, Schreck noted, were exceptionally different from European plants,[75] and several were completely unidentified. To each of these illustrations Schreck added a brief morphological description and some color notes.

Since Recchi had provided no comments on the medicinal purposes of each plant, Schreck had none to offer either. But it is clear that this whole section was prepared in a rush, perhaps to meet Schreck's own deadline (he was anxious to begin his mission to China), or because Cesi was as anxious as always to send the work to the press.

To no avail; the sunspots supervened. Faber's attention was distracted. But it is to him that we owe what is perhaps the most interesting and novel part of the *Tesoro Messicano*: the almost four hundred pages of his commentary on just thirty-five animals, titled "Recchi's Images and Names of the other animals of New Spain in the Exposition of Johannes Faber, Lincean of Bamberg."[76] Once more the title points to the problematic and complex fraternity of pictures and taxonomy. But Faber himself did not have the theoretical capacity to do much with this conjunction.

What he produced was certainly the most discursive, lively, and wide-ranging section of the whole work, even though its contribution to natural history was far less precise and accurate than Schreck's drier botanical and ethnobotanical work.

In each case, Faber gave a *descriptio* of the relevant animal—bird, beast, or worm—below its illustration. Here he provided the basic taxonomic and coloristic data. But where the *descriptiones* are succinct, the *scholia* that follow upon them are prolix and expansive. They contain digressions that are wonderful, fantastic, and almost always of high interest. The range of reference to both old and contemporary authorities (particularly the latter) is breathtaking. With humor and insight Faber writes about a multitude of natural historical matters, some important, others trivial. Even when he is long-winded he often remains stylish. Some of this material is dense and substantial; other parts seem altogether more trivial, such as the excursuses on the song of parrots (cf. fig. 10.5) and the domestic habits of dogs. Some pages seem quite rigorous and not at all credulous; others are strained, gullible, far-fetched, and old-fashioned. But on the whole, Gabrieli was right when he commented how neglected these "dense and attractive" pages have been, so rich in "erudition, experimentation and personal observation."[77]

Faber's pages allow us to piece together a good portion of the evidence for the otherwise forgotten researches of the Linceans. As with Stelluti's *Persius* of 1630, one often has the feeling, in reading the *Tesoro Messicano*, that it takes only the slightest pretext, the most inconspicuous suggestion, to provoke some engaging digression or another, on either scientific or on personal matters. But there, for the most part, the comparison ends.

Barely had the comet of 1618 passed through the skies when Cesi wrote from Acquasparta to Faber in Rome: "I am waiting here and hoping for better news about the Privilege, and soon, God willing, we will give the final handshake to the Mexican book."[78] By then both Cesi and Stelluti had been working on Schreck's text for several years, as well as on the illustrations for his and Faber's contributions. Then followed a long hiatus, during the period when Cesi, Cassiano, and Faber above all were devoting themselves to encouraging Galileo and preparing the *Assayer* for the press. Finally, with that work safely off to the censors, and with the Jubilee year right before them, Cesi wrote to Galileo, just two days after Christmas 1624: "I'm pushing as hard as possible for the printing and completion of the Mexican [book] and other works too, prior to the end of this Holy Year."[79]

Some hope this! Faber had ever more difficulty in persuading the morose and irritable Petilio, still alive and now jealously guarding his manuscripts in his museum in Rome, to show him the illustrations. He needed to see their colors as much as anything else. Throughout 1624 and 1625 he complained about the problem, not just about the pictures of the animals but also about the difficulties of editing Schreck's section.[80]

Then there was another, perhaps even more pressing issue from 1624 on. There was simply too much to do for the Jubilee. By the end of that year, the Linceans had already decided to publish their great panegyric to the symbol of the Barberini papacy, the bee, in the form of the *Apiarium*. Both Faber and Colonna were actively engaged with Cesi in preparing it and devising its contents. Besides, the printing of so large a work as the *Tesoro Messicano* threatened to take a longer time than was left to them.

"Rome, from my Museum at the Pantheon of Agrippa, on the calends of January in the solemn year of 1625." This is how Faber concluded the introduction to his "Exposition" (as it was called) on the animals of Mexico. He had just recorded that

> This little [sic!] work of mine I completed at the command of his Excellency Federico Cesi, Prince of the Linceans, at whose expense all of this Mexican book has been done, for the common benefit of the Republic of Letters. I do not need to review here how much he has always been esteemed for his erudition, and for his recondite knowledge of mathematics and the true and (so to speak) real philosophy.[81]

Indeed not.

Faber probably felt that he had in fact completed his work by then. After all, he and his colleagues had been working on the project for many years, and they were desperate to have it ready for the Jubilee year. True, their attention had constantly been diverted by the need to support Galileo; but Cesi, if not the others, never doubted that the Mexican Treasury was their prime project.

Perhaps it was just as well that the Jubilee year brought a delay, since it produced two important benefits for the work. Along with the many visitors, important and less important, who flooded into Rome for the first Jubilee of the Barberini pope were three missionaries who had spent long years and gained much experience in America. Faber was able to consult them, and over the next year or so included much information obtained from them in his text. Perhaps the best known of these was Gregorio de Bolivar, the Franciscan who had written a brave memorandum to the pope about clerical abuses in the American church. From Bolivar, who had himself conducted many anatomies of animals, Faber obtained much important data for his entries on a whole variety of American—and especially Peruvian—species.[82] Two Dominicans, Pietro de Aloaysa and Bartolomeo de la Ygarza, also provided important information, with Aloaysa contributing descriptions of American birds,[83] as well as on the mysterious two-headed beast known as the amphisbaena.[84]

## CASSIANO IN SPAIN

The second benefit of the Jubilee year was perhaps even more important for their project. In that critical year, Urban felt that he had to be perceived as an impartial figure in Catholic politics and as a guarantor of peace, if not in Europe then at least in Italy. This ambition was threatened by the continuing tensions between France (or rather Richelieu) and Spain over the possession of a small but troublesome piece of land in northern Italy, the Valtellina. In order to patch up the differences between France and Spain, the pope sent his nephew, Cardinal Francesco, on a diplomatic mission to France and Spain. A number of notable and learned figures accompanied him, including Cassiano dal Pozzo, who then occupied the position of "maestro di camera" in the household of Urban VIII.

As the papal legation traveled through France in 1625 and Spain in the next year, Cassiano noted down both antiquarian and natural historical material that might be relevant to his and to his colleagues' work.[85] Never one to let observational opportunities pass, or to neglect the possibility of acquiring new sources of natural historical information, he collected much data relevant not only to his own ornithological interests[86] but also to the material being prepared for the book on Mexico. Knowing, moreover, that the work was not quite ready for publication, and that it was still in the process of being printed, he recorded a great deal of material to be passed on for inclusion in the manuscript that the poor German doctor Faber had thought to be complete.

But it was especially when Cassiano arrived in Madrid that he discovered how much could still be added to the Mexican Treasury. Through a Scottish friend, David Colville, he succeeded in getting permission to visit the Escorial, and there, to his great delight and amazement, he was able to see Hernández's originals.[87] Immediately he asked the librarian of the Escorial, Andres de los Reyes, to copy out those parts of Hernández's original manuscripts that had not been transmitted to them via the papers of Recchi they had acquired so many years earlier. As de los Reyes wrote in a letter to Cassiano of August 10, 1626, though the intention was to transcribe all of the Hernández material, he was only able to devote two hours a day to this laborious task. So he had to be content with much less.[88] What he actually achieved, as far as we know, was the transcription of five small treatises on animals (quadrupeds, birds, reptiles, insects, and aquatic animals) and one on minerals (in his rushed letter he calls these four or five small books), as well as a rather ample "Alphabetical Index of the Plants of New Spain by Francisco Hernández, first physician of the King of Spain."[89]

When Cassiano returned home in October, he brought these transcriptions with him. "I'm delighted that the Mexican Treasure is expanding every day," he wrote to Faber on October 26. Faber was probably not so delighted; but he gamely incorporated Cassiano's contributions. "I've brought back from Spain the text of the Natural History of Hernández, which will serve you in revising the text that you've printed," Cassiano continued.[90] Needless to say, this was by no means

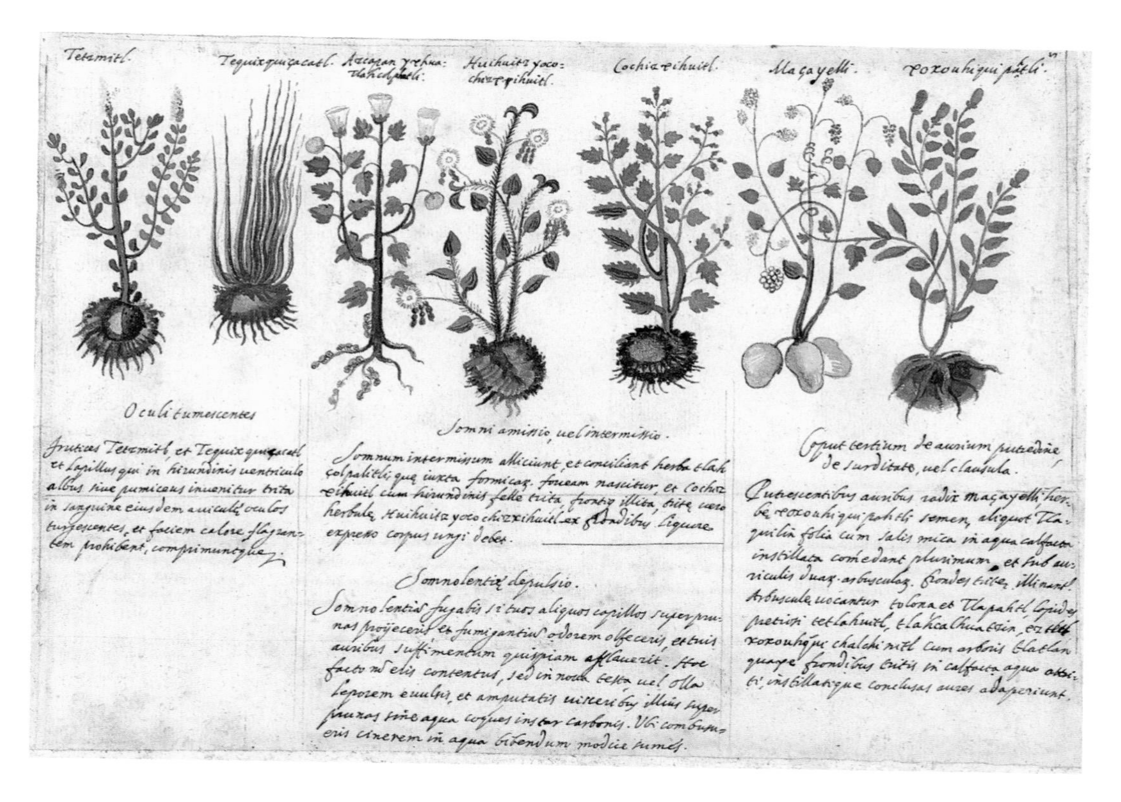


Fig. 9.3. Aztec plants and their uses (Cassiano dal Pozzo's copy of the Aztec herbal known as the "Codex Badianus"). Watercolor and bodycolor over black chalk, inscribed with the Aztec names of each plant above and descriptions of their pharmacological properties below, 208 x 343 mm. Fol. 7. Windsor, RL 27914. The Royal Collection © 2000, Her Majesty Queen Elizabeth II.

the whole of Hernández's text, but rather the five treatises on animals and one on minerals that together made up the *liber unicus* separately paginated at the end of most copies of the *Tesoro Messicano*.[91] This book is entirely unillustrated, and is the only direct transcription we have from Hernández's originals in the Escorial (rather than from Recchi's selection). It thus contains a great deal of new material not to be found in Recchi's compendium at all. No wonder that the Linceans decided to have it printed too.

But Cassiano brought back another item in his baggage. It was both touching and directly relevant to the Mexican project. When he visited the gardens of the "speziale" Diego Cortavila in Madrid in 1626, he saw a marvelous manuscript, an Aztec herbal, containing the illustrations of more than 180 Mexican plants, with their names in Nahuatl, and a description of their medicinal uses in Latin. This was the manuscript that subsequently became known as the "Codex Badianus," named for Juan Badiano of Xochimilco, the Indian translator of the original Nahuatl text into Latin.[92]

Little in the history of ethnobotany is more moving than the preface by its author, Juan de la Cruz, the Indian physician of the College of Santa Cruz in Tlatelolco (where Bernardino de Sahagún had once taught). "Endowed with no theoretical learning, but taught by experience alone," he was offering his "little book" to Don Francisco de Mendoza, son of the viceroy of the Indies,

> for no other reason . . . than to commend us Indians to His Sacred Majesty, even though we are unworthy. Would that we Indians would make a book worthy in the King's sight, for certainly it is most unworthy to come before the sight of such great Majesty. But you will recollect that we wretched and poverty-stricken Indians are inferior to all mortals, and for that reason our innate smallness and insignificance surely merit your indulgence.[93]

For Cassiano the value of the work was self-evident. De la Cruz's special pleading was unnecessary. Though Cortavila presented the original codex to Francesco Barberini,[94] Cassiano had a close copy made, so close that one could almost call it a facsimile (fig. 9.3).[95]

Although Gabrieli believed that this copy of the Codex Badianus was made in the 1650s, on the grounds that there is no sign of its influence in the *Tesoro Messicano*, we now know that it was made just after the embassy to Spain. But there was probably no time to make further additions to the section on plants by Schreck. Colonna, who was preparing his own "Annotations and Additions"[96] to Recchi's commentary on the plants, was in Naples and does not seem to have known or taken notice of the Indian codex. Nevertheless, like the material transcribed by de los Reyes for Cassiano, the Windsor copy of the Codex Badianus is yet another piece of evidence for the Lincean interest in the ethnobotany and ethnopharmacology of America; and, in the light of the most recent researches on all these manuscripts, provides a further instance of the providentiality of Cassiano's occasionally manic campaigns of copying, reproduction, and facsimile making.

## LATE ADDITIONS

It is possible that when Faber laid down his pen to his introduction at the beginning of 1625, he thought that his work was done, and that the book on Mexico would soon be published. But the material kept on coming in, and the Linceans felt that most of it was too important to be ignored. They always had great difficulty in letting a project go. Faber was tired out, and Cesi unwell and worried about his wife's gynecological problems. Besides, he wanted to devote his remaining energies to the part of his work that he most desperately wished to complete—the *Tabulae Phytosophicae* or Phytosophical Tables that would supplement their *Thesaurus*.[97]

The printing process itself was hampered by the need for all these additions. Nine months after Faber signed his preface, Cesi wrote a weary letter—typical of these days—to Galileo. He was enclosing a copy of the *Melissographia* to serve, as he put it, as an example of their "particular study of natural observations."[98] With regard to the Mexican book, he added, "more than fifty folios [i.e., around three hundred pages] have been printed, with the addition of many new things."[99] He realized that many more folios were still to follow.

Just over a year later, on October 7, 1626, Faber wrote to Johannes Thuilius in Padua, saying that the summer had been so hot that he could barely move his hand to the pen; but just that week he had at last finished his section in the *Tesoro Messicano* on the American merganser.[100] It had been his good fortune, he noted, to have had good conversations with no fewer than five missionaries, two born in America, and three almost entirely educated there.[101] Father de Aloaysa, whom he had met in the course of the Jubilee year, had provided him with a long description of the bird, and several drawings of it and some related birds had been sent to him by Cassiano from France.[102] No wonder Faber's section on mergansers had had to be expanded to eleven pages! New information was always too tempting for Faber. At this point, he wrote, almost 70 folios had been printed (that is, about six hundred pages); and it was increasing to one hundred.[103]

At least Faber now realized what they were up against: the scale of their own ambitions for the work. Progress was steady but continued to be slow. There always seemed to be something to add; and both Colonna and Cesi had yet to complete their own contributions. With the *Apiarium* out of the way, Colonna could turn his full attention to Schreck's text on the plants, and bring it up-to-date in the light of the most recent and advanced researches. His "Annotations and Additions" were added at a very late stage in the printing and constitute some of the most valuable portions of the whole book.[104] Mostly brief supplements to Schreck's and Recchi's text, they lend a tone of precision that often seems lacking elsewhere in the book (and especially in Faber's contribution).[105]

Things now began to move swiftly—or so, for a while, it seemed. After all, there was not much time. As always, Stelluti kept Galileo abreast of their project. In mid-August 1627, he wrote that he and Cesi were making a final push to have the Mexican book printed. They hoped that it would come out in a few months. Colonna's and Faber's annotations were complete, and Cesi, though taking his summer break, was working on his own contribution to the volume.[106] The reference seems an innocuous one; but it is, in fact, to Cesi's great plan to order the whole of the botanical world in tabular form, the *Tabulae Phytosophicae*. This is the first time that we hear, at this late stage, of the plan to include Cesi's tables in the work. It was yet another disastrous decision—at least in terms of the Linceans' plans for publication.

Three weeks later Cesi himself wrote to Galileo. As always, his domestic tribulations were getting in the way. For twenty-five years, he reflected, they had been plaguing his work. Here, of course, he was alluding to the difficulties his father and the rest of his family had been making for him ever since the foundation of the Academy in 1603. But at least, he wrote, "the labors of our publications are at the forefront. . . . I am very busy with these things. . . . the first volume of the Mexican book will come out very soon; its riches are such that it will demand a second and third volume."[107] Anyone would have wilted at the prospect of producing two more volumes. Surely what they had achieved so far was enough?

In the meantime Riquius, their panegyrist and rhetorician, went to work on polishing the prose of his colleagues and coordinating the prefatory material and the laudatory epigrams. On January 20, 1628, Cesi announced to Galileo that "the Mexican book is almost finished" and that they were working on completing his tables (as well as Stelluti's *Persius*).[108] All they had to do now was to arrange for the final privileges, or so they thought. A frontispiece had been commissioned from their favorite engraver, Matthias Greuter, who had already proved his skills on the *Melissographia*. Finally Faber signed the last words of his "peroration to the benevolent reader," addressing it from "my museum at Santa Maria sopra Minerva," on March 16, 1628.[109] Colonna signed his preface on the calends of June, from Naples. In his appendix on the *Narcissus serpentarius*, printed on the very last page of his additions (fig. 10.11), Colonna commented on the ruddy colors of that plant, which he had observed, he recalled, in September of that year "while these pages were to be printed."[110]

But their work was still not done. Although the Linceans were desperate to send their book into the world before Cesi died and before Galileo got into further trouble, there remained one more stumbling block: Cesi himself. He was not yet done with his *Tabulae Phytosophicae*.

So the agony continued. Though Stelluti reported to Galileo that the Tables were in press at the end of 1628,[111] on November 10, 1629, Colonna wrote disconsolately to Stelluti that the Mexican book was "already decrepit, even before its public birth."[112] This must have seemed all too true. Five days later Colonna acknowledged to Stelluti the reason for the last-minute delays: "Judging from what you tell me, our most excellent Prince is proceeding much too tardily with the printing of his Tables, distracted as he is by his domestic affairs"[113] (and, as we also know, his declining health).

The *Tabulae Phytosophicae* are one of the most elusive yet important parts of the whole of the *Tesoro Messicano*. Realizing that the *Tesoro Messicano* would reflect the range of their researches better than any of their other publications, Cesi and his friends decided to append the most complete of his frontispieces to the great "Theatre of Nature" he had been preparing for so many years. As ambitious as Jean Bodin's very different *Theatre of all of Nature* of 1597, it was intended to offer a specimen not just of his particular insights but of how he conceived of the world as a whole. So they decided to print "The first part of the Phytosophical Tables taken from the Frontispieces of the Theatre of Nature by Federic Cesi, a prospectus of the whole Syntax of Plants toward the training of Students and the Science of plants."[114]

Cesi did not live to see the completion of his tables. He died in August 1630, leaving only twelve and a half of them ready for publication and the remaining seven and a half in draft form alone. But still they offered critical testimony to one of the most important aspects of Cesi's own work.

What were the Linceans to do? Schreck had long since left for China, and Faber had predeceased Cesi in September of the previous year. Galileo was writing and preparing the way for his *Dialogue of the Two World Systems*. Only Stelluti, Colonna, and Cassiano remained active members of the Academy. Some copies of Faber's portion had already been printed up, with a cover title, in 1628, and some of these probably circulated too. The many pages of Schreck's text and its woodcuts, which Cesi had managed to proofread in the two years before he died, were also ready. Greuter, too, had prepared his frontispiece for the work as a whole. The question, then, was this: to print or not to print, and in what form? The *Tabulae*, after all, were still not complete.

A few copies of the book survive with frontispieces by Matthias Greuter dated 1628 or 1630 and twelve and a half *Tabulae Phytosophicae*. The full twenty tables, as reconstituted by the faithful Stelluti, only appear in copies with frontispieces by Mattias Greuter's son, Johann Friedrich, dated between 1648 and 1651. It is clear that the *Tesoro Messicano* had an unusually complicated publishing history. But whether one can speak of six separate "editions," on the basis of the frontispieces dated 1628, 1630, 1648, 1649, and the two different versions of the work dated 1651, is moot (the second version has a second altogether new title page).[115]

Cesi's death, and the struggles that led to the trial of Galileo, meant that everything had to be put on hold. A chill had settled upon everything. It took almost twenty years for Stelluti, much helped by Cassiano, to assemble all the different pieces of the text, make some additions, obtain all the right privileges, and find a sponsor for the publication of the closest to what could be called a complete version of the *Tesoro Messicano*.[116] In the slightly differing volumes dated between 1649 and 1651 there appeared not just Hernández's *liber unicus* on the animals of New Spain, brought back by Cassiano in 1626, but also the twenty tables of Cesi's *Tabulae Phytosophicae*. Stelluti had pieced together the last seven and a half of them from Cesi's papers. In this way, the work Cassiano and Stelluti finally succeeded in having published gives us the richest sense we have of the entirety of the Linceans' researches. Full of defects, lacunae, and last-minute changes though these versions of the book are, they still give us the best overall picture of what Cesi and his fellow Linceans had for so long hoped to achieve.

## SPONSORSHIP AT LAST

On February 20, 1649, Cassiano communicated some sensational news to his friend Nicholas Heinsius in Leiden: the "book of Mexican things" had found a new patron. At last it could be revived from the dormant state in which it had lain for so many years. Heinsius should tell Johannes de Laet, the great editor of geographical and natural historical works (who had already been nagging about its progress some thirteen years earlier),[117] that the work was finally on its way. Funds had at last been found for its publication.

Cassiano explained that Paolo Sforza, second husband of Olimpia Cesi, the daughter and heir of Prince Federico, had sold all the printed copies of the *Tesoro* in her possession for an extremely low price. These consisted of "one thousand good copies of the text, and three or four hundred spoiled ones (since these had not been looked after as they ought to have been)." All were now being sold for the risible sum of one thousand *scudi*—that is, less than one *scudo* a copy!

Cassiano had found a buyer for the lot: Alfonso de Las Torres, Philip IV's agent for Spanish, Sicilian, Neapolitan, and Milanese affairs in Italy. For a work containing a "huge quantity of paper and illustrations"—over four or five hundred, wrote Cassiano (in fact there were over eight hundred!)—this was extraordinarily cheap. At last Cassiano and Stelluti could proceed. The copies that Cesi had had printed up in the few years before his death could now be bound for publication. To them could be added the valuable indices of the work, the remaining "synoptic tables" by Cesi, and, above all, the *liber unicus*, or "the history by Francesco Hernández in the Library of the Escorial, and of which I had a copy made during the Cardinal's legation and which was then given to Prince Cesi."[118]

Within less than two years the work was ready for publication. But the *Tesoro Messicano* rarely cast a happy spell on its patrons. By the time it appeared in its definitive form (or in as definitive a form as it would ever attain), de Las Torres had gone to Spain and died. Fabio Colonna had died as well. The name of Vitale Mascardi appeared as the printer on the title page, rather than that of his father Giacomo, with whom Cesi had originally worked. The same with the title page, where Johann Friedrich Greuter replaced his father Matthias as designer (fig. 9.4).[119]

The various title pages of the different issues (both actual and planned) reveal a great deal about the checkered publication history of this noble work. The first two by Matthias are dated 1628 and 1630. There is no difference between them. In both cases the publisher (or printer) is given as Giacomo Mascardi. The earliest form of Johann Friedrich's title page, still bearing Giacomo's address, carries the date of 1648. But this never served as the title to a complete version of the work. The text of the titles of 1628, 1630, and 1648 does not vary:

> Treasury of the Medical Matters of New Spain, or History of the Plants, Animals and Minerals of Mexico from the accounts of Francisco Hernández, First Physician of the New World, written in Mexico City itself; collected and brought to order by Nardo Antonio Recchi of Monte Corvino, General Archphysician of the King of Spain and of the Kingdom of Naples, at the command of Philip II, King of Spain; illustrated with notes by Johannes Terrentius, Lincean, Physician and Doctor of Konstanz in Germany; now first published

Fig. 9.4. Johann Friedrich Greuter. Engraved title page, *Thesaurus* (1649–51).

for the benefit and use of students of Natural History through the efforts and expenses of the Linceans. The reverse of this page will give the contents of the rest of the volume. With the privileges of the Pope, the King of France, and the Grand Duke of Tuscany.

This is no mean title; and those that followed were not much different. The title pages with Vitale Mascardi's imprint—in other words, the one sponsored by de Las Torres—were dated 1649 and 1651 (fig. 9.4). They introduce what can now be considered the definitive versions of the *Tesoro Messicano*, and contain a number of telling differences from the earlier title pages. They announce that other material had been added to the texts already printed; that a synopsis of the work as a whole appeared on the next page; that the work had been divided into two vol-

Fig. 9.5. Second title page for the *Thesaurus* (1651).

NOVA
PLANTARVM, ANIMALIVM
ET MINERALIVM MEXICANORVM
HISTORIA
A FRANCISCO HERNANDEZ MEDICO
In Indijs præstantissimo primum compilata,
DEIN A NARDO ANTONIO RECCHO IN VOLVMEN DIGESTA,
*A IO. TERENTIO, IO. FABRO, ET FABIO COLVMNA LYNCEIS*
*Notis, & additionibus longe doctissimis illustrata.*
Cui demum accessere
ALIQVOT EX PRINCIPIS FEDERICI CÆSII FRONTISPICIIS
Theatri Naturalis Phytosophicæ Tabulæ
*Vna cum quamplurimis Iconibus, ad octingentas, quibus singula contemplanda graphice exhibentur.*

ROMAE MDCLI.
Sumptibus Blasij Deuersini, & Zanobij Masotti Bibliopolarum.
Typis Vitalis Mascardi. Superiorum permissu.

umes (the second, of course, being Hernández's *liber unicus*); and that the work was now dedicated to "Philip IV, Great Catholic King of Spain, the two Sicilies and the Monarchy of the Indies."

Many copies of the 1651 issue carry an additional title page as well (fig. 9.5), much more pedestrian than Johann Friedrich's resplendent allegory. It shows a belaureled ship and simple personifications of history and fame. Its purpose was to indicate that part of the expense of publication was borne by the booksellers, Blasio Deversini and Zanobio Masotti (who helped de Las Torres by buying three hundred copies from him just before he returned to Spain),[120] and that the printer remained Vitale Mascardi. The letterpress gives a new and different title than the previous one: "New

History of the Plants, Animals, and Minerals of Mexico, first compiled by the most distinguished Doctor Francisco Hernández in the Indies, and then digested into a volume by Nardo Antonio Recchi." It also has two further, significant emphases. Rightly crediting not just Schreck but also Faber and Colonna for the notes and learned additions, it proudly declares that at last "Some of the Phytosophical Tables of the Frontispieces of the Natural Theatre of Federico Cesi" have been added. Neither of these important signals appear on any of the Greuter title pages. Then comes a critical additional statement, which one might possibly have expected to find on one of the earlier title pages, but in fact appears now for the first time: "along with very many images, about eight hundred, in which individual species are graphically presented to the viewer."[121]

The illustrations of the book were indeed something to be proud of—but proud *enough*? The question haunted Cesi ever since he and his friends decided to present the Recchian compilation to the world. From the beginning the matter of the illustration of the work preoccupied them, and they constantly worried about the accuracy and quality of the drawings and woodcuts they commissioned. It was an issue that impeded their progress as much as any other.

The long trajectory of the book is clearly indicated by the sequence of privileges, approbations, and imprimaturs over the years. The earliest privilege dates from 1612; the last approbation from 1651.[122] The most complete version—"definitive" if one wants to call it that—of the *Tesoro Messicano* thus appeared a quarter of a century after Faber signed off on his preface of August 1625, forty years after Galileo first saw the copy of Recchi's manuscript with its illustrations in Cesi's house in 1611, and exactly eighty years after Hernández first began his research on which the book was based.

Each of the privileges acknowledges the Linceans' great labors in preparing the Recchian material and their role in publishing it. Each mentions the extraordinary expenditure of both effort and money involved in the publication. Each makes clear Stelluti's prime role in having the work printed, right from the beginning. Over and over again the privileges emphasize the medicinal value of the work and the illustrations. Copyright, furthermore, is made to apply not just to the content but also to the abundant images. In the very last of the approvals (the most definitive of all), the Jesuit censor, Baltasar de Lagunilla, omits all reference to the Linceans, and is content with a simple approval of the usefulness of the medical information the book contains.

Louis XIII's 1626 privilege is particularly interesting. It was owed, as it plainly states, to the intercession of Cardinal Francesco Barberini. Once more the recently elevated cardinal was able to help the Academy he had just joined. The matter, of course, was clinched at the time of his legation to France, almost certainly at the suggestion of the ever-enterprising Cassiano dal Pozzo.

Practically no two copies of the *Tesoro Messicano* are the same.[123] Any combination of its various parts seems possible. Almost every time a new title page appears, a few new pages of text are added or substituted, while the prefatory matter fluctuates quite considerably, according to the most prominent or important patron of the work at any particular juncture over the course of its long history. Although the core texts remain basically the same, sections, privileges, letters to the readers, indices,[124] and errata all appear in different places in the work; pagination changes slightly; additions such as the final seven and a half Phytosophical Tables are identifiable by the different paper on which they are printed; and changes are signaled not only by omissions and additions of parts of the text but also by a host of much lesser—sometimes minuscule—typographical alterations and devices.[125]

Just as can be judged from Cassiano's letter to Heinsius of February 1649, and from a comparison of the various title pages, privileges, and permissions, the major change—and a real break—comes with the volumes made up to go with the titles of 1649 and after. These carry the names of Philip IV as chief patron of the work and Vitale Mascardi as typographer (fig. 9.4).[126] There is one striking difference between the copies of the *Tesoro Messicano* that carry the title pages of 1649 and 1651 and the earlier ones. It emerges with particular clarity in the prefatory material to the various sections. This is the almost total suppression of the name of Francesco Barberini as a patron and enabler of the work.

## EXPUNGING THE BARBERINI

An exchange of letters between Stelluti and Cassiano in November and December 1650 provides critical details with regard to the final adjustments to the work and offers useful figures about the size of the printing. On October 27 Stelluti had written to Cassiano saying that he was sending him "the index of medicaments of the plants of Mexico, which has finally been completed, as well as the treatise on the animals which you brought from Spain." After this, he continued, there was nothing more to add "except for the dedicatory letter to the King of Spain, Mascardi's letter to the readers, and the summary of the privileges; when these are printed I'll send them too."[127] One thousand copies of the work were destined for Spain (to which de las Torres had just set out); four hundred were to remain in Rome.[128] Many of these copies were defective, as the newly reprinted pages did not seem to have been bound up with them; and these would now have to be packed up, sent to Spain, and added to the remaining copies.[129]

But at the very beginning of this letter—appropriately the last one in Gabrieli's great corpus of Lincean letters—Stelluti clarified exactly what these newly reprinted pages were. They were none other than "the three dedicatory letters,"[130] in other words, the dedicatory letters to Faber's *Exposition*, Colonna's *Additions and Annotations*, and Cesi's *Tabulae Phyto-*

*sophicae*. They had had to be completely redone. Since the earlier printings of these letters, an urgent change had become necessary. The name of the Barberini, especially that of Cardinal Francesco, had to be removed.

On the basis of the most complete and final copies of the *Tesoro Messicano*, it would be hard to tell how much the Linceans wanted to express their thanks to their great patron, or, indeed, how much they felt they owed to him. The earlier "editions" had made this altogether clear, but the later ones contain barely a sign of such acknowledgment. What happened to cause this strange—but all too characteristic—obliteration of one of the prime supporters of the work?

When Innocent X was elected to the papacy following the death of Urban VIII in 1644, the Barberini family fell into disfavor and disgrace. They stood accused, among many other things, of amassing unbecoming quantities of wealth and property, of profiting from Urban's infamous nepotism, and of embarking on disastrously ambitious projects. *Quod non fecerunt barbari, fecerunt barberini* rang the pasquinade. Francesco himself had to leave Rome on the night of January 16, 1646, and went to Paris. Although he returned from France just over two years later, he remained in high disfavor with the Spaniards and the now-strong Spanish faction in Rome, led by the pope himself. His name became well-nigh unmentionable, and his supporters were made to suffer one kind of humiliation or another for their association with the now besmirched family of Barberini.

No slight coincidence, in this climate, that de Las Torres should have had the new edition dedicated to Philip IV himself. This was done in terms that praised the king's grandfather for having sent Hernández to the Indies in the first place, and the Linceans for having saved the work from near oblivion (though in this de Las Torres had some role too). In his letter to Philip, de Las Torres recalled how, happening to be in Rome on Spanish business, he had decided that no expense should be spared to ensure that the Linceans' edition of Hernández and Recchi saw the light of day.[131] At the same time, Stelluti and Cassiano—perhaps in concert with Mascardi—took the no doubt prudent step of dedicating the completed set of Cesi's *Tabulae* in this volume to Don Roderigo de Mendoza, Philip's special legate to the court of Innocent. How could they *not* have attempted to expunge the name of Barberini from such a work?

First of all, they removed the separate title page to Faber's section on the animals. Not only did it carry the dedication to Cardinal Francesco, it actually bore the Barberini coat of arms, the now notorious trigon of bees.[132] On the verso of the second title page was an epigram by Riquius supposedly in honor of Francesco Barberini himself. Although it actually praised the Linceans and Faber in particular, it too had to be removed. Keeping most of Faber's introductory letter to this section, Mascardi and his Lincean advisers Cassiano and Stelluti changed its dedicatee from the cardinal to the general reader, and removed all references to Francesco on its final pages. They also took out the two references to his uncle, Urban VIII. At the same time, they retained Faber's glowing reference to Cesi's achievement. In the earlier "editions" Faber's section ended with three laudatory poems, one from Holstenius to Francesco Barberini Μουσηγέτης, one from Holstenius to Faber, and another from Müller to him. The first, of course, was removed.

As a way of dealing with the long dedicatory letter to the cardinal that had initially come at the beginning of Colonna's *Annotations and Additions*, they changed it to seem as if it had been written to Cesi. In fact, they even gave this new version the same date as the original one to Francesco Barberini—June 15, 1628. The original letter had made no bones of Colonna's admiration for and gratitude to the cardinal, referring to him effusively as the chief patron of the Linceans. Following the disgrace of the Barberini, this was no longer on—especially if they still wanted to insist on the role of the Linceans. Perhaps, too, Colonna's letter was felt to be still too well disposed to the discoveries and achievements of Galileo. And so most of it came out too, to be replaced by the factitious letter to Cesi (perhaps composed by Colonna himself in the last year or two of his life). Actually written many years after Cesi's death, it is a tortuous piece of work, says very little, and shows all the signs of hasty ex post facto fabrication.

Finally, Cassiano and Stelluti decided to replace the long and poetic letter to Cardinal Francesco that had served as the introduction to the *Tabulae Phytosophicae* (and that also emphasized the cardinal's sympathetic erudition and support for the Linceans). In its stead came the much terser piece dedicated to Mendoza, but which still uses some of the charming horticultural imagery of the original letter to Francesco.

Once again—as so often in these years—expediency prevailed over reality, this time in the form of the futile attempts to banish the name of a key player from a record that was pretty much already set. The decision to do all this, paradoxically enough, was taken by those once much favored Barberini clients, Stelluti and (even more so) Cassiano. But we have come to expect paradoxes in this situation. Ironically, the name now removed was that of the very family that had first supported Galileo and then caused him so much trouble. They, in the end, surely hold much of the responsibility for the concomitant oblivion into which the Linceans' work sank after the trial of 1633. How bitter all this was! And how little, in a way, can the expunging of the Barberini name make up for their miserable desertion of Galileo!

## TEN *The Doctor's Dilemmas*

### DESCRIPTION, DISSECTION, AND THE PROBLEM OF ILLUSTRATION

#### FABER'S EXPOSITION

There are important and compelling moments in the *Tesoro Messicano*. It is a work that has been much neglected and underrated, perhaps because amid so many words and illustrations it gives the impression of being little more than a farrago of information, useful and useless at once. Recchi may have thought that he put the Hernández notes in order following the instructions of Philip II, but the book still gives a strong impression of disorder. The 1790 editor of Hernández bemoaned the "wretched and calamitous fate" that befell the manuscripts in the hands of Recchi;[1] while Sprengel's description in his classic history of botany is typical enough of the general opinion: "An untidy work, from which—because of the too short descriptions, the rather crude illustrations, and the unspeakable Mexican names—there is very little to learn."[2]

But this could hardly be more incorrect. The taxonomic, ethnobotanical, and pharmacological aspects of the contributions by Schreck and Colonna are extraordinarily precious. Some of the species illustrated by them are described for the first time ever; several more are now lost or endangered. When Sprengel wrote he found thirty-seven entirely unknown species; many more are now on the verge of extinction. As a record of a turning-point in the history of botany these contributions are without parallel; as documentation of much that is still—and fast—being lost, it is invaluable. Even that most apparently discursive and anecdotal part of the book, Faber's *Expositio* on the animals of Central America, is of exceptional interest and value. Few other books give so rich an account, largely through the prism of strangers, of a now vanished culture of zoology and botany.

Pharmacology and medicine may seem to be the main focus of much of the work, but this is a native medicine made compatible with and equivalent (or even superior) to the medical and healing systems of the West. The delicacy and tact with which native names and native uses are preserved never flags. Abundant and organized according to an unusual variety of purposes, the indices themselves are uncompromising in their commitment to Nahuatl names and terms. They are the silent but eloquent traces of an unprecedented operation of cultural respect.[3] In fact, Nahuatl is the paramount language of the bulk of the indices; and in them, unpretentiously, the Lincean commitment to the vernacular achieves one of its most unexpected and noblest moments.

But it is Faber's section, certainly the most discursive of all, that provides us with the evidence and the invaluable traces of the Linceans' researches. It contains much that is critical for the history of science, and not a little for natural history itself. In it may be discerned the successes and failures of the Linceans, as well as what they left complete, and what only inchoate. Here and there—but running like some sad leitmotif throughout—are the fragmentary indications of Cesi's passionate and enterprising commitments in other domains.

The *Exposition on the Animals of Mexico* contains the fundamental texts on the earliest Western use of the telescope and the microscope. They offer expansive and moving descriptions of how Galileo first made his sharp-eyed discoveries in the sky. With the aid of the telescope, Faber reminded his readers, Galileo discovered that there were "innumerable hitherto unknown fixed stars, beyond those already known"; that the surface of the moon was not smooth and polished, but rough and pitted, with deep craters and high mountains; that the sun was the major source of light—and reflected light—in the heavens; that the galaxies were composed of thick clusters of stars; that there were four more planets besides those already known; that Saturn had two rings around it; and that Venus had phases like the moon.[4] Here too Faber gave credit to both Della Porta and Kepler for their respective roles in the development and use of the telescope, and recalled the famous meeting on the Janiculum in April 1611 when Galileo demonstrated its use to Cesi and his closest friends.[5]

Then, taking up the theme of the sharp-eyed lynx, Faber emphasized the clarity and sharpness of vision enabled by the microscope, the term he proudly says he invented.[6] While the texts on the telescope have long been known, those on the microscope are far less familiar. Faber tells of the examination of the smallest insects and animalcules, so small that they escape the range of normal vision.[7] Later on come useful details about the skilled Germans Nerlinger and Schad, who worked for Duke Maximilian of Bavaria and who ground lenses "with the sharpness of lynxes' eyes."[8] Faber described Cesi's discoveries of seeds on the undersides of the leaves of ferns (plants previously believed, he reminds his readers, to be seedless);[9] of Stelluti's work on the bee and the louse;[10] and of how all this inspired their examinations of other kinds of seeds, eggs, and reproductive systems in general.[11]

Here, in short, may be found much of the scarce printed evidence for Cesi's cosmological researches, for his work on ferns, fungi, fossils, and for projects such as that promising but oddly uncritical work the *Taumatombria*.[12] This was his strange but often remarkable treatise on unusual rains and other phenomena of the upper and lower air, that we know only from the brief allusions to it and drafts for it in his manuscripts,[13] and which he had to abandon in the wake of the struggle for Copernicanism. But above all Faber's section fleshes out what we already know of Cesi and his friends' concern with the anatomy of things and with the great mysteries of biological reproduction.

It would be hard not to admire the extraordinary discussions of the anatomies Faber conducted in his hospital of Santo Spirito in Sassia (he says he dissected well over 100 humans alone);[14] yet no one has as much as noted, let alone studied, them. With his fellow Linceans Müller and Cesarini, Faber conducted one animal anatomy after another.[15] He provides detailed accounts of his very careful dissection of an eight-month-old male fetus[16] as well as that of a two-headed calf born in April 1624 in a villa belonging to Marcello Sacchetti (the calf had been given to him by Francesco Barberini) (fig. 10.1).[17] Over and over again he examines the digestive tracts of ruminants. Here too are his and Cesarini's examinations of a tortoise and a turtle, as well as their extraordinary investigation into the pulmonary systems of both the turtle and the chameleon. These examinations are remarkably precise—even to the extent of taking the temperature of the chameleon's blood. To read the description of their palpitation of the breathing apparatuses of these animals is to be impressed not just by the intensity of their observations but also by the supple finesse of Faber's Latin.[18]

There are further references to anatomies of a boy killed by lightning,[19] and of a mole and a rat done with Müller (one of these is certainly recorded in the drawing of a hermaphroditic rat in MS 978 in Paris, fig. 8.40).[20] Indeed, Faber makes a particular point of describing hermaphroditic anatomies, not only ones such as those of the rat and the mole, but of humans too.[21] He even anatomized a wolf sent to him by Cesi in 1628.[22] If one could not decide what sex an organism was on the basis of external or surface characteristics, then one could always, Cesi and then Faber began to realize, cut down to the organs of reproduction themselves. These could at least provide the basic elements, the constant and consistent clues, to the classificatory puzzle.[23]

Faber refers to his own book of *Juvenilia*, in which he says he recorded the many anatomies he conducted in Santa Spirito in Sassia as well as in the hospital of Santa Maria della Consolazione on the other side of the Tiber. It is clear from this passage that a draftsman was often present at these anatomies. In the course of his discussion of the crania of tigers and lions, he recalls the specimens in his own museum of animal curiosities (with its more than one hundred animal skeletons).[24] He also mentions Filippo Liagni's now very rare book of 1621, which contains twenty-one strange etchings of animal and human skeletons, a work dedicated to Faber himself (fig. 10.2).[25]

Here too are close discussions of the embryology of various kinds of

Fig. 10.1. Anatomy of two-headed calf born in April 1624. Woodcut, *Thesaurus* (1651), p. 626.

626 ALIA ANIMALIA NOVÆ HISP. NARDI ANT. RECCHI.

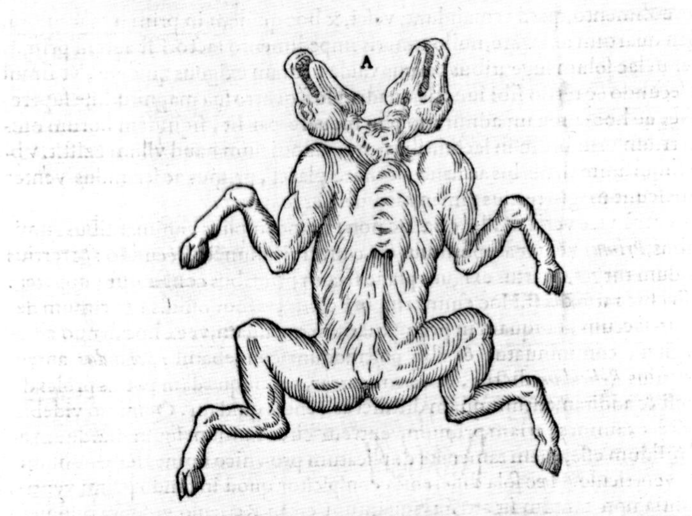


a Vitulus *monſtroſus biceps, menſe Aprili, Anno M. DC. XXIIII. propé* Romam *in Villa quadam Illuſtriſs. D.* Marcelli Sacchetti, *natus, detracta pelle hic ſtudioſo Lectori exhibetur. Cuius alterum caput vt paulo fuit elongatius, ita pes alter & coxa fuit retractior. In quo quid monſtroſi repertum fuerit, partim ex historia iam ſupra data clarum eſt, partim in ſequenti etiam figura delineatur.*

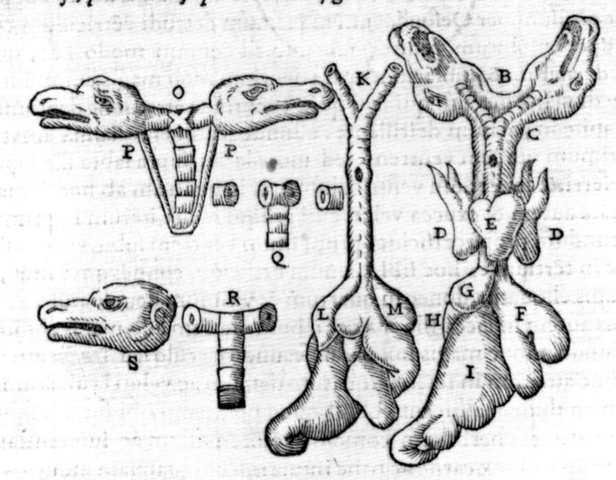


b *Duo capita cum ſua Aſpera arteria,* Pulmonibus *, Corde , & Ventriculis quatuor.*

*Vtriuſque*

eggs and of the implications of the relationship between albumen and yolk (fig. 10.3). Faber even thinks of writing a book on the intrauterine life of humans.

In keeping with the Linceans' exceptional interest in habitat and behavior, Faber writes about mergansers, cormorants, cranes, flamingos, and pelicans. Often he draws on the ornithological material that Cassiano had sent him, such as the picture of a merganser from France[26] and the *discorsi* on pelicans and the ruby-throated hummingbird from Canada (cf.

Fig. 10.2. Filippo Liagno, skeleton of a heron. Etching, *Liagni Scheletra* (1620–21).

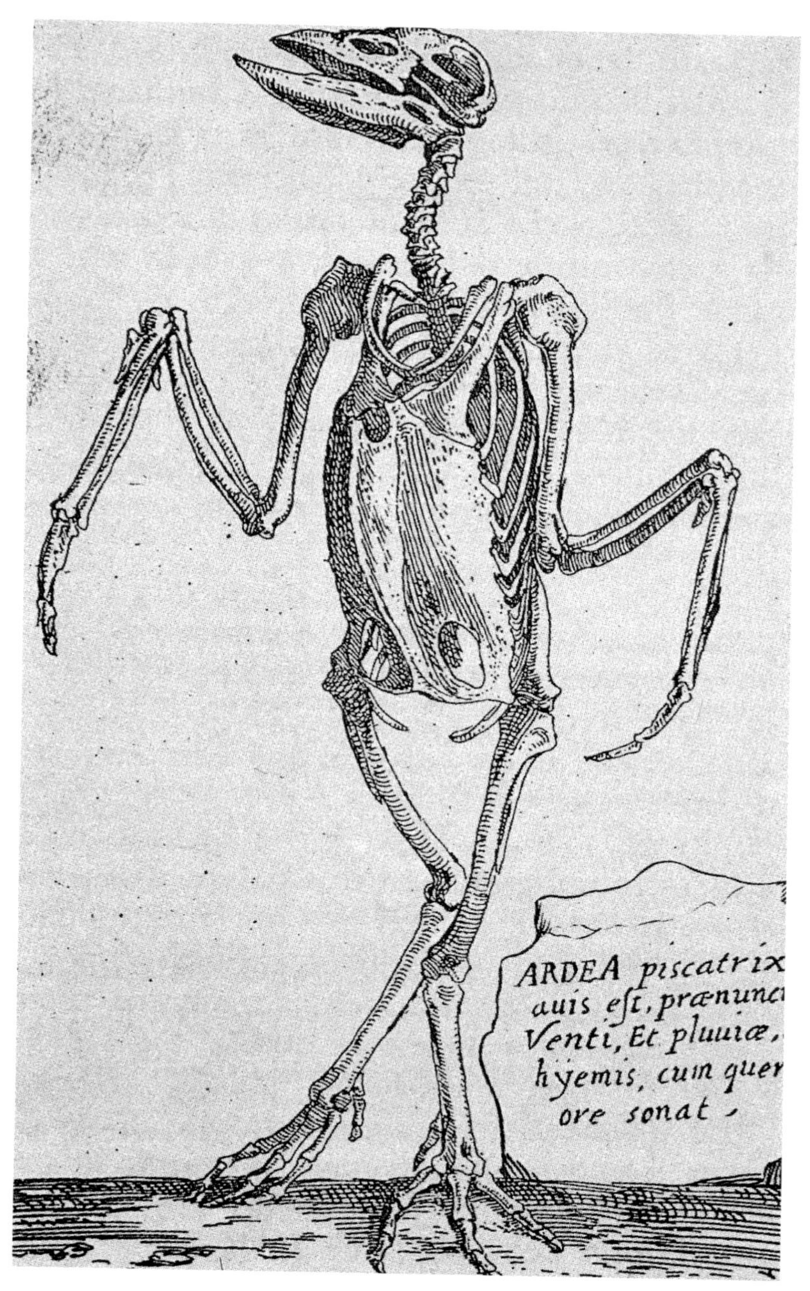
ARDEA piscatrix
auis est, prænunc
Venti, Et pluuiæ,
hyemis, cum quer
ore sonat.

figs. 1.5–7).[27] He augments Hernández's description of the toucan with the help of a report sent to him by Cassiano (via his brother Carlo Antonio) from Paris, where, in 1625, Cassiano was able to study the toucan sent from Brazil to Louis XIII (fig. 10.4).[28] Faber devotes a surprising amount to the behavior and mimicry of the parrots of the grandees of Rome, from Scipione Borghese's to Cesi's wife's. He notes down the Dutch songs sung by a parrot belonging to a Flemish merchant in Rome (fig. 10.5).[29] Faber shares his fascination with the details of mimicry with

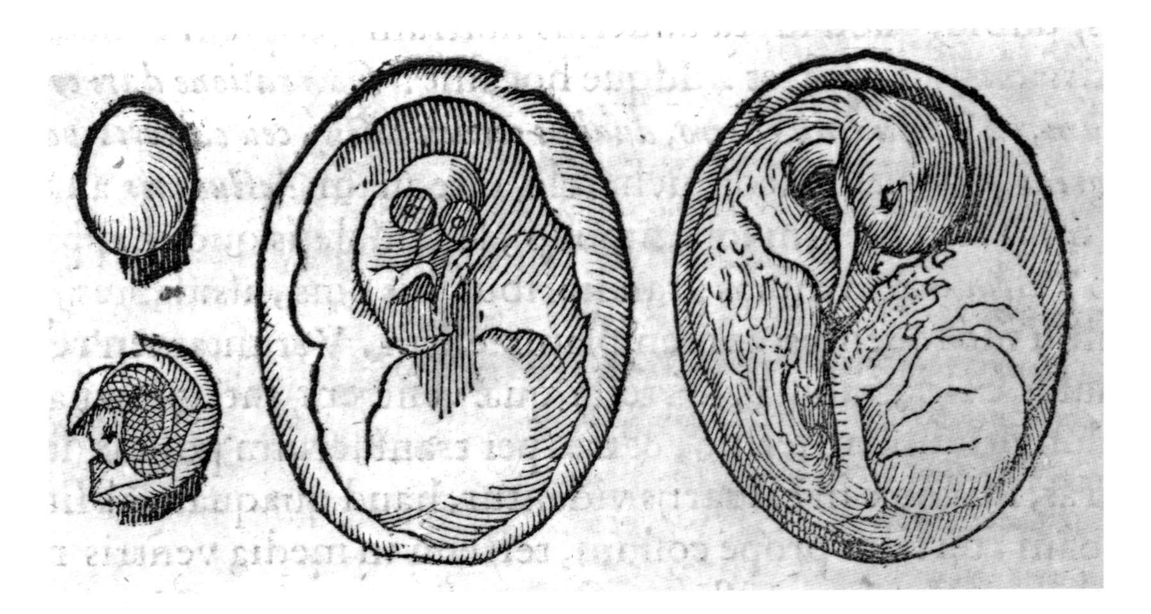

Fig. 10.3. Lizard and chicken eggs. Woodcut, *Thesaurus* (1651), p. 766.

Stelluti, who wrote on the subject in his commentary on the poems of Persius in 1630.[30] Just like Cassiano, Faber was fond of dogs, and he provides a delightful account of the different degrees of domestication and the peculiar talents of dogs belonging to the Medici in Florence, comparing them with the information he obtained from the missionaries about the *Canis mexicanus* (dog trainers will find much on the history of their subject here). There is barely an area of human interest to which Faber does not have something to contribute.

These are all tours de force of discursiveness; but time and again they yield up quantities of new material for the history of natural history. "If you are interested in botany or natural history," wrote Giacomo Mascardi in his initial letter to the reader,

> if you are a doctor, philologist, or phytosopher; if you are the florist of Princes, or of noblewomen, and flatter with sweet things; if you are a seller of drugs, a trainer of pharmacists, a druggist, or aromatist, and you seek health, delights, or money from the newest herbs, you will find myriads of new things, new images [over one thousand of them, on eight hundred pages, as Mascardi had just remarked] and new words too, which from every point of view will fill your eyes, mind, and desire.

The work is also, of course, a rich resource for the history of medicine. Faber, after all, was one of the leading medical practitioners in Rome, and he tells of his own experiences, such as the demands placed on his energies in 1600, when he had to care for more than two thousand sick people and foundlings in his hospital of Santo Spirito. He outlined his treatments for rabies and dropsy, his diets for the sick and sedentary, his views on the usefulness of fasting, and the right and wrong times to do so. He

Fig. 10.4. Toucan. Woodcut, *Thesaurus* (1651), p. 697.

Fig. 10.5. The song of a parrot. *Thesaurus* (1651), p. 717.

named many doctors who merit further investigation, from Giovanni Battista Ferrari's friend Marsiglio Cagnati[31] and Francesco Barberini's personal doctor Taddeo Colicola, to the Roman obstetric surgeon Francesco Mambriano and the writer Giulio Mancini, as well as a large number of now forgotten and half-forgotten *medici* and *archiatrae:* Marc'Antonio Luciano, Pietro Moretti, Quinto Sereno, Pompilio Tagliaferri, Francesco de Almeida, Pompeo Caimo, Ludovico Cascarino, Prospero Cichino, Angelo Colio, Prospero Marziano, and hosts of more or less obscure others.[32] Faber seems to have known almost everyone.

His text is full of cases broadly related to his and Cesi's concerns with reproduction. He records the case of the Florentine woman who bore no fewer than thirty-six children, including a set of septuplets. He describes his treatment of a boy with an abdominal hernia, and writes a great deal about medicaments, mineral springs and baths, and the salubrious uses of certain kinds of food (sheep meat is one unexpected example, raised in connection with a discussion of the llama). Over and over again Faber mentions contemporary pharmacologists and rhizotomes, such as the forgotten Polish expert on antidotes, Johannes Davidowicz, for instance, and the Linceans' friend and colleague from Flanders, the pharmacist and phytognostician Enricus Corvinus.[33] Then, too, Faber discusses the dangers of fumes and other noxious odors, and writes—as the Linceans always had—of poisons and their antidotes.

If the more rigorous parts of Faber's discussions were often contributed by Fabio Colonna, it was Cassiano who provided him with large quantities of information about more anecdotal matters, and about the ethnographic and the folkloric aspects of animal behavior and use. There is much information on the contents of Cassiano's laboratory-museum (as there is on several other museums in Rome too, including that of Francesco Gualdo), and a great deal on his ornithological researches, only inadequately reflected in Olina's *Uccelliera*. Just as Faber's references to fossils and ferns flesh out the scraps of information we have on Cesi's work on these subjects (and the illustrations that went with them), so too do the many ornithological passages help situate Cassiano's barely known treatises on birds.[34] They also cast light on some of the more splendid drawings—of the pelican, cormorant, and stork, for instance—from the paper museum and now in the Royal Collection and elsewhere (e.g., figs. 1.3–8).[35]

Faber is particularly interested in the problem of musk and musklike odors. He spends many pages worrying about how exactly the American civet produces its distinctive odor, and from which part of its anatomy it comes. Colonna helps him out with this, sending him two long letters on the subject and three illustrations, including one of its anal region (fig. 10.7).[36] Between Faber's original illustration of this animal (fig. 10.8) and the letters and illustrations sent by Colonna comes one of the German doctor's longest and most distracting excursuses. In it he moves from the question of the production of the civet's scent to that of various kinds of musk. He speculates at endless length about the origins of ambergris and amber and offers a number of rather careful observations on insects and insect parts in amber.

The subject of ambergris, of course, generates a long discussion of whales and whaling, and recent cases of beached whales. He notes that Cassiano has a set of whale's teeth at his house, as well as of that of the fearful anthropophagous fish, the *lamia*. As always, Faber is perhaps a little too credulous when it comes to Cassiano, but after a further divagation, now on the subject of dolphins, Faber returns to more basic matters.

Whales raise the question of categories: are they mammals or fish? How do they reproduce? This, as we have seen, was the basic question underlying so many of Cesi's researches. From fossil amber (one of the examples of mixed species over which the Linceans obsessed the most) Faber turns to other kinds of fossils as well; and so, once again, the issues come full circle. Motivating all these studies is the problem of the relationship between reproduction and classification. Faber confronts the problem of the status of fossil woods and bitumen more than once, only to move indecisively away from them. He knew that these were the special domain of Cesi's own researches and that the latter would eventually find a way of determining the classes to which they belonged. Were the fossil woods and "metallophytes" that so absorbed Cesi for many years plant or mineral? And how exactly was one to account for the imperfect plants and animals, those phenomena of nature whose principles of reproduction were obscure or which stood on the borderlines between one class and another? The problem extended in many directions, including odd phenomena such as *lyncurium* and fungiferous stones.[37] Classificatory concerns of a similar order also underlay Faber's many references to hermaphroditism.

Faber gave a long and sympathetic account of the botanical excursion the Linceans made in 1611 on the slopes of Monte Gennaro and which none of them ever forgot.[38] From his voluminous papers we now know just how fundamental a role this trip played in the beginnings of binomial classification.[39] Problems of classification also drove Faber's apparently old-fashioned interest in a variety of monstrous forms, notably the peculiar and incomprehensible skeleton of the so-called *Dracunculus* in the possession of Cardinal Barberini (fig. 12.10). Was it really a monster? Could it ever really have existed at all? And if so, how could it possibly be fitted into the order of nature?[40] Then too, there were less improbable but just as puzzling cases, such as the two examples of two-headed calves, one dissected by Faber (fig. 10.1) and the other, with four eyes, possessed by Cesi himself. How did such phenomena arise in the first place?

And what, we often wonder, as we encounter ever more frequent references to Cassiano's museum of natural history, were its principles of organization? The museum on paper seems to have had a variety of such principles; but how specifically did Cassiano order the natural historical specimens in it, as opposed to those from antiquity?[41] What places were they supposed to hold in the chain of being or the order of nature? Perhaps this question was never resolved; perhaps the specimens Cassiano owned were merely there to offer clues, nothing more, to nature's order—even when, in the case of anomalies, they offered clues *ex contrario*.

But Faber's *Exposition* contains much else too—and much, on the face of it, that has practically nothing to do with natural history. Several of his artist friends are movingly commemorated in these ample pages. His discussion of the Mexican lizard reminds him of a lovely nocturne by his friend Elsheimer, which shows the boy Stellio mocking Ceres for the eager drafts she takes in the course of her tiring search for her daughter Proserpina (fig. 10.8). The fact that Stellio was then transformed by Ceres into precisely that lizard (*Stellio mexicanus*) for his impudence is what provokes Faber's recollection—not what is seen on the painting itself. Even so, his recollection of the painting, and of Elsheimer himself, prompts Faber to a fine piece of ekphrastic writing, in which he praises the painter's skill in depicting natural phenomena of all kinds.[42] Faber pays tribute not only to the much-loved German painter but also to the Flemish landscapist in

Rome, Paul Bril,[43] and, above all, to his great friend Peter Paul Rubens, whom he cured of pleurisy in 1602.[44]

Moreover, there are many richly detailed references to the missionaries from New Spain who provided him with so much information, both in person and in their written accounts.[45] Faber seems to have been especially fond of Gregorio de Bolivar; and he knew the famous missionary to China, Nicolas Trigault, who returned from there in 1624. Cornelis Drebbel, the Dutch inventor who lived in England and made such a notable contribution to the history of the microscope, was another acquaintance;[46] and Faber offers a picturesque account (in the course of his discussion of whales) of the submarine Drebbel constructed for James I to help in the defense of La Rochelle.

Sometimes it seems as if almost anything could turn up in these pages: popular Roman games, elaborate marble furniture studded with precious stones, the varieties of buffalo mozzarella, the voraciousness of moles, pearls and pearl fishing, the mineral springs known as the *Acque Albule* at Tivoli and the dermatological ailments they cure, women with beards, poisonous vapors, rattlesnakes, and, as always, any number of museums and cabinets of curiosity throughout Italy.

Often Faber's scholia on the animals of Mexico seem to be little more than farragoes of information, only remotely or ostensibly related to the animals themselves. Yet the whole is more coherent than may at first appear. There are brilliant and erudite observations on both local and exotic flora and fauna. Faber is interested in everything to do with sex—and not only because of the Lincean fascination with problems of biological reproduction. Irrelevantly, for example, he records the story of a lunatic who, maintaining that anyone who did not cease from sexual activity would not go to heaven, cut off his genitals. But what is really at stake throughout is the study of reproductive process and anatomy; and this, in turn, is a critical index of the profound Lincean concern with the order of things.

The Linceans—and above all Cesi—may have tried to illustrate everything they encountered and observed, but they came to understand that illustration could never be entirely adequate to a number of the basic tasks they set themselves. The whole process of description by visual means had severe, inherent, and constitutive limitations. What use was the mere reproduction of surface? Single species changed too often in the course of their lives to be capable of being captured in their essence. What one saw on the surface could vary from hour to hour, from season to season, and from age to age. Surface description could never provide adequate clues to order. Mere pictures would never be enough.

Nor, in a sense, would observation either—at least, not on its own. Time and time again one admires the Linceans for their firsthand observations of things. Repeatedly Faber and Stelluti use the phrase *oculo teste*, according to the testimony of the eyes.[47] But reliance on ocular evidence alone could often be misleading. It was precisely this that flawed Stelluti's

Fig. 10.6. Hendrik Goudt, engraving after Adam Elsheimer, *Stellio Mocking Ceres.*

and Cesi's researches on fossils and on other subjects.[48] Through his friendship with Galileo, however, Cesi knew that observation was not everything. He was aware that the reliability of firsthand observation was often suspect. So was experiment. Hypothesis was at constant war with observation, and in important ways had to precede it.

This is precisely what lies at the heart of the double tension in the

*Tesoro Messicano*. On the one hand, the long pursuit of illustrations; on the other, the provision of the great synoptic tables, with their hesitant but ever-present emphasis on the order provided by geometry.

It is true that the microscope gave the Linceans the possibility of investigating more deeply, of looking beneath the surfaces of things. Dissection and anatomizing constituted the more practical, the more tactile side. But here too lay a fundamental difficulty. There were always more surfaces to be found, and ever a beyond. When to stop, when to know that one had reached the last innard? However much one described and portrayed as one penetrated more deeply, whether with scalpel or microscope, there was always more below, something ever smaller. The aim of the Linceans was to go as deep as they could in order to find what was essential in things—and that essential they found in the elements of biological reproduction. Soon, however, they realized that such elements were best and most efficiently described in terms of mathematics and mathematically ordered components.

Only in his Phytosophical Tables did Cesi begin to set this out. It is not surprising, then, that the Linceans should have decided to add them to the *Tesoro Messicano* (even if they could not wholly follow them). After all, these were tables that were predicated on the desire to bring order to nature by finding the right places for things. No wonder that the whole illustrative project was at first found wanting, and then failed. In his Letter to the Reader at the beginning of the work, Mascardi acknowledged that Cesi's contribution filled the need for logic and clarity amid the profusion of material he was faced with. To a much greater extent than his friends, Cesi intuited the need for a system of classification in which pictures would have to take a backseat—if not disappear altogether.

## ILLUSTRATION PROBLEMS, YET AGAIN

The illustrations were indeed what held things up from the beginning. But not at all for the more theoretical and abstract reasons just outlined—for much more immediate and practical ones.

It may be difficult for us now to appreciate how consequential the issue of illustrations in books of natural history once was. After all, Linnaeus hated them,[49] while our own age has become accustomed to the easy availability of images, of every description. Perhaps because he was brought up in early seventeenth-century Holland—where one could have pictures of almost anything—Johannes De Laet seems to have been exceptionally attuned to the prospective value of the *Tesoro Messicano* (though he also had his own more selfish motives in mind):

> By chance I have received a book in Spanish printed in Mexico City on *The Plants, Animals, and Gems of New Spain* by Frate Francisco Ximénes,[50] who took so much material from the books of Dr Hernández . . . ; but because Ximénes has provided no pictures, his book can hardly be of much use to us. For several years we have

> been awaiting the compendium of that great book, as put into order by the most learned Nardo Antonio Recchi (as I learned from Fabio Colonna). It began to be printed a while back, and I saw its printed title a few years ago already; last year, I asked my cousin Elzevier to enquire in Rome what hope there might be for the book. I understand that the work has either been interrupted or stopped, because a similar book by Nieremberg has been published in Flanders;[51] but . . . he has provided no images besides those which Clusius and others have already given. It is for this reason that I am surprised that so desirable a work [i.e., the *Tesoro Messicano*] has been passed over.[52]

De Laet was writing to the Linceans' friend Lucas Holstenius on October 10, 1636, asking him to check whether all this was true, and whether there was any possibility of acquiring the illustrations for his own books (produced on behalf of the Dutch East India Company and for his publisher friends the Elzeviers). De Laet accurately identified one of the main aims of the Linceans in editing the work of Hernández: to provide the world with copies—at least a good number of them—of Hernández's pictures of the flora and fauna of Mexico. But it was precisely this that proved to be one of the chief stumbling blocks to the expeditious publication of the work, right from the very beginning.

Already in 1582 Juan de Herrera, as we have seen, had worried about how best to reproduce Hernández's originals.[53] Later on, both Recchi and Petilio were reluctant to let others see their copies. Throughout the 1590s, natural historians from Clusius to Aldrovandi were anxious to lay their eyes on the illustrations, whether the originals or the copies. The Linceans were always discussing them, even after Cesi actually managed to acquire some of them. Galileo himself puzzled over them at Cesi's home in May 1611.[54] From that moment on Cesi never stopped trying to get the best kind of illustrations for their edition of Recchi's compendium.[55]

At a meeting of the Academy on October 29, 1611, it was reported that "the printing of the Mexican Treasury is going ahead more and more, since the engraver and the painters are working hard on the images."[56] On January 13, 1613, Cesi, Faber, Valerio, and De Filiis met for various items of business and to pay homage to one of their admirers and prospective patrons, Johann Gottfried von Aschhausen, prince-bishop of Faber's home city of Bamberg. Johann Gottfried asked De Filiis, their librarian, to arrange for all the books they had written and had yet to write to be sent to him. "And because he takes a special delight in botany," the minutes of the meeting continued, "the Prince [Cesi] thought that it would be a good idea if the librarian could put together a book made up of the printed images of the Mexican plants to be offered to him, with a dedicatory inscription added and with laudatory epigrams prefixed to it by Faber and by Valerio."[57]

This book, factitious though it is, still exists. Preserved in the Barberini collections in the Vatican Library, it contains 167 woodcuts of plants, with neither names nor text—except for the dedicatory title and the two epigrams, one in Greek by the unfortunate Valerio, and one exceptionally vacuous piece of Latin by Faber.[58] We know that from the minutes of the next meeting of the Academy that the book was assembled by Faber and then presented to the prince-bishop. These minutes, however, conclude with a matter of greater moment: "The just-printed *Letters on the Sunspots* by Galileo were presented at the meeting, and it was decided that they should be distributed to the Linceans and their friends."[59] Thus did the maculate sun outshine the splendors of the Indies.

As usual, though, Cesi wasted no time. Desperately making use of the slightly quieter period between the preparation of the sunspots book and that of the *Assayer*, he turned once more to the *Tesoro Messicano*. It was clear that the illustrations would not stand comparison with those in works such as Besler's *Hortus* of 1613,[60] or even with the second part of Fabio Colonna's *Ekphrasis* of 1616, with its sprightly and precise illustrations of lesser-known plants.[61] Of course Cesi wanted the images in the *Tesoro Messicano* to reflect Hernández's originals—or Recchi's copies—as closely as possible; but there were so *many* of them! Recchi himself had at least six hundred. Cesi thus faced two problems: on the one hand, by remaining faithful to the originals, he would have to eschew the kinds of precision and finesse they may not themselves have had (despite the constant praise of their colors); and he had to have his artists produce as large a quantity of images at as fast a pace as possible.

Unfortunately, the evidence for much of this process is sparse. Who, for example, were the engravers and the "painters" of the images already referred to at the meeting of October 1612? Aside from the Greuters (the engravers of the title pages) there is no indication in the book of who the artists or engravers of the woodcuts might have been—whether the botanical and zoological illustrations, or the wonderful head- and tail-pieces of Linceans, Lincean flowers (*Stanhopea tigrina* Batem.), and landscapes with Mexican plants and animals.[62] But Giovanni Baglione's 1642 *Lives of the Painters, Sculptors and Architects* offers a precious clue. "Isabella Parasole of Rome, wife of Leonardo, did a book engraved[63] with different lace designs of her own invention and other kinds of ladies' work. . . . also by her hand are the woodcuts in the book of plants by Prince Cesi of Acquasparta, that most learned nobleman." To which Baglione adds: "and the works which Isabella could not complete were engraved by Giovanni Giorgio Nuvolstella."[64]

These are two crucial names. That of Nuvolstella[65] occurs several times in Cesi's household accounts for 1618–19 as an engraver of woodcuts in the Mexican book. In a series of receipts and brief contracts between July 20, 1618, and November 30, 1619, Stelluti recorded payments to Nuvolstella and to an assistant, Nicolo Martini, for the woodcuts of "the Indian plants and other things" (though when Nuvolstella arrived at

Acquasparta on July 20, Stelluti recorded that he did so "in order to engrave the pictures of the *animals* of the Mexican book").[66]

None of this, however, tells us as nearly much as we would like. Did Isabella do all the drawings, and Nuvolstella—only known as an engraver—the woodcuts? Or did Isabella also make some of the woodcuts herself, as Baglione's text suggests? What role, if any, did her husband Leonardo play? Baglione records that he was responsible for the woodcuts in Castor Durante's famous herbal of 1585; and it has sometimes been supposed that she was responsible for the original drawings for that work too.[67] But the woodcuts of the plants in the *Tesoro Messicano* are very different from those in Castor Durante's book. On the other hand, the fact that the woodcuts in the *Tesoro* were based on the existing images in Recchi's manuscripts makes the whole matter of the actual draftsman of the woodcut images rather moot. Does it matter? Nuvolstella and his assistant Martini could easily have made their cuts directly from the Recchian illustrations without any intermediate drawings by another artist at all. In the absence of any further evidence, there is no clear solution to this conundrum.

Fortunately the evidence for the progress of the illustrations, and the endless complications and difficulties they caused, is much clearer. In several respects (though unfortunately not with regard to authorship) the letters corroborate the accounts. On February 25, 1619, Cesi wrote to Faber to say that he was making his engravers work on the Mexican book. With his usual optimism he added that "soon I want to bring the work to a conclusion; in the meantime please put in the request for the privilege."[68] Two months later he wrote: "I am hoping for better news about the privilege; in the meantime . . . please compare the figures made here with the originals, and then send them back."[69] His concern for the accuracy of the illustrations remained unabated. On May 18, he was still awaiting the revised figures of the animals and still hoping to expedite the printing of the book.

From these letters, Gabrieli rightly concluded that by the middle of 1619, book 9 of Schreck's section—the book on animals[70]—was ready to be sent to the printers.[71] Then came a pause. Faber and his colleagues were too involved in helping Galileo with the *Assayer* to have time for much else. As soon as that book was out, Faber could turn again to the remaining tasks on the *Tesoro*. The letters from late 1623 on provide an insight not just into his dealings with Petilio—long since moved to Rome—but also into the many problems of the illustrations: the stalling, the interruptions, the refusals, and—above all—the preoccupation with getting the reproductions just right.

First Faber had to attend to the part of the manuscripts that Schreck had been unable to finish editing. This was the section that came just before Faber's *Exposition*, titled "The names and images of other plants of New Spain"[72]; and on December 19, 1623, he went to Petilio's, where he began to work on it,[73] and to add a few comments on Schreck's minimal observations.[74]

But Petilio was not an easy collaborator. The day after this visit, Faber wrote to Cesi that "We should be a little patient with the grumpy old man; but at least I'm in a good place with him."[75] This was a little smug, perhaps; but they needed Petilio. Faber had to get on with what came next, namely his own section on the animals of Mexico. For this he had to consult the Recchian originals so jealously guarded by Petilio, who had made it a condition that he be consulted at every stage about the description of the plants and animals in the book.

A year or so passed. On February 3, 1626, Faber sent Cesi a corrected and supplemented quire. It had taken a lot of effort, he recalled, especially seeing that the prickly old man ("il Spinoso," they called him) had made one difficulty after another for him. He was both petulant and drunk. Faber had not had a chance to look at "his" animals; the session had been too short. In a postscript Faber added that Petilio wanted to be sent the quire then being printed (probably the section by Schreck immediately preceding Faber's *Exposition* on the animals), and to look it over.[76]

Was Petilio being fussy about the illustrations too? Did he too want to be sure, in his senescent way, that justice was being done to the much-touted quality of the Recchian illustrations? To judge from Faber's subsequent letters to Cesi this does seem to have been the case. Always anxious to make his treatise move ahead, Faber went to Petilio's home and studied the illustrations there one by one. He worried about their identifications. Sometimes he seemed remarkably ignorant:

> As for the first figure, I doubt whether it's a wolf, since its tail is not hairy, which is almost a trademark of wolves—unless they are different in Mexico. I do not know for sure what animal the third figure could refer to; the fourth differs not a little from the third, as is clear from the description.[77]

And so Faber kept on going back to Petilio, one day after another at this stage, to consult his manuscript on the animals. He moved with as much speed as possible on his project, but it was a constant struggle to see what he needed.

All this is vividly recounted in the correspondence of 1625–26. The old man gets odder and odder, he is a Cynic, he eats a dog of a meal, he has a continuous discharge from his body, he is increasingly fastidious, he really doesn't want to have anyone around at all.[78] "Yesterday," wrote Faber on February 18, "I got four other animals out of him; but because Petilio "was drunk and disconnected I couldn't do anything else."[79] In June he waits for Bolivar to come so that they can go and see Petilio together;[80] in January 1626 he spends the whole day at the Barberini Palace, which, he notes, prevents him from visiting the old man.[81] This he will do the next day. In the meantime he has a drink with Cardinal Francesco, who lets him have the catalog of the animals in the Dutchman

Paludanus's museum in far-off Enkhuizen. Faber mentions this too in the *Tesoro*.[82] Sometimes it seems that he has to put everything in.

But at least things were moving along. By February 6, the woodcuts of the animals were in full production. "I'm also sending the drawing for the woodcut," he writes to Cesi,

> the other I'll have this evening. I also wanted to send the large one, so that he who draws it on the wood can, if he wishes, take something from it; the second cut will be of the innards and the head and other details, where has to be a little more careful in showing everything in a small picture. Now, thank God, the bull and the calf are complete; next I'm coming to the boar, and your Excellency should send the painter to Petilio to take the drawing of the navel it has on its back, which Maggi forgot to do.[83]

These references are to the unhappily inaccurate illustrations of the anatomy of the two-headed calf (fig. 10.1), the "Mexican bull" (a bison, fig. 12.7), and the "Mexican boar" (a peccary) in the *Tesoro*[84]—but who is Maggi, here named for the first time? It is almost certainly Giovanni Maggi, whose life and works are described by Baglione immediately before those of Parasole and Nuvolstella, and who had made a number of illustrations of birds used by Cassiano in his *Uccelliera* of 1622.[85] But aside from this elliptical reference we know nothing else about his role in the illustration of the *Tesoro Messicano*.

Petilio, who died shortly afterward, is never mentioned again. Faber's book on the animals was now almost complete. It remained for Colonna to get on with his *Annotations and Additions*. But at the same time Colonna kept an eye on the work of his colleagues.[86] How could he not? He was too skilled a natural historian. All the while that he worked on his updating of Schreck's contribution, he kept Faber informed about relevant material elsewhere and sent him new drawings to update those he found inadequate.

In October 1625, obviously concerned about the decidedly odd illustration of a civet in Faber's long discussion of that beast in the *Tesoro*, Colonna had sent him three much better images to serve in its stead (e.g., fig. 10.7; cf. fig. 10.8). And since part of Faber's discussion was about the scent the animal produced, Colonna added a detail of its anus for good measure.

The accompanying letter from Colonna describes both the male and the female civet in some detail. He elaborates at length on the production of the "odoriferous humor," more abundant in the male than the female. Finally he reflects on how different his own images are from that of the Mexican civet illustrated some forty pages earlier by Faber (fig. 10.8), as well as from those of others in the world. It is true, he acknowledges, that regions and climate can affect the way species look (just as Westerners look different from people in the East); but still, he continues, painters

Fig. 10.7. Civet and detail. Woodcut, *Thesaurus* (1651), p. 580.

Fig. 10.8. Civet. Woodcut, *Thesaurus* (1651), p. 538.

often make mistakes. *Verum & pictura fallax est,* he recalls, citing a well-known dictum of Pliny the Elder's.[87] Of such doubt was their growing skepticism about pictures formed.

But skepticism was also mitigated by optimism. Time and again both Faber and Colonna made clear not only their commitment to the usefulness of pictures, but also their strong sense that pictures could always be improved. This attitude also held things up. They constantly circulated pictures among themselves and discussed them. Faber reproduced the Recchian image of the Mexican amphisbaena (fig. 12.8), and worried whether it showed this puzzling worm as two-headed or not. Then Cassiano sent him another drawing of an amphisbaena, clearly depicted with two heads, and allegedly made from the life (fig. 12.9). To our eyes it could hardly look more improbable; yet Faber adduces it as key evidence in favor of the actual existence of the two-headed beast. Though he has his doubts, Faber concludes that it must, on balance, exist. After all, he has a picture of it from Cassiano![88]

A similar doubt—all too easy to understand when we look at the illustration—creeps into Faber's description of the preposterous Mexican dog, the Aztec *Ytzcuinteporzotil* (fig. 12.5); yet he concludes that it is a faithful reflection, but for the splendid colors, of the Recchian original.[89] On the other hand, the illustration of the Mexican leopard (the *Tlacoozelotl* or *Tlalocelotl*) is so precise—*especially* in the colorful original—that it is easily distinguished, Faber says, from the preceding illustration of the Mexican tiger, and from the panther so diligently described by the ancient writers (whom, of course, he cites at length).

In cases such as these taxonomy and illustration go hand in hand. In sorting out the problem of leopard versus panther, Faber has yet another card up his sleeve: he notes that while not separately illustrated in Recchi's manuscript, the panther is actually shown in his pictures of the palace of the Montezuma kings of Mexico, its skin often spread out on the floor at their feet.[90] Image, ethnography, and classification could not be more closely intertwined than in passages such as these—except, perhaps, in those two strange masterpieces of integration, the *Apiarium* and the *Tabulae Phytosophicae*.

Colonna had fewer doubts about the possibilities of illustration than Faber. More direct and to the point, he had a keen eye for the defects of illustration in those parts of the book he was assigned to revise, namely Schreck's books on the plants. Not only was he a superior botanist to Schreck, he was also able to do many of the new and revised drawings himself. There was much, therefore, to be done. Colonna went through all of Schreck's text very carefully indeed—even though, of course, it had already been printed.

But he was less than satisfied. To judge from his letters of 1626–27—and from his own text too—he continued to have ready access to the Recchian originals. In correcting Schreck's (and sometimes Hernández's) morphological and taxonomic mistakes, Colonna studied the Recchian drawings more closely than any of his colleagues. Along with Cesi he was more preoccupied than any of their colleagues with the accuracy of the illustrations.

Colonna's is the only section of the work to refer to the page numbers of the Recchian manuscripts. If anyone wished to reconstruct these, the evidence would be found here.[91] Comparing Schreck's descriptions with the Recchian originals, Colonna does not hesitate to point out where the woodcuts go wrong.[92] But he also alerts the reader to inaccuracies in the Recchian illustrations themselves (though, like Faber, he prizes their colors and their taxonomic utility). He unhesitatingly adds color details where Schreck has none, and very quickly notes where the latter fails to reproduce something actually available in Recchi.[93] And if something is *not* there at all, or the illustration in Schreck is unduly defective, he tells the reader where to go for a better one. For example, he sends the reader to specific works by Monardes and Clusius for more elegant illustrations

Fig. 10.9. *Lysimachia*. Woodcut, *Thesaurus* (1651), p. 882.

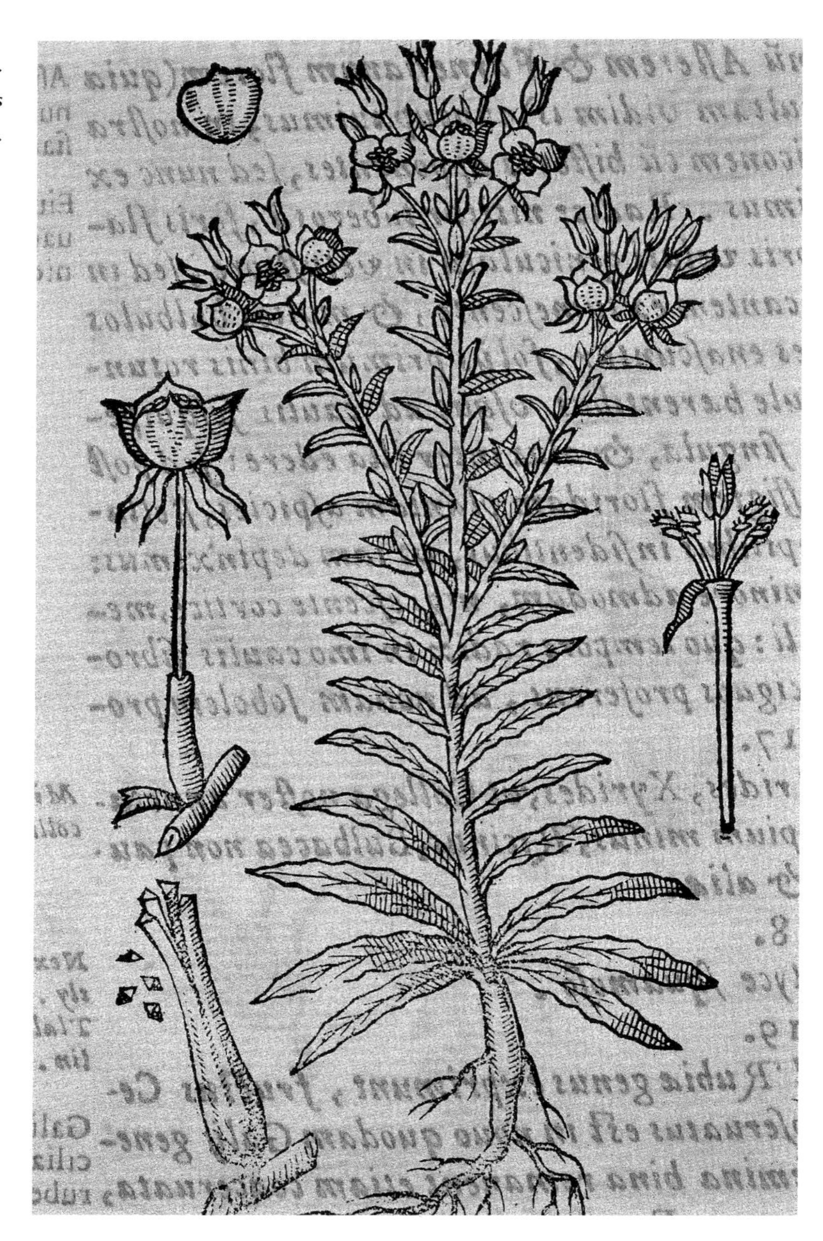

of the Aztec *Pycielt*, the tobacco plant (which he also describes in the most nuanced coloristic terms).[94] Colonna knows the vast botanical literature like the back of his hand.

## ILLUSTRATION AND IDENTIFICATION

Throughout his comments, but particularly in a dense set of three pages toward the end,[95] Colonna worries about the illustrations and about whether Schreck actually got his identifications right. Time and again, he observes that one plant is actually another, illustrated on some other page in Recchi's manuscripts. Schreck's *Axochiatl*, for example, is actually a *Lysimachia:* he knows this from a specimen brought back from Virginia and sent to him by his Veronese friend, the botanist Johannes Pona. To accompany his argument he provides a much superior illustration of the plant, along with details of its stem, flower head, and sexual organs (fig. 10.9).[96]

Colonna's correspondence of 1627–28 is full of references to plants that particularly interest him, and he discusses them in ways that reveal much about how closely he and his colleagues worked together. On November 28, 1627, he wrote to Stelluti that once more he has written a short piece about the *Planta caesia*:

> Do improve it by correcting both the concept and the language. Add and subtract as you wish, if you think this is the right thing to do; otherwise let me know and tell me what better to do. I can't find anything about the *Barberina* among the Mexican plants that satisfies me.

He reminds Stelluti that in the supplement to the *Tesoro* they should point out that the bees that habitually circulate around the flower of the *Flos cardinalis Barberini* [*Lobelia cardinalis* L.] are yet another tribute to the Barberini family.[97] Once again natural historical observation is pressed into the service of panegyric. Not surprisingly, Colonna discusses and reproduces this plant on the very page on which he alludes again to the *Apiarium* and the scale of its achievement (fig. 10.10).[98]

With new botanical names like these, others must follow. In the same letter to Stelluti, Colonna also analyzes the appropriateness of the name *Columnia* for the Aztec *Coyolli* (*Areca catechu* L.). Since it vaguely resembles a column, he notes, the name is especially apt. The same for the Mexican plant that he says he is about to name *Stelluta* in honor of his friend, and that happens to have a flower shaped like a star, *stella*.[99] And Colonna drifts off into other preoccupations. He hasn't yet been able to have a drawing made of the granadilla or passion fruit, a plant that especially interested him;[100] his illness has prevented him from seeing the fruit of the [*Narcissus*] *serpentarius* that he has had drawn on several occasions; and could Stelluti please send him some news of the little dragon belonging to Cardinal Barberini,[101] and a few seeds of the *Flos cardinalis*?[102]

All this is typical of Colonna's intensity. At the very last minute, he was still pushing ahead on every front, particularly with regard to the illustrations. Of course he received his drawings of the fruit of the granadilla, and of the fruit and bulb of the *Narcissus serpentarius*. These were intended to supplement a first set of drawings of these plants, which had already been cut and printed up; now, at the very last minute, he had to get the new drawings to the printers.[103] In fact, the woodcut of the *Narcissus serpentarius* appeared on the last page of Colonna's contribution, just before Cesi's *Tabulae*, as a final appendix to the book (fig. 10.11). In it Colonna noted that he collected the fallen fruits and seeds at the end of September

Fig. 10.10. *Flos cardinalis*. Woodcut, *Thesaurus* (1651), p. 880.

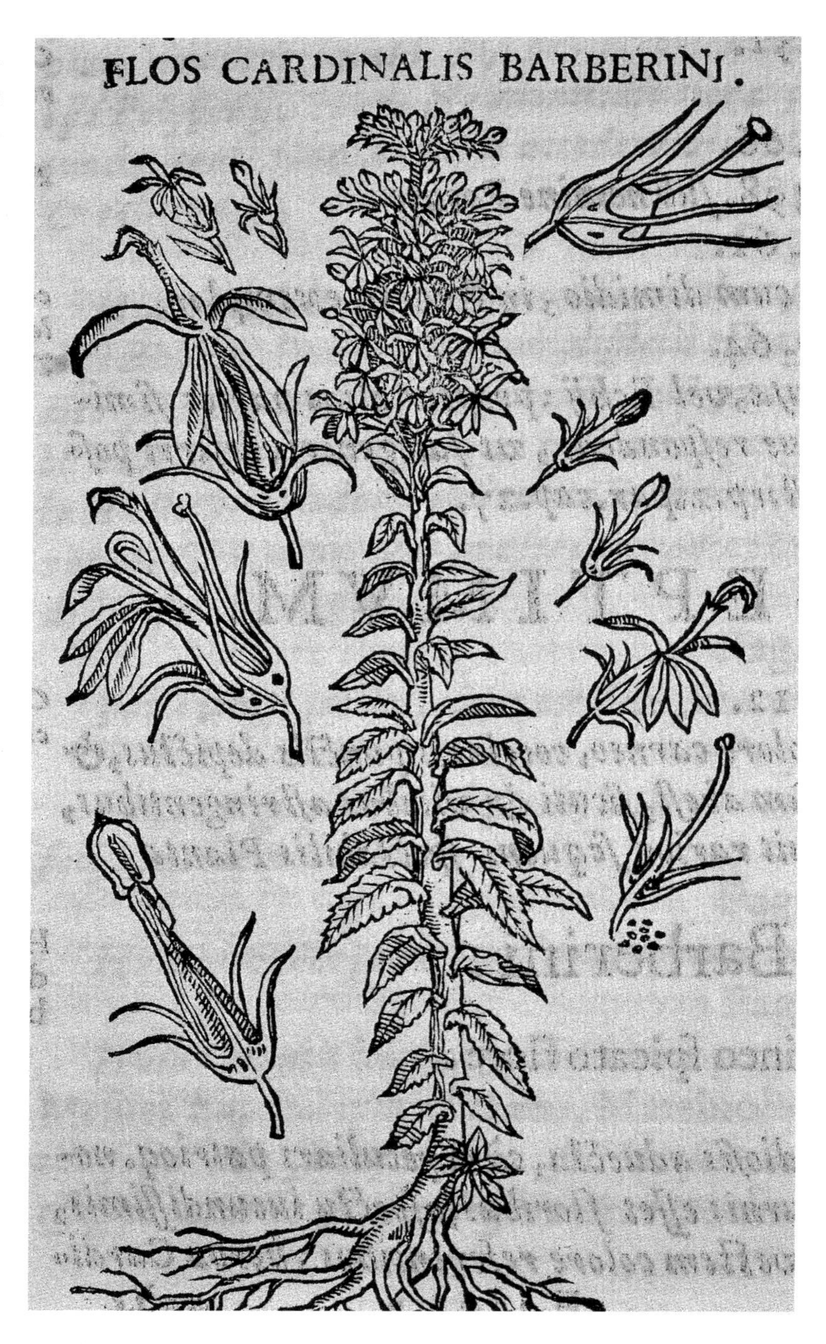


1628.[104] They were represented here too, along with the bulb (fig. 10.11). This was cutting things very fine indeed.

On the same day that he wrote to Stelluti, Colonna managed to write to Cesi too. Suggesting yet further changes and additions to his description of the *Planta caesia*, he turned to yet *another* of his botanical preoccupations, the *cannella*, or cinnamon plant. The woodcutter, he said, had produced a very unsatisfactory illustration of its leaf (fig. 10.12). It was

Fig. 10.11. *Narcissus serpentarius*. Woodcut, *Thesaurus* (1651), p. 899.

FAB. COLVMNAE LYNCEI ANNOT. ET ADDITIONES. 899

Appendix ad Serpentarij Narcisi descriptionem.

*Ne immemores verborum in calce descriptionis elegantis huius nouaeq. plantæ videamur, & ne quis nobis ignauiæ notam inurere forsan possit, pro virili enixi sumus, vt quid amplius huius plantæ historiæ adderemus. Hoc anno mense Septembri floruit. Illustris, autem* D. Bernardinus de Corduba *(de quo suprà) vt & nos, & rei herbariæ studiosos solito fauore prosequeretur, florem iussit custodiri, donec fructum perficeret, quem etiamsi minus ad maturitatem protulit, ex eodem tamen imperfecto multa licuit obseruare, & notare, ita vt plantam hanc rarioris, nouiq. planè generis existimare debeamus, nullis nostri Orbis plantis nec fructu vel flore, nec specie similem.*

NARC. SERP. BVLB. ET SEM.

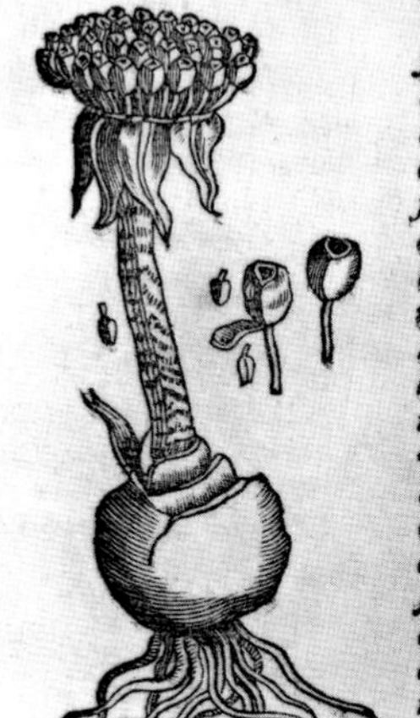

Proprium *habet, vt fructū ferat copiosum in summo caulis, & floris ouali plana basi, duabus vncijs longa, & paulò minus lata. Post Solis in libram ingressum marcescentibus, & dependentibus floris folijs, & longis petiolis ex albo virentibus, & maculosis etiam nonnullis semiunciam & plus longis, fructus profert Oliuæ nuclei magnitudine trigonos, quibus flosculi insidunt, vt diximus, copiosè numero plusquam centum. Flosculis decisis, quod euenit in fine Septembris, quo tempore hoc anno* 1628. *notauimus, dum hæc imprimenda erant, rubescunt, aream trigonam in summo eorum deferentes, ternas cellulas habentes, in quibus singula semina ab imo ex breui petiolo trigona dependebant, & quia adhuc immatura, & ex albo virentia, an plura emittat nescimus; sapore erant ex acerbo amaro. Bulbus, vt ablata parùm terrà licuit obseruare (non enim vltra permissum, ne bulbo vis auferretur) iam folij cacumen vix exerebat extra terrā ex bulbo: magnus est, Alabastri modo candidus, solidus, & squamis crassis tribus distinctus, quæ, vt existimamus, ora fuerunt, ex quibus aliàs caulem, & folia producat. Vtinam propagata in sobolem planta vlterius obseruare permisisset, plura sanè cum exactiore historia proferre nunc nobis licuisset. Iconem quidem, quoad fieri potuit, accuratam exhibere voluimus, vt pro viribus Botanicæ studiosis satisfaceremus.*

both inaccurate and a poor reflection of the Recchian prototype. In fact, it had been made by pressing an actual leaf to the paper and drawing its outlines. So just to be sure, he was having another illustration made of a cinnamon leaf (fig. 10.13), on the basis of an example in the collection of their mutual friend, the indefatigable traveler in the Middle East, Pietro della Valle (who had already provided the Linceans with so much information).[105]

On February 17, Colonna wrote to Stelluti once more, reporting in typical detail on three drawings he had just received: one of the strange *Dracunculus* in Cardinal Barberini's collection (fig. 12.10),[106] one of the leaf of the cinnamon plant, and another of that of the pepper. Colonna scrutinized these drawings closely and found fault with each of them. The cin-

Fig. 10.12. Cinnamon plant and leaf. Woodcut, *Thesaurus* (1651), p. 35.

namon leaf was a little too broad and not oval enough, that of the pepper a little too narrow, and so on and so forth. He says he will have *still* better drawings made of all these.[107] How late Colonna was working on these apparently simple illustrations, and how intent on getting them just right!

The difference between Schreck's truly feeble drawings of the cinnamon and pepper leaves and those that appear in Colonna's *Annotations*

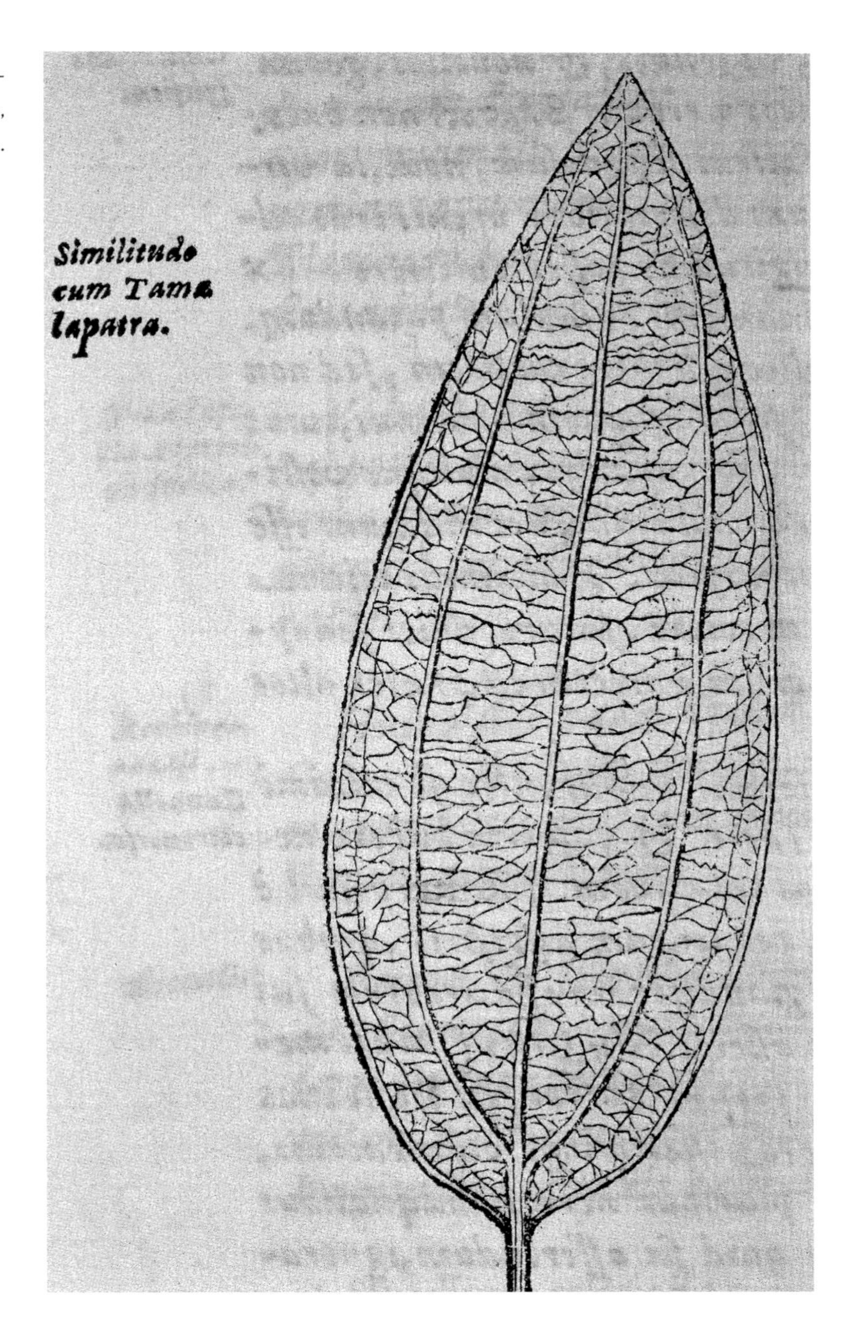


Fig. 10.13. Leaf of cinnamon plant. Woodcut, *Thesaurus* (1651), p. 864.

*and Additions* is vast. Schreck's leaves could hardly be more rudimentary, and they are shown in a wretched and inconsequential way beside the plants concerned;[108] Colonna's are noble in their enlarged isolation.[109] To us they may seem strikingly simple; but the detail of their venation (which Colonna also comments on in his letter) is remarkable (cf. figs. 10.12 and 10.13). None of this appears in Schreck.

In their unadorned concentration on a single part of a plant, Colonna's pictures of leaves stand out from all the other illustrations in the book.

Once again Colonna used them as launchpads for two of his typical taxonomic excurses. What, he wanted to know, were the possible distinctions between the *cannella* and the plants both ancient and modern writers called *cassia* and *cinnamon*?[110] And between the leaf of the pepper and that of the betel nut (both of which he had from Pietro della Valle)?[111] Even today there are few who could bring such skills in morphological analysis to bear on the resolution of taxonomic problems like these.

Though one might not want to. The interest of these illustrations has to do with how they reveal Colonna's growing awareness of the insufficiency of color as a taxonomic marker, and his sense that the concentration on particular *parts* of plants might indeed be more useful. Hence the unexpected illustrations of only the leaves of the cinnamon and pepper plants, and the extraordinarily fine-grained descriptions of their distinctive characteristics.

In this manner Colonna used the occasion of his annotations and additions to Schreck's text to expand his brief. Where he could, he showed the flowers and fruit-bearing parts of particular plants either overlooked or too cursorily represented in Schreck. There are new discussions and better illustrations of plants such as the passion fruit so beloved of the Neapolitan botanists;[112] and wholly new illustrations of a large number of other plants.[113]

Thus ends the long struggle to illustrate the *Tesoro Messicano* in such a way as to do justice to the famous illustrations Hernández brought back with him from New Spain. Even though Cassiano and Stelluti added the *Liber unicus* about the animals and minerals to their edition of 1649–51, they did not manage to have it illustrated. Just getting the new version prepared for publication—and completing the indices and Cesi's *Tabulae Phytosophicae*—was hard enough. Besides, there simply would not have been enough money to provide and print still further illustrations; and there were already plenty—almost a thousand.

## THE EVIDENCE OF THE PROOFS

Finally we come to two more documents that are relevant to the story of the *Tesoro Messicano*. The first dates from the earliest days of the campaign to illustrate the book. It provides clear and striking testimony to Cesi's two main preoccupations: accuracy and precision in illustration on the one hand, taxonomy and classification on the other.

Bound together in two volumes in the rare print division of the Vatican Library are two hundred and seven proof woodcuts for the plants in Schreck's section of the *Tesoro Messicano*.[114] They are inscribed with numbers ranging between 1 and 273 in the first volume, and between 278 and 753 in the second. These numbers relate to the folios in Recchi's manuscript on which the woodcuts were based. Many of the cuts are the same as those in the little volume Faber presented to the prince-bishop of Bamberg in 1613.[115] Quite a few were not used in the final printing of

Fig. 10.14. *Coacihuizpatli.* Pen and ink. Rome, BAV, RG Stamp. I, 81, fol. 88.

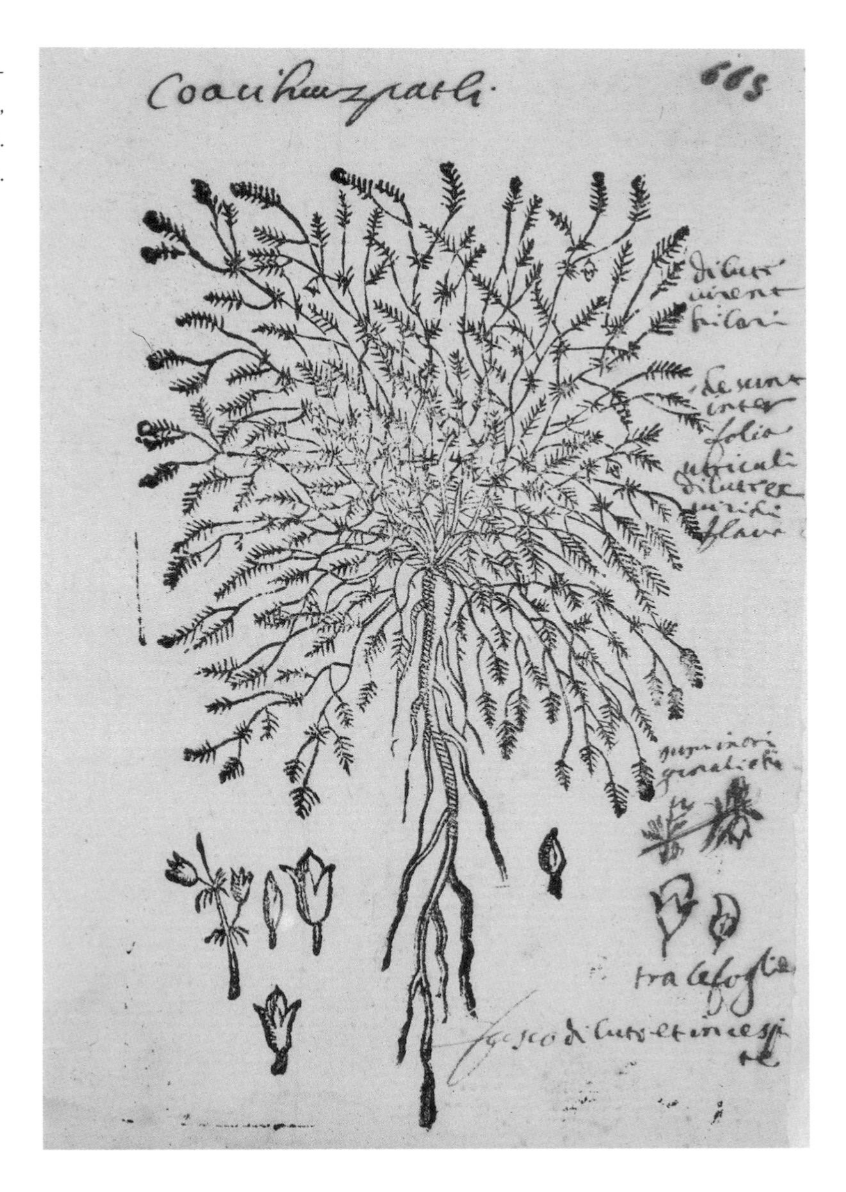

Schreck's contribution, since they were replaced by new and more accurate ones.[116]

Some of the woodcuts, such as that of the *xicama*, are clearly proofs, as may be judged from the much better versions that actually appeared in the *Tesoro*.[117] Many, such as the much-used *Coacihuizpatli*—here noted as a remedy for tooth ailments—have corrections or additions drawn directly onto them (fig. 10.14). These were then incorporated into the final version of the book.[118] Whether these corrections were by Cesi, Schreck, Colonna, or possibly even another among them, like Winther, we cannot be sure (though Colonna, the only one with artistic skill among them, is the most likely candidate). They also made manuscript notes on many of these woodcut proofs.[119] The three or four friends passed these images

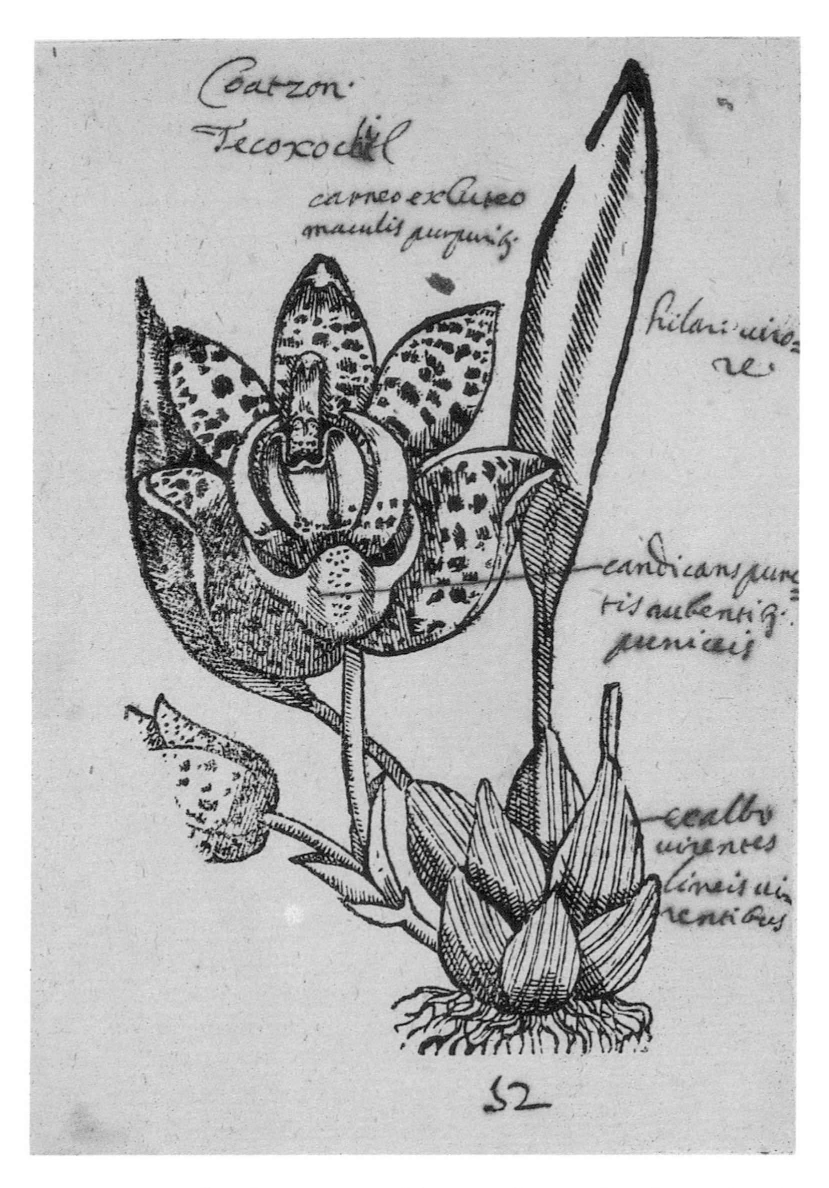


Fig. 10.15. *Coatzontecoxochitl* (*Stanhopea tigrina* Batem.). Rome, BAV, Stamp. I, 80, fol. 22.

among one another for comment, debate, and correction, in a truly collaborative spirit.[120] Most of the color notes were made by Cesi himself (e.g., fig. 10.15).[121] Generally they were based on the Recchian originals but often were taken from actual specimens. Cesi and Colonna were responsible for the bulk of the taxonomic observations and queries. In the pages of these oddly factitious volumes, taxonomy and pictures meet with unusual intensity. The two friends obsess about getting the names of the plants right, whether in Nahuatl or in Latin. Wherever they can, they provide the more common names in Latin.[122] They make notes comparing the woodcuts with the originals. They decide that some have been misnamed, and make the correct correlations and identifications in the *Tesoro* itself.

For example, the several "caryophilli" in Recchi are sorted out, and the woodcut supposedly of the *Atepocapatli* rejected in favor of one corresponding more appropriately with the description of that plant.[123] Cesi and Colonna agree that only one of the two attempts to illustrate the *Xazaro* is right, and settle on which is to be placed in the chapter on that plant in the *Tesoro*.[124] Schreck had worried about the different Aztec names for the plant known as the *Caragna*, and had accordingly left his chapter on it unillustrated.[125] But Cesi and Colonna were more certain: "this is the Caragna and it wasn't put in its place because it didn't correspond well [with the description in Hernández]" runs the firm inscription on one of the proofs.[126] And Colonna decided that it could indeed be placed with Schreck's chapter on the *Tlahueliloca quahuitl*.[127] Such was their knowledge of both Aztec and European plants!

The two friends agonized endlessly about matters like these. Always they acknowledged how much better it was to see the originals, whether of the plant itself, or of the Recchian equivalent. "I don't like the way these leaves look," Cesi noted on the only drawing among these proof illustrations, that of the *Coyotzin* or *Tozcuitlapl xochitl*, "you have to see them in the original" (fig. 10.16).[128] Just as in the case of the plants supplied by Pietro della Valle, the shape and striation of leaves became critical taxonomic markers.

The second document is Cesi's own proof copy of the *Tesoro Messicano*. Still preserved in Rome, it forms the largest portion of a volume composed of parts already printed by the time of his death, and a few added later.[129] The parts Cesi saw and corrected date from that brief and critical period between the completion of the first printing in late 1628 and his death in August 1630. The labor he expended on it is thus all the more poignant, but perhaps no more so than in the case of many of the other projects that were curtailed by his passing.[130]

Copiously and often Cesi commented on the illustrations—both on the woodcuts in the text and on the Recchian originals themselves. Once more he made abundant notes on color, as if he were somehow trying to reclaim both the splendor of nature itself and of the pictures assembled by Hernández and Recchi. What emerges most clearly is Cesi's determination to get the names of the plants right, and to find the right categories into which to place them. He lists as many nomenclatural possibilities as he can, as well as the relevant taxonomic criteria. Time and again he corrects the names given to the plants in Schreck's section of the work. He replaces one Nahuatl name with another, and directs the reader's attention to their Latin or vernacular equivalent, where available. When in doubt he provides alternative names. Often he refers to what he regards as the distinguishing characteristic of a species identified by him.

Nowhere is the extent of Cesi's taxonomic knowledge more clearly revealed than in his notes to the unfinished second part of Schreck's contribution, that is, to the "Images and Names of other Mexican plants and

Fig. 10.16. *Coyotzin* or *Tozcuitlapil xochitl*. Pen and ink. Rome, BAV, RG Stamp. I, 81, fol. 56. (Preparatory drawing for the woodcut in *Thesaurus* [1651], p. 282).

animals." In this section, Cesi not only makes a whole series of taxonomic observations, but actually identifies a number of "anonymous" plants (that is, plants not identified by Schreck or by Recchi either).

The drive to classification was thus stimulated by the drive to picture. The plant world offered the Linceans the most abundant and complex problems of classification. And the struggle between the need to picture and the desire to classify emerged more clearly and more densely in the *Tesoro Messicano* than in any other work the Linceans produced.

The woodcut illustrations in the *Tesoro* are symptomatic of the problem they faced. Cesi and his colleagues were all too aware of the fact that the woodcuts did not show the colors of the originals; but at the same time they realized that unless the illustrations showed what was beyond the reach of the eyes of ordinary people (but not of Linceans!), unless, in other words, they excluded the vagaries of surface description in order to penetrate to those zones that the microscope could now reach, it would never be possible to grasp the essence of things.

Shortly before he died, as he surveyed the pages of the *Tesoro Messicano* coming off the press, Cesi must have realized that in at least two respects his illustrations had failed. He understood that they could never be fully adequate guides to the kinds of classification and ordering of nature of which he was thinking when he died. Besides, the woodcuts he and his friends had struggled over for so long were much more rudimentary and lacking in detail than the sumptuous works that their enemies and rivals were even then beginning to produce. There remained, however, at least one body of illustrations of which this could certainly not be said. The tragedy is that these illustrations were hardly published at all.

ELEVEN *Fossils*

Fossils were never far from Cesi's concerns, not since his boyhood days when he wandered among the remains of fossil woods in Acquasparta and the small neighboring villages of Dunarobba, Sismano, and Rosaro (fig. 11.1). Sometimes he could see the smoke that rose from the mysterious underground fires, the fumaroles, that burned beneath the places where the petrified trunks lay (figs. 11.2 and 11.10). How did these fossil woods originate, and why did they seem to be at once both plant and stone? Swiftly Cesi realized that one question could not be resolved with-

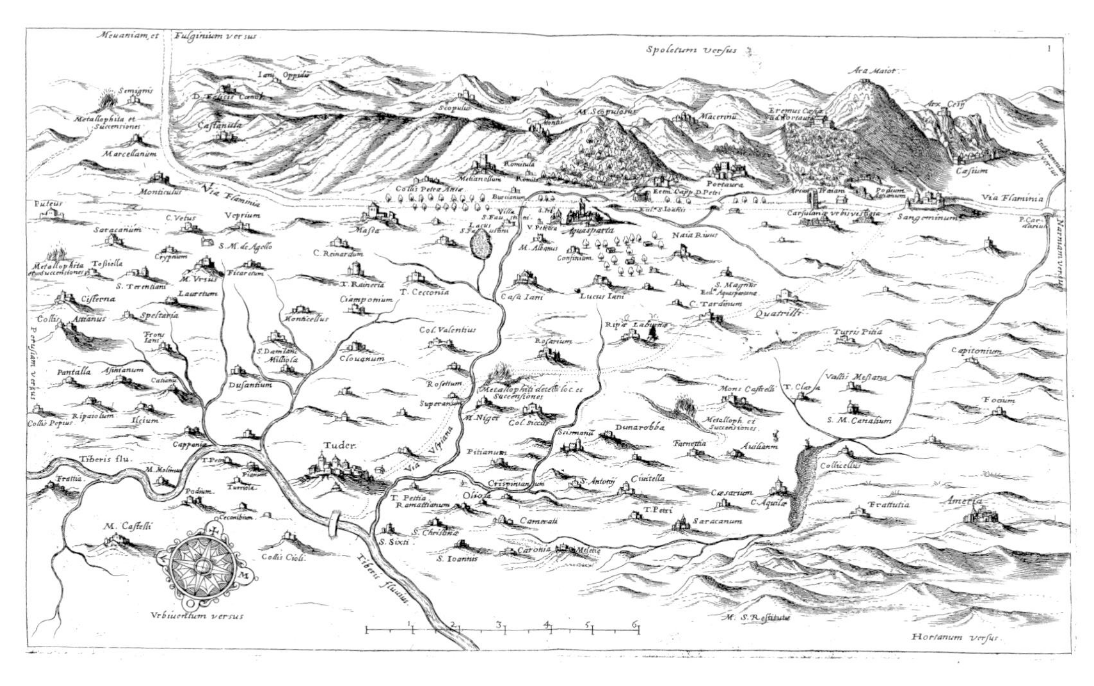


Fig. 11.1. Map of the region of Umbria between Todi and Acquasparta, indicating fossil wood sites and underground fires. Stelluti, *Trattato del Legno fossile minerale* (1637), pl. 1.

Fig. 11.2. Site of fossil woods and underground fires near Rosaro. Stelluti, *Trattato del Legno fossile minerale* (1637), pl. 2.

out finding the answer to the other. In this way two of his central preoccupations flowed together once more: the mysteries of generation and reproduction, and the problem of classification. After all, if you did not know what something actually *was*, or of what substance it was made, where was it to be placed in the scheme of things?

The same issues arose for other kinds of fossils too. Cesi became interested in all of them—and not just because of his effort to arrive at comprehensive knowledge of everything in the world of nature. If the fossil woods were somehow both vegetable and mineral, then what was one to make of other combinations, such as the ammonites? Fossils such as these seemed to be both animal and mineral; but were they also generated from the earth, as Cesi is said to have believed the fossil woods were—improbable as this may seem to us now? Unfortunately, the animal forms that the ammonites resembled did not provide adequate clues to what they actually were, and even less to how they arose.[1]

The old sciences of surface, such as physiognomy and phytognomy, were of little use, since outward form hardly ever turned out to be a reliable or adequate guide to the origins or essence of such things.[2] If only one could establish these, then one might at least *begin* to speculate about some of the answers to the secrets of creation itself, and, at the same time, to embark on the problem of the order of things.

Soon Cesi came to acknowledge that individual specimens could indeed partake of more than one nature (or species or class)[3]—and that their similarity to vegetable or animal forms was not simply a matter of the divine artifice of nature. Of all the natural historical problems that Cesi faced, this was the one that challenged him most.

Fig. 11.3. Piece of fossil wood from Sismano in situ. Pen and ink over traces of black chalk. *Natural History of Fossils* XV, fol. 11. Windsor, RL 25602. The Royal Collection © 2000, Her Majesty Queen Elizabeth II.

## THE DRAWINGS IN WINDSOR

One hundred and ninety-nine sheets of fossil drawings survive in Windsor Castle.[4] They are probably the most precise of all the drawings Cesi and his colleagues ever commissioned. Most are in pen and ink, but some are in watercolor, showing great attentiveness to color and coloristic nuance.[5] Few of the other drawings in the corpus display the same attentiveness to the minutiae of surface, or the same insistence on showing more than just a single aspect of the same specimen. In the case of the fossil woods, for example, one side of a piece of fossil wood appears to be made up predominantly of earth, another is largely metallic, while yet another readily betrays its plant origins.[6]

To go through the pages of the volumes in which these sheets are now bound[7] is to receive the impression that these drawings were made with little sense of the typical. As if obsessed with the desire to establish exactly what each combination was—or might be called—Cesi had his artist convey the very subtlest nuances of the ways in which wood and earth could be combined.

Some drawings show pieces of trunk or branches with bits of earth clinging to them; others depict chunks of soil apparently baked and patterned like wood; others yet show oval-shaped pieces of wood jammed between layers of earth (figs. 11.3–4). Some show pieces of fossil wood actually infiltrated by iron and other minerals (fig. 11.5). Many appear to be petrified pieces of wood, compacted, squashed, and twisted in a variety of different ways (figs. 11.6–7), often looking like some strange

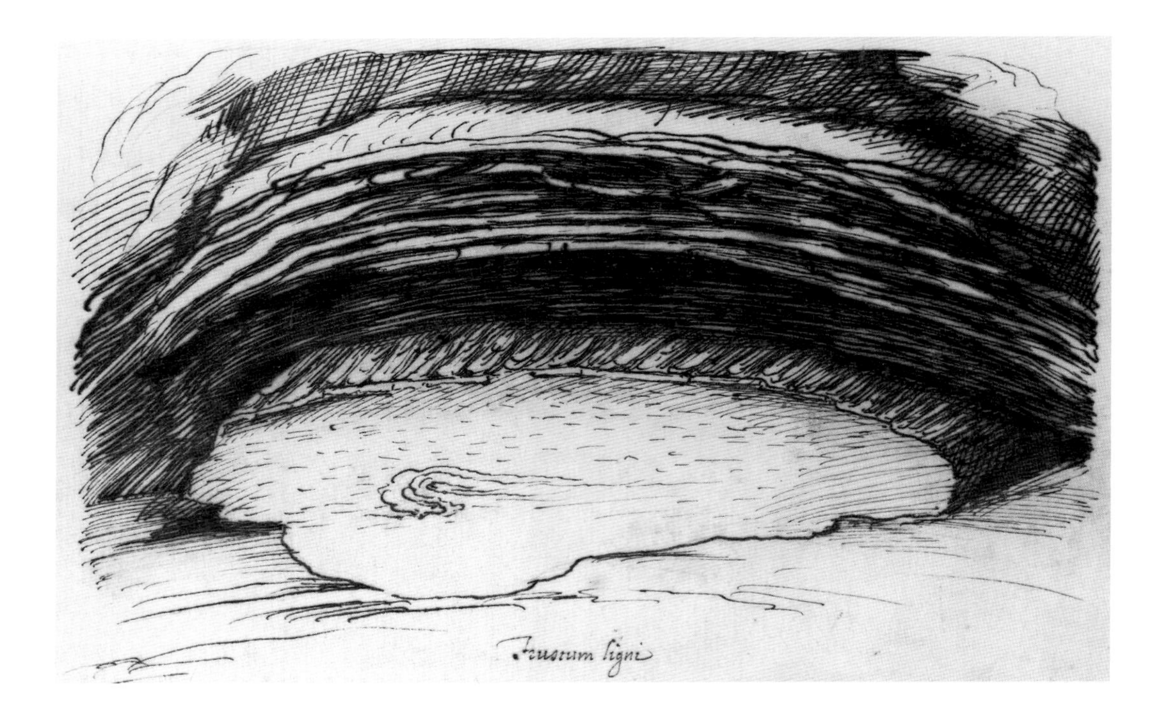


Fig. 11.4. Piece of fossil wood from Sismano in situ. Pen and ink over traces of black chalk. *Natural History of Fossils* XV, fol. 12. Windsor, RL 25603.

Fig. 11.5. *Opposite, above:* Piece of fossil wood infiltrated by various iron minerals. Watercolor and bodycolor and pen and ink over traces of black chalk; retouched in black ink on mount sheet. 180 x 259 mm. *Natural History of Fossils* XVII, fol. 17. Windsor, RL 25711.

Fig. 11.6. *Opposite, below:* Piece of fossil wood, with section. Pen and ink over black chalk, 186 x 251 mm. *Natural History of Fossils* XVI, fol. 52. Windsor, RL 25691. All the Royal Collection © 2000, Her Majesty Queen Elizabeth II.

muscle or tendon of the human body (fig. 11.8). In each case Cesi devised words and names—often entirely new ones—to express the various combinations they revealed. Hence the portmanteaus of *petrilignum, lapifactum, metallophyta, metalligo lignimetallum, ligni metallificati, lapilignea metallificata, lapilignum(aliquantulum metallico), lapiferreum, cretilignum,* and so on and so forth.[8]

Together these drawings provide eloquent testimony to Cesi's almost wholly unknown paleontological researches. They flesh out what little information can be inferred from other sources about the work he for so long planned to write on the subject of fossils and fossil woods. This, like almost every other project of his, was cut short by his death in 1630.[9]

Twelve of the sheets in Windsor represent the often dramatic sites of the fossilized woods and everlasting fires round the Umbrian villages of Dunarobba, Sismano, and Rosaro (figs. 11.9–10). Five show pieces of fossil wood actually embedded in the earth in these locations (e.g., figs. 11.3–4). The largest group—about one hundred and twenty—is formed by exceptionally meticulous and closely observed fragments of fossilized wood (e.g., figs. 11.11–12). Many can actually be paralleled with a series of finds recently made at Dunarobba (cf. figs. 11.13–14).[10]

Almost as compelling as these illustrations are the smaller groups showing aetites or "eagle stones"[11] (the nodular concretions that so fascinated early students of fossils because of their apparently "pregnant" appearance, figs. 11.15–16);[12] the eight sheets with fossil animals em-

Fuscum in Medio latorritium
long: pal: 2. lar. pal: 1. d2. alt: pal: 4.

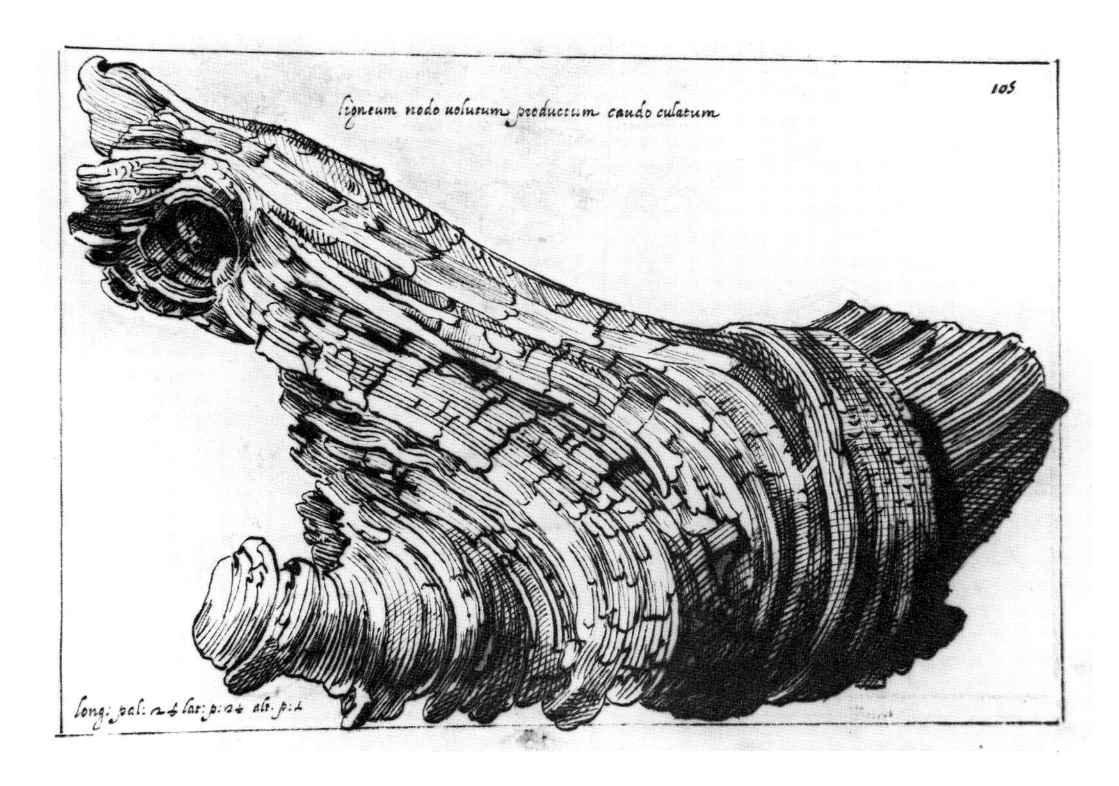
ligneum nodo uolutum productum caudo culatum
105
long: pal: 2 4 lat: p: 2 4 alt. p: 2

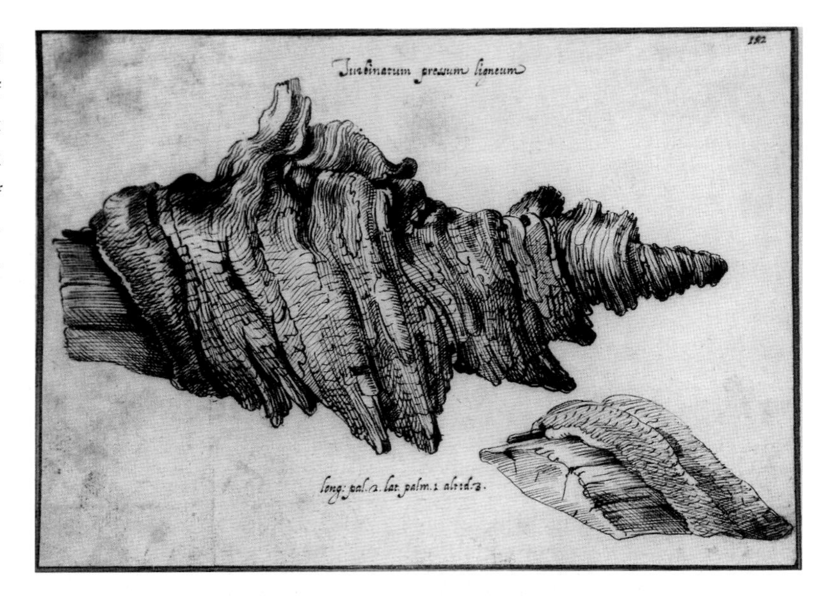

Fig. 11.7. Fossil wood branch with fragment of main trunk. Pen and ink over black chalk, 164 x 229 mm. *Natural History of Fossils* XVI, fol. 5. Windsor, RL 25644.

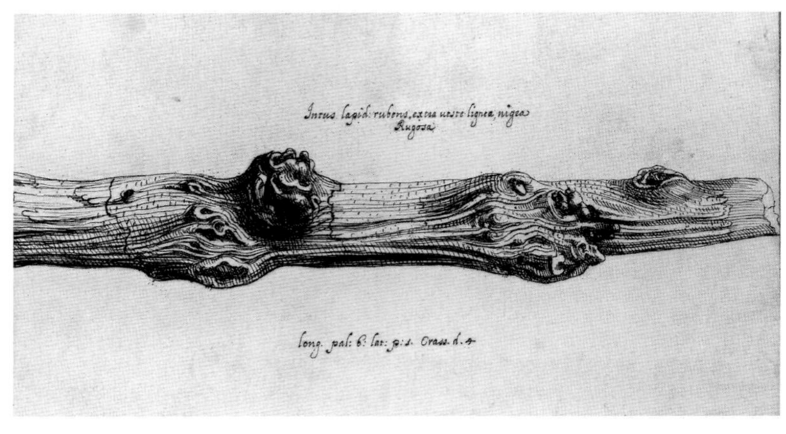

Fig. 11.8. Piece of fossil wood with groups of branch scars. Pen and ink over black chalk, 187 x 267 mm. *Natural History of Fossils* XV, fol. 25. Windsor, RL 25616.

Fig. 11.9. *Opposite, above:* Fossil wood deposits at Rosaro. Pen and ink, 180 x 260 mm. *Natural History of Fossils* XV, fol. 9. Windsor, RL 25600.

Fig. 11.10. *Opposite, below:* View of fossilized wood site and fumaroles between Sismano and Rosaro. Watercolor and pen and ink over black chalk, 182 x 257 mm. *Natural History of Fossils XVI*, fol. 44. Windsor, RL, 25683. All The Royal Collection © 2000, Her Majesty Queen Elizabeth II.

bedded in clay or in one kind of concretion or another; the thirty-seven drawings of baked or burned clays (e.g., figs. 11.17–18); and, last but not least, the three splendid sheets of drawings of ammonites (fig. 11.19).

With the exception of the beautiful and subtle drawings of baked clays, the drawings are almost all in black and white. Together they constitute the first extensive set of field drawings of a fossil site or set of geological features ever made—and they are of remarkably high technical quality.

Both Cesi's correspondence and the frame lines that surround forty-three of the drawings leave no doubt that Cesi intended to publish at least some of them.[13] Similar frame lines appear in all but two of the thirteen plates in the *Treatise on Fossil Mineral Wood* that Stelluti went on to publish in 1637.[14] This is the disappointingly meager book that Stel-

62
Prope Magna feusta Rosarij

194
Inter Scismanum, et Rosarium

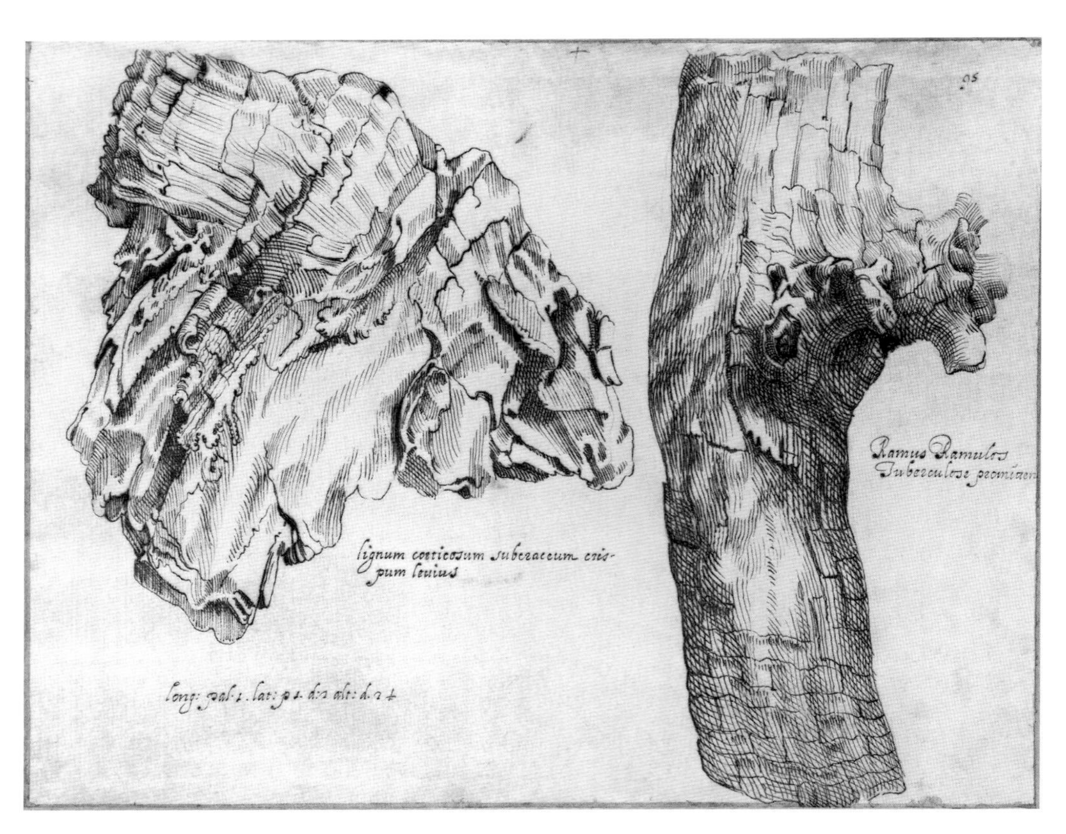
95
lignum corticosum suberaceum crispum leuius
Ramus Ramulos Tuberculose promittens

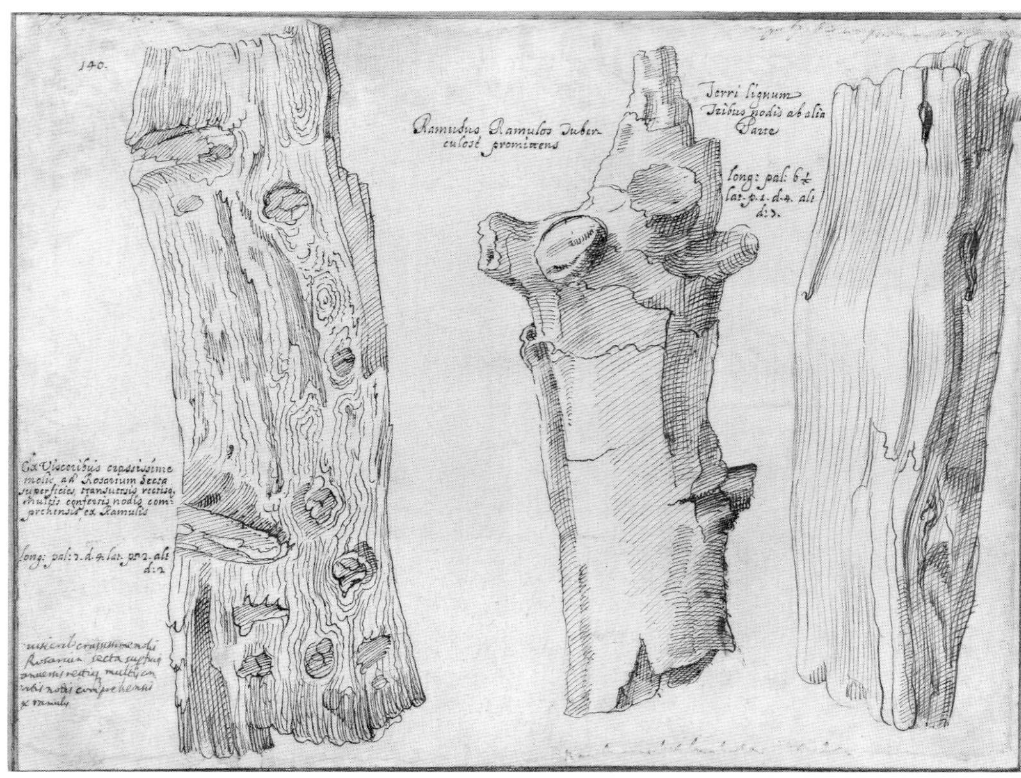
140.
Ramulus Ramulos Tuberculosè promittens

Fig. 11.11. *Opposite, above:* Two pieces of fossil wood, right-hand specimen with branches. Pen and ink over black chalk, 204 x 257 mm. *Natural History of Fossils* XV, fol. 43. Windsor, RL 25634.

Fig. 11.12. *Opposite, below:* Three specimens of fossil wood, one from Rosaro. Pen and ink over black chalk, 209 x 268 mm. *Natural History of Fossils* XVI, fol. 40. Windsor, RL 25679. Both The Royal Collection © 2000, Her Majesty Queen Elizabeth II.

Fig. 11.13. Piece of fossil wood found by Andrew Scott at Dunarobba in 1995.

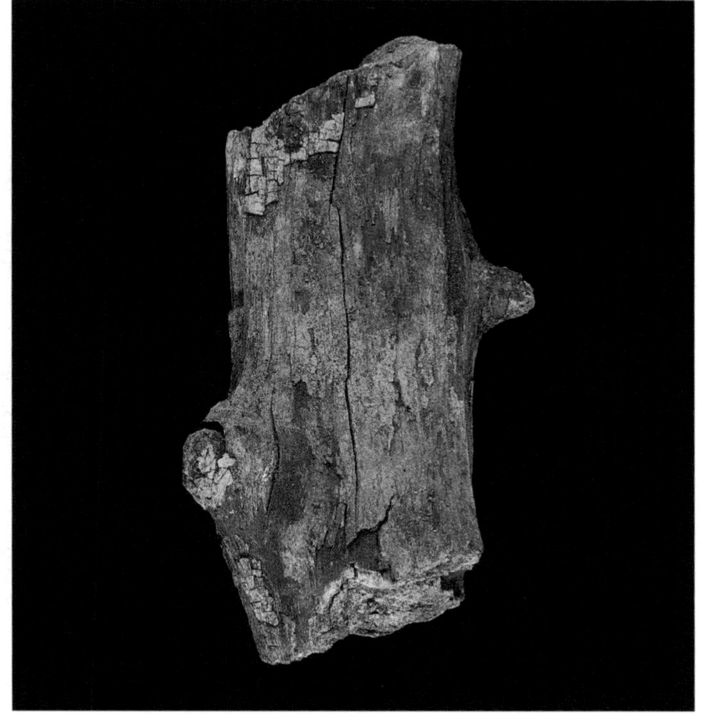

Fig. 11.14. Piece of fossil wood showing sulphur precipitate from decay found by Andrew Scott at Rosaro in 1995.

luti managed to pull together on the basis of Cesi's notes and drawings for the much more ambitious work that he once had planned on the subject of fossils.

## THE BOLOGNA STONE: LIGHT AT DAWN

Let us turn, first, to the more promising days when this project was still in its infancy. From almost the beginning the Lincean preoccupation with generation and classification may be discerned; but within a few years another set of issues emerged.

When Galileo went up the Janiculum on the evening of April 14, 1611, in order to demonstrate his discoveries with the telescope to Cesi and his friends, he concluded his discussions the following morning with an experiment that none of them could have anticipated. Just as dawn was breaking, he invited them to leave the terrace and follow him inside Monsignor Malvasia's villa. He wanted to show them something that his old interlocutor and opponent Giovanni Antonio Magini had sent him from Bologna.[15]

Cesi and the rest of the group must surely have been exhausted and excited by what they had been studying all night in the heavens, but Galileo had one more experiment to share with them. He took them into a small room, and there, in the crepuscular light, opened a little box, a *scatolino*, he had brought with him. To the amazement of those among them who had not yet heard about what it contained, the stone inside it glowed briefly, and then began to fade. They called it the "Bologna stone," or the *spongia solis* (we now know it to have been a piece of calcined barium sulfide).[16] A few moments later, Galileo walked out onto the terrace once more, exposed the stone to the weak light of the dawn, closed the box, brought it inside, and opened it again. Once more it glowed for a while, and then petered out.[17]

The importance of this demonstration can hardly be overestimated. Simple though it was, its implications were profound. Some of these were more directly relevant to Cesi and his recent researches on fossils; others more relevant—perhaps much more—to the newest developments in Galileo's Copernican cosmology. Cesi and his friends could be sure of at least three things about the Bologna stone on that critical morning on the Janiculum: First, the light given off by the *spongia solis* was cold and not accompanied by the generation of heat. Second, it seemed to absorb light not only directly from the sun (hence "*spongia solis*") but also from the weak secondary light of the breaking dawn. Third, this light—even in its weaker forms—could be transmitted in a wholly dark environment.

All this suggested a view of light and of luminous bodies that was wholly at odds with the traditional Aristotelian one. Galileo held that light was quantitative and corpuscular (and could thus be differentially transmitted across a distance, with the atoms or corpuscles of light gradually diminishing in quantity). The Aristotelians believed light to be an incor-

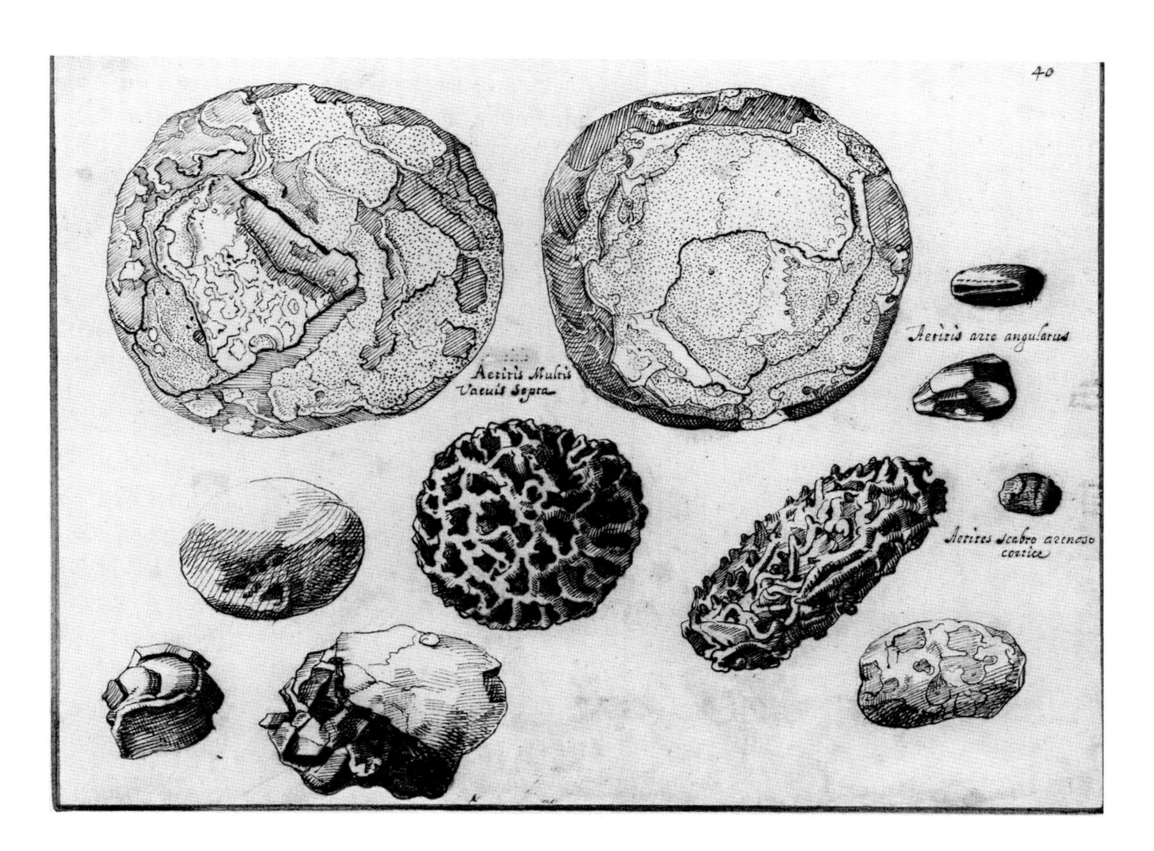


Fig. 11.15. Various types of ironstone concretions (including aetites). Pen and ink over traces of graphite, 192 x 255 mm. *Natural History of Fossils* XIV, fol. 50. Windsor, RL 25579. The Royal Collection © 2000, Her Majesty Queen Elizabeth II.

poreal quality produced by a luminous body consistently diffused across a diaphanous milieu. For Galileo light was something entirely separate from its transparent and luminous context; its action was not such that it could be transmitted in the manner of a rod or tight cord extending to the beholder from the source, as Aristotle maintained.[18] In other words, light from the stars did not reach us directly across a transparent medium; it was not immediately propagated to our eyes; and it could persist, even though subject to diminution, in a dark environment. This proved that there could indeed be things in the heavens that were not susceptible to being seen by ordinary vision, alone and unaided. The naked eye was not necessarily sufficient.

It is not hard to understand the importance of this argument in favor of the telescope. As Galileo emphasized in his letter to Piero Dini of May 21, 1611, the fact that the light of distant stars—say the Medicean "stars"—did not seem to reach our eyes did not mean that they did not exist at all.[19] Furthermore, the claim that light consisted of quanta of corpuscles made it much easier to grasp the differential and variable nature of the received light of heavenly bodies, whether in the form of the scintillations of the fixed stars or the different degrees of luminescence of the planets. Light, quantitative and corpuscular, was unstable, not uniform, and was differentially emitted from its source.[20]

The relevance of all this to the Copernican cosmological positions out-

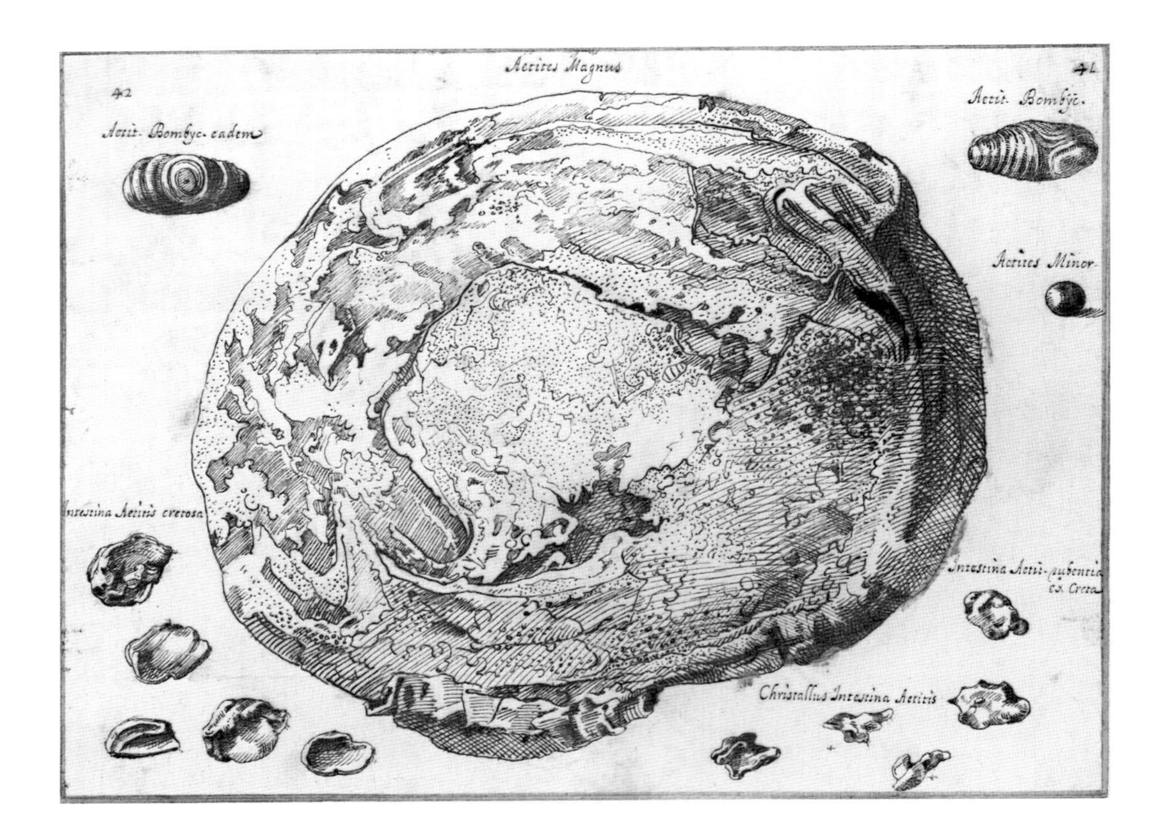


Fig. 11.16. Various types of ironstone concretions (including aetites). Pen and ink over traces of graphite, with corrections in bodycolor, 191 x 256 mm. *Natural History of Fossils* XIV, fol. 50. Windsor, RL 25580. The Royal Collection © 2000, Her Majesty Queen Elizabeth II.

lined in the *Starry Messenger* was great. After all, if light transmitted by the rays of the sun across the very limited light of the dawn could be weakly reflected by the stone in the box, why might not the same apply to the other planets? And if this light was quantitative why might it not be reflected differentially across space, thus accounting for the diversity in planetary brilliance? This all made it much easier to grasp Galileo's proposal that the moon was not naturally luminescent, and that the secondary light of its phases was not its own, but in fact solar light reflected off the earth. Just like the Bologna stone, the moon could receive its light from the sun, absorb it, as it were, and transmit it coldly across the darkness of the heavens. It did not have to burn with its own light in order to do so. Coupled with the application of the same arguments to the phases of Venus, the benefit to the Copernican hypothesis was clear.[21]

But Cesi and his other friends—Stelluti in particular—had slightly different preoccupations. While perfectly aware of the cosmological implications of Galileo's theories of light and heat, they continued to concentrate on its terrestrial manifestations. In so doing, they also hoped to provide additional arguments in favor of Galileo's views on the variability of luminescence and on the separability of light and heat.[22] Just as in the case of the heavenly bodies, there were things on earth that scintillated and glowed in the dark. For example, there were pieces of wood that seemed to ignite when dipped in water, and pyrites and related phenom-

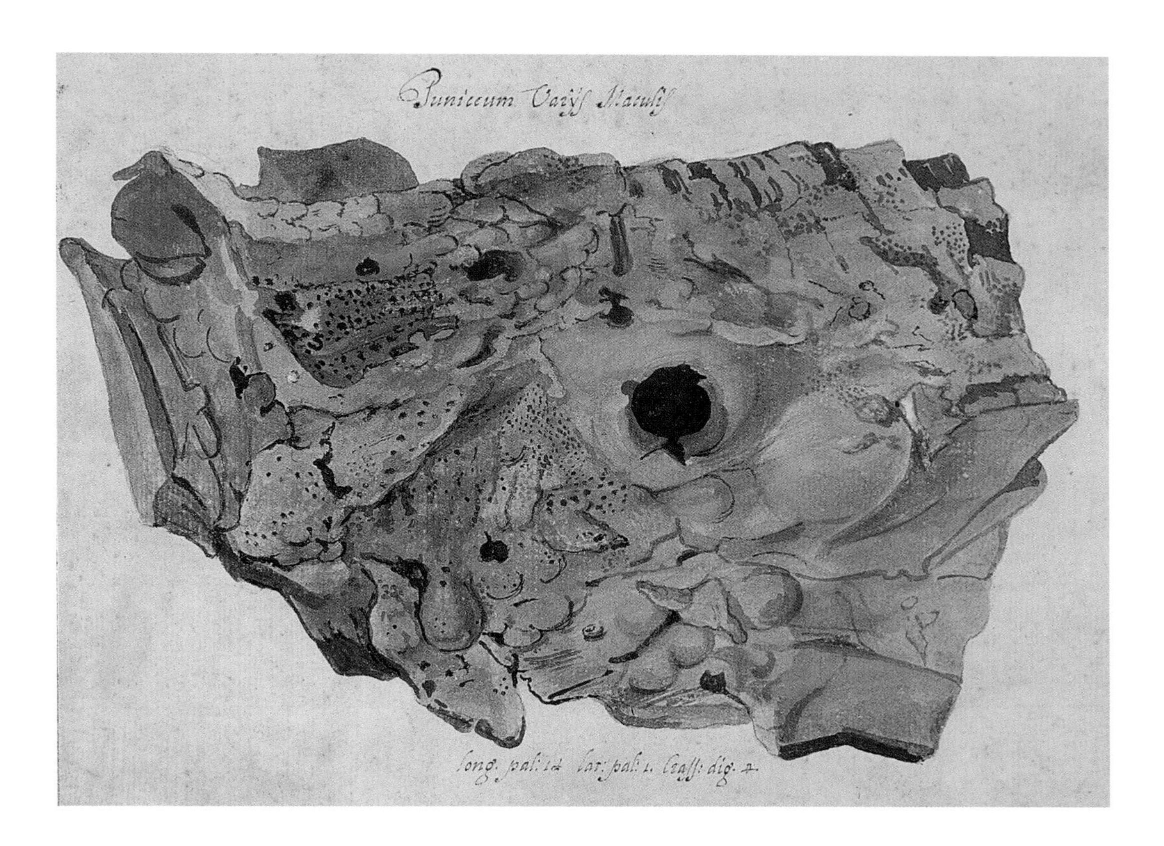


Fig. 11.17. Baked lignitic clay. Watercolor and bodycolor over traces of black chalk, 179 x 242 mm. *Natural History of Fossils* XVII, fol. 34. Windsor, RL 25729. The Royal Collection © 2000, Her Majesty Queen Elizabeth II.

ena that gave off strange kinds of light. The fossil woods, the concretions, the baked clays, the pyrites, and the underground fires themselves all brought to the fore issues not only of spontaneous combustion but also of the spontaneous production of light and the extent to which the generation of light was accompanied by heat.

It is precisely in this context that we may finally begin to grasp the significance of the carbuncle in Cassiano's (and perhaps Cesi's) collection.[23] The Linceans certainly knew of the remarkable chapter in Pliny the Elder's *Natural History* in which many of the issues raised by fossil woods came together with the kinds of problems about light-emitting stones that Galileo was now raising. Pliny begins with fossil woods and moves on swiftly to deal with various forms of luminosity and the problem of the relations between heat and light.

The chapter opens with an account of the "dryites" or the "oak stone," which resembles the trunk of an oak, and which, though stone, burns like wood. It then goes on to discuss the "pyritis" or "fire-stone," which, even though black can scorch the fingers when it is rubbed'; the "phlogitis" or "flame stone" that seems to have a flame burning inside it but is never released; and finally the "anthracitis" or "carbuncle stone" that appears to have sparks running in different directions through it.[24] When touched, however, "it dies away and disappears, but when, on the other hand, it is soaked with water it blazes forth again."[25]

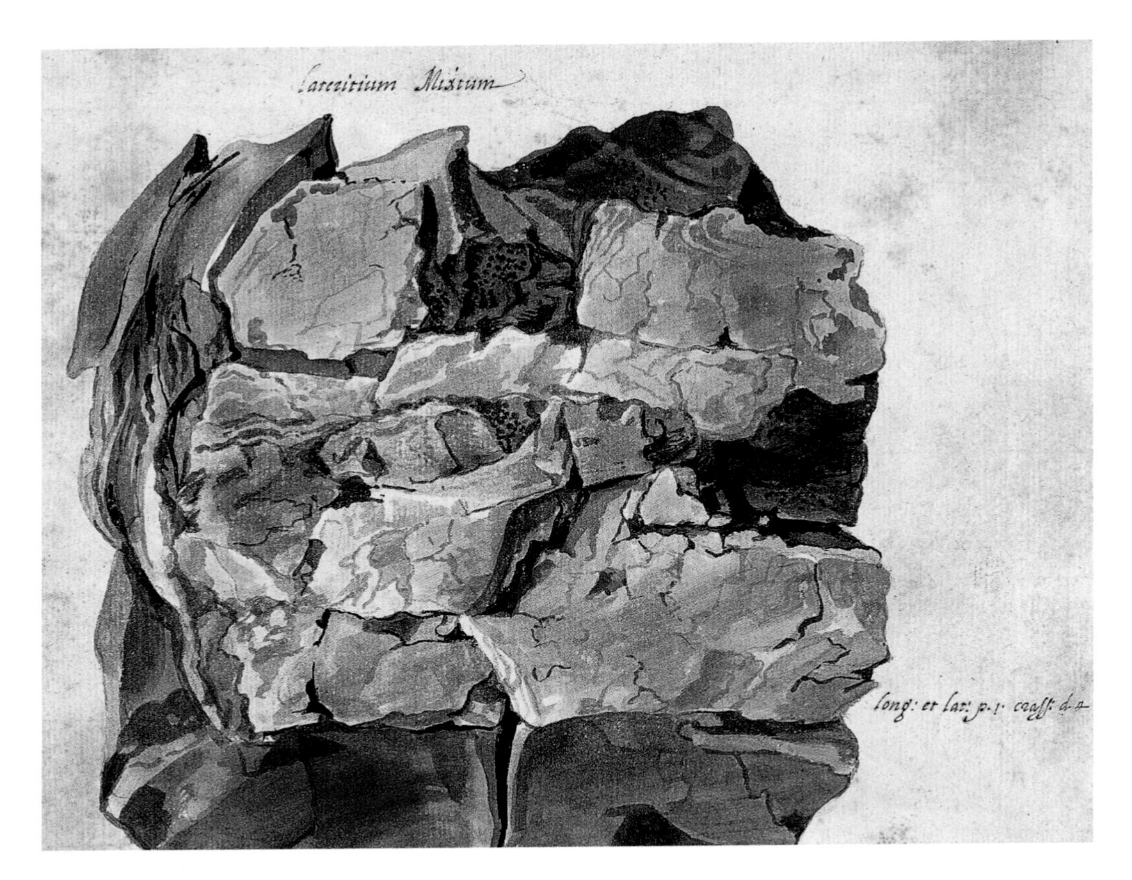

Fig. 11.18. Baked clay. Watercolor and bodycolor over traces of black chalk, 190 x 254 mm. *Natural History of Fossils* XVII. Windsor, RL 25717. The Royal Collection © 2000, Her Majesty Queen Elizabeth II.

All these things appear in one form or another in Cesi's drawings and in his scattered writings.

Earlier in the same chapter Pliny had noted that the *carbunculi* were so called because of their fiery appearance, even though they were not affected by fire and were therefore sometimes known as acaustoe or incombustible stones. Some of these *carbunculi* (of which there were so many that even here there were issues of classification that needed to be resolved) "shine from deep beneath their surface and blaze with exceptional brilliance in sunlight." Others, though dark, "light up more intensely than the rest when they are viewed by firelight or sunlight, and at an angle." The "male Carthaginian stones [another kind of carbuncle] seem to have a blazing star inside them, while the female stones shed all their radiance externally."[26]

There was every reason to focus on the phenomenon of the carbuncle. Pliny's discussion of *carbunculi* was a little thesis on different kinds of scintillation, brilliance, and luminescence, a kind of proleptic illustration, if interpreted rightly, of Galileo's atomistic view of light as quantitative rather than qualitative. The Linceans' researches offered the possibility of providing still further proof of the relevance of the kinds of earthly phenomena described by Pliny to the phenomena Galileo was now so boldly bringing to the fore.

## MORE ON LIGHT AND HEAT

Can it be a coincidence that the only date to be found on any of the fossil drawings is 1611? The inscription above this piece of contorted pyritized wood, or "metallophyte," as Cesi called it, records that it had been found by him and taken to Rome in that year.[27] Almost all the drawings of fossils were, I believe, made around this critical time, when the Linceans hoped that the fossil finds might benefit not only the Galilean hypotheses about the nature of the relationship between heat and light, but also serve their own researches into the nature of the generation of life on earth. In the months immediately following Galileo's demonstration on the Janiculum, the Linceans, with Faber leading the way, did everything they could to obtain pieces of the *spongia solis*, variously known as the Bologna stone, the *pietra lucifera*, or the *litheosphorus*, as it was called by that great Aristotelian expert on fire and light, Fortunio Liceti, and sometimes by Cesi too.[28]

Already on June 10, 1611 Ferrante Imperato wrote to Faber to thank him for having written to him about "those little stones shown by Signor Galileo, with their quality of taking in and retaining light, something certainly unknown to me." But Imperato was skeptical: "I don't believe this to be a natural effect," he added, "but rather an artificial one."[29] Perhaps it would have helped him to adopt the kind of position that Pascal did many years later, precisely with regard to the *spongia solis:* "When we see an effect always turn out to be the same," he wrote in one of his *Pensées*, "we conclude from this that it is a natural necessity, such as tomorrow it will be day, and so on; but often nature deceives us and does not subject itself to its own rules."[30]

Imperato continued that he would love to see at least one of the stones if Cesi could manage to send him one. He regretted that he was unable to discuss the matter with Della Porta, who was out of town. And then, quite typically, he added that he was sending some seeds to Terrentius (who was then working on the *Tesoro Messicano*).[31] Yet once more, botany went hand in hand with their other studies.

For a while they did not make much progress with the Bologna stone. On October 21, Cesi wrote to Galileo that Giulio Cesare Lagalla (the doctor and philosopher to whom we owe the fullest account of Galileo's demonstration at the Villa Malvasia)[32] "was writing on the subject of light as a result of the stones which you showed him."[33] But "the material is difficult and it is always extremely hard to find causes without starting off from received opinion." The problems surrounding the Bologna stone, just like the new botanical discoveries from America—to say nothing of the implications of what Galileo had revealed to him in the heavens—only increased Cesi's doubts about the old authorities. Every day the protagonists of the old order were exasperating him more.

The letter begins with Cesi apologizing for not having replied sooner. He has "spent the last few beautiful days minutely visiting and researching my Monte Gennaro nearby" (he was writing from Tivoli). As always, he and the Linceans had been out in the field, collecting and listing botanical specimens.[34] But the pleasant tone soon dissipated as he realized how much he had to attend to. He had been trying to find a copy of a recent letter of Galileo's to Grienberger. He had to read Galileo's *Discourse on Bodies on and in Water*. Lagalla was worrying about the Bologna stone. Terrentius and the other Linceans were going to write about it too. The Mexican Treasury was being readied for the press.

Clearly, Cesi was under some pressure. Galileo's opponents were gathering their forces, and every one of the Linceans' fields of research seemed to undermine the bases of Aristotelian science. Cesi could contain himself no longer. At this point in his letter he delivered himself of one of the most acerbic of all his utterances about traditional learning: "The old philosophers, more enemies of novelty than friends of the truth, do not cease to give me material to laugh at in their calumnies; and these we must uncover and attack as much as we can."[35]

It was a moving moment.

For several months more the Linceans continued to worry about the Bologna stone. In April of 1612, Della Porta wrote to complain that "those pieces of stone which retain light in the darkness aren't really working. Could you please let us have a good piece for the Academicians down here."[36] Faber, on the other hand, seems to have made the most of the specimens he had. His distinguished compatriot and supporter of the Linceans, the prince-bishop of Bamberg,[37] visited him at his home in December 1612, when he was doing an experiment in which he demonstrated that "even without the light of the sun, air can be included in a dark place, and give off light there." "A fine *artificium philosophicum*!" the prince-bishop noted.[38] In February Cesi wrote to Galileo that the prince-bishop was a great admirer of his, and that having seen "a small piece of stone which receives and conserves the light in the hands of Signor Faber," wanted a piece for himself. Unfortunately, Cesi continued, his own specimen had lost much of its original vigor.[39] Within a couple of months, Galileo stepped into the breach, and sent him a new box of luciferous stones with which to start working again.[40]

But all this was swept away by the commotion over the sunspots. Many of the issues about the relations between light and heat were clearly relevant to that controversy, but the Linceans had too much else to do now. Galileo needed their support more than ever. For the moment he himself set aside the matter of the retention and emission of quanta of light—with the exception of one important aspect of the problem. He found that it could be used to bolster his interpretation of those vexed and controversial lines in Psalm 18:5–6:[41]

> In them [the heavens] he hath set a tabernacle for the sun, which is as a bridegroom coming out of his chamber, and rejoiceth as a

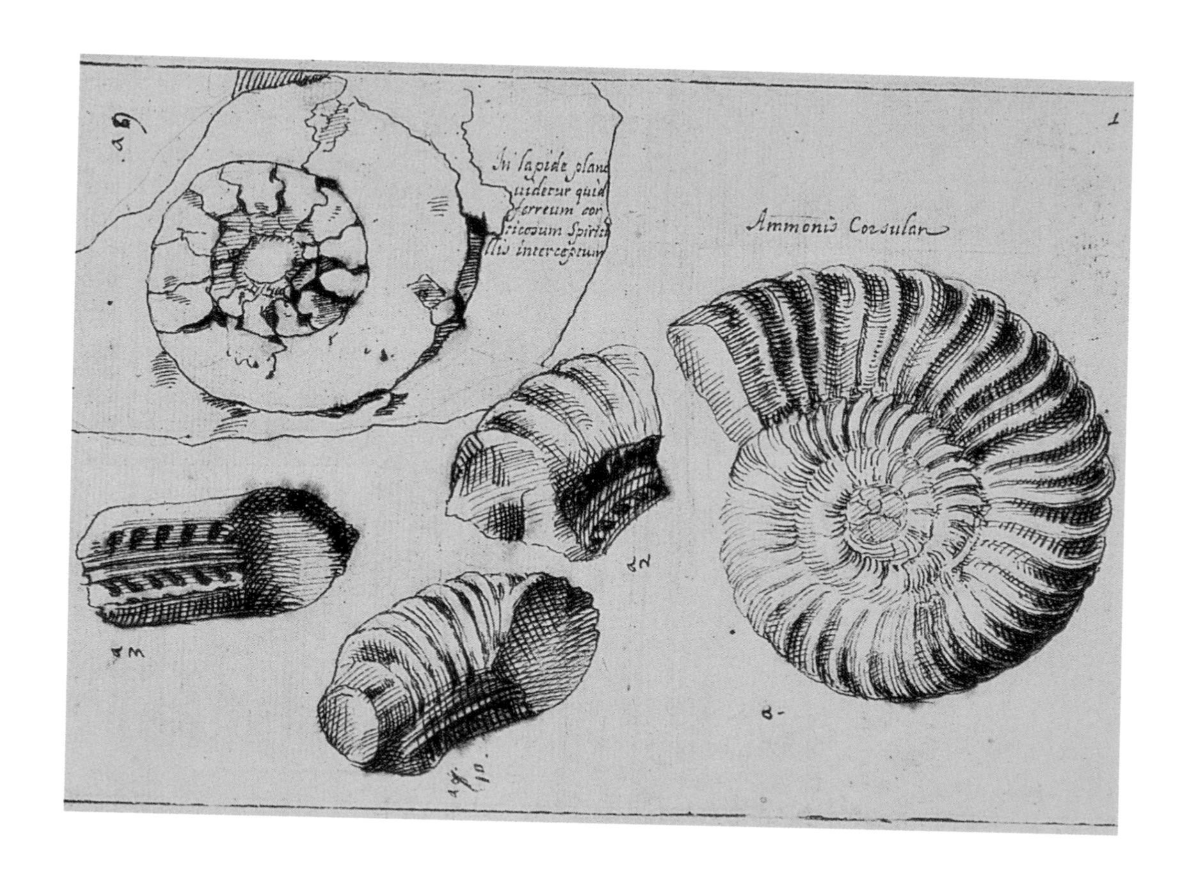


Fig. 11.19. Ammonite, three fragments, and a cross-section of ammonite in rock, probably from Carsulae (all belonging to the genus *Asteroceras* from the Lower Jurassic). Pen and ink, over traces of black chalk, 175 x 239 mm. *Natural History of Fossils* XIV, fol. 7. Windsor, RL 25536. The Royal Collection © 2000, Her Majesty Queen Elizabeth II.

> strong man to run a race. His going forth is from the end of the heaven, and his circuit unto the ends of it; and there is nothing hid from the heat thereof.

These verses, as we have seen, were by taken by Galileo's opponents as an argument in favor of geocentrism. But in his second letter to Piero Dini of March 23, 1615 (intended to be passed on to Bellarmine for his judgment), Galileo argued—using the lessons of the Bologna stone—that the universe as a whole was vivified by a special caloric spirit, distinct from light. This substance or spirit, he maintained, originated from outside or from the edges of the universe and then focused in the sun. There it was

> considerably strengthened, and then re-emitted to warm and vivify all bodies, living and non-living. Thus the sun must be at the center of the universe, and its axial rotation, attested to by the motion of the sunspots, somehow moves all the other bodies in the heavens.[42]

Galileo adopted a similar position in the *Letter to the Grand Duchess Christina;*[43] while in the controversy over the comets soon after, Galileo was forced to reflect once more on problems such as the behavior of light in a transparent context. By this time, however, the Linceans had other matters on their mind.

Long after his condemnation, Galileo still continued to worry about the nature and relationship of heat and light. In the 1640 *Litheosphorus*, his old interlocutor, Fortunio Liceti, once more defended the Peripatetic positions on heat and light.[44] He even argued that the Bologna stone could be used to prove that the moon had a natural luminescence of its own. Galileo, of course, would have none of this, but in order to refute Liceti's interpretation, found himself obliged to renounce the corpuscular theory of light that had served him so well in the second decade. By 1640, in any case, Cesi had long since passed on, and almost the only evidence we have of *his* views is provided by the paltry *Treatise on Fossilized Mineral Wood* that Stelluti published in 1637.

But this treatise barely mentions the problem of light, and skates over what is perhaps the still more critical aspect of Cesi's concern with fossils—that of their middle nature, their "mezzana natura," as he called it. This was an issue—with its clear implications for the problem of classification—that Galileo left almost entirely to Cesi (though what he wrote in the *Assayer* had a profound influence on it).[45] The problem of sorting out how best to order Nature lay in Cesi's own hands.

## MIDDLE NATURES

Let us return to Cesi's ambitious outline, preserved in the Biblioteca Nazionale in Naples, for the ordering of all human knowledge—natural historical, physical, mathematical, and cultural.[46] It also gives an account of the various treatises that Cesi planned to write. Among these were several that, had he completed them, would substantially have expanded our sense of the relations between geological phenomena and Galilean astronomy, optics, and atomism. The summary of the contents of his *Coelispicium*, for example, has headings for comets, novae, and sunspots, as well as for various problems relating to the transmission of light in the heavens. The *Taumatombria*, "on marvelous [celestial] rains," was to discuss another general topic of central concern to Cesi: that of extraordinary liquefactions, saps, juices, and resins. All these phenomena are to be related to the old theories about the petrifying juices instrumental in the formation of fossils, as well as to the even older issue of meteorites (which demonstrated, as it was then thought, that things could be petrified even in the upper atmosphere).[47] Above all, this whole section of the Naples manuscript points to the problem of light that is miraculously derivative or "stolen"—*de luce furtiva admiranda*—and of "nocturnal light and the art of the light-bearing stone."[48]

It is at just this stage that the manuscript moves on to describe a pair of works that are of even greater relevance to the problem of the fossil finds: the planned treatises "On Middle Natures in General" and "On Imperfect Plants." The first of these, as we have seen, proposed to deal with

> things of doubtful nature, or doubtful species, or ambiguous things. On [the nature of] ambiguous Nature. Two different natures joined in a single species. . . . species participating in two natures. A very remarkable corner of the Theatre of Nature, all the more hidden, concealed and obscure because passed over.[49]

It would be hard to imagine a better description of the chief problem that Cesi was now studying, that of the so-called middle nature of the fossils finds of Acquasparta. And indeed, following amphibians and the "whole class of oviparous quadrupeds," but before things like bats, giraffes, mergansers, and so on,[50] Cesi noted that he intended to include discussions of "zoophytes," "lithophytes' or "metallophytes," and "zoolithophytes."

At this point in his outline, Cesi moved on to the subject of "imperfect plants," that is, plants that were somehow or other "mutilated, deficient, inordinate." The most important of these were the fungi, and for the rest of his life Cesi devoted his attention to these and other problematic instances of the crossover between classes, where the principles of reproduction were so often not clear. No wonder that in his treatise on imperfect plants Cesi also wanted to include the lithophytes, the zoolithophytes, and the omnipresent metallophytes. They all had to find a specific place within the order of nature, not outside it.

Cesi concluded his sketch of the treatise *On Imperfect Plants* with a section on what he called the *officinae naturae*, workshops of nature (or, as we might say, the laboratories of nature): the *lapificium Tiburtinum* (the travertine quarries of Tivoli), the *ignificium Puteolanum* (clearly the *Solfatara* at Pozzuoli seen by Cesi when he went to Naples in 1604), and, of course, the *lignificium Acquaspartanum*.[51] One can almost sense the pride with which Cesi noted that he had visited them all himself. Never, even when it came to the abstract processes of classification, did Cesi renounce his primary commitment to firsthand observation.

It is clear that the chief significance of the Naples manuscript lies in the evidence it provides for Cesi's belief that no adequate system of classification, taxonomy, or ordering grid, could ever be produced without first confronting the problem of what threatened to confound such a system. The borderline cases threatened to do just that, since they participated in more than one class, and could therefore not easily or satisfactorily be classified within the grid of things, or within the boxes provided for them. No test case (other, perhaps, than that of the fungi) occupied Cesi's attention more closely than the fossil woods and concretions from the hills and villages round Acquasparta. The bulk of these fell into the category of "metallophytes," a term that underlined their status as problematic combinations of mineral and vegetable. They too were examples of those phenomena of nature that Cesi repeatedly described as being of a middle nature; and they too dramatically posed the problem of how to establish fundamental criteria for a classification that would not be subverted by apparently overlapping or intermediate classes.

It is precisely this problem, that of the *mezzana natura*, that emerges so clearly from the drawings in Windsor. While there are admittedly a few rudimentary etchings in the Paris codices, where the problem of objects of a double or triple nature also seems to be addressed (e.g., figs. 8.35–36),[52] the drawings in Windsor represent another kind of faith in representation altogether. It is not just that the drawings are of a much higher standard than those in the Paris manuscript—or, indeed, than those in most of the codices in the Cesian corpus. In the drawings of fossil woods, Cesi and his draftsmen went to extraordinary lengths to emphasize, both verbally and visually, their overlapping—and therefore ambiguous—nature. Even a superficial survey prompts the question as to whether one is dealing with wood or with stone.

The aetites—fossil forms that often contained mollusks, gastropods, and bivalves—raised a similar question. Even if one did not stumble over the definitional problem of animal *versus* mineral, there was still the issue of how such combinations arose in the first place. Did the impression they gave of being pregnant with whatever else they contained within them offer a clue (cf. figs. 11.15–16)?[53] Like "the jokes of nature," apparently pregnant forms were a central focus of the Linceans' researches.[54] The Linceans took seriously all such instances of nature apparently gone awry, nature against nature, as it were. Pregnant fossils, like pregnant fruit, seemed to offer at least some clues to the secrets of generation and reproduction. Once again it was impossible to resolve the matter of classification without first addressing that of reproduction.

The inscriptions on the Windsor drawings not only declare the double or triple nature of these finds, they also reveal Cesi's search for appropriate terms for the phenomena he was seeking to describe. They testify to the intensity of his nomenclatural drive, and the need to establish an adequate classification for his fossil finds. Moreover, the terms he uses for the texture and surface accidents of his samples and of what he saw in the field is just as various. He has a whole vocabulary to describe the lignosity of the Acquaspartan finds: barky, knotty, full of whorls, rootlike, encrusted, baked, caked, tuberous, turbinate, convoluted, anfractuous, ferruginous, spongy, pyritic, sulfurous—to say nothing of the different ways of describing the "undulating waves" (that is, the growth lines), veins, and grains of the fossil woods.

All this is reflected in the drawings that document every subtle difference between the specimens of fossil woods. The draftsman—whoever he or she may be—does his or her best to match Cesi's verbal descriptions in pictorial terms. Measurements are always given, and the same specimen is often shown from several different angles. Thanks to Cesi's faith in representation, we are provided with clear visual evidence of how wood is pyritized, baked, turned into stone, composed half of clay and half of wood; how specimens that look no more and no less than a branch or knotty twig of a tree resemble nothing so much as pure stone. We can almost imagine the earth itself turning into these woody forms (as Stelluti later said Cesi claimed). The drawings show not only the inexhaustible varieties of woodiness but also the veins, undulating growth lines, grains, filings, metallic fibers, sumptuous speckles and twinkling pyritic elements.

In these drawings, even more than in their inscriptions, Cesi still clings to the old beliefs in surface description and subjectively perceived qualities. Within a few years, however, he and the Linceans would be reading—first in the *Assayer* of 1624 and then in the *Dialogue on the Two Great World Systems* of 1631—that such qualities could not tell one about the essential nature of things, or be of much use in the matter of classification.[55] Although Cesi soon discovered that in order to classify things one had to be able to discover what was essential about them, these drawings make it almost impossible to escape the problem of overlapping and borderline forms. There could hardly have been a more persistent attempt to record the minutiae of difference, as the Aristotelians would then have put it, until the remarkable drawings of citrus fruit that Cassiano commissioned a few years later from Vincenzo Leonardi for his friend Giovanni Battista Ferrari.[56]

1624 was indeed the critical date. Maffeo Barberini had just become pope and his nephew Francesco a cardinal. It was really only from then on that Cesi and his friends could devote more of their time not just to the never-ending issue of classification, but to thinking about the fossil remains that so absorbed them. By now their work on the *Assayer* had been completed. It was the fullest statement of Galileo's Copernicanism before the publication of the *Dialogue on the Two Great World Systems* and the trial of 1633. At last they could return to their most cherished projects.

## FOSSIL WOODS

On May 18, 1624, Giovanni Battista Winther wrote to Faber from Acquasparta, where he had been staying at the Cesi palace with Cesi, Stelluti, and Galileo, who was now finally resting from the intense labors of the previous few years. It was a pleasant caesura.[57] They had more time than usual for quiet reflection, and they could return to less urgent matters than the grand cosmological controversies that had beset them for so long. The conversation took a rather more aesthetic turn than usual, a turn that would be of some consequence in this very year:

> Twice I have been in a very mountainous and clayey place, where a large quantity of very hard woods are to be found beneath the earth, black and perfumed, with very beautiful veins, regarded by our

Prince Cesi as minerals [i.e., not as wood]; we've had many discussions about the site and the petrified substance of some of them.[58]

Winther says that he had read Cesalpino's chapter on ebony and discussed it with Cesi. As anyone who looks through the Windsor volumes can judge, the interest in veining and in ebonylike fossils is intense. But how did such phenomena, the apparent results of the underground fires and the burying of the wood—if that is what it is—arise?

On August 18 Cesi himself wrote enthusiastically to Cassiano that "daily I discover wonderful things in the natural laboratories of these parts, about which I shall write to you in due course."[59] He was referring to the same *naturae officinae* mentioned in the Naples manuscript, and especially to the *lignificium Acquaspartanum*.

Five days later Stelluti, himself still in Acquasparta, wrote to Galileo about more finds of extraordinary and unexpected fossil woods. It was the height of summer, but even in those hot days, he noted, Cesi did not cease to make his "very beautiful observations" of that puzzling intermediate class, the *legno-minerale*. Cesi had found

> very large pieces up to eleven palms in diameter, and others with iron fibres in them, or similar metallic pieces within such woods [c.f. fig. 11.5], and others which secrete a kind of resin [cf. figs. 11.3–4], like incense, even having a similar odour; and then too he has found a very large quantity of woods which have been turned into stone or pyritized,[60] of most extravagant forms.[61]

If Galileo were to come by, Stelluti concluded, he would be able to see both the woods and the underground fires.

It was at this time that Cesi seems to have thought that one innocuous way of enlisting the support of Cardinal Francesco would be via their fossil discoveries. *Curioso* that Francesco was, he probably would have been interested in them anyway; but who did not know how important aesthetic considerations also were when it came to the family of the pope? So Stelluti sent the cardinal an artistic product made from a large piece of the *legno-minerale*: a small table that Cesi had succeeded in having made from "a solid piece of the mineralized wood which I have found in these parts." Flattering the cardinal, Cesi wrote that "the learned and virtuoso delight which you take in natural observations has encouraged me to send it to you, as something which can best show you the nature of these things, especially the variety of its veins and concretions."[62]

By the beginning of December Cardinal Francesco was caught up in it all. Cesi wrote him an extraordinary letter. Obviously he felt that at this favorable juncture he could take the cardinal into his scientific confidence. His letter deserves to be cited at length, since its contents could hardly be more revealing about the nature of his paleontological concerns:

> In the few moments of leisure left over to me from the troubles of my affairs, I have been working on my physical observations, and particularly on the new type of middle nature between plants and metals, the veined and varied mineralized wood discovered by me in the last years in these parts and which I have already brought to you. I have studied and seen many differences, full of curiosities and excesses so immense in size that in the family of similar composite things they even exceed the proportions of the cetaceans amongst feeling creatures. . . .
>
> I have found a great variety of petrified objects—in other words, those which have become totally petrified . . . and others which are mixed, where some parts remain of wood, the others of stone, or of a similar condition, with both the interior and exterior mixed in many different ways and shapes.
>
> That which gave me the greatest occasion first for wonderment and then for contemplation was the discovery of metallized woods, and of those that were also wholly or partly transformed into metal and strangely altered. I believe this phenomenon to be something completely unexpected by natural historians, as the mixtures described above are very pleasing and at various stages of petrifaction. But for many the matter—being more of a natural problem than a scholastic one—has remained extremely doubtful and difficult.
>
> Moreover I believe that the relationship between these terrestrial natures will open up the way to a full understanding of them, until now very limited and confused. Because aside from the above-mentioned middle natures, I have found bodies which are very ambiguous and which show very different combinations: wood and earth, wood and stone, wood that is both stony and metallic. I have found wood and mineral juices drawn together and participating equally of each substance, in a body almost neutral, or rather combined and reduced to an apparent unity, whence the very obscure nature of those terrestrial liquids which we call bitumens will be greatly clarified.[63]

This, then, was the crux of the matter. Avoiding the potentially dangerous problem of the generation of fossils, Cesi went on to discuss a whole series of "mixed" phenomena. All appear several times both in the drawings and the correspondence. They include such things as gagate, anthracite, "lithanthracite" (a low grade anthracite), fossilized ebony, aetites,[64] and finally the fragrant agalloch. Having already commented that he had found specimens of gagate far from the Ganges, where it was supposed to arise, Cesi concluded of agalloch that its origin was also unknown, but that "it is said to be transported by floods," and that it too was not "likely to be very far from this marvelous new kind of metallophyte."[65]

How proud Cesi was of this discovery! He noted that he was attempting "*as much as it is permitted to me* to survey them in writing, to arrive not only at the full historical point of view, but also at every natural consideration and discussion, in searching for causes and examining as much as can be found about these subjects or concluded from them."

But why this reticent tone? It is barely perceptible, but unmistakable nonetheless. What constraints underlay Cesi's words? The question soon became urgent (though never clamorous). It may seem that Cesi gives us no help with it here, at least none that we can immediately detect. But is there an allusion, in the passage that follows, to the controversy about substances that are pregnant with light, and to the whole issue, raised by Galileo himself, of the dissemination not of earthly but of celestial light? This, after all, was a risky domain.

> I have also observed, on such occasions, furnaces lit by nature, their manner and their fumaroles. I also saw in this new class of things not only sulphurous efflorescences, but also saline, vitreous and crystalline ones; and I confirmed to my particular satisfaction many of my theories, made over many years, about the essence and causes of these fires, which although very little known and hidden to human intelligence, always seemed to me for the most part to be full of light.

He will come and give the cardinal a summary of his views in person the following week. In the meantime he sends a few small gifts as illustrations of what he has been describing: "a piece of petrified trunk in which both substances are very clearly evident; a piece of simple fossilized ebony; and a splinter of another kind of this [ebony], much more solid and varied, broken up by bits of metal; similar to those in the attached fragments, rather fragile and brittle. "[66]

In those days, when seemingly private correspondence often served public purposes, Cesi's letter to Barberini was immediately circulated, and recognized as central to his work. Several drafts for it survive,[67] as well as at least six copies, generally in the company of manuscript versions of Stelluti's *Trattato*.[68]

In rendering Cesi's letter to Francesco Barberini into English, it is hard to avoid the word "fossilized." One wants it, because it would so easily account for the phenomena described by Cesi. But although the term "fossil' was certainly current, it was generally used in the sense of anything dug up from the earth, and without the evolutionary and paleontological connotations it now has. Indeed, although there was nothing in the existing literature on fossils to offer an adequate explanation for the fossil woods of Acquasparta, Cesi himself never provided anything really convincing either—at least not in the few surviving indications of what he once intended to write.

## THE ORIGINS OF FOSSIL WOOD

Many of Cesi's friends knew about his fossil researches, even when they were not directly involved in them. In his commentary on Mexican animals[69] in the *Tesoro Messicano*, Faber referred to these researches on several occasions. Faber himself realized that many animals, insects, and even fossil forms like amber raised the problem of phenomena that participated in more than one class of things. In an important passage on the subject, concluding with a discussion of the double nature of amber, he explicitly referred to the significance of Cesi's finds: "Our most illustrious and excellent Prince Cesi was the first to discover and reveal this middle nature between plants and minerals which he treats specifically in his books on metallophytes." "Last year," Faber continues, "he sent both a specimen and different species of these from Acquasparta to Rome to the most illustrious Prince Francesco Cardinal Barberini."[70]

In an earlier passage Faber commented on the beauty and elegance of pieces of furniture made from fossilized wood from Acquasparta. Not only had he sent a table of this substance to Cardinal Francesco, Cesi "recalled to us the ancient use of working terebinthine wood into tables, caskets, small chests, and various other pieces of furniture."[71] Just such pieces are also mentioned by Cassiano in an undated letter to the antiquarian Lorenzo Pignoria in Padua;[72] while among the curiosities listed in the postmortem inventory of Cesi's museum are two inkwells and a rather crude and broken table all made from fossil wood.[73] Time and time again we are reminded of how closely aesthetic and scientific issues were linked—and not just as a way of ensuring patronage (as must have been the motive for the gift to Cardinal Francesco).

In any event, as Faber noted, it was Cesi who was

> the first to uncover and to see to the digging up of this substance. Not only was it sprinkled with the most beautiful flecks and waves, it was also adorned, as it were, with all the other qualities of such concretions. These were so attractive that they seemed to compete—and even to outdo—the elegance of the veins of the *Arundo* or Indian shot. You could even say that these were tiger-like tables, since they glistened with a dark yellow color and were marked with wavy lines. . . . But the interested reader will learn all these things much more clearly from Cesi's writings themselves, which will very soon be published, and in which many new things discovered and observed by him will be presented.[74]

How much they all kept on hoping that Cesi would deliver on the promise of his researches, right until the very last moment!

In the notes to his *Persius*, published the year after Cesi died, Stelluti also made much of the aesthetic aspects of the fossil finds (at least when he found a reasonable pretext in Persius). Thus, just as in the case of some of the ancient beds made of cedar mentioned by Persius, the wood found by Cesi "has a hundred of the most beautiful and diverse forms of undulating waves, that are rather like camlet and other forms of watered and shot silk of the kinds that one uses today. When well seasoned it can be beautifully worked and acquires a lustre like ebony."[75]

Then Stelluti moved on to rather more critical matters. He recorded that according to Cesi, the Acquaspartan wood was nothing but "the trunks of trees, born beneath the earth where there are other minerals." They began as earth and gradually acquired

> the form and nature of wood; and for this reason he wished to call it by the name of metallophyte, finding it of a middle nature between plants and minerals. This was first discovered and invented[76] by Prince Cesi.

Stelluti recalls how

> On many occasions I accompanied him when he went to observe the place where this wood originates, and not without amazement I saw the earth transformed into wood, and there I found large pieces of trunks up to thirty and more *palmi* in circumference.

The size of these pieces was an important factor in the two friends' speculation about the origins of the fossil woods: in their view, they were so large that they could not possibly have been naturally occurring forms. They must have been formed from the earth itself.

This is an extraordinary position, one that points to the shortcomings of relying too much on observation and too little on a reasoned hypothesis about the nature of petrifaction.

At this point Stelluti says that Cesi was fully engaged in writing about the fossilized woods,

> both from a historical and a physical point of view, with his observations about it, and a variety of figures and causes of these matters, such as one will be able to see in his books *De Metallophitis*, which will shortly be published, and in which he also writes about other relevant objects, and about the said [middle] nature discovered by him, and in particular of stones not observed or described by anyone else.[77]

It is clear that Stelluti wrote these notes—and these lines in particular—in the course of the last days of Cesi's life; later on, when he refers to Cesi's discussions with Ferrante Imperato about a kind of *Cytisus* very similar to the ebonylike minerals he had discovered, he says, poignantly, that Cesi's planned books on the metallophytes "ought soon to be printed."[78] A vain hope! Nothing at all appeared for seven years—and even then what did appear did so under Stelluti's name, and was not even a shadow of what the prince had intended.

## STELLUTI AND PEIRESC ON FOSSILS

Every project—and not only the book on fossils—was thrown into disarray by Cesi's death. Stelluti later wrote that in taking Cesi away before he could complete his books, it seemed that Nature herself envied his great knowledge.[79] The work on fungi—the other archetypal instance of things of middle nature—also never saw the light of day. The publication of the *Tesoro Messicano* was held up for almost twenty years. Practically none of the Linceans' plant investigations ever appeared. Faber died in 1629, and when Cesi died a year later, the surviving Linceans were naturally discouraged. In any case, the events that led up to Galileo's trial in 1633 further contributed to an atmosphere that had become decidedly uncongenial for all of them.

At this point the fate of the Linceans' investigations into fossils and the problem of middle natures lay in the hands of two men—Cassiano and Stelluti. Both were determined not to let the memory of Cesi's researches in this domain fade, but Stelluti, for better or for worse, was now the real guardian of Cesi's scientific heritage. He knew he could count on the support of Cassiano, into whose hands Cesi's material had passed, and who remained the closest link to Cardinal Francesco. Despite the latter's abandonment of Galileo, Stelluti felt that they could still perhaps count on his patronage, financial or moral. He and Cassiano must also have thought of invoking other support, from the grand dukes of Tuscany, for example, who remained faithful to Galileo all along. In 1635, as if to remind the Medici of Galileo's fossil researches, Cesi's widow sent them gifts of some fossil woods as well as a pair of large and rather fragile tables of fossil wood. These were still so crude that they had to be polished by an ebony-worker and strengthened by the addition of regular wood.[80]

At the end of that year, Stelluti wrote to Galileo, now more of a Medici protégé than ever, to express sympathy for his troubles, and to cheer him up with some news about the progress of an old project of theirs:

> Your Lordship ought to know that when Signor Balì Cioli [secretary of Grand Duke Ferdinand II of Tuscany] was here recently [in May 1635][81] he went to visit our Lady the Duchess [Isabella Salviati, Cesi's widow] on several occasions, and when he left she gave him several pieces of the fossil wood that is born near Acquasparta. . . . and at the same time he wanted to know where this fossil wood was found, and how it was generated—since he had read in the commentary to my *Persius* that Prince Cesi was writing about it. Her Ladyship the Duchess ordered me to write something about this, which I did, and this was sent to their Serene Highnesses along with a box of different pieces of the said wood.[82]

This, in fact, was the year in which scholarly interest in the fossil finds came to a head. During its course, yet another player came onto the scene—that most polymathic and stimulating of all Cassiano's friends,

Nicolas Claude Fabri de Peiresc, the antiquarian from Aix. For several years he too had been preparing a work of his own on shells and fossils. Although he had been informed of the Linceans' work before then, it was only now, five years after Cesi's death, that he decided to take a personal interest in the fossils of Acquasparta.

To judge from the correspondence of Peiresc, Cassiano, and Stelluti himself, sometime in the course of 1634, quite possibly at the behest of the grand duchess of Tuscany, Stelluti began putting together his summary of what he knew of Cesi's views on fossils. He could himself recall Cesi's thoughts about them, and he had probably studied one or two of Cesi's draft notes on the subject. He also owned one of the copies then circulating of Cesi's remarkable letter of December 1, 1624, to Francesco Barberini.[83] In the correspondence of the time, Stelluti's "summary" of Cesi's views was referred to as a *discorzetto* or *libretto*, and at least five copies are known to survive.[84] One was sent to the Medici, another to Francesco Barberini, another to Cassiano, and another to Peiresc.[85]

Peiresc did not like what he read. Nor, indeed, did he like what he had heard of Cesi's views. To him they did not ring true—especially not those on the generation of fossils.[86] Already in early 1635 Peiresc began writing to Cassiano about the fossil remains of Acquasparta and about Stelluti's "little book." He asked Cassiano to send him pieces of the fossil finds

> which have the crust on them; and those which have a mixture of metal; and those which have a mixture of resins and gums; and those which retain some of the nature of wood and of stone together; and above all those which retain the nature of clay, susceptible to fire and bricklike by nature. These seem to me easier to understand than the others, unless they contain mixtures of different natures together, and appear to be incorporated and entangled one within the other.[87]

Peiresc was clearly interested in the problem of middle and combined natures too, but his foremost concern was to discover how the fossil woods came about in the first place. So the immediate task was to get to southern Umbria and make the required firsthand observations.

But he could not go to Italy himself. Instead, he decided to send his old friend Jacques De La Ferrière, the personal physician of the cardinal of Lyon, to examine the site for him. "When one cannot go to the sites oneself, it is a good idea to send practical people who also have insight into the most recondite mysteries of nature," he wrote to Cassiano.[88] Both De La Ferrière and another French expert, Claude Menestrier, were already in Italy and had just gone to Sicily on behalf of Peiresc in order to examine some of the alleged fossil remains of giants there.[89] After several adventures and mishaps, they arrived in Acquasparta in November 1635, and within a few weeks De La Ferrière sent Peiresc his detailed report. It too is preserved in a neat copy in the Library of the Old Medical School in Montpellier.[90]

In his report De La Ferrière gave an account of the fossil woods and baked clays of both Acquasparta and Rosaro that could almost be a summary description of the drawings in Windsor—"large lumps of baked brick-colored clay of rounded shape. . . . chalk or clay little by little turning into a substance approaching that of wood, and then effectively into wood itself" and so on; but then, after describing the site at Rosaro in some detail, his tone changed:

> And as for Prince Cesi's views (in support of his theory of generation) that he had seen pieces which at one end were like hard clay and at the other wood, here I think he's made a mistake. He assumed that the beginning of petrifaction was rather the beginning of the generation of the wood. Indeed, I have a piece given to me by the Cavaliere dal Pozzo which is wood at one end, and like a half-baked brick at the other. Now, there's nothing more natural than the conversion of wood to terra cotta or to stone—but not the making of wood from the earth.[91]

Indeed not. De La Ferrière was correct, and Cesi—if his views were being correctly reported—wrong.

At about the same time, Gabriel Naudé visited the site. One of the true dissimulators and cynics of the century, he was one of the most brilliant of the libertine antiquarians, and had long served as librarian to the famous, from Antonio Barberini, through Cardinal Guidi di Bagno (himself interested in the results of De La Ferrière's researches), and on to Queen Christina and Cardinal Mazarin. Clearly, anyone who was anybody in the Republic of Letters recognized the significance of what Cesi had brought to the fore. But almost all of them—and certainly Peiresc and Naudé—were dismayed by his apparent failure to acknowledge other possibilities concerning the origins of fossils.

Like Peiresc, Naudé believed, or came to believe, in the organic origins of fossils. Both felt that in his eagerness to reject the Aristotelians, Cesi had too swiftly discounted a number of alternative positions, such as the view that fossils, especially marine fossils, were left in high places as a result of waters receding after the inundation of the earth, or as a result of a natural cataclysm, such as an earthquake (the Aristotelian position).

Naudé did not hesitate to criticize Cesi's views. Years later, in his stinging *Mascurat*, he used them as the basis for a merciless satire on the credulousness of Cesi and his associates. "In deferring too much to the judgment of others," he wrote, "[Cesi] trusted almost nothing to his own.

. . . Just as there are some spirits who have the marvelous disposition to believe nothing, so too there are those who even more vigorously doubt nothing at all."[92]

Such an opinion, about a man who could not have been more resistant to received opinion all his life, is certainly too harsh. But what more precisely were Cesi's views about the origins and generation of fossils, and were they really his to begin with? There was widespread interest in them; even Galileo in his exile was kept abreast. Copies of Stelluti's *discorzetto* or *trattatello* were circulating in both Rome and Tuscany—to say nothing of France—and disseminating what purported to be Cesi's views on the origins of these woods.[93] Stelluti decided to have it sent to the printers, along with the few engravings after Cesi's drawings that were already in circulation—a mere fraction of what he once planned to publish, and with hardly a mention of the whole problem of middle natures.

## THE TREATISE ON FOSSIL WOOD

The overlong title, in a way, says everything: "Treatise on newly discovered fossilized mineralized wood; in which it is briefly treated of the various and mutable nature of the said wood, represented in several figures, showing the place where it arises, the diversity of the undulations to be seen in them, and its such various and marvelous forms; by Francesco Stelluti, member of the Academy of Linceans and native of Fabriano." Not a word of the classificatory and taxonomic problems that so preoccupied Cesi; a strong emphasis on the aesthetic interest of the fossils; and—as we see so clearly in La Ferrière's letters as well—a preoccupation with the geographical location of the fossils. The treatise was certainly brief. It contained only eight pages of text and thirteen engraved plates.[94]

In his dedication to Francesco Barberini, Stelluti explained that the copies of his manuscript that were circulating had aroused such interest that he had decided to have it printed, along with a number of "very curious illustrations." He was dedicating the work, he added, to the cardinal, not only because of his interest in "contemplating the hidden parts of nature,"[95] but also because he was offering him a subject, as he put it, that was "wholly new and ambiguous."[96]

But what did Stelluti actually have to say about all this? Less, perhaps, than we might expect; and rather different, perhaps, from what Cesi might have said.

As so often with the Linceans, things were seldom what they seemed. Would Francesco Barberini, the very man who had so coldly abandoned Galileo at the time of his trial four years earlier, not have recognized the kinship between the design of this title page (fig. 11.20) and that of Galileo's *Letters on the Sunspots* of 1613, the real beginning of his rift with orthodoxy (fig. 5.1)?[97] Let us examine the contents of the treatise on fossil wood a little more closely.

Stelluti began by declaring that he would be discussing the fossil wood

Fig. 11.20. Title page, Stelluti, *Trattato del Legno fossile minerale* (1637).

TRATTATO
DEL LEGNO FOSSILE MINERALE
NVOVAMENTE SCOPERTO
NEL QVALE BREVEMENTE SI ACCENNA LA VARIA,
& mutabil natura di detto Legno, rappreſentatoui con alcune
figure, che moſtrano il luogo doue naſce, la diuerſità
dell'onde, che in eſſo ſi vedono, e le ſue così varie,
e marauiglioſe forme.
DI FRANCESCO STELLVTI ACCAD. LINCEO
DA FABRIANO.
All'Emin.mo & Reuer.mo Sig. Card.
FRANCESCO
BARBERINO.

IN ROMA, Appreſſo Vitale Maſcardi, MDCXXXVII.
CON LICENZA DE'SVPERIORI.

discovered by Cesi. This was "by nature a metallophyte, partaking both of metal and of plant; but much more plant than metal." He emphasized not only its localness but above all its remarkable aesthetic qualities—the variety of its beautiful growth lines, and its many diverse forms. Its closest antique parallel, he said (for one always had to begin with the classical references), was probably the cedar wood of Mauretania, from which the ancients had also made furniture, as he had already mentioned in his Persius.[98]

Then he turned more directly to the matter in hand. He specified the location of the finds in the area once occupied by the west arm of the ancient Tiberine Lake, and had this nicely illustrated by a map showing the region extending from Todi to Amelia and up past Acquasparta into the mountains to the east (fig. 11.1).[99] Then, moving on to make one of his basic points, he embarked on the most improbable part of his discussion. "As far as I have been able to see," he asserted, "the wood is not generated from the seed or root of any plant whatsoever, but only from a piece of earth, containing much clay, which is slowly transformed into wood, nature thus operating until all the earth is converted into wood."[100]

This odd and unlikely proposal is the central element of the Stellutian theory of the generation of fossil wood. At the same time, it is not so difficult to understand how Stelluti (and perhaps Cesi too) might have arrived at such a view. If we look at the Windsor drawings, the fossil wood often seems to emerge from the clay—but can Stelluti really have been so naive? Or was it a case, once again, of not being critical enough about what he thought he had seen and observed firsthand?

Stelluti elaborated his position on the basis of the subterranean fires or "fumaroles" of the region:

> This conversion into wood I believe to take place with the assistance of the heat of subterranean fires. There, snaking underground, they emit thick and continuous smoke, and sometimes flames, especially when the weather is wet, and with the further assistance of sulphurous and mineral waters. And when the heat is sufficient, the wood is browned, or somewhat scorched, and becomes like coal. If then, the earthen material is as yet unconverted into wood, these flames fire it, and it becomes like vases of terracotta in a kiln, or like bricks.

This was a critical moment in the text. The actual observations were unassailably accurate.[101] But Stelluti seems to be saying that the underground fires helped transmute the clay into fossil wood, and that when the heat was more intense, any unconverted clay was baked like terracotta. On the basis of observation the Linceans had inferred "that the essence of this wood is nothing other than earth, because I myself have seen pieces of which one part was hard clay, the other wood and the rest coal."[102] So much for close observation! The Linceans were better at it than almost anyone else; but this still did not prevent them from sometimes drawing the wrong conclusion, just as Stelluti did here.

There is a indeed a lesson here about how empirical and observational evidence may be used to bolster a wholly incorrect hypothesis.[103] In support of his view that the wood could not have plant origins, Stelluti drew a further set of conclusions. Emphasizing the fact that the fossilized trunks always lay prostrate, he pointed out that they were of such a size and weight that they could never actually have stood upright, as trees must. Moreover, they never seemed to reproduce any part of a plant besides the trunk; and so they could not have originated in the plant—or more specifically, the arboreal—kingdom. "The raw material," he goes on to demonstrate, "is nothing other than clay-rich earth, in confirmation of which one may adduce its great weight and the fact that even a small flake can sink in water." Worse and worse.

But it could not be said that Stelluti was unaware of the opposite (and certainly more correct) theory—which he promptly refuted: "Neither can it be believed that these pieces of wood are the trunks or fragments of other trees buried in this area, or fallen and covered by earth, and then fashioned with these waves by the mineral waters which have their springs there, or created by subterranean fires," he said, because "no trees have ever been formed by Nature like those in the following illustrations."[104]

If he came so close to acknowledging the organic origin of fossils, why then did Stelluti attempt to suppress, or sideline, this much more plausible account of the origins of fossil wood? There is no hint in this treatise of the possibility that the "waves" on which both he and Cesi commented so much were in fact growth lines (c.f. fig. 11.21); instead, he harps on their beauty and on other pretty particularities of the grain. He describes in some detail the different ways in which the fossil woods burn; and then discusses how some samples may seem to have wood inside and earth outside, and some the other way round. In this and many other ways, he asserts, nature jokes—*scherza la natura*, the typical phrase of much natural history in those days—and makes us marvel at the wondrousness of its works.[105] There is no sense at all here, as we certainly know there was with Cesi, of the problems that such "jokes" presented to the task of classification.

Immediately before moving on to discuss the illustrations of his treatise, Stelluti produced what he regarded as one of his trump cards: "In the absence of any other proof of the generation of this wood, this may suffice: a quantity of damp earth having been removed from around a piece of the wood, and placed in a room in the Palace of Acquasparta. . . was found after some months to be wholly converted into wood."

Even a schoolboy could have put forward the alternative explanation that upon drying the clay cracked and separated from the wood inside, and fell off. According to Stelluti, however, Cesi was so amazed by this that he had no doubt "that the earth itself is seed and mother of this wood, the earth of these parts being most suitable for its generation."[106]

Why, then, did Stelluti protest so much in favor of this explanation of the generation of fossil woods? And why did he continue to insist that Cesi's finds demonstrated that the woods passed through various stages, beginning with shapeless pieces of earth right down to those that were wholly wooden? True, many of the drawings do show pieces of wood with more or less soil actually clinging to them, some seeming to be pure earth, others mixed, and others pure wood, just as he described them in his book; but can he really have been satisfied with so

Fig. 11.21. Section of fossil wood showing growth lines. Stelluti, *Trattato del Legno fossile minerale* (1637), pl. 4.

superficial a conclusion? Did he and his colleagues really trust so naively to a prima facie explanation of what they had observed? No wonder Naudé was so critical.

## ILLUSTRATING THE TREATISE

At the end of Stelluti's *Trattato* come its thirteen plates, a truly meager selection in comparison with the massive body of visual documentation that survives in Windsor and that was put together by Cesi in preparation for his own treatise on the subject. For no immediately apparent reason, all the illustrations in Stelluti's book (for the most part based on drawings in Windsor) are of specimens from the area around the little village of Sismano near Acquasparta.[107]

The series begins, however, with two striking double-page maps. The first, which shows the territory of Todi and Acquasparta, with the four main sites where metallophytes were discovered (fig. 11.1), is followed by another double spread showing the site close to the Castle of Rosaro—the very site described so closely by De La Ferrière—where a large quantity of the fossil woods had been found, and where the fumaroles (nicely illustrated in figs. 11.2 and 11.10) had been burning for a period of ten years.

Then come four pages showing the meandering growth lines (the "waves") of the fossil wood (fig. 11.21).[108] No drawings for these plates survive. They offer yet another instance of the Linceans' sensitivity to what then would have been construed as the marvelous artfulness of nature, equal at least to the artifice of man. But aesthetics, as often, was combined with science. Not realizing that these wavy lines were in fact growth rings, Stelluti took them to be evidence that the specimens could never have been real trees, for in that case they would have been regular concentric circles, rather than wandering waves or other patterns.

An illustration of a large specimen of fossil wood embedded in the earth follows (fig. 11.22), and another of two more pieces of the white resinous kind, mentioned several times in the correspondence.[109] All are based on the equivalent drawings in Windsor (cf. figs. 11.3–4). Stelluti thought that the oval shape of these fragments was a result of their having been unable to grow upward: he believed that the great weight of the earth in which they were embedded meant that they could not grow into the circular or cylindrical forms of ordinary tree trunks.[110] This he took to be further proof of his view that these fossil forms were actually generated from the earth itself.

Three pages of interesting specimens of fossilized wood and of the pieces of baked earth called laterite come next.[111] One of these Stelluti describes as follows:

> Trunk D . . . also excited great wonder, its innermost part being totally petrified, and near red in colour, but the outside rough, and knotty, exactly as drawn. Its entire length was six palms, and its width approximately one palm in diameter.

Little could be plainer than this. The emphasis on color, texture, size, and the relative state of fossilization occurs throughout Stelluti's spare comments on the plates. Aside from an occasional reinforcement of the view that fossil woods are generated out of earth, there is barely a theoretical moment amid all this high description.

The last plate shows some ammonites (fig. 11.23; cf. fig. 11.19). Stelluti's text about them is significant, not because it adds anything to the discussion so far (indeed, it comes as a kind of distraction from the relatively concentrated discussion of fossil woods, concretions, and baked clays), or because it is in any way an accurate account of the origins of these fossils (he held that all fossils belong to the mineral kingdom and grow within rocks!).[112] Rather, it reminds us of how much more Cesi

C
E
F
G
H
A
B
D
Questo pezzo di legno ouato nel modo che si uede era di misura dalla lra A. à
B. palmi tre; da D. à C palmi tredici; da E ad F palmi undici; da F à G palmi
dieci e mezzo. da G poi sino ad H fu il legno dalla terra ricoperto, e non si po=
tè uederne il fine.


Fig. 11.22. Piece of fossil wood from Sismano in situ. Proof illustration for Stelluti, *Trattato del Legno fossile minerale* (1637), pl. 7. Rome, BAV, MS Barb. Lat. 4355, fol. 27r.

himself planned to discuss, and what his own treatise, had it ever appeared, might have illuminated. This plate, he said, shows several stones—looking like snails or snakes turned in upon themselves—of the kind called Ammon's horn:

> I would always have believed them artificial, and not natural, had I not discovered them to be like this, not only in the territory of Acquasparta, but also in my native territory of Fabriano. My Lord Prince Cesi was writing about their generation and nature, but was prevented by death from finishing the treatise about this material. . . . He was writing not only about the generation of the said plants and woods, and of aetites, which are also generated in the said parts in large quantities, but also about all the other stones known until now, and of still others not yet observed or described by other authors. By long and diligent observations he discovered how the

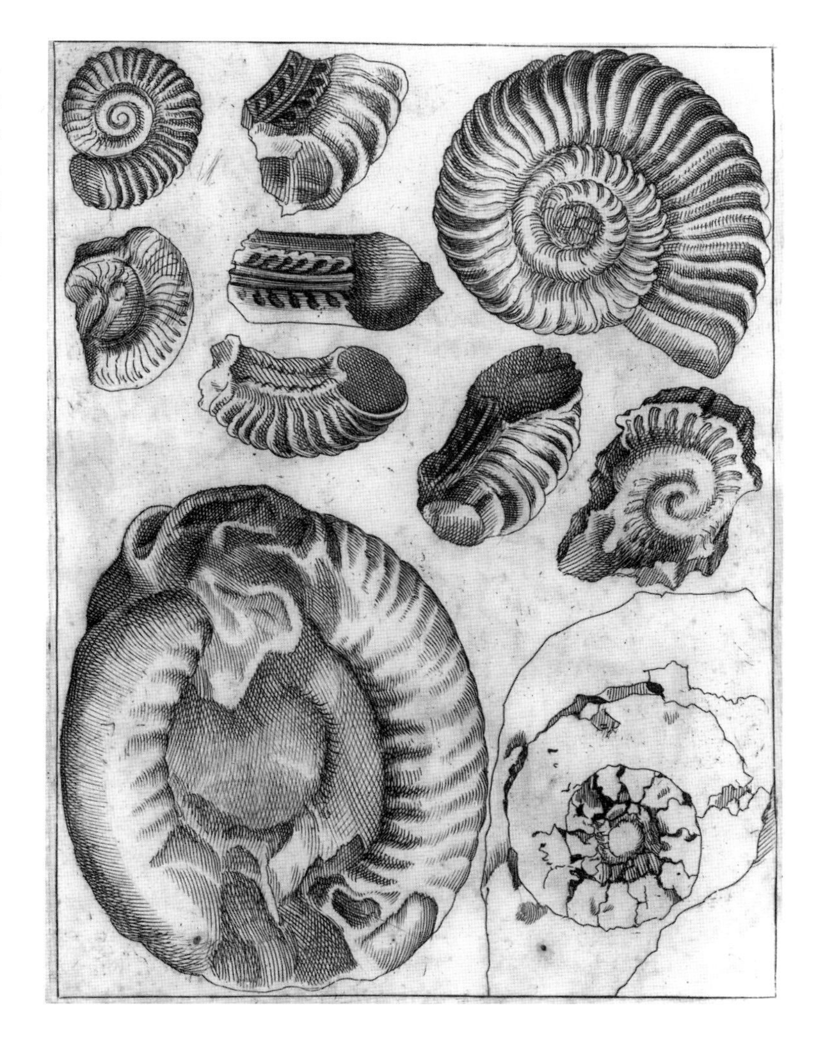

Fig. 11.23. Ammonite, three fragments, and a cross-section of ammonite in rock. Proof illustration for Stelluti, *Trattato del Legno fossile minerale* (1637), pl. 13. Rome, BAV, MS Barb. Lat. 4355, fol. 33r.

> pregnant aetite is generated;[113] and how those little trees with branches and leaves are formed in the arboreal stones (which Imperato calls "imbosked"), so closely resembling nature that they seem to be artificially painted.

At this point Stelluti draws up a list of stones that were of interest to Cesi, including corals, petrified fungi, mixed (i.e., variegated) marbles, bucardites, lyncurium,[114] ostracites, fossil worm tracks, fossil sea urchins and fossil sea-urchin spines.[115] Many of these are also to be counted among the *figulapides* referred to in the first of the *Tabulae Phytosophicae*. A few are illustrated in the Cesian manuscripts in the Institut de France (as in figs. 8.35–36),[116] while others appear in the beautiful volume of mineral specimens, gems and other curiosities in Windsor (e.g., figs. 1.18–19 and fig. 14.3).[117] It would have been a most ambitious project, had Cesi survived.

Passages such as these reveal much more than the aesthetic dimensions of what passed for scientific description in those days. Even in their relatively simple way they went beyond the connections between art and nature, between *naturalia et artificialia*, so favored by *virtuosi* like Cassiano (Cesi himself, like many of his friends, loved the "mixed marbles" illustrated in the fifth volume of the *Natural History of Fossils* in Windsor Castle, and paralleled by plates 6–9 in Stelluti's treatise). They remind us of the fundamental Lincean emphases on observation, discovery, taxonomy, and, above all, the interest in problems of generation and reproduction. But the book ends in a lower key. After the tantalizing glimpse into the further areas of Cesi's paleontological researches, Stelluti laments how Nature herself had shortened the days of his life "only so that he should not come to discover the manner in which she keeps her operations hidden." Again he is referring to the *arcana naturae*, which would eventually yield their mysteries to science.

But how much more would Cesi actually have discovered had he lived? The indications, at least in the field of paleontology, are not especially promising. Was he obliged to hold back in his researches on fossils because of his relationship with Galileo? Probably.

## CESI'S VIEWS

We know from two of the surviving drafts for the *Trattato*—and to some extent from Faber as well—that Cesi intended to produce three books on metallophytes, but that he barely finished writing one.[118] These books would have also contained his views on other fossil forms, such as those mentioned in the letter to Francesco Barberini and in the long paragraph about ammonites and other figured stones at the end of the *Trattato*. But we are in the dark about their more specific contents. Stelluti's book gives almost no hint at all of Cesi's struggle with the problems of classification and with their implications for the study of paleontology. The *Tabulae Phytosophicae*, on the other hand, provide a few indications of Cesi's sense that the project of a fixed system of classification was thwarted by the issue of time. This, after all, presented a major obstacle to the traditional views of fossils.

In all his work, Cesi realized, like Linnaeus long after him, that any attempt at classification had to be accompanied by the study of generative and reproductive principles. But to what extent was Stelluti justified in claiming that Cesi believed that wood was generated from earth, that the seeds of the wood were actually earth? There is little in Cesi's own work to suggest this. Could it all have been a front, to divert attention away from the ever more obvious conclusion that fossils were often the vegetable and animal remains of long past ages when parts of the earth pushed up above seas and lakes, and that phenomena such as fossil woods were indeed the remains of ancient trees that had become fossilized? Such views, of course, stood at odds with the traditional biblical accounts of both the Creation and the Deluge.

As we have seen, the events leading up to and following Galileo's condemnation in 1616 had made the whole matter of biblical interpretation extremely tense. Galileo had attempted to reconcile nature and the Bible in the most unacceptable ways, the theologians alleged, and he had twisted Scripture to fit his new interpretations of nature and the cosmos.

Grassi and his fellow Jesuits believed that Galileo was not sticking to his last when he tried to interpret Scripture in such a way as to suit his own views.[119] This central aspect of the controversy around Galileo found an ominous reflection in the very letter of November 1628 in which Colonna first wrote to Stelluti about the opening page of the *Tabulae Phytosophicae*. Having praised Cesi's "very beautiful and new division" of the metallophytes, he continued:

> How it will be when it comes to the subdivisions of meteors[120] and heaven and earth I really cannot imagine. I am not speaking of corporeal things, about which you need to curtail your intelligence—because if the scholastic theologians are of one opinion today, they will have another much harsher one tomorrow. *Et sapienti basta il motivo*—and to the wise man the motive is enough.[121]

The meaning of this slightly cryptic remark is clear: take the hint! Colonna's letter continued, even more ominously and not a little rancorously:

> As I wrote on another occasion to you, I think it would be a good idea to warn Galileo not to expatiate too much on scriptural matters, particularly with regard to the miracle of the furnace, in which the three [holy] children are seen to walk and to praise God—in case these men find the occasion to prohibit his work, in order to make themselves the prior inventors of all your inventions, as they were unable to do with regard to the movement of the earth, and the moon not shining in and of itself.[122]

Little upset Galileo more than the idea that others might usurp his own work, or claim their discoveries for his. Nor did he like being told not to expound on scriptural matters. What is most relevant here is the way in which Colonna goes directly from the Cesian researches to the vexed situation around Galileo. Indeed, as if aware of the intensely risky context in which they were all moving, Colonna makes a passing reference to the great Dominican heretic Tommaso Campanella; and then expresses his hope that God will grant Galileo a long and healthy life.[123]

Thus, finally, we come closer to what may really have stood in the way of the full publication of Cesi's views on fossils. In so many respects his findings presented a threat to the standard interpretations of the Bible; and these, as most of Galileo's clerical opponents argued, were the preserve of the theologians. The well-known story of the Three Holy

Children in Daniel 3, for example, had repeatedly been adduced by Orazio Grassi as one of the proofs of the transparency of flames at the time of the comets controversy. He set this against Galileo's insistence that the comet could not be a flaming substance precisely because it *was* transparent (for Galileo, it will be remembered, the comet was a matter of light reflected on gases passing through the atmosphere). In the *Assayer* Galileo went to elaborate lengths to demonstrate not just that fire was not transparent,[124] but that the story of the Three Holy Children in the Bible did not imply that either.[125]

Cesi and Galileo had begun to find that almost every natural historical matter, including the problem of fossils, related directly to one or another aspect of the controversies that swirled about them. And if these issues could not be impugned from a mathematical—and what we would now call a scientific—point of view, then they were always vulnerable hermeneutically.

One of the most moving aspects of Cesi's work was his skepticism about the traditional Aristotelian views of nature and its objects (despite Naudé's unfounded allegation of gullible incredulity). However much he may have been steeped in Aristotelian natural history, however much he may have been inclined to accept the charms of its appeal to the purely empirical, he resisted it more and more. His researches on the fossil woods led him to researches on other fossils. What these views actually were he may have had to suppress. At the very time Galileo was in the deepest trouble for proving—over and over again—that the world was not the center of the universe, the suggestion that the world was not made in a week, and that there was no fundamental unity of species across time, was indeed dangerous.

We will never know exactly what Cesi thought about the origins of the fossils of Acquasparta. It is unlikely that we shall ever discover whether he, like Leonardo, formulated some rather more general explanation for the presence of marine fossils on mountaintops, far from the oceans. But what we *do* have, in the fossil drawings at Windsor, is still further visual testimony to his inexhaustible curiosity. Nothing, it seems, escaped the eyes of Cesi and his Linceans, no wrinkle or muddy patch of nature, no specimen of God's creation, large or small. In their commitment to the close visual recording of nature they may have tried to record too much that was not strictly relevant to the project of classification; and they probably devoted too little attention to the problem (often set out by Galileo) of the relations among observation, hypothesis, and experiment. Perhaps their energies were too much absorbed by the amassing of visual and verbal records of so much that had never been documented before.

History may have committed Cesi's ideas to oblivion, but the extraordinarily detailed evidence on which he planned to base them has now finally emerged from the detritus of history. Whatever results he may or may not have achieved, the lasting aspect of his work has now finally been recovered: the first comprehensive study in visual form of the fossil remains of a particular region. There was nothing like this before, and nothing for a long time to come. Cesi's own contribution—massive, committed, and full of remarkable natural historical intuition—fills a major historical gap; and, in its fierce observational intensity, prepares the way for a new kind of commitment to visual illustration as an aid to science.

# PART FOUR

PANTOGRAPHICE
VLVM SABELLIVM ALBANI PRIN
CHRISTOPHORVS SCHEINER SOC. IESV.
S. GAVINVS
Inspice et fac secundum exemplar

# *Pictures and Order*

# TWELVE *The Failure of Pictures*

## THE EVIDENCE OF THE SENSES

On the face of it, Cesi and his friends never gave up their faith in pictures. They believed in their usefulness as records of what they saw, as aids to memory, as a means of comparison and identification, and as a more or less efficient form of disseminating the knowledge they accumulated. Right until the end Cesi worried about the illustrations for the *Tesoro Messicano*. He puzzled over how best to represent the strange combinations of animal, vegetable, and mineral that constituted the fossils. He never stopped having drawings made of his microscopic examinations of the reproductive parts of ferns, fungi, and bryophytes. Every plant and animal that came to his attention was submitted to his draftsmen for illustration.

But things are seldom what they seem. From the beginning Cesi's work was riven by a fundamental tension between the desire to picture everything and the desire for order. Pictures showed too much. They could convey texture and color and irregularity in meticulous detail; but it was precisely this that detracted from their ability to show what was essential and regular about things.

The more accurate pictures were, the more they reflected the untidiness and disorder of nature. Pictures might help in showing what was exceptional in nature; but in striving toward system and systematization, Cesi realized that he had to establish what was regular, rather than what was anomalous.

Color, as we have repeatedly seen, was a particular problem.[1] Time and again the Linceans were faced with the old difficulty that the color of a particular species could vary according to the time of day, the season, or even the age of the specimen. When it came to the determination of class, color was never a reliable guide. By the 1620s—when the Linceans were commissioning their best drawings—they seem to have moved far away from the blithe confidence in color expressed by Aldrovandi when he wrote that color was an "excellent indicator, and, when joined with other accidents of smell, taste, and touch, a very sure means of arriving at a perfect knowledge of mixed species, whether perfect or imperfect."[2] Aldrovandi, that most unsystematic heaper-together of the regular and the irregular, the monstrous and the normal, the anomalous and the typical, may indeed have discerned the centrality of the problem of "mixed" species; but his faith in color (and visual representation in general) seemed misplaced. The same went for texture and smell too.

No one put the basic problem more clearly than Galileo, in a famous passage in the *Assayer* of 1623. This was the book on which Cesi, Faber, Cassiano, and Ciampoli had been working with such intensity for at least five years, the very years in which they first applied the microscope to their investigations. They knew its contents like the back of their hands, and played a critical role in the actual formulation of many of Galileo's positions. "I say," wrote Galileo,

> that whenever I conceive of any material or corporeal substance, I am compelled of necessity to think that it is limited and shaped in this or that fashion, that it is large or small in regard to other things, that it is in this or that place, at this or that time, that it moves or is immobile, that it touches or does not touch another body, that it is one, a few or many; nor can I by any stretch of the imagination separate it from these conditions. But that it is white or red, bitter or sweet, sounding or mute, of pleasant or unpleasant odor, I do not feel compelled in my mind to conceive it as necessarily accompanied by such conditions. On the contrary, if we were not assisted by our senses, reasoning and imagination would never apprehend these qualities. Therefore I think that tastes, odors, colors and so on . . . are nothing but pure names, and reside only in the feeling body, so that if the animal is removed, all these qualities are taken away and annihilated.[3]

This eloquent assault on the unreliable subjectivity of what Locke—and Berkeley after him—later described as the secondary qualities helped bring into focus the Linceans' reservations about color. In the very work which the Linceans had been instrumental in publishing, Galileo made it clear why pictures would never do—nor, indeed, any description that depended on the all too subjective senses. Only those qualities that depended on the objective and mathematically definable properties of mass, space, extension, and number (more problematically) could help to define what was essential about the things of nature.

## DOUBTS ABOUT PICTURES: SUBJECTIVITY, ANOMALY, AND SIMILITUDE

Of course there had been reservations about pictures before. Galen, the ancient medical writer whom everyone (including the botanists) read, had rejected them as being distractions from the substance of what had to be taught. By the middle of the sixteenth century (the first great age of illustrated scientific treatises and manuals),[4] the concerns grew more persistent. The year 1543 was the true *annus mirabilis*—in more ways than one. This was when both Copernicus's *De Revolutionibus Orbium* and Vesalius's *De Fabrica Corporis Humani* appeared, just a few months after the fullest version of Leonhard Fuchs's great herbal, the *De Historia Stirpium* of 1542 (immediately reedited in 1543 and 1545). Even more than anatomy and zoology, botany turned out to be a critical area for the exploitation of illustrations as an aid to research and teaching.

Even then, however, there were many who worried about illustration. Despite his deep commitment to it, Vesalius wrote repeatedly of his anxiety that the subjective hand of the artist might interfere with the clarity and plain detail of the illustrations. Fuchs had a similar concern. "Shading should not obliterate the basic forms of the plants," he wrote.[5] But here, in the introduction to his handbook (and in almost the same breath), Fuchs also invoked the old Horatian ideas about the efficacy of pictures as a didactic aid: what you saw with your eyes sank in more deeply and effectively than what you heard (and by analogy than what you read too).[6]

Of course Fuchs realized, as did every other botanist from Otto Brunfels and Hieronymus Bock on, that illustrations could help immensely in the tasks of identification, comparison, and taxonomy. But when it came to the complex business of classification, the challenge of representing the essence of a plant rather than its mutable or seasonal aspects made the whole matter of illustration all too problematic. It is significant that in the work of the greatest of the early botanical classifiers, Bauhin, there is hardly an illustration to be found. In the *Pinax* of 1623, the culmination of his classificatory efforts, there is none at all. Pictures simply did not go with classification. And if pictures *were* to represent the all too mutable appearances of things, one would need far too many of them, as Linnaeus would caustically assert in the *Genera plantarum*.[7] At what point did Cesi begin to realize the scale of *this* problem?

Certainly not at the beginning. He and his colleagues started to have drawings made of everything they could, not just because of their faith in representation by visual means, but because of the influence of the one figure who was their real predecessor when it came to the collection of vast quantities of visual information, and the picturing of as much as could be found in the world of nature: Aldrovandi.[8] It was he, more than anyone else previously, who had attempted to reproduce the multiplicity and variety of things in visual form. Aldrovandi had had illustrations made of everything that he possibly could. In them, monstrous, anomalous, and totally idiosyncratic lemons featured next to the most apparently normal ones; fishes and plants were turned—we would say distorted—into the most preposterous shapes, often quite improbably anthropomorphized (e.g., fig. 14.2). The most unnatural-looking specimens were set beside the most natural-looking ones.

## Limon Periſtero-cephalos.

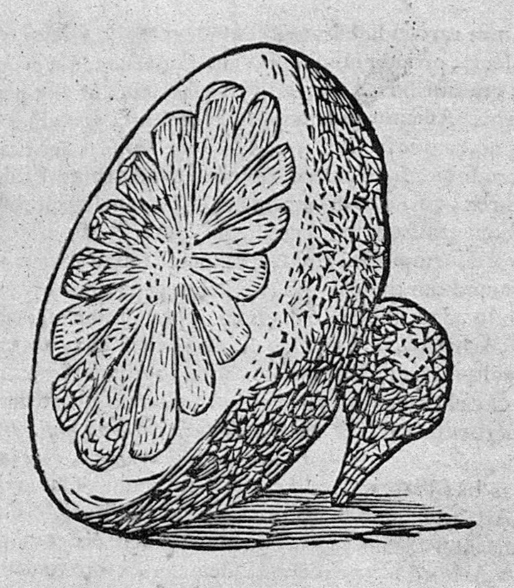

*Cur tenet adnatum Limon caput iſte Columbæ?*
*Nempè vt ſit duplici nomine peſtifugus.*

## Limon Proſopætos.

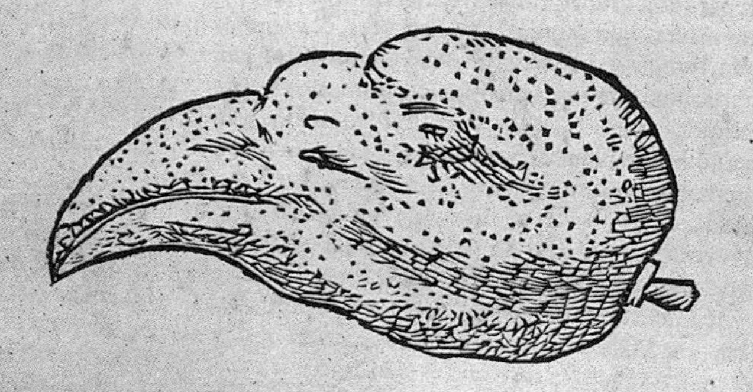

*Nunc Aquila, hic Limon, renouatrix mira iuuentæ*
*Fit tibi, Quo ſumpto ſecla Aquilina trahas.*

Limones huiuſmodi figuræ, magnitudiniſquè duos in diuerſis temporibus obſeruatos adhùc extantes iam ſiccatos aſſeruo, quod non parum mirabile eſt.

Verùm hæ raræ, atq; deformes quodammodo conformationes, præternaturalem Plantarum ſuarum ſtatum oſtendunt, opportunum propterea ſequitur agere de Morbis Aurei Mali, eorumquè remedijs.

Fig. 12.1. *Opposite:* Misshapen lemons. Woodcut, Aldrovandi, *Dendrologia* (1668), p. 515.

Fig. 12.2. Plants for heart disease. Woodcut, Della Porta, *Phytognomica* (1608), p. 223.

When Cesi wrote to Stelluti and De Filiis on April 10, 1605, about how best to achieve the aims and purposes of their new foundation, he specifically addressed the connection between pictures and irregularity. Item 17 on his list of *Propositions* for their Academy emphasized the importance of employing an engraver, "since in printing the compositions of the Linceans, the greatest expense will be in the figures." It was an urgent matter, this proposition continued, to "move forward on this matter, in order to be able to illustrate every observation and *capriccio* of ours."[9]

This was a critical conjunction. In using the word "capriccio"—not far from the English "caprice" or "whim"—Cesi was referring to the "games of nature," as they were also called—that is, to the exceptions, singularities, prodigies, and monstrosities that would so exercise all of them (and Cesi in particular) in the following years. Already by then Cesi was beginning to realize that only by understanding what fell between the interstices of the rules and what subverted them would one ever be able to approach an understanding of the order of nature. And pictures went well with caprice, irregularity, and monstrosity.

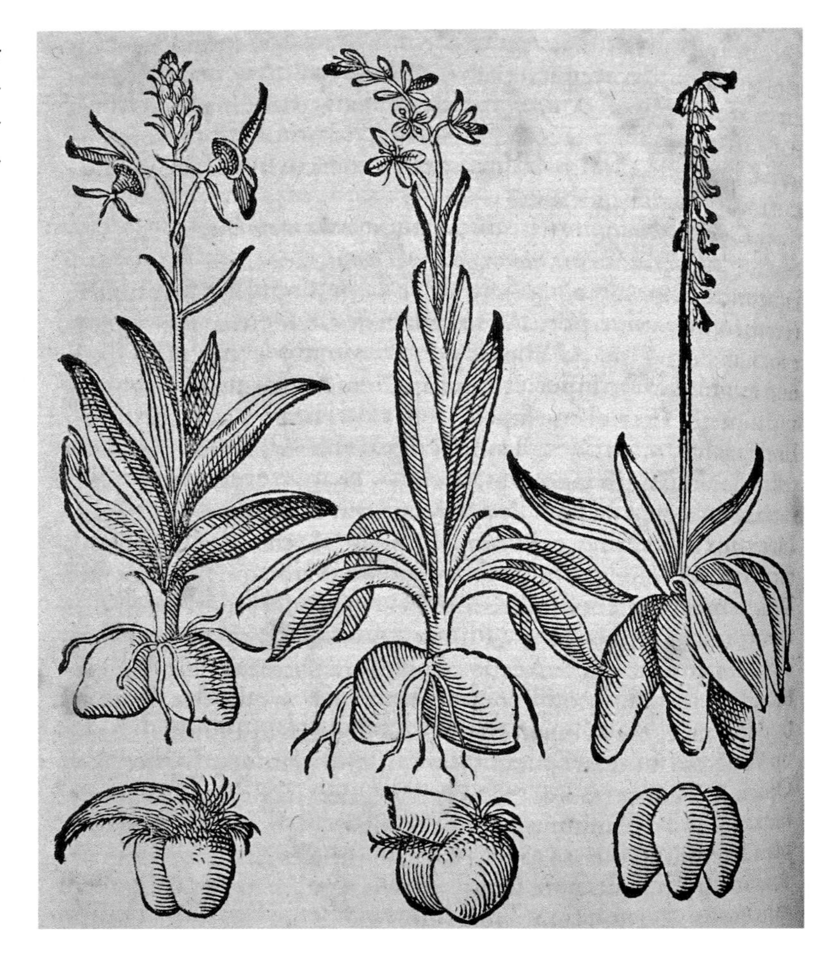

Fig. 12.3. Plants for sexual problems. Woodcut, Della Porta, *Phytognomica* (1608), p. 216.

Here too was a serious problem. You couldn't make things appear to be so abnormal that they never resembled anything else at all. Somehow or other they had always to be made to resemble something else—even if only vaguely. It was not their sheer exceptionality that made the monstrous and anomalous what they were, it was the fact that they always retained at least a minimal relation to that from which they diverged. Foucault rightly observed that the great motor of such "classification" as there was in the great illustrated scientific works of what he called the "Renaissance" was similitude—and Aldrovandi was no exception. Often in these earlier writers, and in much of the early Lincean corpus too, one has the impression that anything could be made to resemble anything else. You could explain the character of something by examining what it seemed to resemble. As a result, similitude—so dependent on surface resemblance—was often forced by wishful thinking.

In particularly puzzling cases, Aldrovandi and his editors had repeated recourse to the apparent solutions offered by anthropomorphizing parallels and the various other strategies of explanation via similitude (fig. 12.1; cf. fig. 14.2). Hence the obvious kinship with the Della Portan sci-

Fig. 12.4. Heads of man and lion compared. Woodcut, Della Porta, *Physiognomia* (1598), p. 167.

ences of phytognomy (where the powers of plants are related to the other kinds of living things they resemble, or the parts of the body they resemble, figs. 12.2–3), and physiognomy (where human temperament and virtues are explained in terms of the surface resemblance of body and face to their supposed animal equivalents, fig. 12.4).

At first sight Cesi and his friends may seem to be heir to this tradition. Remnants of the old Della Portan approach are to be found not only in the endless physiognomic and genital similitudes in the volumes in the Institut de France (e.g., figs. 8.29–33),[10] but also in much finer drawings such as that of the carefully drawn fruit or calabash with an apparently human face (fig. 14.1),[11] which is still of the same order as the remarkable illustration of an anthropomorphized piece of apple bark in Aldrovandi's *Dendrologia*, for example (fig. 14.2).

But when the Linceans began reading the works of Cesalpino and their own Fabio Colonna, they began to realize the problems presented by the exceptions and anomalies of nature to the idea of classification. It was clear how useful pictures were when it came to dealing with the oddness and anomalousness of things, or with borderline cases. Pictures, being richly descriptive (at least potentially so), could show just how much the aberrational varied from the regular. Pictures could embroider where rules could not. They could show texture where rules were firm and untextured; and so on and so forth. But at the same time, in their determination to classify the natural world, and to find the right places for things, both Cesi and Colonna realized that you had to strip away irrelevant detail—in other words, exactly that which presented to the eyes whatever was distinctive about irregularity and monstrosity. Pictures on their own, however useful they might be, were not enough. In this lies the great paradox of Colonna's series of illustrations in his *Phytobasanos* of 1592, and the two parts of his *Ekphrasis on lesser known plants* of 1606 and 1616, all of which so inspired the Linceans (e.g., figs. 9.1–2).[12]

The very title of the two works of Colonna's maturity, *Ekphrasis*, exemplifies the tension to some degree, for it is a straightforward use of the ancient term for the verbal description of works of art.[13] In the 1606 edition Colonna had rather unself-consciously emphasized the importance of the illustrations as a basis for the communication, confirmation, and identification of species; but in the second part of 1616, he made it clear that while he did not intend to give up on illustration, the task would now be to find a still more secure basis for classification. He knew that he and his colleagues would have to begin moving away from their reliance on the external appearances of things. Pictures were too often misleadingly inaccurate—or even downright deceptive—while the colors of things were so mutable that they more often stood in the way of classification than aided it. Even so, despite their many concerns about illustration and its great potential for inaccuracy, neither Cesi nor Colonna could bring themselves to renounce it.

## PROBABLE AND PREPOSTEROUS PICTURES

The problem appears with great force and with melancholic results in the *Tesoro Messicano*. In a sense the book is a failure. Both it and the many surviving documents relating to its genesis and production testify over and over again to the Linceans' commitment to reproduction in visual form. There can be no question of how attached both Colonna and most of his fellow Linceans remained to illustration. But the illustrations in the *Tesoro* are indifferent, and inferior to many others in other natural historical works of the time—admittedly as much a result of practical difficulties as anything else.

Page after page in the book—as well as in the documents associated with it—reflect the Linceans' difficulties with pictures and their doubts about them. When Faber realized that the illustration he had of an American civet (fig. 10.8) was of limited use in his inquiry into the musk-producing parts of that animal, he asked Colonna for help. He wasn't sure whether it was only the male animal that produced a musklike secretion from its genital area, or whether the sacs from which this liquid issued were to be called testicles or not.[14] Colonna obligingly went to look at the pair of specimens in Don Bernardino de Cordoba's menagerie in Naples, and had three drawings made of that animal, including one of its anal region (figs. 10.7).

> As soon as I learned that you still had some doubt about whether those little places from which civet-musk is extracted could be dignified with the name of testicles . . . I immediately went over to look at that animal again, and examine the matter more closely.[15]

Colonna then went on to describe the animals in considerable detail, paying special attention to their colors, as if, once more, to make up for the inherent deficiency of the medium of woodcut.[16]

But there is more. Faber ended his own digression on the civet cat with the concluding words of Colonna's letter. In a discussion in which pictures loom so large, they are striking and paradoxical:

> On this page you can see for yourself just how much my picture of this animal differs from that of the Mexican example and from others in the world. I do not deny that there can often be great differences [between things], because of geography and climate; but I also know this: *painters often make big mistakes*. I have experienced this in other cases too. Remember what Pliny said in book 25, chapter 2: "Painting is indeed deceptive, and with its many colors, is particularly so in the copying of nature, where it also falls short as a result of the varying ability of the copyists."[17]

It was not even necessary to add the next sentence from Pliny: "Nor can one simply paint plants at individual stages of their lives, since they change their appearance according to the fourfold changes of the year."[18] Everyone knew this.

What faith could one then hold in pictures? Here, if ever, was a two-edged sword. No book could testify better than the *Tesoro Messicano* to Colonna's and his fellow Linceans' belief in pictures as an aid to natural historical investigation. He—and they—invested huge amounts of energy in getting them right. But they were also aware of just how misleading pictures could be. Still, Colonna's drawing of a civet cat was clearly better than Faber's copy of the Recchian original. It bore out his argument about the animal and its genitalia, and it was clearly of greater taxonomic help than the all too improbable illustration Faber had supplied for the book.

To go through Colonna's *Annotations and Additions* to the *Tesoro* is to be impressed again by how much, in spite of himself, he kept his faith with pictures. He went through all eight hundred of the woodcuts in Schreck's section with a fine-toothed comb. He compared them with the Recchian originals and noted wherever the copyists—or Schreck himself—made a mistake. He corrected Schreck's identifications and pointed out where better illustrations were to be had in the other botanical handbooks of the day, especially those of the ever-authoritative Clusius. Often he added details of his own, extensively supplementing the pictures with words. He indicated exactly where the illustrations could have been more accurate still, and made refined and careful observations about color. Even with all the faith in pictures in the world, one could always use words too.

Over and over again, moreover, Colonna provided wholly new illustrations where he felt that the originals were lacking in detail or accuracy. If a simple leaf could be of taxonomic assistance, he had it drawn and engraved[19] with a precision unimaginable in the Schreckian woodcuts (cf. figs. 10.12–13). For the most part, however, Colonna's main concern was with the parts of the plant that had to do with reproduction—bulbs, flowers, seeds, and so on.[20]

Fig. 12.5. *Canis Mexicana*. Woodcut, *Thesaurus* (1651), p. 466.

Whatever the defects of pictures, Colonna believed that they were always *potentially* reliable. You could always check them, at least in principle, on the basis of actual observation. This was certainly more than could be said of the second- and third-hand copies in the Recchian manuscripts and the *Tesoro Messicano* itself.

The illustrations in Colonna's *Annotationes et Additiones* are thus much more accurate than those in Faber's commentary on the animals of Mexico. This is not just because most of Colonna's illustrations were taken directly from the specimens themselves, or that he took such care in getting them right. It also has to do with the fact that most of us are likely to be better at discerning improbability in animals than in plants. It is a matter of discerning improbability on a gross scale rather than a small one.

But even so: very many of Faber's animals, if not rudimentary or improbable, do indeed seem preposterous. Who could possibly believe that the "Mexican dog" or the overly Americanized civet cat come even close to the reality of the animals they are supposed to represent (figs. 12.5 and 10.8)?[21] Similarity may all be relative, a matter of convention; but still! The absurd oversimplification of illustrations such as those of the "Peruvian sheep"[22] and the Mexican wolf,[23] tiger,[24] leopard,[25] bull,[26] and boar[27]—to say nothing of the ridiculously anthropomorphic grins with which the hapless woodcutter generally endows them—strains all credulity (figs. 12.6–7).

Not that Faber is wholly to be blamed. After all, the animal illustrations were only copies of Recchi's copies of Hernández's originals (which themselves were probably not too accurate to begin with). The illustrations of most of the plants in Colonna's section—and even many in Schreck's—could be checked against actual specimens (even if they were not actually based on them). In this respect de Cordoba's civet cat was a

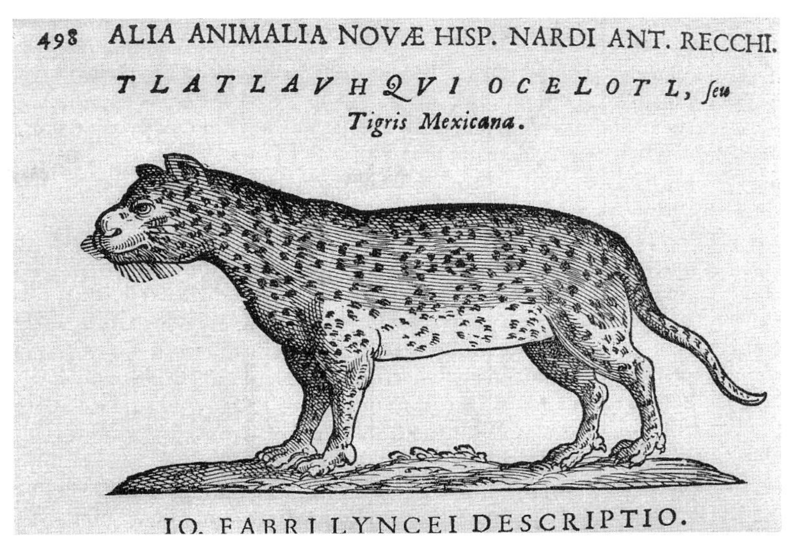

498 ALIA ANIMALIA NOVÆ HISP. NARDI ANT. RECCHI.
TLATLAVHQVI OCELOTL, seu
Tigris Mexicana.
IO. FABRI LYNCEI DESCRIPTIO.


Fig. 12.6. Jaguar. Woodcut, *Thesaurus* (1651), p. 498.

Fig. 12.7. Bison. Woodcut, *Thesaurus* (1651), p. 587.

notable exception. No wonder that Faber spent so much time running back and forth between his home and Petilio's, constantly trying to ensure that the copies had been properly and accurately done! And still, given the second- or third- or fourth-hand quality of it all, no wonder that he had such trouble in determining what species the animals actually were!

There was the problem of the Mexican panther (actually an ocelot, *Leopardus pardalis*), for example. Faber couldn't be sure whether it was the same as the panther of the ancients, or a hyena, or the modern equivalent of the ancient lynx. Was the Mexican squirrel a kind of weasel or not? Was the Mexican bull (clearly a bison) actually a kind of buffalo? After all, the Italian *bufala* was not a buffalo at all. He worried endlessly. The illustrations in his section were simply not accurate or detailed enough to arrive at definitive conclusions.[28]

Faber himself was hardly unaware of the shortcomings of his illustrations. He often corrected them. "Its tail is a little longer than the picture. . . . the hairs are less dense than shown here. . . . the spots are larger than in the illustration" he wrote of the woodcut of the Mexican tiger.[29] Often he too had to supplement the meager black-and-white evidence of the woodcuts with long coloristic descriptions, especially in the case of birds such as the colorful hummingbird or *Picus mexicanus*. Here he pulled out every one of his lexical and rhetorical stops.[30] But once more there's an irony.

In cases such as the hummingbird, Faber was describing the colors not of the bird itself but of the Recchian drawing: further evidence, it may seem, of faith in the fidelity of representation. But the paradoxes multiply. With illustrations such as that of the Mexican bull, for example, Faber could still not be sure that it was sufficient as an aid to classification and taxonomy—and so he made a minute comparison of the illustration with the verbal descriptions of that animal in Thevet, Acosta, and Aldrovandi.[31]

Once more faith wavered. There were serious moments of hesitation. Faber observed that to judge from the illustration, the Mexican squirrel seemed larger than the European version, but then added, "if, however, it is permissible to make a conjecture about this sort of thing on the basis of a picture."[32] Suddenly—at last!—proof is separated from conjecture, at least in the domain of the evidence of the visual. It is a critical moment.

Fifty or so pages later Faber arrives at the strange *Aper mexicanus*, or Mexican boar. He comments:

> to judge from the colors of the illustration, this is a boar, not a domestic pig. It has really coarse bristles and very shaggy hairs, which tend to a dark brown color (if, that is, a conjecture and a picture carry any weight).[33]

Here, if ever, is doubt. Picture and conjecture could not be more clearly aligned. But conjecture was a stage prior to proof.[34] It signaled probability rather than certainty. While pictures might not provide you with the proof you wanted, at least they offered a reasonable basis, sometimes even a good basis, for conjecture.[35]

But what about proof and certainty? You had to know what things really were, how to classify and how to name them. You had to be able to do so even in the case of those things that seemed to go beyond the realm of possibility and any known reality. Could pictures come to the rescue here? Could conjecture somehow be forced closer to proof?

It was precisely at this delicate point that expediency and the doubtful certainty of pictures came jarringly together. Right toward the end of Faber's *Exposition on the Animals of Mexico* are two astonishing instances of the way in which social and patronal expediency could force an assertion of the value of appearance, of what may be seen on the surface of things. It is hard to believe that Faber, Colonna, and Cassiano were not suspicious of what they showed in their illustrations; but the stakes were high.

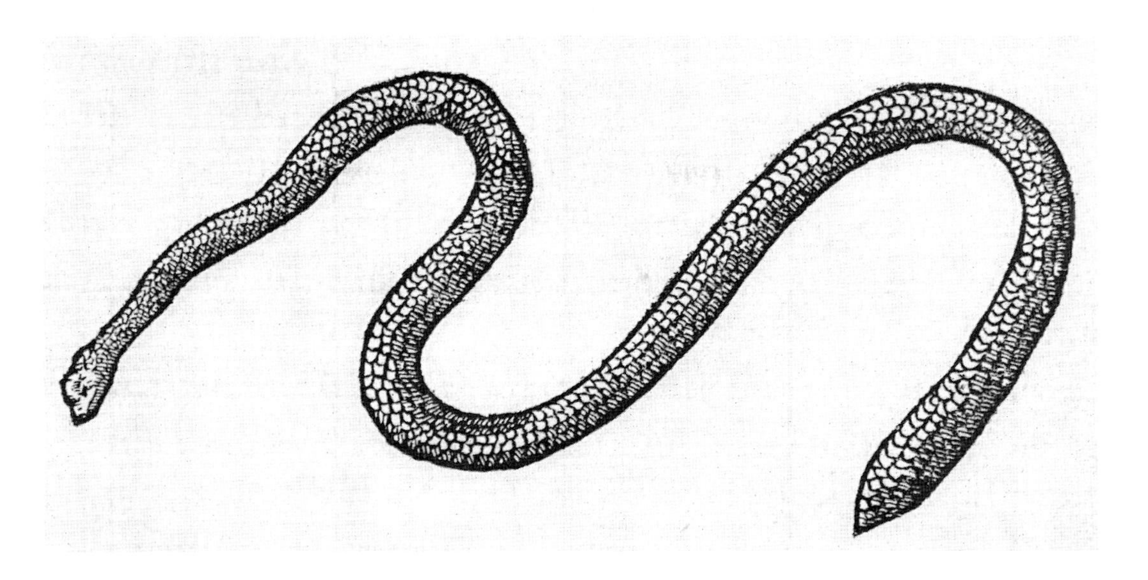

Fig. 12.8. *Maquiztetzauhuatl* or *Amphisbaena Mexicana*. Woodcut, *Thesaurus* (1651), p. 790.

## THE WORM AND THE DRAGON: PRESSURES OF PATRONAGE

The two final animals in Faber's *Exposition*—aside from a pair of uncontroversial scarab beetles—strain all credulity. The first of these is the *Maquiztetzauhuatl*, a peculiar wormlike animal—probably a glass or worm snake—then known as the Mexican amphisbaena.[36] Faber reproduces the Recchian drawing (fig. 12.8). It shows something that is just conceivably a two-headed beast; Faber noted that if you cut it in half, it would grow a new head: from half the thing came a new one—or perhaps a half-new one?[37] In this way it seemed to offer a clue to the mystery of its own generation. But did it really have two heads, as suggested (though hardly proven) by the illustration? Faber reviews all the ancient sources on the subject with his usual erudition (some say yes, others no) and decides that the case does indeed seem unlikely, and that in any event the illustration does not really seem to show two heads either.

But then, in a seeming about-face, he brings in another piece of evidence: "Just as I became convinced that the two-headed amphisbaena was most likely the stuff of myth and fable, rather than of truth, the Cavaliere Cassiano dal Pozzo, one of our Linceans, contrary to every hope and belief of mine, showed me the most truthful image [*verissimam imaginem*] of an amphisbaena, in the form of a drawing with all the appropriate colors." For these men, a good picture, based on actual observation, could turn everything around. Faber had received this drawing just as he was sending off his book to the printers. It had been brought directly from Paris, where a friend of Cassiano's was present when it was drawn from an actual specimen there.[38]

So we turn with some expectation to the reproduction of the drawing Cassiano brought back with him (fig. 12.9).[39] What a disappointment! Though described by Faber as a "very truthful image," it is so utterly im-

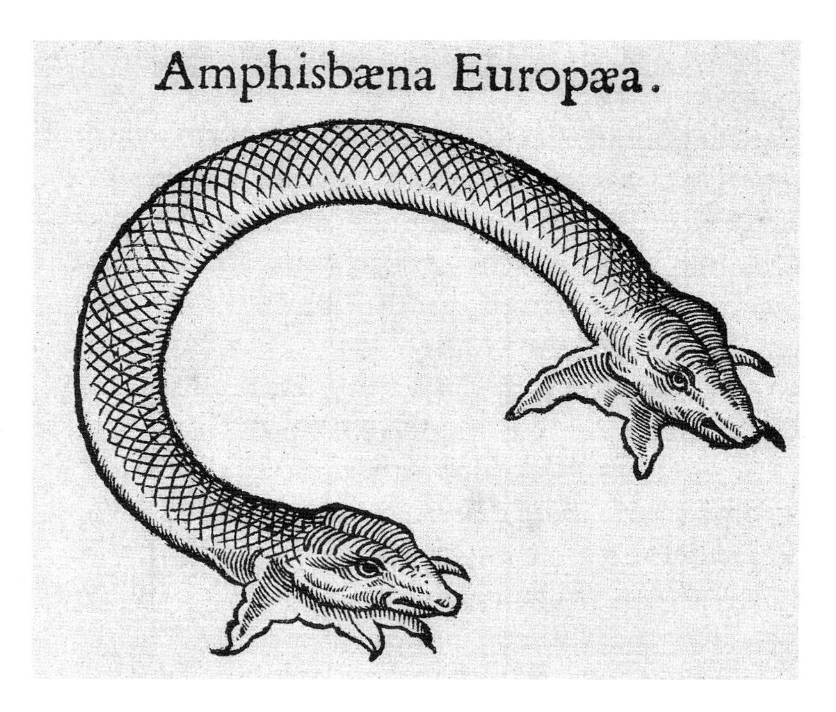


Fig. 12.9. *Amphisbaena Europaea*. Woodcut, *Thesaurus* (1651), p. 797.

probable—and so strangely rudimentary—that it is impossible to believe that it could have had any source in nature at all. How could anyone possibly have believed it, let alone someone with as much experience as Faber? And yet he did (or at least made a good show of doing so). But why, and on what possible grounds?

Anything, Faber concluded weakly, was possible in God's universe; and if God could produce two-headed serpents in the New World, or a boar with a navel on its back (as in the case of the "Mexican boar" or peccary), then one had admit that the New World was capable of producing animals that anywhere else would be regarded as monsters.[40]

But what is *really* going on here? It is all a little hard to credit. Did Faber really believe that Cassiano's implausible drawing proved the existence of so improbable a beast as the two-headed amphisbaena? He said he did. Perhaps it was because he felt that he had to make a gesture of belief in the certainty of *any* drawing offered by Cassiano. In truth, almost all of Cassiano's drawings are exceptional in their accuracy and detail. Could Faber not tell the difference here? He must have been able to do so. So what kind of pressures applied in this case?

It is hard to tell. Nevertheless, in the case of the next (and final) animal in the book, the matter is much clearer. Faber devotes no fewer than thirty long pages to the *Dracunculus Barberinus*, a really strange little beast whose half-fleshed skeleton belonged to Cardinal Francesco.[41] It looked like a serpent but had a horn, a pair of feet, and wings (fig. 12.10). Was this a serpent that could fly, a winged dragon, just like the beasts of myth? Surely not. How could anyone believe in the veracity of so peculiar an illustration?

Fig. 12.10. Alleged skeleton of the *Dracunculus Barberinus*. Woodcut, *Thesaurus* (1651), p. 816.

It appears to show the head of a bat (or perhaps a small ape?), the forelegs of a rat, and a vertebrate tail—as well as wings and two feet. For some reason Faber needed to believe both in it and in the original. After discussing a variety of other improbable beings, he concluded that the *Dracunculus Barberinus* could not be a mere invention. He marshaled his very considerable philological expertise to analyze the distinctions between the various possible Latin words for such a being, like "draco," "anguis," and "coluber." And he referred to what to us may seem to be the nearest parallel, the bat, an animal already much discussed and anatomized by the Linceans, precisely because it too stood on the borderlines of several classes.[42]

Not surprisingly, this gave Faber yet another opportunity to refer to the Lincean use of dissection as a means of determining what things really were, and how best to classify them. Over and over again he and the others had anatomized animals and hermaphroditic specimens of one kind and another in order to establish the right class to which to assign them.[43] As if to demonstrate his clear and long-standing awareness that a better way of getting at the truth below the surface was to cut, to penetrate to, and to magnify the insides of things, Faber here inserted a reference to a dissection he had himself done of a rabbit, in order to establish whether or not it was hermaphroditic (as maintained by Herodotus and many modern authorities).

Once more, penetration to the inside of things in order to find the truth that lay in the reproductive system replaced the old reliance on misleading and all too changeable surface (though of course the Linceans continued to seek to depict what they saw within). In an effort to bolster all this still further, Faber now added an account of Winther's anatomy of a wolf, done in his presence, as a means of establishing more precisely both the accuracy of the Recchian illustration of the Mexican wolf and where to place that animal in the scheme of things.[44]

Faber knew better than to take his and Recchi's drawings at face value. Yet in the case of the *Dracunculus Barberinus* and the Mexican amphisbane he came perilously close to doing just this. Indeed, when he reviewed all the possible evidence from both ancient and modern writers (which he did at even greater length than in the case of the amphisbane), he concluded that the *Dracunculus Barberinus* was a real, if very rare, beast.[45]

Two possible explanations lie before us: First, that for Faber the testimony of pictures did not merely offer the possibility of moving beyond probability and conjecture in the direction of truth, but that visual representation itself was capable of attaining the certainty of propositional statements. Second, that Faber did not believe any of this at all, but rather that he had to declare that he believed it to be true, simply because the *Dracunculus* belonged to Cardinal Barberini. The circumstances of patronage, in other words, forced him to assert that it was a possible animal—and not a fiction, a monster, or merely an invention.[46]

Observing that the *Dracunculus Monoceros Card. Barberini* was not actually one of the usual distant examples of Mexican fauna that illustrate the book, Silvia De Renzi noted that Faber's aim here was to offer (after four hundred pages!) a still more challenging task: to use words and a picture to dissipate all doubts about an animal

> that even the most favorably disposed listener would have judged to be unreal. . . . The whole epistemological framework which the Lincean doctor made use of to distinguish between true and false here collapses in the face of the impossibility of arguing against an object possessed by their patron Cardinal Barberini; and both anatomical expertise and philological skill are made subordinate to the more important task of affirming the advantage for other naturalists of describing, for the very first time, an animal held by everyone else to be nonexistent.[47]

The case is surely paradigmatic. In the critical years leading up to the trial of Galileo, the need to flatter the Barberini drove the Linceans to make repeated show of believing in what appeared on the surface. In this sense they abandoned Galileo—though if one reads deeply enough in the *Tesoro Messicano* one may understand the ways in which they never did at all.

"And this," Faber concluded,

> is a succinct but genuine and wholly true exposition of this animal, not one artificially put together by a peddler of false wares, but one really brought into existence by God and by Nature itself. Just as we have set this out in words on paper, so too we have wanted to portray it accurately in a drawing, to both the eye and mind of the curious reader. And we have done so all the more willingly because we know for sure that a dragon of this kind has never been so exactly described and so elegantly depicted as here.[48]

Did Faber actually believe in this animal? No—as little as he believed in the amphisbane. What is between the lines is clear enough. He did not believe; but he was obliged to make as convincing a case as he could for what he knew to fly in the face not just of conjecture and probability, but of truth itself. In effect, this part of his contribution to the *Tesoro Messicano* was all a cover-up.

Nowhere do we see the reasons for the failure of the Linceans to enter the mainstream of history more clearly. They were too often successful at masking their real achievements and commitments. Good science, so to speak, was buried in strategy. It was so well done that only a very few people would ever appreciate their achievement. After hundreds of pages of the most valuable material (despite the endless, sometimes pitifully anecdotal digressions), Faber had to conclude with a strenuous yet ultimately pathetic defense of what he knew not to be true, something that he knew would hinder his science rather than aid it. This was the tragedy of so much of the Linceans' work.

## AN ESSENTIAL TENSION

And Cesi? For all his sponsorship of the great campaigns of illustration that have formed the subject of these pages, it gradually became clear to him that reliance on surface could not even begin to suffice either for proof or for the great classificatory tasks he set himself. It was not just a matter of the relative subjectivity of the artist's role, or even of the subjectivity of interpretation. Strive as the Linceans might to reproduce the variable colors and textures of things, these were not of much help in Cesi's efforts to sort out where to place the anomalies of nature, to find the essential features of plants, and to find a system that even if not complete would at least offer the possibility of fitting in all the new species that might conceivably come their way.

Along with the effort to reproduce what he found in visual form, Cesi thus strove to produce something that in many respects stood in diametric opposition to this: his famous *Tabulae Phytosophicae,* which he began working on already in the second decade and never completed. He was working on them as he lay on his deathbed—and even then he must have known that what he was attempting to do was only half of what might really be accomplished if he could only engage in some of the mathematical strategies that he only hinted at in the vaguest of ways. In the end, he never came close (except verbally) to the kinds of geometrical diagrammatization or construction of the world through mathematical relationships that he suggested in his tables. This was a task that he might indeed have handed over to Galileo; but by the time Cesi came to his dying realization, Galileo was desperately putting the finishing touches to his own exposition of two ways of viewing the world, the *Dialogue on the Two Great World Systems*.

The essential tension in all of Cesi's work in these years—and therefore in all the Lincean projects—was between the desire to record and describe everything (on the one hand), and the need to reduce and classify (on the other). But he came to realize that the more extensive, descriptive, and veristic visual representation became, the greater the need to avoid what was merely secondary in the appearance of objects.

The aim, in the first instance, had to be to penetrate to the heart of what was essential about things. Otherwise the world would remain infinitely fragmented, with species unrelated and unsystematized. One would understand nothing if one could not see the essential relations between things. Nature would be filled only with interstices, or (an equally problematic state of affairs) with an endless array of species, each partaking of the nature and qualities of other species.

The use of the telescope, and a little later of the microscope, only complicated the matter further. The deeper one looked into inner or outer space, the more one found there, and the more the amount of visual information expanded. One had to do as Copernicus did, and Galileo after him, and reduce the density of pictures to the geometrical simplicity of diagrams, or—more rudimentarily—of tables. Not for nothing did Copernicus have printed on the title page of his *De revolutionibus* the very words that Plato, the greatest ancient embodiment of anti-Aristotelianism, is reputed to have inscribed on his door: *Let no one unskilled in geometry enter here*. Not for nothing did Galileo want to have the following words on the title page of his own collected works: "From this one will understand in infinite examples the usefulness of mathematics in drawing conclusions about natural propositions, and how impossible it is to do philosophy without the escort of geometry, in conformity with the pronouncement of Plato."[49]

Since Cesi was not even remotely the mathematician that Copernicus and Galileo were, his solution was to work not toward the diagrammatization of nature but rather to its presentation in tabular form, in a manner not dissimilar from that of the boxes and tables of the Ramists before him. Even without mathematics, the Ramists had attempted to reduce knowledge itself to tabular form. They too, like Cesi afterward, foundered in their attempts. There was no easy solution to the problem of order.

# THIRTEEN *The Order of Nature*

## THE MOCKERY OF ORDER

All his life Cesi wrestled with the problem of how to find order in multiplicity. It was not just that he and his friends had collected a superabundance of information, nor was it only a matter of inventing some more or less principled form of organization for it. From the very beginning Cesi realized that in order to understand the secrets of nature he would have to devise a system of classification. But nature itself seemed to thwart him. For where was one to place the irregularities and anomalies? Every one of them, in one way or another, subverted the necessities of order and structure. The "jokes of nature," of which he and many other naturalists of the time spoke so often, were just that: they introduced disorder into order and made a mockery of it.

There was a further difficulty too, a common yet grave one. A substantial number of the species[1] the Linceans took as the objects of their study seemed to stand on the borderlines of classes, or to participate in more than one of the then-accepted categories and classes. Regular classes were essential to order; mixed ones—or members of classes that spilled from one class into another—subverted that regularity. Cesi's study of fossils yielded an apparently unending sequence of "mixed" or ambiguous categories, such as the metallophytes and zoolithophytes of Umbria; so did his and Heckius's (and later Colonna's) examination of the world of fungi, lichens, and bryophytes. The Mexican Treasury is full of species that seem to belong to neither one class of things or another; many more are to be found among the drawings in Windsor and Paris, ranging from what is described as a "petrified twig" (but is probably a piece of coral)[2] to an exceptionally careful drawing of that long-standing conundrum of nature, the goose barnacle (figs. 14.3 and 14.4).

Cesi realized the scale of the problem early on. The old Aristotelian notion of "differences"[3] as a means of distinguishing between species was not proving helpful. By early 1604, the Linceans were already seeking advice on the subject from the great botanical experts of the day. On February 17 they sent a letter to Caspar Bauhin, "by far the most expert of all in the differences and powers of plants"[4] and pleaded with him to share his expertise in taxonomy with them. On the same day they wrote to Lobelius in London, telling him of their struggle with "the differences of plants and other miracles of nature."[5] A month later they approached Clusius in Leiden, to learn not only about "the differences of plants, seeds, and other similar things, but also about the arcane and rare things generated by nature."[6]

These letters were written by Heckius on behalf of all four of them. At this stage he was just as preoccupied as Cesi with the problems of classification—but also more confused. A few years later he sent one of his illustrated notebooks back to his friends in Italy and wrote that that the butterflies and annelids it illustrated were the "playthings or mockeries of nature, and miracles of mixtures." But what were they really: jokes or combinations? Heckius did not know. Nor did he know where to place such things. They were so varied that he constantly seemed to encounter exceptions to the order of nature. Surely God could not have intended a world full of exceptions? This did not seem likely. Heckius was stuck. He knew that each exception seemed to have an inner, structural order that related it to other beings apparently in the same class; but for the rest, all he could do—and he was a medical doctor, after all—was to concentrate on the effectiveness of plants, and try to relate their powers to their particularities. This seemed a useful and practical way out.[7]

To concentrate on the particularities of things, however, meant overlooking their essences and the possibility of grouping specimens into species on the basis of underlying similarities. Other than noting, presciently perhaps, a few structural similarities between things, he made no headway at all with the problem of classification. All too aware of the problem of exceptionality, Heckius now resorted to another concept, that of species "which we call imperfect." Thus he introduced a term that from then on would be fundamental for all of them, and for Cesi in particular.

At this stage Heckius's conception of the relationship between particularity and essence was still vague. He could see that difference was related to effectiveness; but then he gave up. All such matters, in the end, had simply to be referred to as "the hidden and neglected things of nature."[8] Once more, a central problem for the Linceans was left at the level of a commonplace of the time.

Cesi himself, however, was not content to leave matters there. He could not have been more keenly aware of the need to reveal exactly what was hidden, and to seek the inner structure of things, just as Heckius had begun to do. At that point he felt he might come closer to the elusive essences—or rather the distinguishing characteristics—of things. Only then could he and his fellow Linceans proceed to work out adequate and systematic forms of classification.

This is what Cesi now turned to do. Aware from the beginning of the particular classificatory problems presented by the fossil world, Cesi consulted Ferrante Imperato's massive 1599 work—all twenty-six books of it—on natural history, and then his son Francesco's 1610 book on fossils, to see if either offered any help. But these books turned out to be of far less assistance than those of the great botanists.

He studied with particular care Cesalpino's *De Plantis* of 1583 and Colonna's 1592 *Phytobasanos* and 1606 *Ekphrasis*. In Cesalpino he found the argument that the bases of classification—the "characters" as they were sometimes called—should be the parts of a plant, such as the seed and fruit, that were of functional importance.[9] Aware of the weaknesses of systems based on extraneous criteria (such as history and habitat), or on artificial ones (such as the alphabet), he realized the need to found a system that would be predicated on signs *within* the plant. Such signs, he understood, would have to be unchanging across different specimens of the same species (thus excluding the outward morphological character of parts of the plant such as leaves and roots). Cesalpino grasped, for the first time, the value of a classification based on the reproductive system of the plant—in this case the fruit and the seed. Cesi knew his Cesalpino well, even before he met Faber, one of Cesalpino's star pupils.

But the figure who really provided the link between the great Italian herbalists of the sixteenth century, and who even exceeded the scientific authority of Faber after he became a Lincean himself, was Colonna.[10] In his *Phytobasanos* Colonna revised the traditional Dioscoridean material not just in terms of his own fieldwork, but also in light of the new writings about botany and the new specimens that were daily pouring in from home and abroad.[11] From Colonna Cesi learned that in order to make any progress in the matter of classification, one had first to strive to determine which parts of the plant were constant and unchanging, rather than just attend to the variations and vagaries of surface, or to what could easily be seen with the naked eye.

## PRECEDENTS

The Linceans were faced with an immense problem, not just because of the multiplicity of information they were gathering and the many phenomena that simply did not seem to fit any kind of classification, but because their precedents were so inadequate to the tasks they set themselves. First among these, of course, was Aldrovandi. Heckius had written to him for help even before they sought the collaboration of Clusius, Bauhin, and the others.[12] Hundreds of Aldrovandi's manuscripts still survive in the University Library in Bologna. They testify to his omnivorous recording of every imaginable kind of fact about natural history and to his attempts to record them by visual means. They include not only watercolors of hundreds of specimens of flora and fauna, but also a huge set of woodblocks—more than thirty-six hundred were produced by 1599[13]—intended to be used in the publication of the material Aldrovandi assembled.[14] Foucault was right to claim that Aldrovandi came at the end of the Renaissance epistemological tradition, in which history was the inextricable and unitary fabric of all that was visible of things, and of the signs that had been discovered or lodged within them. For Aldrovandi—and many of his contemporaries, even Conrad Gesner—historical diachrony had the same explanatory and categorical status as a comparative anatomical or taxonomical grid. The assigning of names was done on the basis of an arbitrary mixture of historical, personal, and physiognomic motives. Mythology, history, and crude judgments about resemblance all featured in his haphazard, nonsystematic, and thoroughly random and inconsistent taxonomic scheme. Occasionally the manuscripts offer a vague attempt at presenting material in hierarchically graded form; but the books eventually published on their basis proceeded in the old-fashioned way of dividing up each subject into a seemingly endless range of random headings.[15] The overall impression is of a profusion of such headings, lacking logical order, almost as promiscuous as the encyclopedias and bestiaries of the Middle Ages and the great compilations of antiquity.

This is also the impression one has of most of the encyclopedic collections (as it has now become the fashion to call them) of the sixteenth and seventeenth centuries. Sometimes there does seem to be a kind of struggle to order or some form of methodical arrangement; but by and large it is hard to discern a system that goes beyond the glorification of the owner and his relation to the divine.

For Cesi, the link with Aldrovandi was an important one. Aldrovandi provided him with a model for the single-minded recording, in both verbal and visual form, of *as many* of the facts of nature as possible—through reading, a wide correspondence, exchanges of specimens, and commissions to artists and collectors of flora and fauna as well as through first-hand observation. Together with his students Aldrovandi made many excursions in central and northern Italy in search of new and rare plants, animals, and minerals. He put together the largest assemblage of natural historical information until his time; but the drive to order was far less committed and consistent than what Cesi envisaged. For Aldrovandi, as for all his Renaissance predecessors, such classificatory principles as there were were all too often based purely on resemblance, as Foucault emphasized; and everything could be made to resemble everything else. History and natural history were mixed together in an apparently seamless fabric of knowledge—just as may sometimes seem to be the case with Cesi. But only sometimes. Almost always misunderstood by modern scholars, the natural historical work of the Linceans, and of Cesi above all, is to be distinguished from that of almost every one of their predecessors on the basis of their recognition of the importance of classification in understanding the world of nature, and of the imbrication of the classificatory endeavor with the even more critical investigation of the processes and mysteries of reproduction by way of dissection and examination by microscope.

## FABIO COLONNA, AGAIN

There was thus good reason to have written to Bauhin, Lobelius, and Clusius. Much more so than Aldrovandi, by 1604 these men had made considerable contributions to the study of botany in the field and were acknowledged leaders in the taxonomy of plants. Most of the older writers on botany whom they were inclined to use offered too few significant clues to the problem of classification, especially Mattioli, whom Cesi in particular used as a kind of guide to plants in the field.[16] But Bauhin was already working toward a more satisfactory classification of plants than hitherto available and was developing a concordance of names and a taxonomy that would eventually culminate in his *Pinax* of 1623. He was already preparing his 1614 book on hermaphroditism, the archetypical problem of mixed species, with its clear and significant relation to issues of biological reproduction. Clusius and Lobelius had written a series of works, largely published by the house of Plantin in Antwerp, which were more profusely illustrated than anything before. Because of their abundant firsthand experience both at home and abroad, and their knowledge of plants from the New World, they offered texts that could finally begin to break free from the straitjacket of Dioscorides, Aristotle, Theophrastus, and their ilk.

But soon Colonna, building on the basis of his study of Cesalpino, published his 1606 *Ekphrasis*, which more than fulfilled the promise of the *Phytobasanos*. Not only did the *Ekphrasis* contain more plant descriptions but the old Dioscoridean and Theophrastan names were now firmly collated with their modern equivalents. Here botanists could find a secure basis for the comparison of species that had previously gone under a variety of different names. Special attention was devoted to the structure of the reproductive parts of plants, and to the flower in particular. In order to aid in the task of identification and dissemination, as well as to illustrate his discoveries about the anatomy and physiology of plants, and—not least—to present the hitherto unknown plants he had assembled, the work was adorned with no fewer than 156 illustrations (as opposed to only 37 in the *Phytobasanos*). But it met with little public success, so it was republished with additions in 1616, as the *Minus cognitarum stirpium pars altera*, a book now sponsored by the Linceans.

In Colonna's work, illustration and the drive to classification sat uneasily together; but it still provided the Linceans with a key model. Cesi could turn to Colonna for advice on the crucial issue of those many beings that seemed to stand on the borderlines between plant, animal, and mineral life. Indeed, in the additions to the second part, Colonna had given just a sample of his expertise in this area. In the little treatise on the murex attached to it,[17] for example, Colonna showed that fossil shells were not just stones with peculiar shapes but were animal in origin;[18] while in a text that anticipated the discoveries of the Danish paleontologist Steno some fifty years later, he demonstrated that the *glossopetra*, or "tonguestones," as they were known, were not in fact stones at all, but rather fossilized sharks' teeth.[19] Colonna profoundly subverted the old doctrines of similitude by pointing out that with fossilized specimens it was insufficient to classify them on the basis of their morphological similarities to extant species, or parts of them. What seemed to be fossil fungi, for example, were more likely to be fossilized marine accretions, such as coral.[20]

But Colonna's real achievement, as we have seen, lay in the field of botany. Now, with the work of observation and description firmly grounded, and with the disorder of the old nomenclatures somewhat tidied up, he could turn more fully to the problem of classification. Whereas in the first part of the work on "lesser known plants,"[21] he had emphasized the importance of the illustrations as a basis for the communication, confirmation, and identification of species, in the second part he made it clear that his interest lay in finding firmer bases for classification. This entailed moving away from reliance on the external appearances of things. Taking up some of Cesalpino's ideas, Colonna insisted that genera ought not to be based on similarities of leaf form, but rather on the characters of the flower, the receptacle, and especially the seed. They all revealed their affinities more clearly than any obvious external form.[22] In following Cesalpino, Colonna thus made a significant step toward the Linnaean system of classification based on the reproductive features of plants. Investigation of the principles of reproduction, as well as of the elements of reproduction, came to be of crucial importance, and this necessitated the ever closer and more detailed scrutiny of the innermost parts of things.

## THE REDUCTION TO ORDER

In October 1611 Cesi wrote to Galileo from Tivoli about the botanical excursion he had recently made to Monte Gennaro in the company of Faber, Schreck, Müller, and Henricus Corvinus.[23] In his field notes on this and subsequent trips, Faber reduced the usual descriptive and polynomial names of many of the plants they saw to the simplicity of pure binomials.[24] In this way too he and his colleagues once again anticipated by over a century the work of Linnaeus.[25]

Those, of course, were the days in which botany and astronomy went intensely hand in hand. It is hard not to recall, in considering Faber's simplification and reduction of the complexity of the old botanical names, Cesi's letter of July 21, 1612, to Galileo, in which he wrote that there was "not the slightest doubt" that one of the great satisfactions of the Copernican system was "the removal of the multiplicity of movements and spheres, and their too large and complex diversity." [26]

Here lay the crux of the matter. How could one ever be sure that such a reduction and simplification would not just be an arbitrary construct of the human mind, rather than an adequate reflection of the order of nature? No one (except of course Galileo) was more aware than Cesi of the danger of imposing the ordering needs of one's own mind on the order of nature itself—whatever that order might be. In the same letter, Cesi had written that it was hard to imagine that the great "farrago of spheres and revolutions" in the Ptolemaic system was a purely natural thing—it must, he felt, have been just the fantasy of some overly cerebral intellectual ("*aborto d'huomo cerebroso o miscuglio di strani fantasmi*"), "who instead of acknowledging the pure and simple proofs believed it [the Ptolemaic system] to be the work of Nature itself."[27]

In fact, Cesi was replying to a letter from Galileo three weeks earlier, in which Galileo had put the matter clearly:

> I have heard with pleasure that this time you are occupied in thinking about the Copernican system, and that you are inclined to prefer it to the Ptolemaic one—especially if the eccentrics and epicycles could be totally removed. About which in particular, I only want to tell you what you already know much better than I, namely that *we should not wish nature to accommodate itself to that which seems better disposed and ordered to us; but that we should rather accommodate our intellect to what nature has made*, in the certainty that this is the best and only way.[28]

Cesi never resolved the problem of classification; but the manner in which he brought his work to its conclusion (however tentatively) had serious consequences for every one of the Linceans' strategies: not least for the ways in which they decided to record nature—or rather, for how they conceived of recording it.

## THE MIRROR OF REASON AND THE THEATER OF NATURE

Just as Colonna was completing the second part of his *Ekphrasis*—and very shortly before the Linceans began using the microscope for the first time—Cesi's anxieties about classification seem to have come to a head. One indication of how he struggled is provided by the complex manuscript in Naples outlining his project for organizing all knowledge in the world.[29] Its opening section, headed *Mirror of Reason*, was intended to outline a "universal synoptic scheme [*typus*] for every form of contemplation and reasoning, methodically adapted into the fullest encyclopedia."[30]

This was to be no typical encyclopedia, however. At the outset Cesi declared that its organization had to be based on new and methodical principles. Knowledge would be presented in as logical and systematic a way as possible, in order, as Stelluti put it later, to "manage all the disciplines."[31] In his 1633 survey of the illustrious literary and intellectual figures of Rome, Leo Allacci described Cesi's still unfinished *Universal Mirror of Reason* as containing "the universal art of the sciences."[32]

Like so much else of Cesi's, however, this work never actually appeared. Nevertheless, the Naples manuscript provides a clear sense of his mission to present all of human knowledge in as ordered a fashion as possible. In attempting to cover every area of thought, from metaphysics to physics, Cesi realized that he had to provide a legitimate base for the artifice of the mind *from the things themselves*.[33] He was all too aware that even though the ordering constructs of the mind were likely to be artificial, they still had to be derived from nature itself.[34]

This, he knew, would be a difficult, if not impossible, task. The heart of the problem lay in the exceptions and anomalies of nature. He could not just proceed in what may then have seemed one of the most efficient ways available to him: the arrangement of subjects into tables of hierarchically graded and bracketed "topics" (or "places"), such as could be found in any number of the encyclopedic works produced by Peter Ramus and his many followers.[35] This way of breaking all branches of knowledge down into their supposedly constituent parts still dominated the study of the arts and sciences.

Nor could Cesi simply adopt the old Aristotelian approach, elaborated by Ramus, of taking the "differences, divisions, and all the members of the sciences, and of the conceptions themselves, and ordering them from the broadest groupings down to the last little branches of the ultimate particles of things," as Cesi himself put it.[36] He could just as well have been describing a traditional pedagogical handbook. But in observing that the subjects of his *Speculum* were divided into classes on the basis of "invention," he was not just acknowledging the artificiality of mental constructs, he was using a term derived both from traditional rhetoric and from the work of Ramus himself.[37] Expanding this approach, he then noted that his synoptic tables would not only contain classes devised by others, but also new classes, derived from his own researches.

Cesi achieved this by confronting the problem of anomaly and mixed (or borderline) categories head-on. He did so after announcing the publication of what he called the frontispieces of his immense *Theater of Nature*. Many of these, as we shall shortly see, were actually published—unlike so many of his other projects. In these frontispieces the idea was to set out a compendium of the whole of nature ("*universa natura*"). They were supposed to enable the reader to "complete the classes, from the first to the last," under headings that ranged from the highest things to the lowest. These in turn would be so organized as to correspond both to the concepts and to the differences that distinguish things from one another. Just how he intended to accomplish this yoking of the empirical and the theoretical turned out to be extremely problematic. But it represented a radical departure from the old Ramist way of doing things. For all that, however, Cesi would pay the price of never escaping entirely from the empirical habits of traditional science. It was precisely this that eventually limited—kept down to earth—his conceptual advances.

Cesi had already been brought up short by some central obstacles in the course of his study of fossils and fungi. These obstacles all revolved around the problem of phenomena that seemed to stand on the borderlines of classes, that were intermediate in some sense or another, or that seemed to participate in several classes. For where were *such* things to be placed in the ordered system he planned? If anything, they threatened to subvert it. How was he to integrate antistructure into structure? And how was he to indicate participation in more than one class? He had to face the two chief difficulties that necessarily confront anyone who tries to classify the phenomena of nature: first, the problem of the series, where gradation militates against placement under headings, in boxes, or in any kind of grouping to which a boundary can be imagined; and, second, the problem of the borderline case, where participation in more than one class plays the role of spoiler in the establishment of a clear structure of separate classes.[38]

Cesi planned two treatises on these problems. The first was to be titled *On Middle Cases in General* (*De mediis in universo*) and would deal with the whole issue of ambiguity and of things "of a doubtful nature," namely cases where "two different natures joined in a single species. . . . species participating in two natures."[39]

The second treatise was to be devoted to the related problem of imperfect plants. It was intended to address "the whole genus of plants in which monstrosities and marvels occur as a result of some particular defect or peculiar characteristic, arising rather from the rudiments of the plants than from the plants themselves."[40] At the heart of the problem lay plants with no visible signs of seeds or reproductive organs, such as the ferns and bryophytes and the broad group later called cryptogams, where the usual kinds of indicators for classification were somehow absent—and therefore "imperfect."[41]

It is precisely from his jottings on this subject that we get a feel for Cesi's sense of the close relationship between the parts of reproduction and classification, and for the urgency with which he wished to resolve such questions. In this he was perhaps spurred on by his own anxiety about his wife's repeated miscarriages; but it surely also had to do with the actual researches on which the Linceans embarked so intensely, and which they had so vividly illustrated in the pages of the manuscripts now in the Institut de France. In these researches they constantly faced the problems not just of the mystery of generation and reproduction, but also of the apparent absence of any of the traditional parts with which to attempt an organization into classes.

With the fossils the Linceans entered one of the most critical of all the realms in which species that seem to participate in more than one class are to be found. In an age before Darwinian evolution and the geological theories of Lyell, borderline cases—say the ape-men so much discussed by Gesner and others in the sixteenth century and later—presented almost insurmountable problems. They were difficult if not impossible to tolerate because they could not be placed in the right category.[42] Only once an evolutionary view of nature was accepted did it become possible to accept the idea of gradual transitions, rather than big ruptures and clear boundaries.[43] As long as a commitment was maintained to a world that only existed after the perfect creation of God, fossil remains could not adequately be explained, and the problem of the borderline continued to be a thorn in the side of those who expected to find a place for everything in God's universe.

## ENCYCLOPEDIAS AND FRONTISPIECES

To Cesi's mind, the old encyclopedias did not make the grade. They offered no rational guidance through the multiplicitous profusion of nature, nor any sense of how oddity might be related to regularity—let alone inserted into it. In any case, they were too unsystematic, haphazard, and unguided by any of the overarching conceptual principles he constantly sought.[44] In the moving statement of Lincean ideals titled *On the natural desire for knowledge* that Cesi read to the session of the Academy on January 26, 1616, in the presence of Galileo, he made plain that he knew all too well what he was up against:

> The field of knowledge is truly vast—vast because of the abundance of contemplations and investigations[45] and vast because of the abundance of written works. Some think that even without help and convenience they can derive great profit by gathering together for themselves an immense machine of undigested material in their minds. . . . But there are indices and very full repertories, dictionaries, and lexica of all the professions. The best writers are digested into books of commonplaces. There are florilegia, maxims, deeds; theatres, anthologies, gardens, and any number of workshops.[46] There are the libraries which give us all the books which have been read and judged; and we order them according to author and to subject.

But all this was not good enough. Instead, he recommended:

> method and the synoptic art itself,[47] with its types, conjunctions, divisions, unions, and conditions, as in my *Mirror of Reason* in particular, where I set out the whole [of the subject] before the eyes of the investigator, so that at any one moment, with his memory more alive and his mind more acute and awake, he may freely and unobstructedly move forward both in invention and composition. Unfortunately, only a few seek the convenience of this approach, and embark on their studies with weak means and little order. . . . It is hardly surprising that of the few people who study, even fewer—actually, very few indeed—arrive at any noteworthy degree of knowledge.[48]

While the old encyclopedic systems might indeed aid memory,[49] they were too artificial and incapable of adequately showing the relations between things.[50]

Cesi continued to fret. But Galileo offered further inspiration. He had shown the way to simplifying the complexity and overabundance of the Ptolemaic spheres and eccentrics. Cesi himself, in his famous letter of 1618 to Cardinal Bellarmine,[51] made his own plea for the kind of simplification that had become thinkable only as a result of Galileo's successful penetration of the crystalline spheres with the aid of the telescope. Now, inspired by Cesalpino's and Colonna's emphasis on the importance of establishing the basic principles of reproduction in the plant world, he realized the necessity of the ever closer and more detailed scrutiny of the innermost parts of things.

This, of course, is what the Linceans began to do with the aid of the microscope. They used it separately from and in conjunction with their many dissections of animals. They examined the magnified sori and sporangia on the underside of fern fronds; they did their anatomies and cut to the insides of things, there to examine the tiniest and slenderest things within. But the more they looked, the more they found. The problem of multiplicity would not go away. Even the microscope could not help them as much they had hoped with the relationships *between* things. It is not surprising that Cesi should so often have expressed his frustration at his inability to see order in chaos.

At the end of 1622, clearly overwhelmed by the tumult of things, Cesi wrote to Faber:

> I find myself in the midst of that grand chaos of the methodical distribution of plants, and seem to have almost totally overcome it. This will form a good part of my *Mirror of Reason* and *Theatre of Nature*.

He was still working on these old projects, and they had become ever more daunting.

> You know how large this undertaking is, even aside from the turbulence of my affairs. . . . The Tabernaemontanus[52] and Bauhin's commentary on Mattioli[53] you lent me have been of great use and assistance in the course of my revision of this uncountable army of plants; but the most convenient in terms of placing and ordering so huge a plantation would be Lobelius's two volumes of pictures of plants, the *Icones Lobellianae*,[54] as an aid to memory; [I would like to] cut up all the pictures and place them in the manner you know and as Terrentius already did in his way; so please get me two unbound copies of these books.[55]

Visual representation and methodical order still went hand in hand, as Cesi once more took up his efforts to find the right places for things.

These were intense and satisfying months. Work on the manuscript of the *Assayer* had been completed, and he could switch from botany to fossils and back again, without too many interruptions. A couple of months later, on January 21, 1623, Cesi thanked Faber for Bauhin's *Phytopinax*[56] and asked him whether there was anyone

> who has distinguished between the fossils in an orderly fashion, and has enumerated them into their classes, and especially the metallic ones, and those which are called half-minerals; and similarly if anyone has synoptically reduced the sciences into trees and tables. I find myself, dear Faber, in the middle of the distinctions and differences of all things. . . . Particularly with regard to stones I have had real trouble in finding among the authors even the shadow of a system of order worth anything at all, but still I think I have recovered their conjunctions, differences, divisions, and affinities, to a certain satisfaction of my own, and I do not think it will displease you either.[57]

The terms are very similar to those used seven years earlier in his *Discourse on the Natural Desire for Knowledge*.[58] This is the language of someone who has thought hard about order and method. By this time Cesi had become even more intensely aware than before of the need to *include* the ambiguities of nature in the table of classification, and to find the right place for them in the scheme of things.[59] Precisely the same preoccupations emerged in the much circulated letter on his fossil finds to Cardinal Francesco of December 1, 1624, where, it will be remembered, he had noted the high difficulty of dealing with the problem of *mezzanità* in the case of fossils.[60]

But it was beginning to seem as if there were corners of nature left untouched by such problems. From then on Cesi turned his attention, finally, to the production of the first of his great frontispieces to his *Theatre of Nature*. In them he would set out both the problems and the principles of classification, in as orderly and organized a manner as possible.

The closer the Linceans came to having their Mexican book ready to be sent to the printers (and the day was long enough in coming), the more Cesi realized the scale of the problem he had on his hands. The book would contain an immense quantity of new information, and more than a thousand illustrations; but it was desperately lacking in structure. Cesi worried that it would give the appearance of an old-fashioned encyclopedia. It could hardly be described in any of the terms he had used in his description of traditional reference books and encyclopedias in the *Natural Desire for Knowledge*. If anything it was a kind of miscellany (however stuffed with valuable information) in the haphazard sense in which this term would have been used in England at the time—Baconian, one is tempted to say, rather than Galilean.

Cesi's decision to append some of the frontispieces of his *Theatre of*

*Nature* to the *Tesoro Messicano* thus made a great deal of sense. He rushed to have as many as possible ready before the work was printed. Time, after all, was running out (though no one could have predicted that he had less than two years to live). In these frontispieces he set out some of the principles and guidelines by which he proposed to order first the plant world and then the rest of creation itself. The *Tesoro* would have seemed incomplete to him had he not included at least an overview of his position on the structure and order of nature. Alas, it turned out to be less of an overview than a series of suggestions, a set of insights, an array of samples and soundings.

On October 14, 1628, Colonna wrote to Cesi from Naples:

> I have received [via Stelluti] the ten *Phytosophical Tables*; and having glanced at them, I was amazed by the sublime intelligence of your Lordship, by the acuity of your observations, and by your disposition of things. It is certainly an invention wholly worthy of Your Excellency—and you are rightly called Prince of Sciences.[61]

Observation, disposition, and invention: there could be no three more fitting terms to describe the basic processes that underlay Cesi's thinking about his system of classification (though not necessarily in this order).

Six days later Colonna wrote to Stelluti. Just as in his letter to Cesi, he observed that he was so overwhelmed by the brilliance of the tables that he really had nothing else to add. Cesi, he wrote, was dealing with things that no one had touched on before.[62] Moreover, he had gone far beyond the plant world,

> so that in these tables he has dealt not only with earthly things, but also with heavenly and ideal ones; the result is that the tables will be less understood by *semplicisti* than by metaphysicians. In this I have nothing to say except that I am full of admiration for his Lordship.[63]

Colonna realized the extent to which the tables were abstract and theoretical, rather than empirical, as they may at first sight seem to have been. In them Cesi worked out the metaphysics, as it were, of classification.

## CESI'S TABLES: GEOMETRY AND THE AMBIGUITY OF CLASSES

Twelve and part of the thirteenth of the frontispieces were printed in the version of the *Tesoro Messicano* ready in 1628.[64] The remaining seven and a half (edited by Stelluti) were added to the edition of 1649–51. In the long and arduous struggle for adequate systems of classification in the history of science, these frontispieces, or tables, mark a turning point. Not for nothing did Cesi call them his Phytosophical Tables (by evident analogy with "Philosophical"). When Galileo spoke of philosophy he meant what we generally call science, and Cesi wished to put the study of growing

Fig. 13.1. Federico Cesi, *Tabulae phytosophicae. Thesaurus* (1651), p. 905.

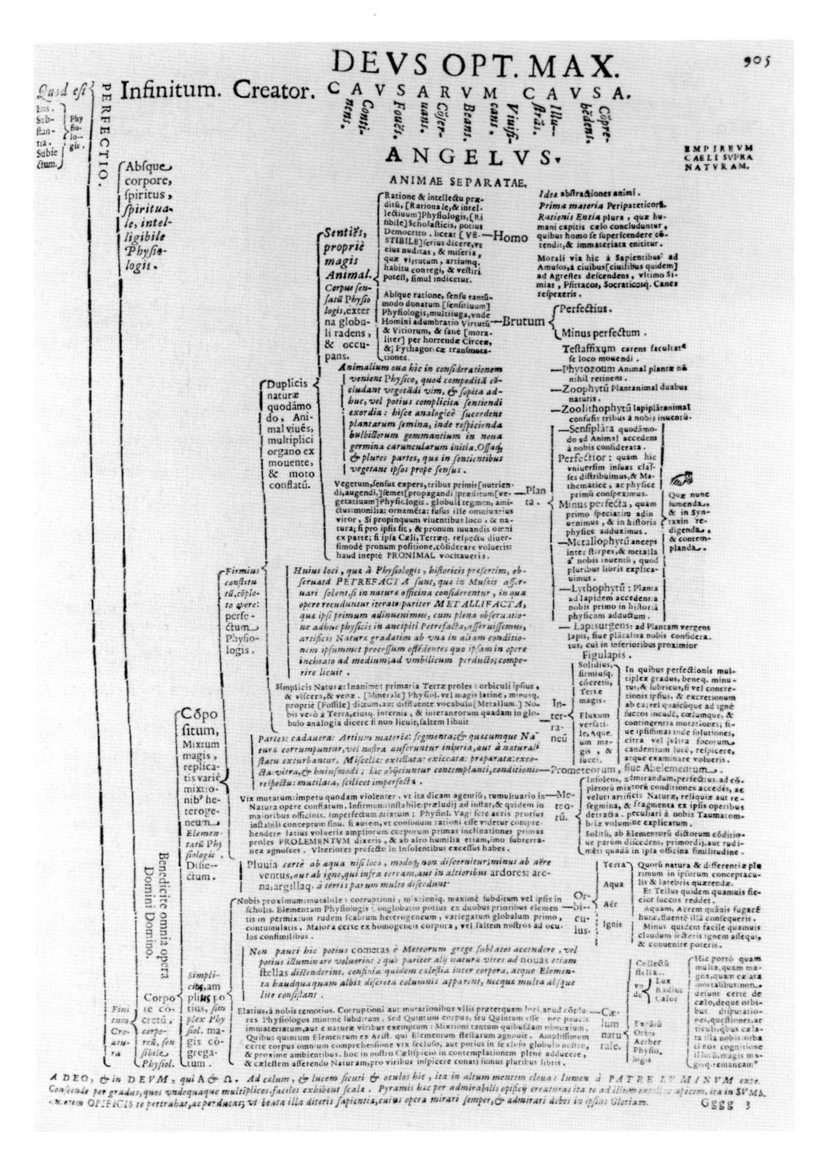


things—"*phyto*sophicae"—on the same level of rigor and authority as Galileo's study of bodies in space.[65]

Even before Ramus, many had attempted to reduce their field of natural historical specialization to tabular or hierarchical order. Francesco Imperato for example, had done just this with the fossil and mineral world in his significantly titled *Universal Method of Fossils* that preceded his huge treatise on natural history of 1628. This was published in the same year as the first printing of Cesi's tables. But no one had ever attempted to reduce quite so much of the *whole* of the field of nature to tabular form—and certainly not accompanied by so much detail.

To turn to pages 905–950 of the *Tesoro* is to be confronted with almost fifty pages of densely packed information (cf. figs. 13.1–2). Its precedent,

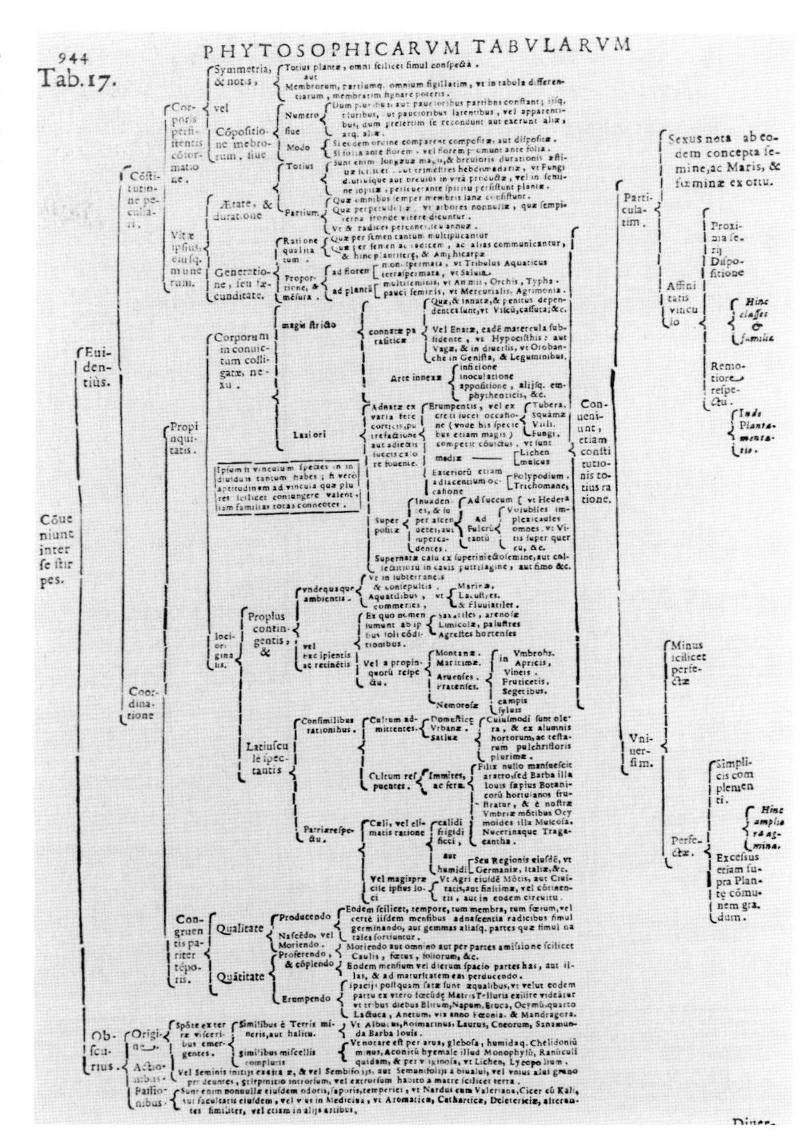


Fig. 13.2. Federico Cesi, *Tabulae phytosophicae. Thesaurus* (1651), p. 907.

of course, was the *Apiarium* of three years earlier. And just as in the case of *that* extraordinary attempt to reduce one small example of the natural world to order, the *Tabulae Phytosophicae* are so dense that it may seem impossible to recover any serious kind of ordering principle within them. At times the desire for control and order gives way to a kind of uncontrolled expansiveness about those things Cesi knew best—hence the constant disruption of the more conventional forms of hierarchical bracketing by the irruption of smaller and larger boxes, of subsidiary brackets, of pointing fingers, and of plain columns of text. Even so, these are pages that constitute one of the most extraordinary deployments of vertical *and* horizontal bracketing, cross-referencing, and hierarchicalization ever made in the service of the reduction and classification of knowledge.

But Cesi did not really succeed. He tried to do too much. He wanted to divide up nature in such a way that everything in it might somehow find a place in his tables. He hoped to offer a conspectus of all knowledge in which the relationship between things would immediately be clear. But often enough he never quite decided in which direction the relationships should go. On the one hand, there were relationships *across* space (surface relationships, rather as in a map); on the other, relationships *within* space (from surface to depth); and then there were the hierarchically graded relationships, defined by larger and smaller brackets, which he could not ever bring himself entirely to renounce. In all of this the necessity of detail warred against the demands of simplicity.

At the top of the first table—at the top of everything—comes the infinite, at the bottom the finite (fig. 13.1). The Creator is infinite; his creatures, finite. God is the cause of causes. Below him come the angels and the empyrean; the separate souls; reason and intelligence; and then *homo*, man. This is all fairly standard. So too is the next stage down, that of the *bruta*—not just animals, but all things lacking in reason.

At this point we enter the first area of distinct interest. It is also one of the keys to the Cesian view of things, and where he begins to diverge from writers like Bodin. The *bruta* are divided into more perfect and less perfect. At the top of the subdivisions that follow come the *testafixa* (animals lacking the capacity to move themselves). Then *phytozoa*, (animals retaining something of a plant—a sharp and perceptive division, still present in the modern sense of the phytozoa), *zoophyta* (partaking both of the nature of the plant and of the animal), and the *zoolithophyta*. Already we are in the territory of mixed and ambiguous things.

No wonder Colonna was confused. In the third of his letters of 1628 on the subject of Cesi's tables—the same letter in which he went on to warn both Galileo and Cesi to be more prudent in announcing their discoveries[66]—he acknowledged receipt of the first printed page of the tables:

> The first page, a table of the whole of the *Theatre* and the basis of division—or better, distinction—has pleased me very much. . . . To tell the truth I do not know what the stoney plant-animal called *zoolithophytum* is. I would like to know about it, since it is something I haven't yet seen—probably because it hasn't yet been sent to me. The same with the plant-metal—unless it is the branchy silver described by Imperato.[67] Certainly this is a very new and important distinction. But how it will be when it comes to the subdivisions of heavenly phenomena and heaven and earth, I really cannot imagine.[68]

Colonna was right to be worried. The next part of the tables does indeed move back up to the heavens, beginning with what Cesi regarded as the most elemental of heavenly phenomena, the *prometeorum* or the *abelementum*. Then come earth and sky, with the elements of fire, earth, air, and water; and then the stars and lights of the sky. These matters could be left to Galileo; Cesi's contribution had to be in botany.

So he proceeded with his tables. Following the zoolithophytes comes an even more problematic set of groupings. Once again Cesi announces that they are of his own devising. They go from the *sensiplantae* (plants "in some way" approaching the status of the animal), through the perfect and imperfect plants, and on to his particular specialities: metallophytes (on the borderline between plants and metals), lithophytes (plants approaching the status of minerals), and the *lapisurgens* (a stone on the verge of planthood). Then Cesi moves right on to the problem of fossilization and petrifaction.

As if the possibility of classification were not sufficiently subverted by the division into intermediate and transitional phenomena clearly incapable of division and separation, Cesi thus introduced yet another complicating factor. It further threatened the notion of fixity on which all systems of classification then depended. This factor was change over time. How could any species be securely placed when species were likely to change by some long drawn-out process of evolution? Cesi had no clear answer at all; perhaps he would not have been able to express it even if he had. No wonder that the problem is almost wholly absent from the rest of the tables; no wonder that the project stalled. He had neither the philosophical skills for such problems, nor the mood for another fight.

But if there was one issue on which Cesi never lost his focus it was that of ambiguous or borderline phenomena, those he described as "anceps." For every classifier, then as now, they posed an almost insurmountable problem. At a time when every system of order, from the *Kunstkammer* and the cabinets of curiosities to the dialectics of Peter Ramus, from the purely physical to the purely metaphysical, depended on the idea that one could really divide *everything* into boxes, where was one to put those things that could fit into more than one box? Everything was supposed to have its proper place, as most graphically demonstrated, for example, by the chest of drawers or "arca" designed for the mid-sixteenth-century fossil collection of the Dresden doctor Johann Kentmann, where every drawer was carefully marked and correlated (fig. 13.3);[69] or by the "capsulae" in which Lobelius placed roots, seeds, and nuts in his *De Plantarum seu Stirpium Historia* of 1576 (e.g., fig. 13.4).[70] Each specimen could be stored in a marked drawer, or box, or receptacle—the exact *place* for each specimen thus so clearly and fixedly classified.[71] But what if such specimens partook, as Cesi's seemed to do, of more than one class?

The answer surely lay in attempting to discover the unchanging signs or *notae* of what might be identified as a distinctive and possibly nonambiguous class. You could make dissections or use the microscope in an effort to arrive at the smallest parts of things. But the use of the micro-

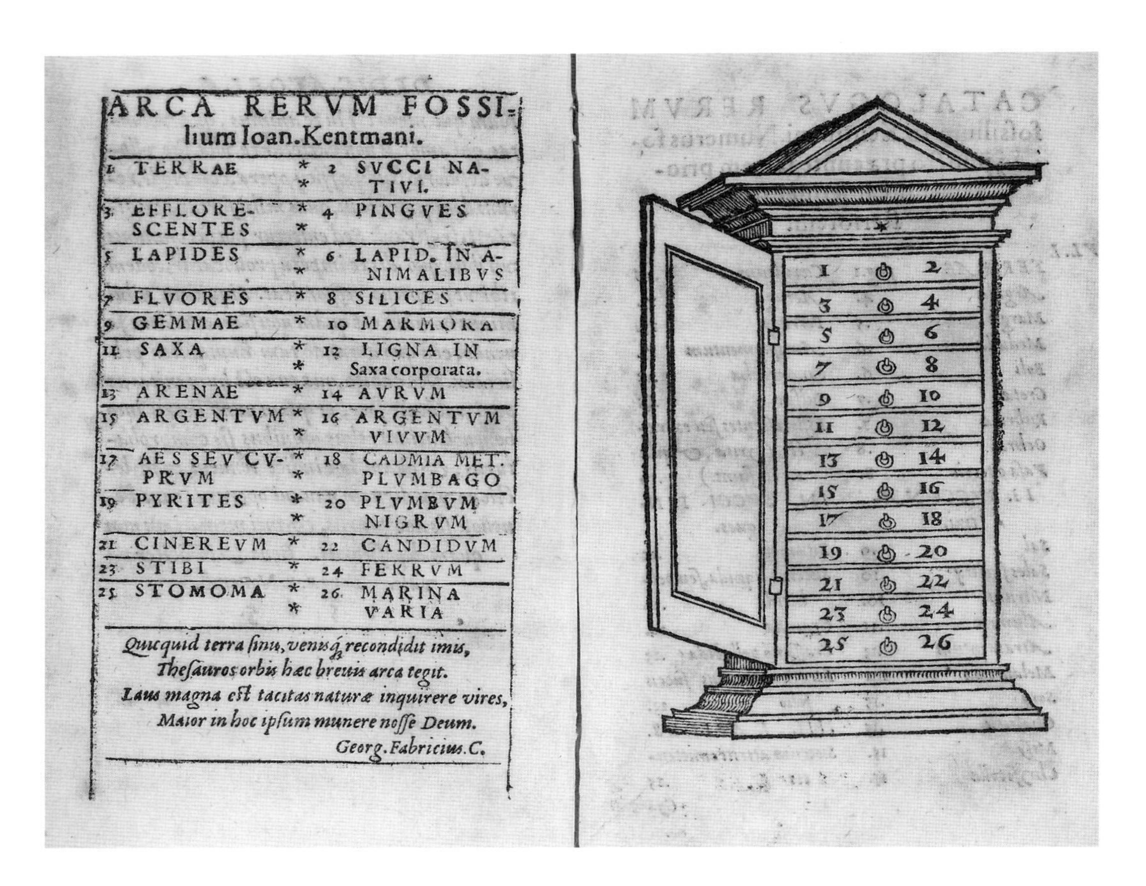

ARCA RERVM FOSSI-
lium Ioan. Kentmani.

| | | | |
|---|---|---|---|
| 1 | TERRAE | 2 | SVCCI NATIVI. |
| 3 | EFFLORESCENTES | 4 | PINGVES |
| 5 | LAPIDES | 6 | LAPID. IN ANIMALIBVS |
| 7 | FLVORES | 8 | SILICES |
| 9 | GEMMAE | 10 | MARMORA |
| 11 | SAXA | 12 | LIGNA IN Saxa corporata. |
| 13 | ARENAE | 14 | AVRVM |
| 15 | ARGENTVM | 16 | ARGENTVM VIVVM |
| 17 | AES SEV CVPRVM | 18 | CADMIA MET. PLVMBAGO |
| 19 | PYRITES | 20 | PLVMBVM NIGRVM |
| 21 | CINEREVM | 22 | CANDIDVM |
| 23 | STIBI | 24 | FERRVM |
| 25 | STOMOMA | 26 | MARINA VARIA |

*Quicquid terra sinu, venisq; recondidit imis,*
*Thesauros orbis hæc brevis arca tegit.*
*Laus magna est tacitas naturæ inquirere vires,*
*Maior in hoc ipsum munere nosse Deum.*
*Georg. Fabricius. C.*

Fig. 13.3. Johann Kentmann's fossil cabinet. Gesner (1565), sigs. a5v–a6r.

scope and the telescope, as we have repeatedly seen, also brought to the fore the basic tension at the heart of all looking: the "big picture," as we now say, with all its unclarity, its unshadedness, its hints of a multitude of components making up the variegated whole; versus the close-up, the consequences of drawing ever nearer to the picture, which now disappears from view, in favor of an awareness of its parts, and of exactly those parts that make it distinctive. The microscope excludes all beyond the parts; the telescope (at least when viewing the heavens) not only magnifies distant and vague surfaces, it brings more and more into view, so that the picture seems larger and more multiplicitous than ever before.

This, then, is the struggle that played itself out in Cesi's tables. On the one hand, he wanted to make the whole of the plant world available to the eyes, a *conspectus* offering a syntax, a putting-together, of everything. On the other hand, each part of the individual plant had to be made available too. A syntax must have rules; and the clues to these were to be based on the essential elements of things to be found within their depths. Thus could surface be related to depth. Cesi had to present both a conspectus of the whole of the plant world, ordered as best he could, *and* (as he put it in the Naples manuscript years earlier) an ordered indication and exposition of everything that could be found in a *single* plant.[72]

These were not easily reconcilable aims. The struggle to resolve them

Fig. 13.4. Compartments (*capsulae*) for roots and seeds. Lobelius (1576), p. 667.

CAPSVLA LIGNORVM.

| | | | |
|---|---|---|---|
| 5 | Citrinum | Xylobalſamum | Scobs Buxi |
| 4 | Rubrum | Surculi Lentiſci | Ebenum |
| 3 | pro Santalo albo | Aſpalathus | Guaiacum |
| 2 | | Ligno Aloes | Echinæ radic. |
| 1 | | | Salſaparilla |

Purgantiũ.

| | | | |
|---|---|---|---|
| 1 pro Scammonio | 2. Diacridium | 3 Colocynthis | 4 Ammoniacum |
| Sena | Semen Chartami | Folia Braſſicæ marinæ | 3 Sal gemmæ |
| Foliis Scammonij | Thymelææ | Chamelæa | 2 Sal petræ |
| Semine hederæ | Sem. Aſari | Semen Geniſtæ | 1 pro Sale nitro |

| | | | | | | | |
|---|---|---|---|---|---|---|---|
| 4 | | | | | | Coccus Cnidius | |
| 3 | Thymus | Chryſocolla vel fortè Chryſolith. | Elleborus albus | | Alhandal | Semen Cataputiæ | Rhaponticum |
| 2 | Epithymus | Lapis Armenus | Niger | | Diacridium | Rad. Tithymali | Myrobalani 5. |
| 1 | ℘ Caſſitha | Lapide Laſuli | Colocynthide | | Scammonio | | Rhabarbaro |

| | | | |
|---|---|---|---|
| Polypodium | Rhabarbarũ | Chamelæa | Eſula |
| Turbith album | Lapatio | Thymelæa | Turbith |
| Turb. cineritio | | Braſsica marina | Polypodio |

Seminum.

| | | | | | | | |
|---|---|---|---|---|---|---|---|
| 5 | Capita papaueris | Acetoſæ | | | | | |
| 4 | Semen Mãdrago. | Portulacæ | Citrulli | | | | Semen violarum |
| 3 | S. Hyoſcyami | Scariolæ | Melonũ | Biſmaluę | Fœnigręci | | Cnici |
| 2 | Papaueris nigri | Endiuiæ | Cucumeris | Maluarũ | Lini | Pſylij | Vrticæ |
| 1 | S. Papaueris albi | Lactucæ | Cucurbitæ | Bombacis | Seſami | Cythoniorum | |

emerges on every page of his tables. What has not been noticed is the extent to which Cesi began to move away from reliance on similitude and resemblance, in the direction of number and geometry. In this he was stimulated by his awareness that the geometrical diagram, more abstract than pictures, could tell one more about the essential aspects of things than pictures ever could. The more the Linceans experimented with the microscope the more they came to the same conclusion. Like the great microscopists later in the century—Hooke and Grew, for example—the Linceans looked into the depths, and saw the geometrical structures of things. Thus—paradoxically enough—while the geometric diagram might seem essentially to have to do with surface, it emerged still more clearly when one cut and penetrated into depth and interior. It also had the great advantage, as was obvious enough, of being able to show the relationships between the structures themselves (whether newly revealed or not).

Why, then, did Cesi not present his tables in more purely diagrammatic forms, rather than in the tabular and gridlike way he published them? Simply because he could not bring himself to be radical enough. He could not entirely renounce the tradition of presenting knowledge—or particular bodies of knowledge—in tabular and hierarchical form; nor could he bring himself to pare away all the detail that stood in the way of clear diagrammatization. He was too timid to so; he wanted to leave too much in. After all, he had observed and found so much!

At the same time, Cesi seems to have had an inkling of the role of time in the formation of species, and of the possibility that it might be used to account for some of his fundamental borderline and "middle" cases, such as the zoophytes, lithophytes, and zoolithophytes. There are indications that he felt that there was indeed something about this process that could be suggested by representation in hierarchical and tabular (if not gridlike) form. The first table does in fact hint at Cesi's awareness not just that these are middle cases but that they are *transitional*. But Cesi could not yet imagine—nor could he have imagined—a purely geometrical way of presenting time. And so he stumbled, interestingly, over how to reconcile the conflicting relationships between surface on the one hand, and density and essence and interiority on the other; between scanning and cutting; between a tolerance for the generality of things and a haunting awareness of the need to find the rules of creation.

The actual terms Cesi used are indicative enough. Even, occasionally, are those used by Stelluti and Colonna. In their suppressed foreword (dedicated to Francesco Barberini) to the *Tabulae Phytosophicae* in the earlier editions of the *Tesoro*, for example, they expressed their awareness of the problem of dealing with so rich and meandering a mass as the *Tesoro Messicano*. So much new botanical material had been introduced into the field that a new kind of guide was necessary. Their solution, they wrote, was to offer just the general "diagraph" of Cesi's "Syntaxis," as a kind of prospectus to so vast a "garden" of botanical information. The messy density of the growths in this garden required a syntax, which would have general rules; and these would be conveyed by what they called the "diagraph."[73] In this way epistemology met with the apt metaphor of horticulture.

The details of all this emerge, haphazardly, in the tables themselves. But gradually one gains a sense of the contradictions and attempted reconciliations between picture making, abstraction, and cutting into depth. Already in the second table Cesi sets out analytic principles for the study and ordering of plants. His model for the presentation and division of botanical material for the purposes of cognition in general (*omnifaria cognitio*) is firmly geometrical. It is not only in this respect that the tables seem to bear the imprint of the recent lessons of Galileo. Also present in these dense and densely thought-out pages is a view of the relations between experience and concept that is very close to the one that we have come to call Galilean.

At the head of the second table comes *Phytichnografica*, that is, the means by which plants are *designated* (Cesi's term) and made available by means of representation for full cognition (*plena cognitio*)—in other words, made available to the *intellect*. Phytichnography is a form of representation that depicts the plant in purely geometrical terms. Cesi makes it sound as if plants can only be fully grasped by means of what we might now call schema,[74] a very different form of picturing from that which the Linceans had hitherto used. Phytichnography would be much more concise than this, and would be conceived as a way of comprehending the relations between looking and the mental processes that underlie perception.

Phyticonography thus turns out to be a process that is basically theoretical. The Cesian view is that the mind (*intellectus*) scans both the surface and the interior of that which is (re)presented to it by phytichnographic designation; but everything presented to contemplation is dependent on the theoretical antecedents that institute the processes of looking in the first place.[75]

Following phytichnography comes what Cesi calls phytoscopy (*phytoscopica contemplatio*). This is a much more practical procedure. What he intends by it is that practical considerations ought to underlie the phytoscopic modes by which the surface of plants are subjected to examination. In keeping with his increasing concerns about the unreliability of surface alone, Cesi views this practical stage as somewhat superficial, because it only makes observations and is not concerned with the *principles* by which things are examined and known. The rules of *this* process, by which the mind actually grasps ("takes hold of")[76] the principles of cognition, is called phytognostic. In other words, while looking as a form of study cannot be renounced, it must be followed by the phytognostic rules (*phytognosticae regulae*).

Then comes the necessary analysis, or what Cesi here calls *dignosis*. This is an ambitious procedure, at least in its conception and aim, since it suggests that Cesi believed that the structure of knowledge itself is revealed by the diagnostic processes by which the parts of the plant are taken apart and separated.[77] At this stage, the geometrical paradigm is never very far from the surface of analysis, and the passage from cognitive processes that have to do with the irregularities of growth to those that are motivated by the need to reduce irregularity is precisely marked by the transformation of the prefix: phytognosis becomes dignosis.

After these three stages come the processes of *loganatomia* and *phytotomia*. They divide the parts revealed by diagnosis into their constituents, right down to the smallest particles, to what Cesi calls "the parts of parts." And these, in the final stage of *phytotomia*, can then be counted.[78]

It is precisely the transition from visual contemplation, in the form of phytoscopy, to the establishment of the *phytognostic* rules that allow true apprehension, or grasping, as Cesi puts it, that is of critical interest now; so too are the *diagnostic* processes of separation whereby the deeper levels of correct and thorough knowledge are attained.[79]

The terms used to describe each of these processes are again significant: *phytichnography*, with its connotation of theoretical representation, for the designation of that which is singled out for presentation to the intellect; *phytoscopy* for the scanning of and looking at those parts to be considered; *phytognostics* for the stage whereby the rules of apprehension are aroused; and *diagnostics* for the separation into parts that reflects the ordering processes that result in correct knowledge (if properly done). The following stages—*loganatomia* and *phytotomia*—are purely corroborative. By dividing and descending even further into the parts and particulars, one penetrates, by means of vision, into the final stages of discernment that enable numeration. In both these stages the idea of *cutting* for the purposes of analysis is emphasized, not only in the major terms derived from the root *tomē*, but also in the use of Latin terms such as *praecisius*, from which, of course, we now have the word "precision." But the cutting in each case is in depth, rather than transverse or planar. It goes with microscopy but threatens geometry.

The radical nature of Cesi's transformation of the terms of vision and knowledge, of the scopic on the one hand and the gnostic and gnomic on the other, is not easy to grasp. What he did was to separate firmly the regime of vision from the regime of analysis. One has only to consider the relevant suffixes: throughout these pages *-gnosis*, pertaining to rules, dominates *-scopia*, the processes of looking. The opposition is crucial, since it takes the proper domain of knowledge out of the field of the visual into that of the intellectual. It directs the evidence of the eyes (that is, observation) into the field of theory and principle. Cesi decisively moved the sphere of diagnostics away from the rule of the scopic. And when we consider how these processes are conflated in the work of the man who preceded Galileo as a Lincean, Giovanni Battista della Porta, Cesi's effort seems all the more remarkable. It cannot have been easy for him to free himself from the approach of that older and venerable figure.[80]

Let us return to the critical opposition between diagram and picture, schema and plenitude, between clarity and propositionality on the one hand, ambiguity, shade, and color (including rhetorical color) on the other. Cesi's tables evince yet a further struggle. It concerns the reluctance, perhaps not even willed, to go below the surface of forms. But when Cesi does so, it is crucial. On the whole the bases for his search for affinity and similarity between things seem to be external (for he could not entirely escape the surface paradigm of his time); but the exceptions, even when apparently cursory, are fundamental.

In the fourth and fifth tables, for example, Cesi speaks of the necessity of examining not only the general morphology of plants but also the special morphology of their parts, such as stalk, leaves, flower, and seeds.[81] He acknowledges that these must be examined for both their external and their internal structure. While similarity and affinity may indeed be sought on the basis of external form, the distinguishing marks of plants, the *characteres* or *notae* as they were so often called, had to be discovered deep within their recesses.[82] However subversive the microscope may have been for general morphology, and for the geometricization of classification, Cesi needed it.

The primacy of mathematics and number in these tables can hardly be doubted.[83] It emerges in the very first of the tables, in Cesi's critical distinction between perfect and "less perfect" (or imperfect) plants. For Cesi, this distinction is defined by the fact that the perfect plants are to be considered "mathematically and physically," and the imperfect ones under the category of physics, or, as he puts it, of "physical history" (in both cases physics is to be understood as referring to physiology in its nonmathematical form). Cesi then spells out exactly what this distinction entails: the more perfect plants, he announces, "we have here universally distributed into their classes," while the less perfect ones, "which we first discovered in their particularities [*speciatim adinvenimus*], have been brought into our physical history." Mathematics cannot be applied to the imperfect single item, and the single item cannot be a category. Take away the possibility of mathematics and one must renounce the rules of order that underlie classification. The point is not whether this entails a failure to find a criterion for the establishment of class; rather, it is the principle of the indispensability of the relationship between mathematics (in the form of number and geometry) and the possibility of classification.

In the fourth table Cesi sets out those factors that are essential for the investigation and knowledge of plants. At its head is mathematics. Only following it is the realm of "physics" (by which Cesi here means biological and physiological factors), and the fields of medicine, morals, philology, and (last of all) metaphysics. For him—at least in these tables—mathematics consists first of *quantity*, then the *kind of form*.[84]

In the next table, Cesi sets out the parts of plants as well as a great variety of the analytic categories susceptible to quantification (such as height, length, width, density, symmetry, asymmetry, paucity, multiplicity, and so on, both with regard to the whole and to the parts). No more detailed and careful proposal for the analysis of the parts of plants had yet appeared, not even in Cesalpino. While the general mathematical and quantitative thrust of this table could hardly be clearer, the exposition, written in a strangely careless Latin, is immensely complex and unresolved. "Quantity" embraces what we would understand by both arithmetic and geometry. Arithmetic (under the heading of "number") precedes geometry, which uses what Cesi calls "external terms" to describe the "figure" and "species" of plants. They might be round, oblong, or more awkwardly shaped. In any event, they possess angles or curves that can only be described geometrically.[85] Such terms apply to the wholes; but when it comes to the parts, "almost all the figures are mixed—which very nicely shuts the door on all our rules."[86]

Nothing could be more revealing of Cesi's awareness of the tension between the rules of order and the resistance to geometry. As long as one stays with the *appearance* of the parts, rule will be subverted. This is why pictures fail: outward appearance is conveyed by pictures, and pictures are not amenable to rule. On the other hand, if one considers the *order* of the parts, one can establish the rationale of order and thereby succeed in completing the figure. This is essential, since (as Cesi puts it in one those boxes typically reserved for specific illustrative material that did not fit with the internal logic of the table) "all figures are in all plants, and are disposed in multifarious ways."[87]

The need for the order of quantification, both arithmetical and geometrical, could not be more explicit in this strange version of the old Anaxagoran position that everything is in everything. Cesi clearly opposed appearance to the abstract form of order and number—to the distinct disadvantage of the former. Appearance could only be contained by the abstract forms of geometry and not by any of the old ways of following nature by the visual equivalences of outward appearance. "You have lines in rushes, spheres in bulbs, in fruits, and in tubers, the circumference of an oval in the heads of the cabbage, and the aspect of a pyramid in the pointed and florid stalk of a campanula."[88]

In one of the most famous of all passages in the *Assayer*, Galileo asserted that the "book of the universe" was "written in the language of mathematics, and the characters are triangles, circles and other geometrical figures, without which it is humanly impossible to understand a word of it."[89] Cesi himself wrote that characters of this book were "mathematical figures and physical experiments."[90] A common and overly hopeful paraphrase of the passage in Galileo is that the book is the "book of nature," and the implication generally drawn is that this is the book of natural history;[91] but it was Cesi, preparing his Tables at the same time as Galileo was composing the *Assayer*, rather than Galileo himself, who effected—or rather, attempted to effect—the reconciliation between the language of natural law and that of natural history. This is the broad context for Cesi's statement that "all figures are in all plants, and are multifariously disposed." There was every reason to conclude the box containing this claim with the injunction: "if you seek the use of the compass, go to the fungi."[92]

This apparently casual remark, so much at odds with the studied omission of mathematics from the definition of imperfect plants in the first table, illustrates one of the essential tensions in all of Cesi's work. However much he struggled to find a distinct place for anomalies and borderline cases, however much he and his friends agonized about the status of the hybrids they grew in their gardens, his greatest difficulty was to figure out how to fit them into the logic to which his work was tending. It was time itself, then, and the anomalous case, the case that participates in more than one class or seems to inhabit the space between classes,[93] that were the obstacles Cesi never succeeded in overcoming. But to deny his awareness of the problems they posed would be to do him an injustice.[94]

All these issues come together in the thirteenth table, perhaps the most important and methodical of all of them. It follows the table devoted to the problem of nomenclature, and deals directly with the classification of plants. In it Cesi sets out the rudimentary principles of similarity and difference underlying the division into the groups he calls species and genera. Although there is a great deal that is old-fashioned in the tables as a whole,[95] Cesi's commitment to the view that "syntactical reason" is predicated not only on external relations but also on internal ones[96] reveals him as a true intellectual precursor of the similar developments in eighteenth-century classificatory thought that culminate in the work of Michel Adanson.[97]

It is in the second part of the thirteenth table, however, completed by Stelluti on the basis of Cesi's manuscripts, that the significance of intermediate and transitional forms for the basic principles of classification becomes clearest. "Each and every plant relates to every other plant. There is no plant which in its total coordination with all the others does not have its exact place, by which it receives its location in the series and chain of the plants themselves."[98] But this statement is no simple reflection of the doctrine of the great chain of being. Rather, it is a remarkable anticipation of the idea of the series that, left unarticulated by Buffon, receives its classic formulation in the zoological work of Lamarck. This is clear from the further classificatory principle stated and exemplified here: "Any plant, placed in its own class for a number of reasons, may in one way or another tend away from it, and by certain distinctive characteristics (*notae*) of its own, have the propensity to relate to another class."[99]

Plants thus relate to one another within each class, and each class relates to each other class. The proof lies not only in the variations but also in the so-called intermediate and transitional plants, whose status partakes of more than one class—in other words those equivocal forms called "ancipites," or "doubtful." These are the plants that Cesi himself described as "intermediate"; they are the ones that seem to link the classes by means of the "uncertain or double nature of the affinities they show between themselves."[100]

At this point, unfortunately, the table stops short of providing examples (there was evidently an ellipsis in Cesi's notes). The overall idea is clear enough: proof of the general series of separate classes is provided by the intermediate forms that serve to link them. This idea recurs in

Diderot's *Rêve d'Alembert* as well as in his *Lettre sur les Aveugles*. Georges Canguilhem rightly noted that the existence of monsters seriously cast into doubt the ability of nature to provide the evidence of order,[101] but at the same time he credited the eighteenth century with having banished the eccentricity of monstrosity—that is, with having rid itself of the idea that hybrids and monsters *transgress* the rules of order.[102] Aberrant organisms came to be believed to offer the possibility of providing, paradoxically, the evidence for the understanding of the *regular* phenomena of systems of organization, just as Diderot had implied.[103]

Long before this, however, Cesi began to go down the same road. For him too irregularity, monstrosity, anomaly, and the equivocal and borderline case served, in the end, to prove the regularity of phenomena that form the basis of the continuity of the series. In fact, it was precisely in the domain of intermediate forms that the Cesian view profoundly differed from the old Aristotelian one. For Aristotle, "things that are different in *genos* [as opposed to *eidos*, or appearance] are too distant and without common measure."[104] For Cesi, everything had something of everything else in it, even if only to a minuscule degree. As transitional and anomalous forms proved, the *genos* was absolutely not to be defined, as in Aristotle, as "that enclosure beyond which there is nothing but pure otherness."[105]

This might have been a convenient doctrine for Ramist topical logic[106]—but not for Cesi. At every stage in Cesi's tables, Aristotle's insistent demands for closure were refuted, precisely on the grounds of the existence of intermediate and anomalous forms.[107] And while his view of intermediate phenomena as transitional was never very clearly integrated into his classificatory principles (such as they were), it was more clearly articulated and empirically investigated than in any other writer before him—and indeed for many years after.

In their attacks on the notion that the Linnaean system reflected the original plan of Creation, both Buffon and Daubenton used recent observations attesting to the existence of "zoophytes" and "lithophytes" as further proof of the minute gradations of the continuity of nature.[108] With Cesi it was never just a matter of the great chain of being; it was the idea of continuity *between the terms of the series* that underlay his classificational heuristics. He made this clear in an important box in the seventeenth table, where he discussed the connections between what he called classes and families. Here he observed that "if you look at the chain itself, then you can only babble about individuals; but if, rather, you look at the aptitude toward chains, which clearly serve to join many things together, then, indeed, you connect whole families."[109] This was a heuristics that explicitly proceeded from the phenomena themselves, but it was not one that he ever claimed to be natural; it was avowedly artificial and conceived of in terms of the operations of the mind or of reason.[110]

Hence the consistent emphasis on rules as well as on observation and looking. Rules must be found because order is predicated on them. Looking is one thing; but outside the regime of cognition it goes nowhere. In other words, empirical processes must be subject to the laws that underlie the organization of mind. Cesi, of course, never used *this* phrase—except by high etymological implication, in his various derivations from the Greek *-gnosis*.

There remained a serious and fundamental problem, however It loomed ever more threateningly over almost all the Linceans' projects over the years. Cesi was really the only one among them—including Colonna—who thought hard and unceasingly about classification. While the rest of them simply went on making pictures of what they saw or discovered, and sought with varying intensity to perfect their pictures, Cesi began to realize some of the basic problems presented by picture making to the systems of classification he was developing. With all their descriptive possibilities, with their marvelous potential for the representation of the texture, color, and mutability of things, pictures were excellent when it came to representing the exceptional, the monstrous, the prodigious, and the anomalous; but how could pictorial representation of the kind they were striving for show the regularity of the classes into which all things would ideally be shown to fall? And how could the limitless and open-ended potential of pictorial description be pared down to show the essential relations between things? For this one needed not the constitutive looseness of pictures, but rather the pared-down rigor of geometrical analysis and representation.

## SCHEINER AND HIS PANTOGRAPH

Hard on the heels of his 1630 *Rosa Ursina*, which appeared in the year of Cesi's death and just a couple of years after the first printing of the *Tabulae Phytosophicae*, Christoph Scheiner, Galileo's old opponent, published his *Pantographice*, "or the Art of Drawing anything you like by means of a linear, hollow, moveable mechanical parallelogram." The *Rosa Ursina*, long delayed in its publication, presented at great and extravagant length the evidence not only for Scheiner's claims for priority in the discovery of the sunspots, but also for the superiority of his theories explaining these and other astronomical phenomena. Surprisingly and ironically it also contained one of the few published writings of Cesi himself, namely his discourse on the fluid substance of the heavens, and his argument, in 1618, with Cardinal Bellarmine on the subject. By that year Cesi was able to see the utter untenability of Scheiner's view of a sun that was not subject to the usual processes of age and imperfection. Galileo's true discovery of sunspots had clearly proved this to him and his fellow Linceans (and, of course many others)—but not to the orthodox members of the papal court and curia.[111]

Ever since his first publication on the sunspots, Scheiner had given himself the nom de plume of *Apelles latens sub tabulam*, "Apelles hiding behind the picture." And with the publication of the *Pantographice* the name

Fig. 13.5. Frontispiece to Scheiner, *Pantographice* (1631).

of this great painter of antiquity, already resonant enough, acquired its full resonance. Here Scheiner described his invention of the pantograph—inspired, as he said, by a painter who had claimed (without disclosing any details) that he had invented an accurate copying device. "I will be the painter in this treatise of mine," Scheiner begins his preface to Paulo Sabelli, duke of Albano; and then goes on to offer a traditional statement about the superiority of pictures over texts. A small painting can show you at least as much as many pages of text. By demonstrating the use of the pantograph, Scheiner would no longer be Apelles lurking behind the picture, but Apelles revealing himself and offering everything to view (cf. fig. 13.5).[112]

In his *Pantographice* Scheiner demonstrated the use of the instrument called the pantograph. As the full title of this book implies, its principle derived from that of a parallelogram (fig. 13.6). The *pan*tograph, as the

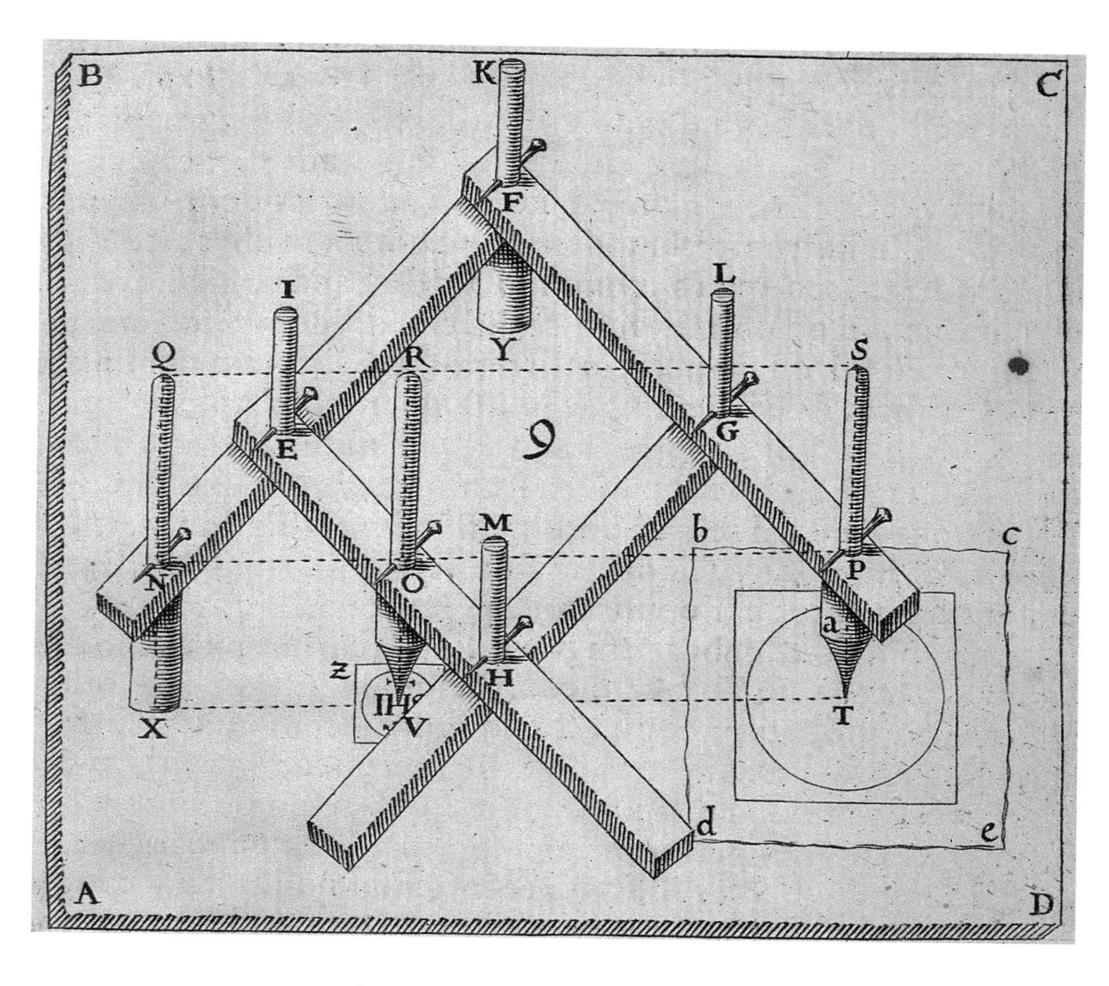


Fig. 13.6. Pantograph. Scheiner, *Pantographice* (1631), p. 29.

word itself suggests, could make smaller or larger drawn copies of everything. But it did not reduce, or cut through the multiplicity of detail, as Cesi's diagraph did. Instead, it transcribed surfaces, in all their vagueness and fullness. Transcription meant exactly that; and so ambiguity was imported from the object of representation. The pantograph could not think for itself; it transcribed, mechanically. And yet it was based on a geometrical figure, the parallelogram. When we think back to the principles of "art" as articulated by Ernst Gombrich out of Karl Popper, we recall Gombrich's view of the schemata that lie at the basis of all picturing; and we realize that for Scheiner too the accurate transcription of reality grew out of the engagement of a geometrical schema. The transition from geometry—on which all schemata must be based—to representation in its flawed plenitude could not be clearer. Almost every illustration in his book makes plain the transition from diagram to picture. Behind the picture (*latens post tabulam*), therefore, lay the diagram; but now the picture could finally come out (*Apelles patens*!) and abandon the mere schemata of things.

Scheiner left history as a reactionary at the time Galileo entered it as a revolutionary. At a comparatively early stage in his career, Galileo wrote on the use of a particular instrument for drawing, namely the military and geometrical compass; but it was never turned into a device for picture making and would only be used for the establishment of order and precision in the realms of battle and geometry. At the end of his career, Scheiner wrote a work on the use of another instrument for drawing, the pantograph. It could represent anything, but only in its visual fullness, not in its intellectual essence. The pantograph could somehow reproduce the sensual surface of things, or if not reproduce it, at least suggest it, via the high susceptibility of the eyes.

The diagram could do none of this. It appealed for its effectiveness to order, logic, and reduction. Each of its marks was significant. Take away a line from a diagram, and it is nonsense. Take away a mark from a picture and all that is lost, in the worst of cases, is some part of its affect. The tension embodied and forecast in Cesi's diagraph and Scheiner's pantograph continued to haunt the study of the world of nature for a long time to come.

# FOURTEEN *The Fate of Pictures*

## APPEARANCE, TRUTH, AND AMBIGUITY

Cesi's doubt lies at the heart of all visual communication of knowledge of the natural world. On the one hand: pictures, with all their potential for descriptive density, and their capacity to record every detail and all creatures in their full anomalousness. On the other hand: diagrams, with their capacity to abstract, to reduce things to their essentials, to show basic relationships between things and within the things themselves.[1] What is at stake is more than a simple polarity. It is a drama that plays itself out over and over again in the history of the reproduction and dissemination of scientific knowledge.

The great irony, given Cesi's reservations about pictures, is that once he died the remaining Linceans commissioned many even better illustrations than before. He himself continued to have them made until the very last minute, despite his hesitations about them. After his death, Cassiano not only incorporated the Lincean drawings into his "museum on paper," but he also commissioned the refined and meticulous drawings of Vincenzo Leonardi. For his own ornithological work, Cassiano ordered the beautiful watercolors of the birds that interested him (figs. 1.3–7); on behalf of his friend, the Jesuit Ferrari, Cassiano paid for the unparalleled series of citrus fruit (figs. 1. 23–29); and to the corpus he acquired from Cesi he added breathtaking sheets such as those of the porcupine and its parts, the oryx, and the remarkable sheets of asbestos and mushrooms on blue paper (cf. figs. 1.1–2, 1.10, and 1.17–19). Such drawings were more intensely descriptive than any of those assembled by Cesi, and rendered with even greater artistry and technical skill. They have a splendor and an attention to detail that exceed most of those commissioned by Cesi himself. They seem to speak to an unruffled confidence in the possibilities of scientific illustration (or what we would now call scientific illustration). This could hardly be called a falling off in optimism about illustration. On the contrary.

Some of the greatest books of illustrated natural history were yet to come. They include such triumphs of picturing as Robert Hooke's *Micro-*

*graphia* of 1665 and Maria Sibylla Merian's *Metamorphosis* of 1705 (to go no further into the next century). The first represents the acme of the illustrated reproduction of things seen with a microscope, the second the apogee of the illustration of the natural history of the New World. Moreover, Merian's spectacular engravings (themselves intended to be colored) were preceded by watercolors on vellum that exceed in quality and technical skill the finest drawings by Vincenzo Leonardi. Many other works (especially Dutch ones) may be inserted into this tradition. Ferrari's two treatises surely count among its most significant harbingers.

Of course there had been many outstandingly illustrated books of natural history before Ferrari. Most of these are well-known and have been much discussed.[2] Several critical works were published in the couple of decades leading up to the publication of the *De Florum Cultura* in that critical year, 1633. Many of them have received attention in the course of these pages. From the work held up as an example to Cesi, Basilius Besler's massive *Hortus Eystettensis* of 1613, to that fine but neglected book of 1625 on the Farnese Gardens allegedly written by another Barberini gardener, Tobia Aldini (but actually the work of the versatile doctor from Messina, Pietro Castelli), garden books generally set the standard of quality for other illustrated works of natural history.[3] In this broad context, both natural historical and horticultural, there was nevertheless little to match Ferrari.

The *De Florum Cultura* contained a series of flower engravings by Cornelis Bloemaert that were unprecedented in their clarity and cleanliness of line. No flower book—or herbal for that matter—had yet paid quite such meticulous attention to the details of flowers, seeds, and bulbs. In his *Flora*, Ferrari attempted to offer a visual basis for the making of taxonomic distinctions, often between "species" that were so closely related to one another that we would now regard them as varieties of the same species. This was the first book to make use of a microscope for a botanical illustration, and to enlist it as a new aid to taxonomy. In so doing, Ferrari showed how the hitherto unseen elements of a plant could now be *counted*, and that *number* could form one of the more secure criteria for determining the distinctions between things.[4] Already there was some shift away from complete faith in pictures.

The situation was not quite so straightforward, however. Ferrari's next book, the *Hesperides*, took the achievement (and particularly the visual achievement) of the *Flora* even further. On the face of it, the *Hesperides* revealed an even greater confidence in the use of illustration. The engraved and etched plates of citrus fruit, once more largely by Bloemaert,[5] were just as meticulously and neatly descriptive as those in the flower book. But there were many more of them, and illustration was yoked to taxonomic drive even more dramatically than before. The book was an attempt to classify the myriad variety of citrus fruit. Many of these are what we would now call monstrous, rather than separate species. But not so for Ferrari. By suggesting that they were more typical than unique, he attempted to insert what to us (and to modern citrologists as well) might seem uniquely monstrous specimens into the order of nature.

Significantly, Ferrari hardly ever used the term "monstrous"—although with the usual Lincean and post-Lincean emphasis on reproductive process he occasionally referred to what he described as misbirths, *abortus*, aborted specimens.

In order to demonstrate his multifold classification of citrus fruit, Ferrari made careful use of Leonardi's marvelous illustrations (figs. 1.40–43). But fine pictures, as we have seen, are not especially well suited to good classification. On the contrary. They have the seemingly inexhaustible capacity to represent the outer limits of the aberrational and the monstrous; but when it comes to showing the normal, the need for graphic reduction stands in the way of the limitless potential for description.

In the *Hesperides*, Bloemaert, following Leonardi and the other citrus artists, expertly represented the most varied, bizarre, misshapen, digitated, pregnant, and cancroid fruits, each one of them assigned a different name. There are few other works in which monstrosity so clearly forms the basis for taxonomy.

But Ferrari did not really know how to deal with the extraordinary monsters his pictures showed. In most copies of the *Hesperides* there was no hand-coloring to help the classificatory effort (though Ferrari attentively detailed the colors in his text); but every conceivable minutia of texture was described in the engraved plates. Not even this, however, was always sufficient for the kind of classification he offered; and so, when he could not fall back on the visual to account for a classificatory or taxonomic turn, he turned to explanation by way of poetic invention and fable, as behooved the good professor of rhetoric he was. He made up strange stories of metamorphosis as a poetic way of describing the etiology of strange fruits.

Hence the long fables and the allegorical plates in his book. They are used as a way of accounting for the kinds of bizarre fruits that he cannot otherwise explain (or that he is reluctant to explain in less colorful terms). The results are nothing that we—or even he—would call scientific, not even remotely so.[6]

For the demonstration of monstrosity the pictorial is without peer; but when it comes to orderly classification the matter is more complicated. In this respect, the *Hesperides*, like the *De Florum Cultura* before it, is a failure. For all its drive to name and to classify, it is taxonomically inept. The spaces the *Hesperides* created for species assignment were too particular, too narrow. Essential relationships between things were not indicated, precisely because there was so little that did not seem to have its own small place. Sometimes it seemed as if everything in this world had to be given its own class—a classificatory oxymoron if ever there was one.

And so the old dichotomy once more raised its head, to detrimental effect.

It had already been a significant factor in the slow progress of the most important of the Linceans' publishing projects, the *Tesoro Messicano*. The pictorial weakness of the book is surely not simply to be attributed to financial constraint. It is at least as much a consequence of the fact that the book is a classificatory jungle, a taxonomic nightmare—as page after page attests. Here is a work that seems to put on display all the irregularity of the world, not its regularity. In the end, the authors of the *Tesoro* could not figure out a satisfactory way of showing the essential relationships between what they took to be anomalous, exceptional, and monstrous. It was not easy to do this sort of thing by means of pictures alone. The book is paralyzed in and by its means. Its illustrations are indeed weak, its scientific results for the most part hopeless (though by no means as hopeless as is often alleged).

The only solution to the quandary, as Cesi came to realize, was to provide a table. In it descriptive picturing would be eschewed and various forms of geometric representation proposed. But the latter would pend. The ideal, as Cesi could only sketch out roughly, was to find a way of representing in diagrammatic form the relationships between things as predicated on their parts. He did not have the time—and perhaps not the ability—to polish this effort, to make it seem even remotely efficient. But to say of his *Tabulae Phytosophicae* that "they fit within a recognizable tradition of botanical taxonomy then in the process of formation," or that "the more we inspect the tables the more puzzling they become and the less novel they appear," or that they simply "reinforce the viewpoint that taxonomy was fundamentally an Aristotelian exercise"[7] is entirely to miss the meaning and value of the Cesian achievement, fractured and incomplete as it is.

The whole issue is brought to the fore by the case of Athanasius Kircher, Ferrari's slightly younger companion and colleague at the Jesuit college. Kircher was a prodigious scholar and one of the best-known authors of illustrated books of the seventeenth century. But what are we to make of his extraordinarily erudite works, with their spectacular and amazing illustrations? It would surely be hard to claim of them that they represent anything but the triumph of illustration. Indeed, they are just that; but as works of science (however we choose to understand that term) they are near to useless. It is no surprise that their appeal should now largely be confined to the bibliophile and the modern student of arcana. One could make no reasonable case (at least not here) for a Feyerabendian tolerance of their strange and oddly reasoned contents. However relativist we may want to be about seventeenth-century science, the contents of Kircher's books are simply too quirky, too odd, to tell us very much at all about the world, except in the most internalized, imaginary, and Borgesian sense (though they are certainly instructive about the byways of the imaginative capacity). A perfect end, one could say, to the unruffled faith in pictures expressed by Cesi and his friends at the beginning of their endeavors.

The dichotomy would never be resolved. Just how little it was is demonstrated by two books published by Stelluti in 1637. Stelluti's struggle, though on nothing like as heroic or intelligent a scale as Cesi, was especially keen. In the enforced pause in his and his friends' work on the *Tesoro Messicano*, in the years immediately following Cesi's death, he turned his attention to two of the projects that stood closest to his heart.

The first was the treatise on fossil woods, the second a reduction to tabular form of Giovanni Battista Della Porta's large, influential, and much republished work on physiognomics, the *De Humana Physiognomia* (which first appeared in 1586). Having just published his *Persius*, that finely imbricated combination of literary, natural historical, antiquarian, and ethnographic knowledge, at once both richly anecdotal and programmatic, Stelluti could now turn to these two items of remaining business. To him they both seemed pressing and urgent. Both appeared in 1637, still within the dead decade that followed the trial of Galileo and that marked the fading swansong of the Barberini papacy.

The little treatise on fossil woods—the *discorzetto*, as Stelluti and his correspondents sometimes called it—was an attempt to do at least some justice to Cesi's unpublished notes on the fossil woods of Umbria. Central to these were his theoretical concerns with things that seemed to be of a mixed—or as Cesi put it, a middle—nature. The real subject of the treatise was thus a profound and basic classificatory problem (it had already been clearly announced in the *Tabulae Phytosophicae*). Stelluti was barely capable of dealing with it at all. Instead, he provided a pathetic selection of illustrations—pathetic, at least in comparison with the stunning corpus of fossil drawings that Cesi had assembled.

Partly because of the climate that followed in the wake of the trial, partly because of Stelluti's own intellectual limitations, and partly as a result of financial constraints, the *Trattato dal Legno fossile minerale* was a failure. It was a slim and cursory production. It is hard to discern the reasons for the choice of illustrations, which are in any case much inferior to the rest of Cesi's fossil drawings, to say nothing of the illustrations in contemporary works such as those by Ferrari. There were plenty of engravers who could have done better; but at this point money was clearly an issue. Although the work was sponsored by funds from Francesco Barberini, by 1637 he was in no mood to overwhelm the closest friends of Galileo with his bounty. Not even all the illustrations Stelluti planned for his book actually made it into the published version—and there were few enough to begin with. How much of this was due to the constraints and circumstances just mentioned, and how much to a failure of nerve and will in the face of the realization that pictures could only be of limited use in resolving the mystery of the classification of things of a middle nature, must remain an open question.

The little-known *On the Physiognomy of the whole of the human body* (thus not just the face) is perhaps the most important work of Stelluti's maturity. It is one of the few published works that may be taken to be particularly representative of his own scientific interests. These interests take us back to the early days of Linceans, and to some of the cruxes of the problems they faced from the beginning.

The book is a very late act of homage to the work of Della Porta, and in it the tension between pictures and tables came home to roost.

It would be a neat and satisfying thing to say that pictures were finally defeated in Stelluti's "edition" of Della Porta. Of course they were not. By any standards, the book is another failure. The very notion of publishing a work on physiognomics without illustrations is too oxymoronic to be sustainable. Not surprisingly, then, Stelluti seems to have begun making plans for another volume that he said would contain illustrations. In the meantime, he reduced a book that depended essentially on its pictures—Della Porta's *De Physiognomia*—to one that consisted of hierarchical, bracketed tables. And he did so both interestingly and efficiently.

Once more the trajectory was typical enough. Already in 1613, when the Linceans were still deeply and, for the most part, unquestioningly committed to picture making, Cesi wrote to Stelluti and asked him to try to procure all one hundred or so plates of Della Porta's treatise on physiognomics.[8] This he did in the long, detailed, and programmatic letter he sent to Stelluti on the occasion of the latter's trip to Naples, a trip that was made in order to stabilize the faltering efforts of the previous five years to set up a branch of the Academy there. Cesi also instructed Stelluti to give Della Porta a copy of their very latest achievement, the just-published *Letters on the Sunspots*, as well as Faber's early printing of the plates of the Mexican plants.[9] At that stage Cesi was proud of the pictures they were producing or having produced for them; and he wanted both Della Porta and his epigones to have a sense of what had already been achieved with the aid of pictures. Della Porta would have been pleased to discover that so far Cesi's belief in the usefulness of the old science of surfaces was untainted by doubt.

But the more Cesi pursued his projects of ordering and classifying, the more intensely he became aware of the incompatibility of the reading of surface appearance as a guide to the inner clues to order. In 1613, Cesi wanted all the plates of the *De Physiognomia;* in 1637, Stelluti, the keeper of the flame, produced a book that eschewed illustration in favor of a set of hierarchized tables. Della Porta had been the great master of the sciences of surface, as demonstrated in his treatises not only on physiognomics but on phytognomics, chirognomics, cryptography, and celestial physiognomy. It was he who had argued with Galileo over priority in the invention of the telescope. And while his own work with magnifying lenses in a tube ought to have undermined his doubt in the indexical status of surface—as it did with Galileo—it still did not.

Visual representation had by now become suspect. It had not yielded the secrets that Della Porta, for one, believed it could. It could not reveal what lay below the surface in any systematic way at all. Ironically, the very invention that Della Porta once thought was more his due than Galileo's had begun to undermine the old confidence in surface; and Stelluti himself began to have second thoughts too. His aim, in his edition of the *De Physiognomia*, was to take the whole of Della Porta's work and have it "succinctly reduced and ordered into synoptical tables," as the title page put it. The idea behind these *Tabulae Physiognomicae*, as one could rightly call them, was taken from Cesi's *Tabulae Phytosophicae*. The historical and natural historical anecdotes would all be removed (although one could hardly say this of Cesi's tables). In order to get to the marrow of the matter, the opinions of other authors (Della Porta had cited a large number) would be cut. In any case, as Stelluti emphasized, the later editions of Della Porta's work contained many accretions and mistakes, which should now be pruned. As much as possible of the "confusion, perplexity and ambiguity" of these editions should be removed. An extensive task of paring down, abbreviation, and simplification lay before him. For Stelluti the solution was to turn once more to the process of reduction to hierarchically organized tables, such as he had learned from both the *Apiarium* and Cesi's *Tabulae Phytosophicae*.

Underlying the physiognomic work, however, was not only the desire to get to the core of the matter but also an emphatic skepticism about surface. As a result of all his paring down, Stelluti declared, everything in his tables would be clear and capable of being taken in at a single glance. This was the opposite of the situations in which the eye wanders listlessly and endlessly over a dense surface, and, never fully satisfied, is unremitting in its search for clues to what lies beneath.

Yet Stelluti still did not get it right. The physiognomic tables are illogical and bewildering; and much less illuminating than Cesi's. They may indeed serve as something of a guide through the Della Portan thickets, but as an explanation of how physiognomics or any of its related sciences are thought to work they are completely inadequate. The problem lay not just in the confusion of material and the lack of logic underlying the tables, or in the putative equivalents they offered between outward appearance and inner character;[10] it lay even more in the constant and ultimately futile attempts to relate everything to everything else (as if Stelluti thought that this was all that classification was about). The book is fraught with analogy and correspondence of every kind, so that everything is indeed made analogous to, or to correspond with, everything else. As a result, the attempt to find the nonvisible essence of things—their propositional essences, so to speak—foundered.

It was not just a matter of establishing a threefold tabulation with the headings *nature*, *signs*, and *relation* (by these, of course, Stelluti meant the basic inner nature of each thing shown, the outward clues to this nature, and the relation to other things like it). This, after all, was precisely the scheme that underlay the old-fashioned pictorializing way of thinking about the world—despite the effort to shrug it off. To grasp this scheme, indeed to use it, you needed a volume of illustrations anyway. Whatever your logical intentions, you could never quite set pictures aside. Stelluti hinted that a second volume of illustrations would eventually be published; but it never was. His energies were soon taken up by his return (along with Cassiano and Colonna) to their largest and most poignant illustrated project of all, the *Tesoro Messicano*. Poignant, because its pictures—and they were poor ones at that!—largely gave them monsters; and the existence of those monsters could only be proven by the evidence of the pictures themselves.

The irony and the tension could not have been clearer. The old sciences of surface had not yielded the Linceans the order they desired; but they were not yet ready to abandon pictures either. They needed pictures as evidence of the anomalies of nature, as proofs of the existence of things no one had even imagined before—from the New World as well as the Old. But pictures gave them little help when it came to placing such new and strange things in the order of nature. The Linceans struggled with this dilemma both consciously and unconsciously. Sometimes their struggle produced exceptional results—even in those domains of classification where pictures, in the end, might have been thought less than useful.

Once more the trajectory is an instructive one. In the volumes in the Institut de France, for example, drawing after drawing reveals the extent to which Cesi's draftsman could not shake himself (or herself) free of the old urge to similitude and anthropomorphization.[11] Puzzling plants and puzzling parts of plants were made to look like human faces, or like other parts of the human and sometimes animal body. If they could not be strictly physiognomized, as we have seen, they were genitalized (cf. figs. 8.29–33, 8.35). As so often in the Aldrovandian corpus and in the older works of natural history, puzzlement in the face of the inexplicable led to forced resemblances: resemblance to things known, and, most of all, resemblance to the human body. Such an approach, in the drawings commissioned by Cesi, fell squarely within the tradition of the sixteenth-century natural historians and received its confirmation from the high physiognomic precedent of his first hero and true father figure, Della Porta himself. The only difference was that the Lincean drawings revealed an even greater preoccupation than he did with issues of biological reproduction, and therefore with genitalia.

But there are many hundreds of finer drawings than these in the corpus left behind by Cesi and expanded by Cassiano. They are so closely descriptive and give such a strong impression of accuracy and verifiabil-

Fig. 14.1 Dried calabash (?). Watercolor and bodycolor with white heightening and gum varnish over black chalk, 151 x 113 mm. Windsor, RL 32569. The Royal Collection © 2000, Her Majesty Queen Elizabeth II.

ity that one would never have thought them capable of being pressured into similitude. And yet, even with them, there are moments when, for all the intense quality of graphic description, they seem to go in just that direction.

Take the strange piece of fruit numbered 388 in Cassiano's paper museum (fig. 14.1). It is executed with great refinement and coloristic subtlety. It looks like a dried apple—or is it a coconut?—with strange markings; but are its markings the product of art or nature? It is impossible to tell. The drawing itself is probably the product of a hand which, when confronted with a "joke of nature," pushed depiction in the direction of similitude, and endowed the object it showed with a physiognomy. This would have served, even if unwittingly, to further inspire wonder at the artistry of nature itself, or at the striking way in which the product of one world (that of plants) could parallel that of another (that of man). And the only way of resolving such a problem would have been by a supplementary text, not by a picture.

This particular problem, however, is a consequence of the fineness of the drawing. If the drawing were weaker, it would be easier to dismiss. An immediate parallel is offered by the ridiculous piece of apple bark in the form of a human face illustrated in Aldrovandi's posthumous *Dendrologia*

Fig. 14.2. Misshapen piece of apple bark in the form of a human face. Woodcut. Aldrovandi, *Dendrologia*, (1668), p. 360.

360 Vlyſsis Aldrouandi

aliquando ad amplam aſcendit, aſt nunquam excelſam ad proceritatem peruenit; ipſa E cortice eſt leui, & ſubatro; ramis quidem oblongis, ſed in latitudinem hinc inde fuſis; folia fert laurinis breuiora, latiora tamen, mollia, & aliquantulum hirſuta, & inordinata; nodos habet faciebus belluarum ſimiles, adeout ex illis concinnari Perſonæ de facili poſſent, quemadmodum ea, quæ delineata proponitur Simiæ ſimilis, &c.

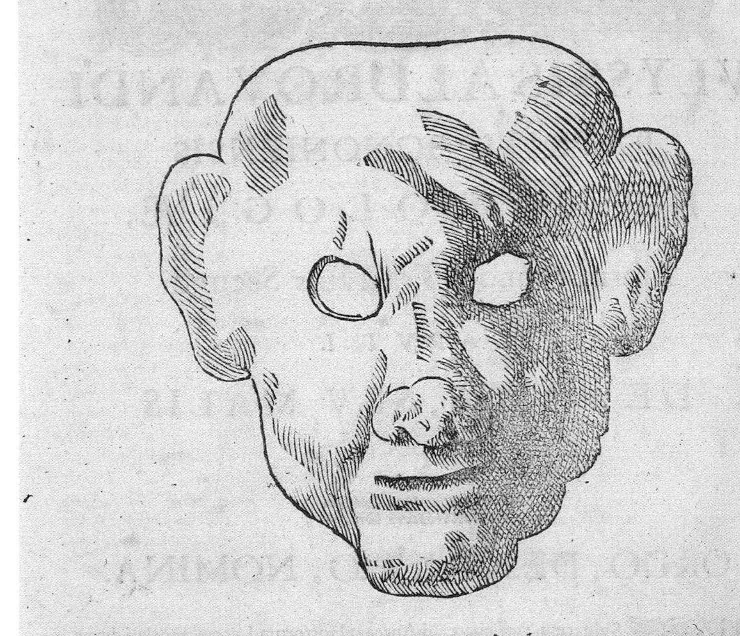

NODOSA MALINI CORTICIS PARS QVÆ LARVAM HVMANAM APPRIME IMITATVR

16\. Radices illi ſunt paruæ ſummę telluri hærentes; Flos ab ramis erumpit candidus, aut parum purpuraſcens; Fructus illi ſuccedit maiori ex parte orbiculatus in ſummo floris, inq; imo pediculi aliquantū depreſſus; illiq; caro ineſt fungoſa, ſuccoſaq; ex quo de facili rugoſus fit; ſapor in maturitate ad dulcem, acidulumq; tendit, in acerbitate verò ad aſperum, & ſtipticū; ex ijs *Melapia* parua, rotunda, pallidiuſcula, valdè odorata, & dulcia quæ H *Melimela* fortè ſunt, quaſi melleamala, recognoſcuntur; *Roſea* colore, veluti roſarum ſubrubenti mediocris magnitudinis, odore roſaceo, ſapore dulci, ſubacido, & diutius

(which has several more examples of this kind) (fig. 14.2). This woodcut shows such evident violence to reality that it only arouses skepticism; whereas Cassiano's drawing, precisely because of its quality, inclines us to believe in its plausibility. Such drawings lend great conviction to what we see as the monstrous, the aberrational, and the borderline; and make us believe (if something like this "apple" is not merely a production of art) that the order of nature must at least occasionally give way to monsters.

Hence the usefulness of the drawings, scattered throughout the Lincean corpus, of deformed nestlings, of pregnant and digitated fruits, of hermaphroditic animals. Eventually the Linceans discovered that even such things could be illustrated in ways that were largely free of the old drive to similitude; and that from such drawings one could tell what fit-

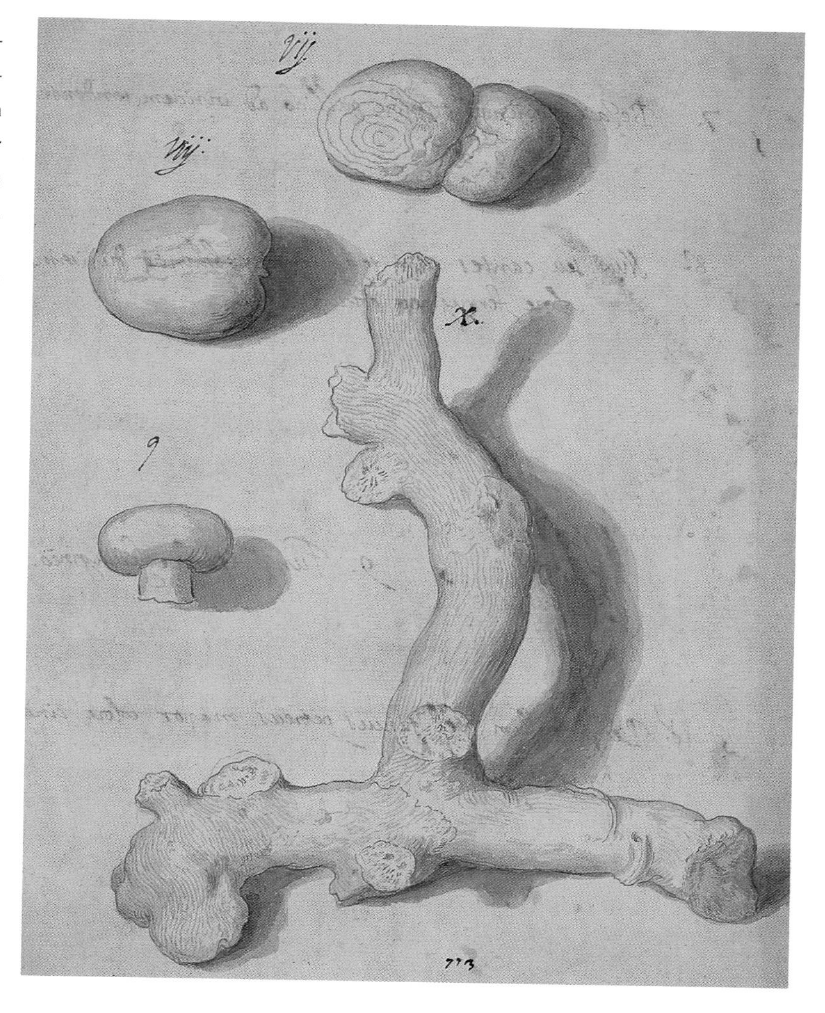

Fig. 14.3. Bezoar, "fungites," and other geological curiosities. Pen, brown ink, and brown wash over black chalk, 233 x 178 mm. Windsor, RL 25526. The Royal Collection © 2000, Her Majesty Queen Elizabeth II.

ted into the order of nature and what did not—however much the natural logic of such things might continue to escape one, or to escape the best of schemes and systems. Drawings were not only *fitted* to the representation of the monstrous, they could actually serve as *proof* (and not just as indicators of probability) of what was genuinely aberrational.

How much could the best of these drawings actually resolve borderline issues—or at least the more nettlesome cases? One of the most interesting pages in the beautiful volume of precious stones and mineralogical and other oddities in Windsor Castle known as *Natural History of Fossils V* shows a set of very puzzling mixed species (fig. 14.3). At the very top comes the troublesome bezoar; at top left is one of those strange types of pregnant stones—or is it a fossilized nut?—that resemble the unresolved case of aetites; below this, something described as a "fungites" in the inscription on the verso—in other words, a putatively fossilized fungus (but probably just some dried out mushroom); and finally, dominating most of the page, an object that looks like a fossilized twig of some kind, or maybe just some type of coral. In the inscription on the verso of the page it is described as "pseudo-coral or stony [petrified] twig, very rare."[12] Not much help here!

These are most precise drawings. They give the impression of high verisimilitude. Yet they hardly enlighten us as to what these strange mixed species actually are. On the basis of the drawings on this sheet alone, we would have considerable difficulty in identifying the specimens shown. It is the inscriptions on the verso, only detected by looking at the sheet against the light and with the aid of a mirror, that give us some idea as to what they are. But do the drawings, even thus identified, help us in establishing the exact nature of each of these specimens? Not at all.

The plain fact is that a problem such as this cannot be resolved, in and of itself, by a drawing alone. Certainly, such cases might be helped by multiplying representations—that is, by using other more securely identified specimens as an adequate basis for comparison. It is this idea, the exchange and dissemination of visual information as a means of comparison—in the interests of taxonomy—that first inspired the Linceans' use of drawings;[13] but it was not enough to sustain an unimpaired faith in drawing.

There were a few successes. Some of these issues were resolved by Fabio Colonna, if not by means of drawings alone, then at least with their aid. Perhaps one of the most absorbing drawings in the corpus—and also one of the simplest—is that of a goose barnacle (fig. 14.4).[13] Here was a phenomenon to conjure with! The goose barnacle is actually a crustacean of the family *Lepadidae* (and in this case the drawing is probably an example of the commonest Mediterranean species, the suggestively named *Lepas anatifera*). For a long time, few could quite make out what it really was. Even its molluscular status was uncertain. It was much discussed by natural historians and folklorists of one stripe or another, and had long challenged the classificatory capacities of both botanists and zoologists. The prevailing view, at the beginning of the seventeenth century, was that it was generated from the spume or froth formed on old pieces of wood, say from shipwrecks; this then became some kind of fungus, which gradually turned into a shell. Eventually the shell itself "gaped open," as the noted English botanist John Gerard put it, and the stringlike legs of a bird seemed to appear, gradually growing into "a foule, bigger than a Mallard, and lesser than a goose."[14]

It is not hard to imagine the interest to the Linceans of such a strange combination, such an improbable yet obviously appealing instance of borderline infraction and category overlap. In the first place, the goose barnacle supposedly began life as a fungus, thus offering yet another instance of the group of "imperfect" things that so fascinated Cesi and the early Linceans. Second, it raised interesting problems of generation, notably in the way it seemed to show the transition from a thing of the sea

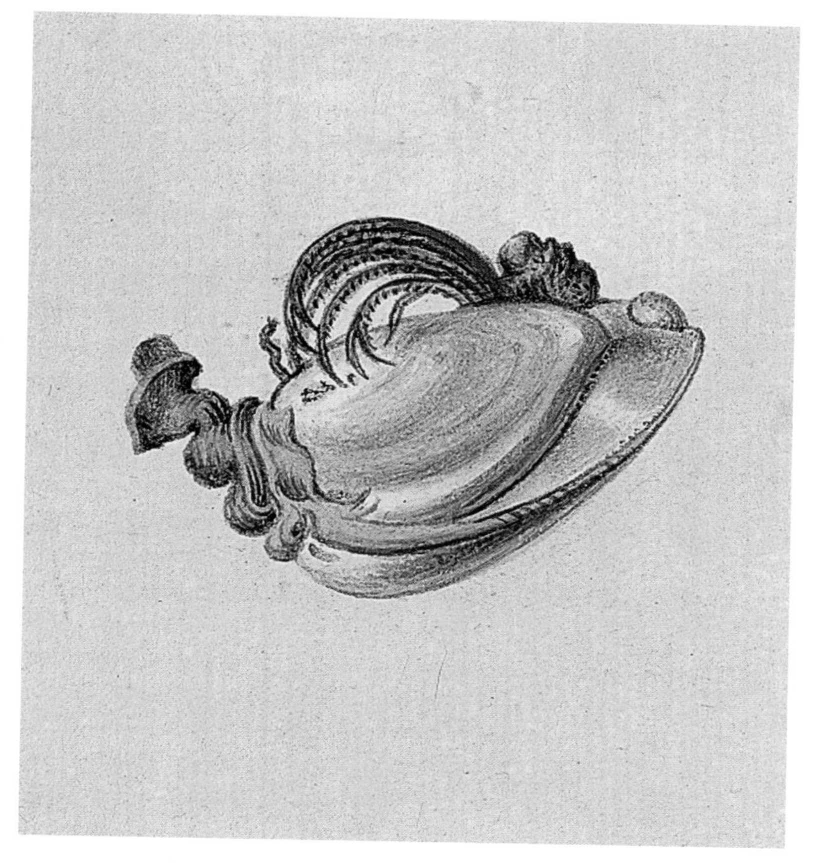

Fig. 14.4. Goose barnacle (*Lepas* probably *anatifera* ). Watercolor, bodycolor, and gum varnish over black chalk, 149 x 110 mm. Collection Cyril Fry, Saxmundham, Suffolk.

to a thing of the air. Third, it posed the problem of deciding just how to classify an object that seemed to partake of both the vegetable and the animal kingdom (if not that of the mineral world as well!). But perhaps the goose barnacle was not a transitional or combinatory phenomenon at all? In this case one had to decide exactly what it was to begin with. This was the sort of problem that all the Linceans, but Cesi and Colonna in particular, had long found especially compelling.[15]

Colonna was the first to see through much of the fantasy that swirled round this strange creature of the sea. Already in his *Phytobasanos* of 1592 he had rejected the fable and the elaborate descriptions that went along with it.[16] It may well have been he who sent to Cesi the surviving watercolor of a goose barnacle (probably long after he had completed his 1616 studies of marine animals). It is a drawing that could hardly be more different from the kind of rough and fanciful illustrations then current, such as occurs in Gerard's 1597 *Herball* (fig. 14.5) or in Claude Duret's *Histoire admirable des plantes* of 1605.[17] It is precise, meticulous, and clear. From the period, it is the least fanciful (and therefore by far the most reliable) record we have of a goose barnacle, and it directly and visibly resolves many of the particular problems of classification this particular case presents. It proves that even drawing, at least in its finer forms, could serve a critical role in resolving some of the old confusions about borderline species.

At the same time, however, in its very precision and close observation the drawing manages to give a visual sense of what it was about this animal that gave rise to the legend that some kind of "fowl" grew out of the shell. One can almost see its incipient limbs. In many respects the drawing actually comes quite close to descriptions such as Gerard's. Even though one could hardly deny that *some* element of exaggeration may have crept in—simply because of the impossibility of an altogether innocent eye, or of wholly escaping from the pressures of tradition—the basis for doubt about the status of the specimen is suppressed as much as possible.

This, then, is what good drawing, forced by anxiety about classification, could be expected to achieve. It could indeed facilitate the determination of species. Paradoxically, it could carry you beyond doubt, and sometimes even resolve fundamental questions of ambiguity. Such drawing reveals an almost totally unprecedented confidence in the possibilities of visual transcription as a means of settling even the most delicate issues of classification—however much faithful transcription of nature may have wandered as a result of the pressures of tradition and the desire of the eye to see what it wanted (and had learned) to see. The point now, if you wanted to go anywhere at all, was to renounce similitude—or at least to make a conscious effort to do so.

But it was not so easy, and the burgeoning confidence in pictures could not be sustained. We can now begin to see why. Pictures can indeed show everything; but how to suppress the temptation to make something look like something else we know? How to avoid investing it with the marks of what we know best—a face, a living being? You may decide to suppress everything except the essential aspects of an object, everything except what you believe to be the determining elements of a class; but in so doing you deny one of the basic virtues of pictures: to show the distinctiveness of things, to show not just what is essential, but what is adventitious or unique. Would it ever be possible to avoid seeing what the eye wants to see, or to show how things are similar, when similarity is so clearly determined by experience, expedience, and desire? Say you strove to achieve objectivity in representation, to show only what a specimen presents to the eye—nothing more: in such a case you would have to show everything.

You describe and describe in great density, and still you are not sure of what it is about objects that might help you to resolve the classification of clearly mixed things, of things that are neither fish nor fowl, neither vegetable nor mineral. The more you describe, the more you show the particularity and oddity, not the regularity, of things. You can produce beautiful drawings such as that of the *Acetabularia mediterranea* and still not know exactly what it is: mineral, fungal, or a just a clod of earth with

Fig. 14.5. Goose barnacle. Woodcut. Gerard, *The Herball* (1597), p. 1391.

mushrooms growing from it. You can make drawing after drawing of fossil woods, and still not know what they basically are.

By any reckoning, Cesi's corpus of nearly two hundred such drawings in Windsor is consistently precise and in many cases superbly sensitive to every nuance of surface. They leave one in no doubt at all about their transcriptional accuracy. Aside from the fact that many can now be compared with actual finds from the area of Dunarobba and Sismano, there are many that can be cross-checked one against another: every test of this kind simply confirms their accuracy. Yet still we cannot tell, even with all the visual information provided by Cesi, and even with such exceptional attention to every imaginable detail, what exactly is what. One perceives immediately and clearly that what these drawings show are seriously mixed phenomena; but we come no closer to establishing what they really are or how they came about. At least not on the basis of the drawings and their inscriptions. Every page suggests another combinatory possibility. No wonder Cesi drowned in a sea of terminological possibilities!

It is true that with animals and plants you could cut into things, ever more deeply if you so wanted, and magnify ever more parts in your search for whatever you defined as species; but if you did this, the problem was that you just found more pictures—picture upon picture—or you saw the geometric elements of things. If you seek the use of the compass, go to the mushroom!

The tussle continued long after Cesi and Cassiano died. Could pictures and classification ever go together? Or did one simply need both? For long into the seventeenth century, handbooks exemplifying the old doctrine of signatures continued to be reprinted. If you wanted the physiognomic or phytognomic works of Della Porta for example, it was easy enough to find them. There were reprints and editions from Rouen, Naples, Frankfurt, Hanau, Padua, Venice, Paris, and Rome throughout the century. In France physiognomics had a kind of heyday as a result of the popularity of the voluminous works of Marin Cureau de La Chambre and the *Conférences* of the painter Charles Le Brun at the newly founded Academy of Painting in the 1660s and 1670s. Before expiring into pure folkishness, the tradition was given a final and powerful shot in the arm by the popular works of Lavater.

Folkishness or not, however, there can be little doubt that the lingering appeal of a science that read the marks of surface to discover the truths that lay below had basically to do with the natural inclination to want to believe in such a possibility in the first place, to believe that our eyes can indeed provide us with evidence we can trust as the basis for further judgment of the world beyond ourselves.

But we are also skeptical beings—or like to think of ourselves as such. Eventually we learn the risks of seeming gullible, of seeming to be capable of being deluded by the evidence of the senses. The more tired the physiognomic tradition in science became, the deeper the split between belief that objectivity in picturing was indeed attainable and belief that pictures were not necessary at all. Linnaeus himself, the greatest of all classifiers, declared bluntly:

> I do not recommend the use of images for the determination of genera. I absolutely reject them—although I confess that they are more pleasing to children and those who have more of a head than a

brain. I admit that they offer something to the illiterate. . . . But who ever derived a firm argument from a picture?[18]

From then on—at least for a while—pictures became ever more divorced from the project of classification. Ironically enough, proof of the failure of that marriage was offered by the one work of Darwin's that was illustrated more than cursorily and could reasonably be considered a failure: his last work, the *Expression of the Emotions in Man and Animals* (1872). In it, after years of standing well aside from the problem of classification (as his theories, in a way, required) Darwin set out to classify something that was generally regarded as unclassifiable: the emotions (and even more so the actual *expression* of the emotions). It is significant that this book is the only work of his to have been extensively illustrated; and yet all we are left with, in the end, is a sense of just how hard to classify the emotions really are. Pictures failed to prove otherwise. By using photography, Darwin clearly thought he could show the emotions in all their raggedness—and yet prove the underlying similarities at stake. This was something he was never able to prove, since the perception of similarity could be made to depend on the evidence of allegedly objective pictures alone. You certainly could do so, as the physiognomic tradition had shown, by forcing similarity in the direction of bodies and faces known; but to do this, as we have seen, does violence to the accuracy of pictures (or at least to what is more or less legitimately perceived as the accuracy of pictures). Precisely because of their capacity for showing detail—all the details—pictures could yield probability, but never proof, just as Faber had discovered in the course of his work on the Barberini *dracunculus*. Unless you took the physiognomic dead end, you needed to cut out detail, pare it down to the minimum, down to a stage when pictures were no longer pictures, in order to see the essential relationships between things. All the *Expression of the Emotions in Animals and Man* could show was that the emotions, ragged to the end (and shown to be just that by means of pictures) were not, in the end, capable of classification.[19]

Though who knows? Perhaps illustration too will have the last laugh, when the time comes—and it seems near—that the emotions too are subject to rule, order, and system.[20]

Be that as it may, the lesson that Cesi learned is one that continues to play itself out from age to age. The technologies of visual reproduction are always capable of lying. They may seem guileless and simple, and they may give the appearance of truthfulness: yet there can never be a reliable correlation between outward appearance and inner truth.

But there are truths, in life as in painting. Stelluti's strange book of 1637 on Della Porta's Physiognomia was written in the immediate aftermath of Galileo's trial. It was dedicated to the very man, Cardinal Francesco Barberini, who had once supported Galileo but who had by now turned his back on him. When Galileo died under Medicean protection in 1642, and people were already beginning to talk about the possibility of raising a monument in his honor, Barberini wrote to the grand inquisitor of Florence:

> His Holiness the pope, with the agreement of the Eminent Cardinals, has resolved that you, with your usual skill, should ensure that it comes to the ear of the Grand Duke [of Tuscany, Ludovico de' Medici] that it would not be a good idea to make any mausoleum for the body of someone who has been made to repent by the Tribunal of the Sacred Inquisition, and who died while he was still doing penance, because this would scandalize good people everywhere.[21]

Cold words from the man who had once so warmly supported the activities of Galileo's closest friends! Had he not encouraged, sponsored, and even joined the Academy of which Galileo was the center and chief ornament? He had. Could anyone have predicted that he would thus turn on Galileo? Perhaps not. He may have been dissimulating his true views on the troublesome scientist from Florence for a long time.

These were years in which you had to be aware of the degree to which people were pretending to be what they were not; or, even more necessarily, to be aware of how much people could cover up what they really were, or really thought. Who knows what strange confluence of psychological motive and scientific reason combined to produce Stelluti's effort to systematize Della Porta? In his dedicatory epistle to Francesco Barberini, Stelluti pointed to the special relevance of the Della Portan science of signatures for the divination of character. After all, he said, princes and potentates had to be able to detect simulation among the courtiers who surrounded them. The features of the face were a good indicator; but still you had to know the science of reading them. Crucial to this science, Stelluti stressed, was the element of choice, *elezzione* as he called it. The word occurs over and over again in the book. You had to know how to choose between good signs and bad signs, and study to achieve that knowledge. The problem was great precisely because appearances were so deceptive. How, Stelluti asked, was one to read the outward signs of the body in such a way as to understand what really lay beneath, to understand the true character of the soul contained in that body? And were external signs such as mouth, forehead, nose, and other parts of the body adequate manifestations of the secrets of the heart? It was impossible to know. To decide correctly was the real choice of Hercules.

The illustrations in Windsor Castle that form the starting point of this book were found in the wake of a dream I once had. In it Anthony Blunt, the famous art historian, pseudo-aristocrat, and spy appeared. This story began with him. In his hands he held the drawing of an orange, and it was this dream that led me to the source of the drawings that Cassiano dal

Pozzo commissioned for Ferrari's book on citrus fruit. As was revealed only a few years before my redicovery of these drawings, Blunt's outward demeanor had long masked his true beliefs. Trusted by many, his apparent allegiances cloaked his deepest commitments. He had learned to construct his life in such a way that that even he may not have known who his real friends were.

In Blunt's case we may never know what he really believed, or what ideals he held true. There may have been none. I remain puzzled by the disjunction between my reading of Blunt's character over the years and what we now know about his beliefs and actions. His enemies believed that his whole life consisted of dissimulation and deception. The fruits of his behavior, while probably seeded by good intentions, turned out to be poisonous. There is no parallel with Galileo—except that both cases are paradigmatic for the problem of interpretation.

Galileo's life was fraught with lies and dissimulation. Some of his most important works were published or otherwise disseminated under pseudonyms or the names of his disciples. Spying and cover-ups became the order of the day, as his enemies sought to discover what he really believed and his friends tried to find out what the next set of tactics of his opponents might be. For a while Galileo managed to veil his Copernicanism; but not for long. He was too frank and outspoken to keep up the mask of appearances his friends sometimes wished him to adopt. His own sense of the truth of things constantly subverted the strategies of those of his intimates who attempted to sustain appearances. The innocent Luca Valerio, as we have seen, could not stand the pressure of doing so. Men and women suffered as a result of the need for dissimulation; and not everyone could resist the efforts made by the Linceans' intellectual, political, and personal enemies to save the appearances.

In the year before Galileo died, Torquato Accetto wrote a book called *On Honest Dissimulation*. In it he held that, for the love of truth, one should remain silent about at least some of one's thoughts. He recommended "the expediency, indeed the necessity, for a wise and free man not to simulate nonexistent virtues (always an abject thing), but to dissimulate."[22]

In the preface to his book on Della Porta, Stelluti warned that slaves and servants could simulate qualities and virtues they did not have; but the highborn—and the truly virtuous—should practice dissimulation when truth was threatened. In other words, there were good masks and bad masks. Once again, you had to know how to interpret them, to see what was essentially true and essentially untrue beneath them.

The energy and skills required to negotiate a moral path under the circumstances described in the early chapters of this book occupied the Linceans all their lives, and Galileo too. In the short run, their efforts failed; in the long run, of course, they did not.

Our trope must take its final turn. Galileo's critical revelations of the fragility of appearance shook the world then known and changed it forever. The sum of his work and the particulars of his observations made it clear that the secrets of nature—like the natures of men and women—could no longer be revealed on the basis of observation alone. What nature displayed to the outer eye of man was no longer a sufficient guide. Indeed, it was often downright misleading. Judgment could no longer rest on any confident assertion of the correlation between appearance and reality (whatever those terms might mean). Skepticism about appearance had to become the order of the day, because the world had become interpretable—with all the perils and profits interpretation entailed.

At the same time, Galileo never wavered in his belief that there *were* truths to be revealed—by abstraction, hypothesis, and experiment. Observation of nature may have lain at the basis of each of these processes, as Galileo and his fellow Linceans showed over and over again; but observation was never enough, even when they closed in on the objects of their study with the aid of technological resources never previously imagined.

At the center of the Linceans' activities stood an unparalleled investment in picture making. Theirs was a deep investment in the worth of appearance and the value of surface. In the course of supporting Galileo, they came to have their doubts, and Cesi above all. As he became more aware, via the new technologies, of the schematic bases of both appearance and representation, he began to realize that what lynxes saw with their keen eyes was not always to be trusted. But the more he turned to abstraction, number, and geometry, the less he seemed capable of renouncing the helpmeet of illustration. He found that he could not renounce either of these poles in his approach to nature.

And so it has continued to be. The tension between the poles remains. The only truth is that appearance does indeed mask that which is essentially defining about things. If we take away the visual description of what we see, we will be left with an unnatural world, one in which everything is regular and subject to rule. In order to grasp such a world we may not need pictures, just as we may never need to explain the lies with which we live. But this would entail a world in which explanation was never required, a dead and dry world free of the incessant chaos of life. We have no option but to live with the tension that incapacitated the work of the Linceans yet never dimmed the sharpness of their vision.

# Notes

## A NOTE TO HISTORIANS OF SCIENCE

1. Shapin 1996, p. xv. Shapin's bibliographic essay on pp. 167–211 offers a comprehensive and fair survey of the literature on the scientific revolution up until 1996; for subsequent surveys, see note 5 of the introduction.

## INTRODUCTION

1. While much is known from documents published by one indefatigable researcher, Giuseppe Gabrieli (cf. especially the almost complete selection of his many articles on the subject in Gabrieli 1989), the story has not yet been integrated with the vast quantities of illustrations of nature that Cesi and his friends commissioned and that have only recently been rediscovered (as recounted in chapter 1).

2. Grafton 1991, p. 179.

3. Ibid., p. 184.

4. For the origins of the name of the Academy (a reference both to the sharp-eyed lynx and to Lyncaeus, the keen-sighted pilot of the Argonauts) see chapter 2, under "The Wanderer of the Heavens."

5. See Freedberg 1992b.

6. For an excellent yet concise overview of current views on the "scientific revolution," see Shapin 1996, summarizing earlier and often heated controversies on the topic. In addition to the works mentioned in his comprehensive bibliographic essay (pp. 168–211), see also Henry 1997; Applebaum 2000; and the new survey by Peter Dear (whose many works on the rhetoric of science in the seventeenth century have been fundamental to my thinking in a number of areas covered by the first half of this book) (Dear 2001).

7. Throughout this book, I shall use the noun "picture" not in the limited art historical sense of a painting in oil or some such colored medium on panel or canvas, but to refer to a visual representation, more generally taken, of something in the natural world. In this sense, I will often refer to the colored drawings Cesi made as "pictures," as a way of signifying their more general status as concrete visual representations of the world.

8. Koyré 1957; and Koyré 1978.

9. As, for example, in the case of the drawing of what he eventually decided was a fossilized fungus (fig. 14.3) and of the strange specimen in fig. 8.20.

10. One exception, of course, is the incomparable Jacopo Ligozzi.

11. Cf. Ong 1958 for the fundamental discussion of the relations between rhetoric and logic in the sixteenth century.

12. *Ageōmetrētos oudeis eisitō*—a very slight adaptation from the motto that according to John Tzetzes was inscribed on Plato's door: *Mēdeis ageōmetrētos eisitō* (Tzetzes Chil. 8. 973). For a further discussion of the implications of this citation, see chapter 12, under "An Essential Tension."

13. Carolus Linnaeus, *Genera Plantarum, eorumque characteres naturales* (Leiden: 1737), fols. ¶¶recto—¶¶¶verso. See also Freedberg 1994a; and chapter 12, under "Doubts about Pictures," and chapter 14 for arguments about the usefulness and legitimacy of pictures.

14. As cited in Atran's marvelous book of 1990 on the cognitive foundations of natural history. See also Meinel 1984. In many respects some of Federico Cesi's most crucial ideas about the role of mathematics in natural history would be taken up by Jungius, himself deeply influenced by both Galileo and Cesalpino. On this aspect of the work of the botanist, mathematician, and physicist from Giessen and Hamburg, see the excellent discussion in Atran 1990, pp. 158–160.

CHAPTER ONE

1. On bottle-imps, see Bromehead 1947a.

2. The literature on Cassiano has grown tremendously in recent years. For a long time, the chief source remained Lumbroso's excellent monograph and selection of letters of 1874. This provided the basis for all subsequent research. The revival in Cassiano studies can really be dated to Francis Haskell's pages on Cassiano in Haskell (1963; reprint, 1980), pp. 98–114. For a while, Haskell and Sheila Rinehart were the only serious contributors to the field (e.g., Haskell and Rinehart 1960, and Rinehart 1961). Finally, in the 1980s, the study of Cassiano really got under way, beginning with the series of *Quaderni Puteani* and the articles by Freedberg, Solinas, and Solinas and Nicolò. This activity was punctuated by the important group of essays edited by Solinas in 1987. Works such as Sparti 1992 and Herklotz's magnificent summation of Cassiano's antiquarian researches (Herklotz 1999) followed, and we now have Solinas's own summation of his many years of work on the subject in the exhibition he organized in Rome at the end of 2000 (Solinas 2000a). For the provenance of Cassiano's drawings in the British Royal Collection and elsewhere, see "A Brief Note on the Provenance of Cassiano's Drawings" following this chapter.

3. For a very good summary, see Haskell 1980; and now, for a review of Denis Mahon's many years of research on Poussin and Cassiano, Mahon et al. 1998.

4. Including, it should be remembered, a notable collection of drawings after Early Christian monuments, mosaics and antiquities. See now Herklotz 1999; and Osborne and Claridge 1996.

5. These volumes he made available to scholars and researchers who came to see him from all over Europe. For an outline of their content, see the remarkable fold-out sheet appended to Dati 1664. For their actual contents, mostly surviving in Windsor Castle (and some in the British Museum), see especially Vermeule 1960 and Vermeule 1966; as well, now, as Osborne and Claridge 1996, with the essay by Haskell and McBurney on pp. 9–23 on how they passed from Rome to England (see also the "Brief Note on Provenance" below). For attempts to identify some of the artists of the drawings after antiquity—including Poussin, Pietro da Cortona, Pietro Testa, and others—see especially Turner 1992 and Turner 1993.

6. On collecting, museums, cabinets of curiosity, *Kunst-* and *Wunderkammern*, and the relationship between *naturalia* and *artificialia* in these collections, there is now an immense literature. See, for example, Impey and McGregor 1985; Schnapper 1988; Pomian 1990; Olmi 1993b (both references to his many earlier articles on the subject); Findlen 1994; Daston and Park 1998 for recent important discussions.

7. Galileo, *Opere*, IX, p. 69 (in the youthful *Considerazioni al Tasso*); cf. Panofsky 1954, pp. 16–20. When you opened Ariosto's *Orlando Furioso*, on the other hand, you beheld, "a treasure room, a festive hall, a regal gallery, adorned with a hundred classical statues by the most renowned masters with countless complete historical pictures . . . in short, full of everything that is rare, precious, admirable and perfect." Thus Galileo wished to suggest that, in contrast to Tasso, the ideas in Ariosto were altogether on a different scale and, more important still, were connected by overarching concepts.

8. "Curious" not because he was strange but because he was a "curioso" (to use the seventeenth-century term), a man interested in natural and artistic curiosities, and possessing what would then have been called a "cabinet of curiosities." On this subject the literature is now vast. See note 6 above.

9. Evelyn, *Diary*, p. 277. On the stones mentioned here, see Pliny, *NH*, XXVII, chapters 11 and 7 respectively, as well as Thomas Browne's *Pseudodoxia Epidemica*, book II, chap. 5, § 8. Pliny noted that when the "enhygros" was shaken, liquid moved about inside, as in an egg; and he wrote page after page about different kinds of carbuncle, a stone that was not just harder than a diamond, but also glowed with astonishing brilliance and intensity. Cf. chapter 11, under "The Bologna Stone: Light and Dawn" and notes 24–26 there.

10. One can only suppose that Evelyn was more fascinated by the drawings after antiquity and by the two strange stones, the "*Enhydrus*" and the carbuncle, about which Pliny the Elder, the most widely read of the ancient writers on natural history, had written at considerable length (cf. the previous note). How could a stone contain water? And what was it about the carbuncle that made it not just harder than a diamond but glow so brightly? Such phenomena of nature often seemed more compelling, even to the most pedantic lovers of antiquity, than collections of paintings—even ones with ancient subjects and archaeologically meticulous settings, such as those by Poussin. On the significance of these stones for the Linceans, see the discussions of aetites and light-emitting stones in chapter 11, under "The Bologna Stone" and "More on Light and Heat."

11. For some of the additions made to it by Carlo Antonio himself, see Haskell and McBurney in Freedberg and E. Baldini 1997; and Sparti 1992.

12. Skippon 1752, pp. 692–3.

13. See note 7 above.

14. They were purchased in Rome by the Adam brothers from Cardinal Albani, into whose collections dal Pozzo drawings had passed. On this see "A Brief Note on Provenance" following this chapter.

15. Windsor, RL 25536–25734; on the drawings of fossils and fossil woods, see Scott and Freedberg 2000, as well as chapter 11 below.

16. Windsor, RL, 25480–25535.

17. Windsor, RL, 27599–27690. As Inventory A in Windsor Castle indicates (as in Freedberg and E. Baldini 1997, p. 31; reproduced in all other volumes in both series of *The Paper Museum of Cassiano dal Pozzo*), there were originally five volumes of bird drawings, four of which were then dismembered, leaving some of the loose sheets in Windsor Castle, while others were sold and dispersed (cf. "A Brief Note on Provenance" above).

18. Windsor, RL 27691–27901. On this "Erbario," see chapter 8 below.

19. Formerly in Rome, BAV, but presented to the Mexican government by Pope John Paul II in 1990 and now in the Museo Nacional de Antropología e Historia de México in Mexico City. Cf. also chapter 9, under "Cassiano in Spain."

20. Cf. chapter 8, under "The *Erbario Miniato*."

21. "Having," just as Skippon recorded, "one fin on the middle of the back, a pair of fins under the gills, a longish snout, wide mouth, a forked tail, and [is] well armed with sharp teeth" (Skippon 1752, p. 693).

22. New York (Sotheby's), September 16–17, 1988; drawings from the estate of James R. Herbert Boone (who had acquired them from a London dealer in the 1930s following their sale by the royal librarian just after the First World War; cf. "A Brief Note on the Provenance of Cassiano's Drawings" below).

23. For details see note 73 below.

24. The same applies to the head and neck of a whooper swan, the study of the foot of a pintail duck, and many other of the finest drawings in the corpus (all these still in private collections in Britain; for excellent illustrations see *Quaderni Puteani* 3, pp. 14–15, and *Quaderni Puteani* 4, p. 170).

25. Though the piece of amber shown here may in fact be a fabrication. I am grateful to Ross McPhee of the American Museum of Natural History for pointing out the significance of this drawing to me: either it shows some remarkable inclusion—the first time such inclusions had ever been depicted in amber—or it shows a fabrication.

26. For a full survey of all of these, see now Freedberg and E. Baldini 1997.

27. On Aldrovandi and his collections, see, in addition to the older literature, the many articles by Olmi, especially Olmi 1976, Olmi 1992, and Olmi 1993b, as well as the useful pages in Findlen 1994.

28. For Aldrovandi's approach to classification, cf. also the brilliant pages in Foucault 1973, notably pp. 34–46, as well as my unpublished "Naming the Visible: Art and Natural History in the Circle of Galileo," Siemens Foundation Lecture, Munich, June 1990; a version published in Italian as Freedberg 2000; see also chapter 13, under "Precedents."

29. In fact, they had already been cataloged at the end of 1970s by the then librarian of the modern Accademia dei Lincei, Ada Alessandrini (Alessandrini 1978), pp. 288–293; cf. also ibid. pp. 68–77 for an excellent summary; Montpellier, BEM, MSS H.

505–507. The three other books seen by Skippon may have been the manuscripts preserved in Rome, BV MS R.57, and Florence, BL, MSS Ashb. 1209 and 1210; but they may also have been one of the several manuscripts of Heckius's still preserved in Rome, BANL. On all of these manuscripts, see the characteristically useful and penetrating descriptions in Gabrieli 1989, pp. 1055–1078.

30. Especially in Montpellier, BEM, MSS H. 506–508; on these, see below, particularly chapter 8, under "Heckius and His Simples."

31. They too had passed into the Albani collections but were probably not regarded as sufficiently interesting for the Adam brothers to purchase for George III in 1763 (cf. below, "Brief Note on the Provenance of Cassiano's Drawings"). Instead they were acquired—along with a large number of other important Albani manuscripts—by the French commissioners in Rome at the time of the French occupation there at the very end of the eighteenth century. A few years later they were all sent to the library of the Medical School at Montpellier, probably around 1804; for the often vague details of this history, see Alessandrini 1978, pp. 48–50.

32. The first published references to these (other than the nondescript entries in Bouteron and Tremblot 1928, p. 223), as far as I know, were Ubrizsy 1980 and Alessandrini, De Angelis, and Lanzara 1985.

33. See chapter 8, under "The Drawings in Paris," as well as the still more detailed discussion in my introductory essay in Pegler and Freedberg (forthcoming).

34. With the exception, perhaps, of the much less comprehensive survey of the mushrooms of Pannonia in Clusius 1601, reproduced with commentary and the original drawings in Aumüller and Jeanplong 1983. But see also Pegler and Freedberg (forthcoming) and chapter 8 below, under "Heckius and His Simples" and note 20 there.

35. Or for that matter in any other form; on the fossils (also of a particular region, namely southern Umbria), see chapter 11 below; and Scott and Freedberg 2000.

36. See chapter 8 below, under "The Drawings in Paris"; and my introduction in Pegler and Freedberg (forthcoming).

37. Cf. chapter 8 below, under "The Drawings in Paris."

38. For the document recording his birth date, see Freedberg 1994b; and Freedberg and E. Baldini 1997, p. 79, note 6; for further biographical data, see as well Freedberg 1989a; and Freedberg and E. Baldini 1997.

39. The best overview of the Roman Jesuit college in these years remains Villoslada 1954.

40. The appointment came immediately in the wake of the publication of his not very distinguished Syriac lexicon, interesting chiefly for his account of how he learned Syriac from the Maronite fathers in Rome, and of life at their college (Ferrari 1622). Ferrari remained a keen defender of the Hebrew language, and of its suppleness and style, as he makes clear in several of his much republished and reedited *Orationes* (see the following note), notably those titled *Hebraicae linguae suavitas, sive De litteris Hebraicis a falsa criminatione vindicatis; Hebraicae musae, sive De disciplinarum omnium Hebraica origine; Hebraicae litteraturae securitas, sive De arguto dicendi genere usurpando; Stylus Hebraicus, sive In hostes brevitatis*; all in *Orationes* 1635, pp. 171–207.

41. Much of his reputation was clearly based on his suave and often florid *Orationes*, first published in 1625 and republished many times in the seventeenth century and occasionally thereafter. See Ferrari 1625; as well as De Backer-Sommervogel, II, 1390–1391.

42. On his work as Barberini gardener, see especially Teti 1642 p. 38, and his own *Aetas Floreae* included in the various editions of his *Orationes* (e.g., Ferrari 1635, pp. 108–20). See the summary of this material now in Freedberg 1996b, and, of course, the authoritative work on the Palazzo Barberini by John Beldon Scott (1991).

43. See Belli Barsali 1981 as well as important but neglected manuscripts in the Vatican Library such as Rome, BAV, Barb. Lat. 1749, 1950 and especially 4265.

44. And also, no doubt, by Pietro Castelli's book of 1625 on the Farnese Gardens on the Palatine (Castelli 1625).

45. For payments see Merz 1991, pp. 326–328 ("*Anhang 4*").

46. Though a few were also done by Claude Mellan and Anna Maria Vaiani. For early admiration—never subsequently repeated, unfairly—of Bloemaert's splendid work for both the *De Florum Cultura* and the *Hesperides* (Ferrari 1633 and 1646), see Evelyn 1662, p. 78.

47. Although most botanists I have consulted concur in identifying these plates as

illustrations of the *Amaryllis belladonna*, it has occasionally been suggested that they in fact show *Cybistetes longifolia*. On Ferrari's struggle to make species distinctions, and the purely external criteria on which he based these, see Freedberg 1994a.

48. Here I use the word "differences" in its general sense (cf. also Balme 1987 for its technical use in the Aristotelian tradition); for the seventeenth-century (and earlier) technical sense, see chapter 13 below and the references in notes 1 and 3 there.

49. See Freedberg 2000, as well as chapter 12 below, under "The Evidence of the Senses."

50. See chapter 6, especially "Cesi Asks for a Microscope," on this subject.

51. Cf. Tongiorgi Tomasi and Ferrari 1986. By "illustration" I here mean a reproductive illustration in a book, to distinguish these plates from the drawings made with the aid of a microscope in the Institut de France just a few years earlier (cf. chapter 8, under "The Drawings in Paris").

52. Even including the highly geometrized illustrations of garden plans on pp. 23, 25, 27, 29, 31, 33, and 216. On these see my "The Failure of Pictures: From Description to Diagram in the Circle of Galileo," forthcoming.

53. Ferrari 1633, lib. IV, chap. 6, "Arte maius Naturae miraculum," pp. 468–502.

54. Ibid., p. 479.

55. The fable revolves around the deity of change in nature, Vertumnus. It is illustrated by the main illustrator of the book, Pietro da Cortona, Poussin's rival in Rome at the time. In the presence of personifications of Nature, Art, and Flora herself, the aging god does his sprightly dance accompanied by three putti, representing dawn, midday, and dusk. Behind them blooms a large Chinese rose, with the Barberini palace rising in the background. For this illustration, its further context, and Poussin's use of it, see Freedberg 1996b.

56. Cf. chapter 12, under "The Evidence of the Senses," as well as chapter 4 below.

57. Even better in Latin, "Verum, ut ingenita foliis asperitas minuta non manibus modo, sed oculis quoque contrectata exploratissima esset; lynceum exploratorem, nempe microscopium, sive conspicillium tubulatum consuluimus," Ferrari 1633, p. 481.

58. Ferrari 1638, p. 479. For the Galilean phrase, see chapter 6, under "Cesi Asks for a Microscope" and note 1 there. It is, of course, an inversion of the even more famous description of the telescope as an instrument that could make "distant things seem near" (cf. chapter 4, under "Della Porta, Galileo, and the Telescope," especially note 2).

59. Ferrari 1633, p. 495.

60. For Ferrari's contribution to the taxonomy of citrus fruit—and to the history of taxonomy more generally—see Freedberg 1992b; and Freedberg and E. Baldini 1997. For centuries it was in fact Ferrari's basic division into three genera—citrons, lemons, and oranges—that dominated the citrus classification (and to some extent still does today, though clearly it has been significantly refined and re-refined).

61. The distinction between what I here, in a broadly modern sense, call species and genera was not, of course, very clear in the seventeenth century. For Ferrari the genera were the three groups of citrus fruit (cf. the preceding note), and the species all his subdivisions of these. In very many cases, what Ferrari regarded as different species would, in the Linnaean sense, have been regarded simply as varieties or hybrids. The same applies to many of the so-called species in the *De Florum Cultura*. For fundamental analyses of the epistemological and historical implications of such terms, see Pellegrin 1986 and Atran 1990. I leave for another occasion a discussion of the natural validity (or otherwise) of such terms, as occurs in the psychological and philosophical literature by writers such as Eleanor Rosch, Brent Berlin, and Frank Keil. See also chapter 13, notes 1 and 3.

62. The question was even further complicated when he had to differentiate species and other kinds of classes that had similar names, as when it came to making a distinction between a *lumia* and a *lima* (not the present-day lime), or between a *limon ponzinus* and a *limon sponginus*, or an "Adam's apple" and an "Apple of paradise," both looking nothing so much as like ordinary lemons.

63. Since they were by well-known artists, art historians had of course already found the preparatory designs for the allegorical plates; but, characteristically, they had not bothered even to look for the drawings of citrus fruits that formed the basis of Ferrari's researches.

64. Ferrari 1646, p. 69.

65. Cf. Turner 1993, pp. 36–37, for a little more information on this still shadowy but clearly very accomplished artist.

66. Olina 1622.

67. Windsor, RL 27628, *Natural History of Birds*, fol. 30.

68. Ibid. verso; signed at the top edge of the sheet; cf. Henrietta McBurney in *Quaderni Puteani* 4, p. 175; *Quaderni Puteani* 4, no. 106, p. 175 and p. 177, pl. 106. The date makes it clear that Leonardi had gone on producing bird drawings for Cassiano long after the publication of the 1622 *Uccelliera*. It should perhaps be noted that this is the only drawing in the whole of Cassiano's Museum on Paper yet to be found with an artist's signature on it.

69. He probably did the best of the mineralogical specimens too, such as those in Windsor, RL, *Natural History of Fossils V* and sheets such as those of the asbestos (fig. 1.19). For his antiquarian drawings (of which there were many too), see especially Turner 1993 and *Quaderni Puteani* 4 (1993).

70. See Freedberg 1993 for a summary.

71. Ibid., p. 391.

72. Ibid.

73. These may well have been the two drawings on vellum illustrated in *Quaderni Puteani* 4 (1993), p. 170, pl. 1 and fig. 2: the first, showing a standing male and female flamingo in the collection of Laurence and Aurea Morshead; the second with details of the tongue, beak, and webbed foot, as well as two flamingos in flight, sold at the Boone Sale, Sotheby's (New York), September 16, 1988, lot 160. A third drawing with details of the head and beak of the flamingo is in the collection of Sven Gahlin, London, and is illustrated here as fig. 1.7. The vellum drawings—at least—are probably those commissioned by Peiresc at the request of the chancellor Méri de Vic and then given to Cassiano by De Vic's son, later archbishop of Auch, when Cassiano passed through Auch in 1625 (see Lhote and Joyal 1989, VIII, p. 59 [letter of February 28, 1629, discussing Cassiano's receipt of a flamingo skin from Peiresc, and the latter's hope that he would eventually be able to send him a live specimen]). See also Solinas 1989a, pp. 102–103; and McBurney in *Quaderni Puteani* 4 (1993), p. 169.

In their letters the two friends discuss everything from the distinctions between male and female birds, to the taste of their tongues, evidently an exceptional delicacy (see also the ex-Boone drawing with details of the tongue illustrated in *Quaderni Puteani* 4 [1993], p. 170, fig. 1.1). The correspondence on the subject of the flamingo runs from March 1627 through May 1628 (Lhote and Joyal 1989, nos. III–IX) and recurs briefly again in October 1634 and January 1635 (Lhote and Joyal 1989, nos. LI and LIX). See especially Cassiano to Peiresc, September 25, 1628, and February, 28, 1629 (Lhote and Joyal 1989, nos. VI–VIII); Peiresc's own treatise on the flamingo, written already in 1605 and probably known to Cassiano, is in Paris, BN, MS Dupuy 669, fols. 39r–42r.

74. Ferrari 1633, pp. 441–442.

75. See especially the refusal to have a monument erected in Galileo's honor, as below in chapter 14 and note 21 there.

76. Ferrari 1646, p. 99.

77. Ibid., p. 360. For more on Pietro della Valle, see chapter 10, under "Illustration and Identification," as well as note 105 there.

78. Rome, BANL, Archivio dal Pozzo, MS XXXIV (32)

79. In Rome, BANL, Archivio dal Pozzo, MS VI (4), fols. 323r–435v. Cf. also Freedberg 1989a.

80. Bound into the front of these notes in Rome, BANL, Archivio dal Pozzo XLII (39), is the detailed contract for the publication of the book, and much else relating to its financing and its illustration.

81. See Jouanny 1911, pp. 110–163, nos. 53–69. Cf. Freedberg 1989c; and Freedberg in Freedberg and E. Baldini 1997.

82. Siena, AS, Balia 795, no. 20, October 6, 1646; cf. also the copy of the *Hesperides* specially bound for the Balia of Siena still preserved in the Biblioteca Communale there, pressmark B.LXXIII.A.7.

83. Cassiano in Rome to Cesi in Acquasparta, August 15, 1622 (*Carteggio*, no. 633, p. 770): "le invio un libro d'Uccelli stampato da un giovine di casa più per prova dei

rami che io vo mettendo insieme, per veder se potessi con un po' di spesa e diligenza dar qualch'aiuto alli scritti di questa materia, che per altro."

84. Cf. chapter 5 below, under "The Publication of the Assayer."

85. Cf. the letter from Cesi to Cassiano of October 4, 1622 (*Carteggio* no. 638, p. 773), in which he says of the *Uccelliera:* "Questo mi goderò con quella maggior sodisfattione che V.S. può immaginarsi, come cosa sua; nella quale sin hora scorgo un esquisitissima diligenza et un'espression mirabile de' soggetti rappresentati, che più niun non credo possa in alcun modo haversi, e sarà senza dubbio di gran giovamento all'Istoria naturale, base e fondamento reali della buona e non mascherata filosofia. E doveriano tutti i studiosi haverne non poco obbligo, con gusto particolarmente de' Signori Compagni, quali meco vedendola fervente non solo nel godimento delle scienze, ma anche nel procurare continuo giovamento a questa parte di esse, tanto più utile quanto abbandonata, sarà di non poco momento per esempio, e di calore all'imitatore."

86. All but one preserved in the library of the old Medical School at Montpellier. Montpellier, BEM, MS H. 170, fols. 85r–87v and 120r–121r; MS H. 319, fols. 108r–115r; and MS H. 267, fols. 207r–212r. The first is only recorded secondhand in the account by Johannes Faber in *Thesaurus*, pp. 697–698. Cf. note 89 below.

87. See chapter 2, under "The Legation to France and Spain", and chaper 9 under "Cassiano in Spain" and note 85 there.

88. Much of the *discorso* on the toucan, originally accompanied by the drawing seen by Skippon, was published by Cassiano's friend Johannes Faber just two years later, in *Thesaurus*, pp. 697–698. The discourse on the toucan was sent by Carlo Antonio to Johannes Faber in Rome and was used by the latter as the basis of his description of the *Picus americanus* in his commentary on the animals of Mexico first published in 1628 and reissued in 1649–51 (*Thesaurus*, pp. 697–704). See Carlo Antonio dal Pozzo to Johannes Faber, February 1, 1628, in *Carteggio*, no. 621, pp. 760–761. On all of this, see also McBurney 1992a, pp. 349–350.

89. Montpellier, BEM H. 319, fols. 108r–115r and H. 267, fols. 207r–212r (dated 1642); H. 170, 120r–121v (probably dating from two years later); and H. 170, fols. 85r–87v (dated 1635; published by Solinas 1989a, pp. 128–129). See also chapter 10, notes 27 and 28.

90. And discovered that it was female because of a number of eggs still attached to its intestine. Montpellier, BEM, H. 170, fol. 87v. Reproduced in Solinas 1989a, p. 129. This insistence on anatomizing in order to examine issues of reproduction was central, in ways we shall soon see, to the Linceans' approach to the study of natural history.

91. London, Courtauld Institute Galleries, Witt Collection, 3218; and Windsor, RL 28739 (leg and feather) and RL 28740 (head).

92. On this trip, see chapter 2, under "The Legation to France and Spain", and chaper 9 under "Cassiano in Spain" and note 85 there. For the passage from Skippon, see under "The Discovery," above, and Skippon 1752, p. 693.

93. For the flamingo drawings and the letters associated with them, see note 73 above.

94. See the important manuscript in Montpellier, BEM, MS H. 267, fols. 1–26, containing extracts from the "Libro originale di Leonardo da Vinci MS che si trova a Milano nella Libreria Ambrosiana o Borromea." On Cassiano's efforts to prepare an edition of Leonardo's *Trattato* (including his collection of drawings by Poussin to illustrate it), see Carusi 1929–30; Stumpo 1986, p. 211; and Bell 1988, pp. 115–121. The planned edition was eventually published in Paris in 1651 by Cassiano's friend Paul Fréart de Chantelou.

95. See also McBurney 1992a.

96. The descriptions cover much more than modern ornithologists might, from ways of sewing their skins together for utilitarian purposes, to details of their flight, their sounds and song, their seasonal behavior, and their habitat; as always there is a good deal on culinary and venational matters.

97. On the Palazzo in Via dei Chiavari and its contents, see now the excellent work by Sparti 1992.

98. For Cassiano's acquisition of Cesi's library and the contents of his museum, see Carutti 1883, pp. 77–82; Gabrieli 1989, pp. 79–90; Nicolò and Solinas 1985; and Sparti 1992.

99. Cf. chapter 8, under "The Drawings in Paris," below, and note 53 there; as well as Freedberg in Scott and Freedberg 2000, p. 13, and p. 71, note 15.

100. Dati 1664, sigs. C4v–D1.

101. Ibid., sig. C4v.

102. The correspondence, first excellently discussed and analyzed by Lumbroso 1874, is largely preserved in the volumes that entered the library of the present Accademia dei Lincei from the collection of the dukes of Aosta in 1973. But see now, for a thorough listing, Nicolò 1991.

103. Dati 1664, sig. C3.

104. Ibid., sigs. C4–C4v.

105. From people such as the scurrilous yet shrewd biographer Janus Nicias Erythraeus to the Frenchman Jean-Jacques Bouchard (on the first, see Garboni 1899, and Freedberg 1992a; on the second see Kanceff 1976–77; on Cassiano and libertinism more generally, see the still authoritative survey by Pintard 1943.

106. For the history and provenance of these drawings, see especially McBurney 1989a–c; and Francis Haskell and Henrietta McBurney in Freedberg and E. Baldini 1997, pp. 9–24.

107. See, for example, Vermeule 1960 and 1966; see also Jenkins 1987 and Jenkins 1989 for the drawings only recently rediscovered in the British Museum.

108. Dati 1664, especially sigs. C3–C3v, referring to his "famous volumes of Natural History" as well as to his volumes of drawings after the antique.

109. With the exception—as so often in this story—of Giuseppe Gabrieli in the 1930s and 1940s (cf., for example, his pursuit of the drawings of mushrooms owned by Cassiano as recounted in Gabrieli 1989, pp. 1436–1445). Another notable exception was Bromehead, who made a useful study of some of the geological drawings before their rediscovery in the 1980s (cf. Bromehead 1947a and 1947b).

110. Cf. Haskell and McBurney in Freedberg and E. Baldini 1997, p. 17.

111. On the history of the drawings, see McBurney (1989a–c); and Haskell and McBurney in Freedberg and E. Baldini 1997, pp. 9–26.

112. New York (Sotheby's), September 16–17, 1988; *Drawings from the Estate of James R. Herbert Boone*.

113. Cf. "The Discovery" above, as well as chapter 9, below, under "Cassiano in Spain"; Gabrieli 1989, pp. 1459–1464; and the longer discussions in chapters 8 and 9 below.

## CHAPTER TWO

1. Cesi in Rome to Stelluti in Fabriano, July 17, 1604; *Carteggio*, no. 15, p. 40.

2. Stelluti in Acquasparta to Galileo in Florence, August 2, 1630; *Carteggio*, no. 1011, p. 1217. For the next part of this letter, about Cesi's unassigned *Nachlass*, see under "The Legation to France and Spain" in this chapter.

3. In Gabrieli (1989, p. 3 (originally published in 1930). In the years between 1924 and 1942, Gabrieli published a remarkable series of articles devoted almost exclusively to Cesi, his friends, and the Academy of Linceans (published chiefly in the *Rendiconti* and *Memorie* of the Accademia Nazionale dei Lincei). They are closely based on documents preserved in the present-day archives of the Accademia dei Lincei in Palazzo Corsini in Rome, and occasionally elsewhere. Few archivists have transcribed so many documents so faithfully, or brought together so much material with such intelligence, acumen, and grace. Thanks to the efforts of librarians at the Accademia dei Lincei—chiefly Ada Alessandrini and Marta Gianni—most of Gabrieli's articles were collected and published as Gabrieli 1989. Together with Gabrieli's massive, accurate, and wonderfully annotated *Carteggio*, published between 1938 and 1942, this work provides the indispensable basis for any study of Galileo and the Linceans—and therefore for the present book too.

4. Gabrieli 1989, p. 3.

5. All the sources for Cesi's life are reliably brought together in Gabrieli 1989, pp. 3–26 and in subsequent chapters.

6. See chapter 11 below.

7. For an excellent discussion of the implications of both the mythological Lyncaeus and the eyesight of the lynx, see now Lüthy 1996.

8. It has been investigated in loving detail by Gabrieli, most devoted and reliable of historians (see note 3 above). The most important of Gabrieli's predecessors were undoubtedly Odescalchi 1806 and Carutti 1883. Gabrieli also had at his disposal the two

large manuscript volumes titled *Memorie dei Lincei*, written by Francesco Cancellieri, on deposit in the library of the modern Accademia Nazionale dei Lincei in Palazzo Corsini.

9. "Arcanarum sagacissimi indagatores scientiarum, et Paracelsicae dediti disciplinae," as Heckius wrote on behalf of the other Lincei to Thomas Merman on February 17, 1604 (*Carteggio*, p. 30).

10. Stelluti and De Filiis seem both to have been born in 1577; Heckius is generally supposed to have been born in that year too; but Rienstra 1968 showed, on the basis of the Deventer Baptismal records, that he was in fact born in 1579, and that for one reason or another he often misstated his own age.

11. Further details in chapter 8 below, under "Heckius and His Simples"; see also Carutti 1883 and Van Kessel (1977).

12. In calling themselves Linceans, or Lynx-eyed, they were also, as noted above, alluding to the legendarily keen-sighted Argonaut, Lyncaeus.

13. Gabrieli 1989, II, p. 1053; cf. also the two manuscripts titled *De Nostri Temporis Pravis Haereticorum Moribus Libri XII*, Naples, BN, MSS IV.H.102 and XI.B.2; as well as Rienstra 1968 for further references to his anti-Protestantism.

14. A typical instance is provided in the foreword to one of his notebooks on butterflies, insects, and snakes, where he describes how "collecting these things was often an inconvenient and difficult task. Indeed my companions often thought me ridiculous as I reined in my horse in mid-course and descended in order to hunt down a single little butterfly, or beetle, or moth" (Montpellier, BEM, MS H. 506, fol. 1v).

15. On Heckius's notebooks, see especially chapter 8 below, under "Heckius and His Simples."

16. This summary of Heckius's travels and vicissitudes is based on Carutti 1877; Gabrieli 1989, pp. 1053–1116; Uhlenbeck 1924; Rienstra 1968; Van Kessel 1976; Schulte van Kessel 1985; Ricci 1988; as well as my own reading of the letters collected in the *Carteggio* and the Heckius manuscripts in Montpellier (summarized in Alessandrini 1978), Florence, Naples, and Rome.

17. Heckius in Prague to Cesi in Rome, January 5, 1605; *Carteggio*, no. 20, p. 52.

18. See the following chapter for a detailed discussion.

19. See chapter 11 below.

20. Cf. chapter 13 below, under "The Reduction to Order."

21. Already at the session of the Academy on October 15, 1603, Cesi was recorded as having given a "Phitosophica lectio," in which he demonstrated that the plant (in general) was to be defined as an "animal vegetativum" (Gabrieli 1989, p. 510). Cf. also Belloni Speciale 1987, p. 62. For the later, critical use of the term "phytosophical" in Cesi's work, see especially chapter 13.

22. Issues of classification will arise in each of the chapters of this book, but especially in chapters 11, 12, and 13.

23. For more on this, see also chapter 9 below.

24. One must, of course, suppose that it was not really epilepsy from which Colonna suffered, but rather some other kind of neurological ailment—perhaps a form of manic depression—that valerian (the Dioscoridean *phu* illustrated on p. 41 of the *Phytobasanos*) helped calm (fig. 9.1). See also chapter 4, under "Botany and Astronomy: The Academy Expands."

25. Cf. chapter 9 below, under "Naples, 1604" and "Hernández and His Manuscripts."

26. On Cesi's visit to Naples see especially Gabrieli 1989, pp. 9–10 and passim, but especially pp. 1497–1548. Much light is also cast on the Neapolitan connections by the *Carteggio* for the years between 1604 and 1610. On the Linceans' work on the material assembled by Hernández and redacted and copied by Recchi, see especially chapters 9 and 10 below.

27. See the useful bibliography of the editions of Della Porta's works in Gabrieli 1989, pp. 702–729.

28. For the editions of his works on these subjects, see Gabrieli 1989, pp. 713–718.

29. The adverbial diminutives are lovely here: "Sed interim quantuluscumque sim, quantulumcumque valeo, vobis me expono et dedo: si non doctrina, saltem affectu merear." Della Porta in Naples to Cesi and the Linceans in Rome, June 25, 1604; *Carteggio*, no. 14, p. 35.

30. Giovanni Battista Della Porta, *De distillatione libri IX* (Rome: Ex typographia Rev.

Camerae Apostolicae, 1608), sig. ai.

31. It was thus the second publication to appear under the sponsorship of the Lincei (the first being Heckius's little treatise on the supernova of 1604 published in Rome in 1605; cf. chapter 3 below for the history of this work).

32. Giovanni Battista Della Porta, *Elementorum Curvilineorum libri tres* (Rome: Zannetti, 1610).

33. Giovanni Battista Della Porta, *De Aeris Transmutationibus* (Rome: Zannetti, 1610).

34. Cf. chapter 4 below, under "Rome, 1611," for a full account.

35. Schreck is another figure who merits vastly more attention than he has received so far, particularly for his work in China; as always, the groundwork has already been laid by Gabrieli (cf. Gabrieli 1989, pp. 1011–1152; cf. also chapter 4 below, under "Botany and Astronomy," and chapter 9 below, under "Schreck and Faber at Work").

36. Already in 1606 Faber had earned the devotion of the Flemish artist Rubens for having cured him of a life-threatening case of pleuritis. On Faber, see not only Gabrieli 1989, pp. 1177–1253, 1577–1584; but also the useful articles by Gabriella Belloni Speciale in DBI 43, pp. 687–689; and De Renzi 1994. Ct. also chapters 4, 9, and 10 below, passim.

37. For a hitherto unidentified portrait of the attractive figure of Cesarini, and for additional material to that supplied by Gabrieli 1989, passim, see Freedberg 1994c.

38. For a gripping summary of the Linceans' preparation of the *Assayer* for publication, as well as of many of the issues involved, see also Redondi 1987. Whatever the shortcomings of the main theses of Redondi's much-discussed book, the section on the *Assayer* remains a first-rate account of the Linceans' involvement in its publication and of their support of Galileo at a critical time.

39. On all this, see Redondi 1987, especially pp. 45–51.

40. Recorded in detail in the still unpublished manuscripts in the Vatican Library, Rome, BAV, MSS Barb. Lat. 5688 and 5689. See also chapter 9, note 85 below.

41. Francis Bacon, *Saggi morali* (London: John Bill, 1618).

42. Cassiano in Fontainebleau to Cesi in Rome, August 1625; *Carteggio*, no. 863, pp. 1059–1060.

43. Though the Englishman was of course aware of Galileo's work from an early stage; for an important example, containing reference not only to critical issues around the perception and behavior of heavenly bodies but also to Galileo's work on the tides, see the *Novum Organum*, bk. 2, Aphorism 46. Cf. Galileo, *Opere*, XII, pp. 255 and 450; as well as Drake 1978, p. 274. I am grateful to Eileen Reeves for these interesting references.

44. Henceforward to be referred to as the *Tesoro Messicano*, and abbreviated as *Thesaurus*; Cf. chapters 9 and 10. The basic article, needless to say, was written by Gabrieli in 1940 (reproduced in Gabrieli 1989, pp. 373–383); note 1 contains references to the many earlier articles he wrote about the contents of and contributors to this central work of the Linceans). See also the articles by F. Guerra, C. Sánchez Téllez, and G. B. Marini Bettólo in Cesi, *Atti* (1986), pp. 307–348; and G. B. Marini Bettólo, "A Guide for the Reader of the Mexican Treasure" (printed to accompany the 1992 facsimile of the *Thesaurus*), Rome: *Istituto Poligrafico e Zecca dello Stato*, 1992.

45. Stelluti in Acquasparta to Galileo in Florence, August 2, 1630; *Carteggio*, no. 1011, p. 1217.

46. For Cassiano's acquisition of Cesi's library and the contents of his museum, see Carutti 1883, pp. 77–82; Gabrieli 1989, pp. 79–90; Nicolò and Solinas 1986; and Sparti 1992.

47. For his chilling words of opposition to the monument to Galileo, see chapter 14 below.

48. Especially in the case of his work on fossils, as discussed in chapter 11 below.

49. Stelluti 1637b.

50. See chapters 9 and 10 for the astonishing history of the publication of the *Thesaurus*.

51. Alfonso de Las Torres; see chapter 9 below, under "Sponsorship at Last."

## CHAPTER THREE

1. See Clark and Stephenson 1977, pp. 172–176, on both this nova and that of 1572.

2. For excellent accounts of this once rather neglected event, see U. Baldini 1981;

Ricci 1988; and now the full discussion (with a useful further bibliography) in Reeves 1997, especially pp. 57–90 ("Neostoicism and the New Star"). Reeves takes a rather different view than my own of the implications of the nova for Galileo's thought. She emphasizes the neo-Stoic dimensions of the discussions about the nova, and in setting out the optical and atmospheric implications of Galileo's view, her findings valuably supplement my own presentation here. She also emphasizes the ways in which Galileo associated the generation of the nova with the phenomenon of the aurora borealis, a view quite explicitly anti-Aristotelian. Surprisingly, however, Reeves does not mention the treatise by Heckius that both motivates the present chapter and forms its central feature.

3. Cf. Ricci 1988, p. 112.

4. Clark and Stephenson 1977, pp. 172–206; Hoskin 1977, pp. 22–28.

5. These were never published, though some of Galileo's notes and a summary description of their contents survive; Galileo, *Opere*, II, 277–284; see also Favaro, *Padova*, I, pp. 275–282.

6. See especially U. Baldini 1981.

7. See under "Hawks and Doves" below, as well as the excellent note in Ricci 1988, pp. 123–24, note 33.

8. On the dissolution of the solid spheres, see Donahue 1981 and Rosen 1985; cf. Favaro, *Studio*, pp. 275–305; Favaro, *Carteggio*, pp. 283–285; and Brahe, *Scripta*, vol. III, especially pp. 279–288.

9. Drake (1957, p. 13.

10. For Galileo's early Copernicanism (from around 1597 on), see, inter alia, Ferrone 1984, pp. 242–3 and notes 19 and 20. For a contrary view, see Drake 1976, p. xi (his astronomical opinions before 1610 were not "dogmatically Copernican") as well as (19pp. 27–31, suggesting an anti-Copernican about-face around 1605, with the second edition of the *Dialogue* of "Cecco di Ronchitti" discussed below.

11. Cf. Ingegno (1978), as well as the important reference in Kepler, *Werke*, I, p. 253.

12. On Bruno's cosmological views, see, for example, Ingegno 1978.

13. "*Maiori forsan cum timore sententiam in me fertis quam ego accipiam*" (V. Scampanato, *Documenti della Vita di Giordano Bruno*, Florence: Olschki 1933, VI, *Documenti romani*, XXX, 202).

14. Galileo, *Opere*, II, pp. 277–279.

15. Cf. the excellent discussion of this topic by Reeves (1997, especially pp. 61–65), who also offers a full account of the neo-Stoic and anti-Aristotelian implications of the view of the cosmos as a single medium stretched from the earth to the most remote stars, as well as of the way in which the moisture constantly exuded from the earth turns first into air, and then into fire. See also Barker 1991, pp. 135–153 for the neo-Stoic dimension of views such as these.

16. Reeves 1997, p. 68.

17. In Galileo, *Opere*, II, pp. 277–284.

18. This useful explanation of parallax is taken from Shea 1977, p. 78.

19. Published with an excellent commentary by U. Baldini 1981—although by avoiding the question of its actual physical composition, van Maelcote also avoided drawing seriously anti-Aristotelian conclusions.

20. Lorenzini 1605; see now also Poppi 1992. On the relative roles of Cremonini and Lorenzini in the composition of the book (it was basically Cremonini's), see now Drake 1976, pp. 9–11.

21. *Dialogo* (1605); available in Galileo, *Opere*, II, pp. 311–334; Favaro 1881; and Drake 1976.

22. Even to those who are expert in Italian, it takes a while to adjust to the quirks and idiosyncrasies of the speech here, while many words—especially the many expletives—are either totally local or simply *hapax legomena*.

23. On the taste for this form, its literary origins, and its contemporary implications, see also the useful assessment by Drake 1978, pp. 48–49.

24. For an opposing view of its authorship (attributing it to Girolamo Spinelli), see Marisa Milani, "Galileo Galilei e la letteratura padovana," in Giovanni Santinello, ed., *Galileo e la cultura padovana* (Padua: CEDAM, 1992), pp. 179–202. Favaro 1881 and Drake 1976 make the Galilean case clear enough.

25. Galileo, *Opere*, II, pp. 311–312. The translation of this and all following passages

from the *Dialogue* is my own. For an alternative, rather more pedestrian one, see Drake 1976 with an illuminating commentary.

26. "Experience," of course, is a critical word here, since the Italian "esperienza" also then doubled for the notion of experiment; cf. U. Baldini 1981, p. 85.

27. Ibid., p. 315.

28. Literally "A cobbler can't talk about buckles." "No seto, que un zavattin no pò faellar de fibbie," or—as Favaro translated it, "Non sai tu che un ciabbatino non può ragionare di fibbie," Galileo, *Opere*, II, p. 315. The proverbial equivalent, of course, is the Plinian "sutor ne ultra crepidam," i.e., "Cobbler, stick to your last."

29. Cf. chapter 5, under "Sunspots," and chapter 11, under "Cesi and His Views" below.

30. Galileo, *Opere*, II, p. 315.

31. Ibid., p. 324.

32. Ibid., pp. 315–316.

33. A clear allusion to the Aristotelian treatise *De generatione et corruptione*; Galileo, *Opere*, II p. 317.

34. Ibid., p. 318.

35. Ibid. In the Verona edition of the work a precaution is taken here by adding that the people who think this "are wild-eyed." Other prudent changes are made in this second edition of the pamphlet too (cf. Drake 1976, p. 42).

36; Galileo, *Opere*, II, pp. 321–322).

37. *Decreta, canones, censurae et praecepta Congregationum Generalium Societatis Jesu* (Avignon: 1830), III, p. 341; repeated in the 1599 *Ratio Studiorum*, regulating instruction in all the Jesuit colleges.

38. U. Baldini 1981.

39. On all these works, see the useful note by Ricci 1988, pp. 123–124; as well as Drake 1976, pp. xv–xvi.

40. Drake 1978, p. 119.

41. For a useful overview, see Drake 1978, pp. 120, 460, and 102–116, on their earlier discussions of mathematical issues. Ferrone 1984 deals with their relationship—as well as with Cesi—in still greater depth.

42. For a useful and clear summary of the complexity of anti-Aristotelian positions—even in the most progressive contexts—in the early seventeenth century, see Mercer 1995.

43. For Galileo's view on this, see now especially Reeves 1997, pp. 65–68, inserting this view into its neo-Stoic context, once also shared by Bellarmine.

44. That Heckius's *De Stella Nova Disputatio* was already being printed by February 1605 emerges from a letter that Cesi wrote to him on April 1, 1605 (*Carteggio*, p. 59).

45. Ricci 1988. Even Ferrone (1984, p. 246) admitted that although Heckius's treatise was a "most precious work," he had been unable to consult it. Until 1998 I knew of only four copies, all in Rome: one in the Biblioteca Corsiniana, one in the Casanatense, and two in the Biblioteca Angelica there. In 1998 Seth Fagen drew my attention to two more copies, one of which, carrying the crest of Robert Bellarmine on its binding, he acquired for the bookseller Martayan Lan in New York (then sold to Jonathan Hill; see his catalogue no. 125, 1999).

46. Cf. preceding note.

47. The problem of how to reconcile the merits of chastity versus the physical need for sex recurs over and over again in his letters and in his many unpublished treatises. On this subject in Heckius, see also the useful summary in Schulte van Kessel 1985. In addition to the concerns expressed in many of the letters, there is also a rather juvenile—but certainly racy—comedy, unstudied by any scholar, about the problem of how to satisfy one's sexual needs while traveling away from one's wife, in Florence, Biblioteca Laurenziana, MS Ashburnham 1209, fols. 46–82 (entitled the *Donna Pudica*).

48. Heckius in Prague to Cesi in Rome, January 5, 1605; *Carteggio*, no. 20, p. 51; cf. his letter to the Lynxes dated December 19, 1604; *Carteggio*, no. 18, p. 46.

49. See, for example—just one out of several—the letter Cesi wrote to Heckius in Prague on July 2, 1605; *Carteggio*, no. 31, pp. 84–87, where love and firmness are almost equally balanced. See now also Biagioli 1995 for the homoerotic bond between Heckius and Cesi.

50. Cesi's words, typically firm and consoling, are worth quoting here: "Quod de

fraternitate reliquenda valedicendo, absit absit absit etiam huius rei minima suspicio. De nuptiis et effeminata requie ac molli, procul procul a tuo pectore et mente cogitatio effugiat. Nulla nulli lyncaeo desideranda requies nisi cum fratribus." In Cesi's letter of March 19, 1605, to Heckius in Prague; *Carteggio*, no. 22, p. 56.

51. Heckius says he saw it on October 11 (Heckius 1605, p. 9; Kepler saw it on October 17—and again on October 21 (Kepler, *Werke*, p. 159 [from the *De Stella Nova* of 1606]); cf. Kepler's *Gründtlichen Bericht von einem ungewönhlichen Neuen Stern, welcher im October ditz Jahr erstmahlen erschienen*, of the year before, in Kepler, *Werke*, I, pp. 390–394.

52. In addition to Heckius 1605, p. 9, see also the relevant passages from Rome, BANL, Archivio Linceo XI, fols. 16–23 (the introduction to his treatise on the new star) published by Gabrieli in *Carteggio*, no. 21, pp. 54–55 (January 24, 1605).

53. Cf. chapter 2 above, under "Four Friends"; and especially chapter 8 below, under "Heckius and His Simples."

54. On this lost work, see also Carutti 1877, p. 13; Gabrieli 1989, p. 1098; and Van Kessel 1976, pp. 119 and 121.

55. Heckius in Prague to Cesi in Rome, January 5, 1605; *Carteggio*, no. 20, p. 51.

56. Ibid., p. 52.

57. Rome, BANL, Archivio Linceo XI, fols. 16–23; see also the relevant passages transcribed and published by Gabrieli in *Carteggio*, no. 21, pp. 54–55 (January 24, 1605).

58. Ibid., fol. 16; Heckius in Prague to Cesi in Rome, January 24, 1605, *Carteggio*, no. 21, p. 55.

59. Cf. Heckius 1605, pp. 15–16.

60. Ibid., pp. 14–15.

61. Heckius 1605, pp. 13 and passim. For Cesi's moderation of Heckius's attacks by deleting the stronger language from the draft manuscript he received (on which more below), see Ricci 1988, 131–132 (cf. also Ricci 1988, note 25, pp. 120–121, for Heckius's views on other worlds and on the Copernican world system in an unpublished treatise *De mirabilibus creaturarum Dei* dated 1601 and preserved in Rome, BANL, Archivio Linceo XXI, fols. 8–146).

62. Cf. Ricci 1988, p. 117.

63. Rome, BANL, Archivio Linceo XI, fol. 21v.

64. Heckius 1605, p. 15.

65. On these, cf. this chapter under "Scripture and the Fluidity of the Heavens"; and chapter 5 below, under "Cesi versus Bellarmine."

66. Rome, BANL, MS Archivio Linceo XI.

67. Compare Rome, BANL, Archivio Linceo XI, fol. 18r, with Heckius 1605, pp. 12–13; the changes summarized in Ricci 1988, pp. 126–127.

68. E.g., Rome, BANL, Archivio Linceo XI, fol. 18; cf. Heckius 1605, p. 13.

69. Rome, BANL, Archivio Linceo XI, fol. 20v.

70. Cesi in Rome to Heckius in Prague, March 19, 1605; *Carteggio*, no. 22, p. 56.

71. Gabrieli 1989, p. 1112, also cited in Carutti 1883, p. 17.

72. Heckius 1605, p. 4.

73. "Non tantum novas in caelo stellas hactenus non visas meus ad vos misit calamus, sed nova etiam in terris genita animantia nunc etiam viscera telluris acutius intuens has nulli descriptas inveni plantas novas etiam prius cognitas," he wrote in the dedicatory letter to the third volume of his *Fructus itineris ad Septentrionales*, preserved in Montpellier, BEM, MS H. 507, fol. 8r. As always, the emphasis on sharpness of sight is central and critical.

74. See Cesi's letter from Rome to Heckius in Prague, July 2, 1605; *Carteggio*, no. 31, p. 84.

75. Cf. chapter 8, under "Heckius and His Simples" and note 19 there; chapter 13, under "Precedents" and note 12 there; as well as *Carteggio*, no. 11, pp. 32–33 (to Clusius in Leiden, March 20, 1604), and no. 42, p. 100 (to Mercuriale in Pisa, April 1, 1606). For the connections with Mercuriale, see Rome, BANL, Archivio Linceo XI, fol. 28 and Archivio Linceo XVIII, fols. 18–20; as well as Gabrieli 1989, pp. 434, 592, 1105, 1107, and 1196. On Mercuriale's support for the young Galileo, see Biagioli 1993, pp. 21–23, 164, 167.

76. See especially letters such as those in *Carteggio*, no. 24, pp. 59–70, as well as

those cited in the following note. See also the *Lynceographum* (Rome, BANL, Archivio Linceo IV, fols. IV-242v, to be published soon by Anna Nicolò) and the *Gesta Lynceorum* (Rome, BANL, Archivio Linceo III, fols. 1–29 and Archivio Linceo III bis, fols. 5–88v).

77. Cesi to Stelluti and De Filiis, April 10, 1605; *Carteggio*, no. 24, especially pp. 66–68; Cesi in Rome to Heckius in Prague, April 30, 1605, the Lynxes in Rome to Heckius in Prague on May 15 and 17, 1605; and Cesi in Rome to Heckius in Prague, June 11, 1605, July 2, 1605, August 13, 1605; *Carteggio*, nos. 26–31, 34, pp. 73–87, 89–90.

78. For the "balbutientes quidam novi philosophi" see also under "Cesi Intervenes" above (Rome, BANL, Archivio Linceo XI, p. 18r, Heckius 1605, p. 13).

79. For a rough early outline, see Naples, BN, MS XII.E.4, fol. 13; for the draft of the *De caeli unitate tenuitate fusaque et pervia stellarum motibus natura* sent in the form of a letter to Cardinal Bellarmine in 1618, see Rome, BANL, Archivio Linceo IV, fols. 552–558; the late, published version only appeared, ironically enough, in Christoph Scheiner's massive *Rosa Ursina* of 1630 (Scheiner 1630), pp. 771–782. See also chapter 5, under "Cesi versus Bellarmine."

80. Cf. the very good summary in Ricci 1988, p. 119 note 23, as well as the useful extracts on pp. 128–129 citing relevant passages from Cesi 1618.

81. For the full exposition of these views, see Scheiner 1630, pp. 771–782; for the early draft, see Naples, BN, MS XII.E.4, fol. 13. See also the reprint, with notes, in Altieri Biagi and Basile 1980, II, pp. 9–35.

82. As Bellarmine would eventually point out to him with much sternness; see chapter 5 below, under "Cesi versus Bellarmine."

83. Archivio Linceo MS 11, fols. 19v–20.

84. Rome, BANL, Archivio Linceo XI, fol. 19v.

85. Heckius 1605, p. 19.

86. Cesi in San Polo to Galileo in Florence, June 20, 1612; *Carteggio* no. 130, pp. 238–239; and then Cesi in Rome to Galileo in Florence, July 21, 1612; *Carteggio*, no. 142, pp. 252–254.

## CHAPTER FOUR

1. Cesi in Rome to Heckius in Prague, June 11, 1605; *Carteggio*, no. 29, p. 81. Cf. Cesi's remarkable letter, more intimate yet firmer than ever, written to Heckius on July 2, 1605; *Carteggio*, no. 31, pp. 84–87.

2. For a recent summary of the well-known Dutch role in the development of the instrument whereby distant things could be seen as if nearby, to use the frequent phrase of the times, see Sluiter 1997. De Waard 1906 and Van Helden 1977 remain the basic authorities here, however. Sluiter briefly traces the much-told history of Hans Lipperhey's presentation of his instrument to Count Maurice of Nassau in The Hague in 1608, and then reminds his readers of the fact that this was known to Ambrogio Spinola at the time, and then available at the court of the archdukes in Brussels well before April 1609, when the papal nuncio (later cardinal) Giulio Bentivoglio wrote about it to Scipione Borghese in Rome (Sluiter 1997, pp. 231–232); soon Galileo's friend from Padua, Paolo Gualdo, and Cassiano's close friend the Paduan cleric Lorenzo Pignoria were corresponding about it—as they would for years to come. Cf. Galileo *Opere* X, p. 250; chapter 5 below, under "Sunspots" and notes 5–8 there; and note 54 below.

3. Although Galileo did not invent the telescope, he was the first to grind long-focus objective lenses and to equip his instruments with aperture rings (see Van Helden 1984, p. 155, with further bibliography).

4. Della Porta in Naples to Cesi in Rome, August 28, 1609, in *Carteggio*, no. 50, pp. 114–115 (p. 115).

5. In fact, Della Porta was so heated about this that he got his own reference to his earlier discussion of the use of telescoping convex and concave lenses wrong. It ought actually to be to the *eighth* (not the ninth) book of his little treatise *De specillis* in the *De refractione optices parte*, or to the *eighteenth* chapter of this work. There is also a further reference to this invention in the section *De catoptricis* of the *Magia Naturalis*, where he again discussed the matter of magnification by means of lenses (as noted by Favaro in *Galileo, Opere*, X, p. 252, note 1). All of these sources were cited by none other than Kepler in a letter to Galileo of April 19, 1610 (*Galileo, Opere*, X, pp. 323–324).

6. Della Porta in Naples to Cesi in Rome, August 28, 1609, in *Carteggio*, no. 50, pp. 114–115 (p. 115). Here it should perhaps be noted that already on October 9 of the pre-

vious year, Della Porta had written a letter to Cesi in which he had drawn a rudimentary support for a parabolic mirror (Della Porta in Naples to Cesi in Rome, October 9, 1608, in *Carteggio*, no. 49, pp. 112–113).

7. Della Porta in Naples to Cesi in Rome, first months of 1610, in *Carteggio*, no. 54, pp. 147–149.

8. For Galileo's considerable exaggeration of the size of the Albategnius crater, see Gingerich 1975; and now Bredekamp 1997 for important suggestions about the reasons why Galileo might have so enlarged it.

9. On his strategy in thus naming the stars, see now Biagioli 1993.

10. Cf. *Galileo, Opere*, X, pp. 323–324.

11. Francesco Stelluti in Rome to Giambattista Stelluti in Fabriano, September 15, 1610, in *Carteggio*, no. 57, pp. 152–154 (pp. 152–153).

12. Della Porta in Naples to Faber in Rome [?], 1612 [?], in *Carteggio*, no. 202, pp. 308–310 (p. 308).

13. For a nuanced view of the critical figure of Christopher Clavius, see now Lattis 1994, with full references to the earlier literature on him.

14. Such were the kinds of arguments put forward by Francesco Sizzi, a supporter of astrology as well, in his *Dianoia astronomica*, published in 1611 but already complete in 1610. Cf. Galileo, *Opere*, III, p. 213.

15. G.B. Lauri, ed. J. Riquius, *Theatri Romani Orchestra . . . . Dialogus de Viris su aevi doctrina illustribus Romae MDCXVIII* (Rome: Andrea Phaeus, 1625), p. 39.

16. Already in the in the first edition (1570) of his commentary on Joannes a Sacro Bosco's basic astronomical text, the *In Sphaeram*, Clavius's support for Ptolemy and his hostility to Copernicanism was clear. Clavius 1611–12, III, p. 301. Texts such as Psalms 19:5–6 and 104:5, as well as Ecclesiastes 1:4–6, made it clear that the earth was immobile and that the sun and stars moved around it (ibid., p. 106).

17. For this very commonly stated objection, see, for example, Galileo, *Opere*, pp. 104–105 and 134–135.

18. As in Drake 1957, p. 75.

19. BAV, Urb. lat. 1079, fol. 292v.

20. In his 1611 edition of Johannes a Sacrobosco's *In Sphaeram* (Clavius 1611–12, III, p. 75). Cf. note 13 above.

21. Several accounts of the meeting—which lasted from 8 in the evening till 7 (despite Gabrieli's claim that it lasted until midnight) in the morning—survive, including those published in Gabrieli, *Carteggio*, pp. 158–162. These include the fundamental accounts in Orbaan, 1920, p. 211 (citing BAV Urb. lat. 1079, *Avvisi di Roma*, fols. 292–293 and Sirturi, 1618, pp. 27–28); for Lagalla's account and his skepticism about its accuracy in showing objects on the moon, see Lagalla 1612, p. 8. Cf. also the reference to this meeting in the eloquent passage by Faber on the Galileo's discoveries with the telescope in the *Tesoro Messicano, Rerum Medicarum*, p. 473. With regard to Cesi and his companion's examination of the Lateran inscription, see now Spiller 2000, containing a useful discussion of skeptical views on the telescope as an instrument of reading rather than just as an observational device.

22. Galileo to Dini, May 21, 1611, *Opere*, XI, p. 115.

23. As in Stelluti 1630, pp. 37–38. Cf. also Lüthy 1996, p. 7.

24. B.A.V., Vat. lat. 9684, ff. 87–88: "Se mai fu tempo che V.S. fosse in Roma è hora. . . . Ogni sera vediamo le cose nuove del cielo, officio veramente da Lincei. Giove co' suoi' quattro e loro periodi, la luna montuosa, cavernosa, sinuosa, acquosa. Resta Venere cornuta e 'l triplice suo Saturno, che di mattino devo vederli. Delle fisse non dire altro: . . . non è però piccola difficoltà se la terra sia il centro del orbi." Also transcribed in Galileo, *Opere*, p. 99; and with some inaccuracies in *Carteggio*, no. 62, pp. 158–160 (Cesi in Rome to Stelluti in Fabriano, April 30, 1611).

25. Cesi in Rome to Stelluti in Fabriano, April 30, 1611, in *Carteggio*, no. 62, pp. 158–160 (p. 159).

26. G.G. *Opere*, II, p. 93.

27. Villoslada 1954, pp. 197–198.

28. Ibid.

29. Galileo, *Opere*, XIX, p. 275.

30. Galileo, *Opere*, XI, pp. 100–101.

31. For the significance of this phrase and its further resonance, see U. Baldini 1985.

32. Cf. the 1599 *Ratio Studiorum*, which set out the rules for the curriculum at all Jesuit colleges: "Our teachers in scholastic theology are to follow the doctrine of St. Thomas; only those who are well acquainted with St. Thomas are to be promoted to the chairs of theology. Those who are not well acquainted with this author, or who disagree with him, are to be barred from teaching. . . . In matters of any importance, philosophy professors should not deviate from the views of Aristotle, unless his view happens to be contrary to a teaching which is accepted everywhere in the schools, or especially if his view is contrary to orthodox faith. In accordance with the Lateran Council, they should strenuously try to refute any arguments of Aristotle, or of any other philosopher, which are contrary to the faith." *Decreta* 1830, III, pp. 339–341.

33. Cf. also *Epistolae* 1635, p. 17.

34. Brodrick 1928, I, p. 485.

35. Cf. Blackwell 1991, p. 142.

36. See the typically thorough account of his life and dispatches in Gabrieli 1989, pp. 1011–1052. But a good biography remains to be written here.

37. *Thesaurus*, p. 532.

38. "I think on Saturday I will make Doctor Faber a Lynx. Perhaps you know him: he is a very good philosopher, doctor, poet, and philologist, especially since I think we are soon going to lose Signor Schreck, at least in part," wrote Cesi to Stelluti in Fabriano on October, 28, 1611 (*Carteggio*, no. 79, pp. 176–177 [p. 177].

39. And *his* immense scientific and literary manuscript remains, formerly in the Archivio degli Orfani in Santa Maria in Aquiro in Rome and now housed in Rome, BANL, have been even less published than those of Heckius.

40. Recorded in the very same magnificent letter to Dini of May 21 (cf. chapter 9 below, under "First Delays" and note 54), in which Galileo defended his discoveries with the telescope; Galileo, *Opere*, XI, pp. 107–108; *Carteggio*, no. 64, p. 162.

41. Faber in Rome to Galileo in Pisa, December 15, 1611, in *Carteggio*, no. 85, p. 181.

42. On Corvinus, "exoticorum diligentissimus scrutator" as Schreck called him in his contributions to the *Tesoro Messicano*, see *Thesaurus* pp. 52, 64, 88, 145, 303–304; as well as the many references in the *Carteggio* and in Gabrieli 1989 on his botanical, pharmacological, and phytognostic work.

43. Cesi in Tivoli to Galileo in Florence, October 21, 1611; *Carteggio*, no. 78, p. 175. More than ten years later, in the very book that published the drawings of American plants seen by Galileo, Faber remembered this same excursion. His account gives some sense of the pleasures that often accompanied their scientific pursuits: "Fourteen years ago, after exercising ourselves in the examination and researching of plants for most of the day, we sat down at sunset. Our bones were almost broken from fatigue (though it was not an unpleasant fatigue), and in the company of our Prince, most devoted and expert in botany as in all the other sciences, we refreshed ourselves with a snack. Finally, when night came, we repaired to the castle of the Prince in the region called San Polo, which is right up against the mountain, and we revived our physical energy at a sumptuous banquet, restoring our minds, too, with sweet nourishment—that is to say, with most pleasant conversation and with philosophical problems" (*Thesaurus*, p. 503, first published in the 1628 version of the *Thesaurus*). Also on this trip to Monte Gennaro, see now Belloni Speciale 1987, p. 59–79; as well as the earlier account by Cortesi in *Annali di Botanica*, VI, 1916, pp. 156–160.

44. Rome, BANL, Archivio degli Orfani di Santa Maria in Aquiro, Filza 420, especially fols. 297–304. On these expeditions, see De Angelis and Lanzara 1985, as well as the brief reference in Findlen 1994, p. 191. Cortesi 1916, pp. 153–156 was the first to publish extracts from the important document in the Orfani archives.

45. See De Angelis and Lanzara 1985, pp. 127–147. Cf. also Cortesi 1916, pp. 156–160. This simplification and reduction of the complexity of the old names also point ahead to the spirit in which Cesi wrote to Galileo nine months later about one of the great "satisfactions," as he put it, of the Copernican system, which was the "removal of the multiplicity of movements and orbits, and their too large and complex diversity" (Cesi in Rome to Galileo in Florence, July 21, 1612; *Carteggio*, no. 142, p. 252). Cf. also chapter 13 below, under "The Reduction to Order."

46. Already in 1577 he had published a book on iatropharmacology (*Theriace et Mithridatia* [Naples: Marino de Alexandro, 1577]). See also Gabrieli 1989, pp. 900–911 for further rather inconclusive speculations on his writings in this domain.

47. Cesi in Acquasparta to Galileo in Florence, February 4, 1612; *Carteggio*, no. 104, p. 204.

48. E.g., Colonna 1606, p. 331 and elsewhere.

49. See especially chapter 8 below.

50. *Lynceographum quo norma studiosae vitae Lynceorum Philosophorum exponitur*, Rome, BANL, Archivio Linceo IV, fols. 1–242.

51. As observed by Olmi 1998b, p. 819.

52. *Praescriptiones* 1624.

53. See chapter 5 below, under "The Case of Luca Valerio."

54. "Every day I find myself liking his [Galileo's] theories more and more, seeing that gradually the most intelligent people in the field are accepting them. As for Venus, this I can easily swallow, but when it comes to the movement of the earth, I would like to be allowed to have a little more time to decide about it [vorrei esser dispensato ancora un pezzo]. In effect, it's something that deserves to be considered maturely and so far I find it quite difficult to grasp" (Welser in Augsburg to Gualdo in Padua, May 20, 1611; Galileo, *Opere*, XI, p. 117).

55. Rome, BANL, MS Archivio Linceo IV, fol. 280v; cf. Gabrieli 1989, p. 521 for the full text of the eulogy of the newly admitted Welser.

## CHAPTER FIVE

1. Odo van Maelcote wrote to Kepler on December 11, 1612, that Galileo had shown him the sunspots, presumably when he visited the Jesuits in Rome in April 1611; while Giovanni Battista Agucchi wrote to Galileo on June 16, 1612, acknowledging that he had told him of the sunspots when he visited Trinità dei Monti at the same time (cf. Galileo, *Opere*, XI, pp. 445 and 329 respectively).

2. Galileo, *Opere*, XI, p. 214.

3. See, for example, Galileo's own letter to Maffeo Barberini on June 2, 1612, in which he said that he had been looking at sunspots with the aid of a telescope for "about eighteen months," Galileo *Opere*, XI, p. 305. Cf. also the letters from Agucchi to Galileo, and from Odo van Maelcote to Kepler cited in note 1 above, both referring to the fact that Galileo had shown them sunspots while he was in Rome the previous year.

4. See the letter from Cesi in Acquasparta to Faber in Rome, January 31, 1612, in *Carteggio*, no. 100, p. 194, in the course of which he also laments the fact that Italian engravers are so inferior to German ones: "mi son si ben doluto di non potermi trovar in Germania, ove fioriscono tutte le professioni et scienze. Ben pol V.S. credermi, che sa quanto io sia germanofilo." On this topic see further Belloni 1983.

5. Galileo, *Opere*, XI, pp. 230–231; for a critical earlier exchange between Welser and Gualdo this year, see chapter 4 above, under "Botany and Astronomy: The Academy Expands," and note 54 there.

6. Galileo, *Opere*, XI, p. 235.

7. Ibid., p. 236.

8. Welser in Augsburg to Faber in Rome and Gualdo in Padua, November 18 and 25, 1611; Galileo, *Opere*, XI, pp. 235–236. It does not, in fact, ever seem to have been discussed whether the references to the German discovery of the sunspots might in fact be to Johannes Fabricius, who published his observations of March 1611 in his *De maculis in sole observatis et apparente earum cum sole conversione* (Wittenberg: Lorenz Seuberlich, 1611). (Cf. also Favaro in Galileo, *Opere*, V, p. 10.)

9. Ibid., p. 237.

10. See now Van Helden 1995 for useful details about Hariot's observations. See also the reference to the discovery made by Johannes Fabricius, probably in March 1611, in note 8 above.

11. It has often been said that the author of the letters, the Jesuit Christoph Scheiner (see below), wished to remain anonymous in order not to bring the Jesuits into dispute by addressing so tendentious a subject (and Scheiner himself would later [Scheiner 1630] write that he was told to publish under a pseudonym by his Provincial); but the full implications of this obviously carefully chosen pseudonym do not seem previously to have been pointed out.

12. Years later the sympathetically inclined Jesuit, Paul Guldin, who had heard him speak in Rome in April and May of 1611, recalled that he had told the professor of mathematics at the Jesuit college in Ingolstadt, Christoph Scheiner, and a number of other

Germans as well, about Galileo's discoveries (including the sunspots); see Drake 1978, p. 175; and Favaro's commentary and references in Galileo, *Opere*, V, p. 10.

13. Galileo, *Istoria*, p. 14.

14. Galileo in Le Selve to Cesi in Rome, May 12, 1612, in *Carteggio*, no. 117, pp. 219–221 (p. 220).

15. Galileo, *Opere*, XI, pp. 332–333.

16. Ibid., pp. 344–345.

17. Ibid, pp. 366–367 and Careggio, no. 142, p. 252. See also chapter 13 below, under "The Reduction to Order".

18. Cesi in Rome to Galileo in Florence, September 14, 1612, in *Carteggio*, no. 159, pp. 267–269.

19. Galileo, *Opere*, II, p. 426.

20. As translated in Drake 1957, pp. 140–141.

21. Galileo at Le Selve to Cesi in Rome, January 25, 1613, in *Carteggio*, no. 209, p. 317–320 (p. 320).

22. Galileo, *Opere*, XI, p. 427.

23. Horace, *Odes*, III, 2 (*Virtus, recludens immeritis mori/Caelum, negata tentata via*).

24. For the censorship of Galileo's draft of the *Three Letters*, see the excellent articles by Shea 1975, pp. 47–49; and Rossi 1978, also pp. 47–49. Cesi's letters to Galileo of November 10, December 14, December 28, and January 26, 1612–13, reveal some of the nuts and bolts of the dispute, while Galileo *Opere*, V, pp. 137–142 has the texts concerned. See also Shea 1986, pp. 118–119.

25. "L'epistola dedicatoria secondo l'avertimenti si smagrirà un po," Cesi in Rome to Galileo in Florence, February 22, 1613 (*Carteggio*, no. 222, p. 330).

26. *Istoria e Dimostrazioni*, p. 1 (but sig. A3).

27. For the earlier and later versions of the Preface, see Galileo, *Opere*, V, pp. 75–88.

28. Galileo, *Opere*, V, p. 17.

29. "non vogliono mai sollevar gl'occhi da quelle carte, quasi che questo gran libro del Mondo non fosse scritto dalla natura per esser letto da altri, che da Aristotele, e che gl'occhi suoi havessero a vedere per tutta la sua posterità," Galileo, *Istoria*, p. 104.

30. Galileo, *Opere*, V, p. 184.

31. Ibid., XI, p. 393.

32. Ibid., XI, p. 405.

33. I base this account of Cesi's role in supervising the illustrations of the second letter on the excellent account in Van Helden 1995, pp. 377–384.

34. Galileo, *Opere*, XI, pp. 326–327; as translated in Drake 1957, p. 84.

35. This was the point made by the friendly Cardinal Conti, when he admitted to Galileo that not only did the Bible fail to support the Aristotelian theory of the incorruptibility of the heavens, it actually implied the opposite.

36. Thus punning on the very word—*helioscopus*—that "Apelles" had applied to his own instrument for looking at the sunspots. For Mascardi's words to the readers of Apelles's letters, appended at the Linceans' request to the 700 supplemented copies of Galileo's book, see Carteggio, no. 214, p.324 (dated February 1, 1613); for Cesi's proposal to Galileo that the printer seem to carry the responsibility, see Carteggio, no. 191, p. 299.

37. See U. Baldini 1984b, especially pp. 14–17 on this; and Galileo, *Opere*, III, 1, 298–307, for the text of the *De lunarum montium altitudine*.

38. Galileo, *Opere*, XI, p. 509.

39. Aguilonius (1613), with illustrations by Rubens.

40. Galileo, *Opere*, XII, p. 28–29. Cf. *Carteggio*, no. 310, p. 419.

41. Galileo, *Opere*, XI, pp. 605–606.

42. Cf. chapter 4 above, under "Galileo, Della Porta, and the Telescope: Early Conflicts"; as well as for the relevant passages from Psalm 104:5, I Chronicles 16:30, Job 26:7, Proverbs 30:3, and Ecclesiastes 1:5.

43. E.g., in Kepler 1937–, pp. 33–34.

44. Galileo, *Opere*, V, pp. 281–8; translation of the whole letter in Drake 1978, pp. 224–229.

45. On this, see *Decreta* 1830, III, pp. 339–341; as well as chapter 4 above, under "Rome, 1611."

46. Cesi in Acquasparta to Galileo in Florence, January 12, 1615, in *Carteggio*, no. 377, pp. 477–481. Galileo, *Opere*, XII, pp. 128–131.

47. Galileo, *Opere*, XIX, p. 308; Blackwell 1991, p. 113.

48. Lorini to Paolo Cardinal Sfondrato, February 7, 1615; in Galileo, *Opere*, XIX, p. 298.

49. Galileo, *Opere*, XII, p. 146; Blackwell 1991, p. 74.

50. Galileo, *Opere*, XIX, pp. 309–10.

51. Ibid., XII, p. 150; *Carteggio*, no. 387, p. 489.

52. Galileo, *Opere*, XII, pp. 160–161.

53. Antonio Foscarini, *Lettera sopra l'opinione de' Pittagorici e del Copernico della mobilità della terra e stabilità del sole e del nuovo Pittagorico sistema del mondo* (Naples: Lazzaro Scoriggio, 1615) ("A letter concerning the opinion of the Pythagoreans and of Copernicus about the mobility of the earth and the stability of the sun and the new Pythagorean system of the world. . . . In which it is shown that the opinion agrees with, and is reconciled with, the passages of Sacred Scripture and the theological propositions which are commonly adduced against it"). See Blackwell 1991, pp. 87–88.

54. On Foscarini's work and its effect on Galileo, Bellarmine, and the whole Copernican debate, see especially Blackwell 1991, "Foscarini's Bombshell," pp. 87–110.

55. Galileo, *Opere*, XII, pp. 171–172. For a translation of Bellarmine's *Letter to Foscarini*, see Blackwell 1991, pp. 265–267; Shea 1975, pp. 50–51. Cf. also Drake 1978, p. 249.

56. Galileo, *Opere*, XII, p. 171; Blackwell 1991, p. 266.

57. Galileo, *Opere*, XII, p. 151.

58. E.g., in ibid., p. 190.

59. Galileo, *Opere*, XIX, pp. 320–321. For the distinction between the judgment that something was formally heretical and the much less severe one that something was "erroneous in faith," see Blackwell 1991, p. 121. See now also Feldhay 1995, pp. 26–27.

60. Along with the wholly obscure *Commentary on Job* (which in the discussion of Job 9:6 had also declared the earth to be mobile and the sun stable) by the Spanish Jesuit Diego de Zuñiga. Both were placed on the Index "until corrected," *donec corrigatur*. Galileo, *Opere*, XIX, pp. 322–323. See now also Feldhay 1995, pp. 28–52.

61. Galileo, *Opere*, XIX, pp. 322–323. For the view that the cosmological theories of Copernicus—and in particular his heliocentric view of the universe—were not a central issue for the Church in the seventy-odd years that elapsed between the publication of the *De Revolutionibus Orbium* in 1543 and the events leading up to the Decree of the Congregation of the Index of March 5, 1616, see also Blackwell 1991, p. 53. For an extension until much later, with the claim that the Galilean controversies were determined by patronage issues rather the clash between Copernicanism and anti-Copernicanism, see Biagioli 1993, especially pages such as p. 274 on the comets controversy.

62. Galileo, *Opere*, XIX, p. 321. Drake 1978, pp. 253–4 has a vivid and plausible reconstruction of the probable events around the actual execution of this order.

63. On Valerio, see especially Gabrieli 1989, pp. 836–864.

64. Rome, BANL, Archivio Linceo XXX, fols. 75r–76v; Galileo, *Opere*, XIX, p. 268.

65. Galileo, *Opere*, XIX, p. 348.

66. For the deliberations, see Rome, BANL, MS Archivio Linceo 30, 75r –76v. For the phrase, cf. Cesi to Galileo, March 7, 1615 (*Carteggio*, no. 387, p. 489); Cesi to Galileo, May 11, 1613 (*Carteggio*, no. 238, p. 353); and Galileo, *Opere*, XII, p. 150.

67. Cesi's words on the occasion of the funeral of Cesarini; Gabrieli 1989, p. 213.

68. Rome, BANL, Archivio Linceo 30, fol. 76r.

69. Cf. chapter 3 above, under "Scripture and the Fluidity of the Heavens."

70. Cf. U. Baldini 1988, pp. 162–163; and U. Baldini 1992, pp. 293–296.

71. *De caeli unitate, tenuitate, fusaque & pervia, stellarum motibus natura, ex sacris litteris*, in Scheiner 1630, pp. 776–780.

72. Ibid., p. 777.

73. Ibid., p. 778.

74. Ibid., p. 779.

75. Ibid., pp. 777–779.

76. Ibid., p. 784.

77. I am grateful to Eileen Reeves for pointing this out to me. On Bellarmine's early concerns about resolving the question of whether the heavens were fluid and boundless, see especially U. Baldini and Coyne 1984; Lattis 1994, pp. 94–99, 212–216. For the neo-Stoic dimensions of these views, see especially Barker 1991, pp. 133–154; and Reeves 1997, passim.

78. Scheiner 1630, p. 780.

79. See now Freedberg 1994c.

80. Galileo, *Opere*, XII, pp. 422–423.

81. Drake and O'Malley 1960, p. 6.

82. G.B. Rinuccini to Galileo, March 2, 1619; Galileo, *Opere*, XII, p. 443. Typically, Biagioli 1993, p. 274, claims that "the dispute [over the comets] from 1619 to 1626 does not seem to represent a clash between Copernican and anti-Copernican positions, but rather the play of patronage dynamics in a field turned much more insidious by the 1616 placement of Copernicus on the Index." This seems to me to underrate the Copernican dimension and overrate the influence of patronage (as always with Biagioli)—aside from the obvious fact that this particular statement seems to want to have it both ways.

83. Biagioli 1993, p. 62.

84. Galileo, *Opere*, II, p. 410.

85. Ibid., p. 409.

86. In contrast to the Aristotelian view that comets were exhalations from the earth that actually caught fire as they moved in the upper air. Cf. Shea 1977, chapter 4, emphasizing the Aristotelian derivation of Galileo's views (as has for a while been a fashion in Galileo studies). For a different interpretation, see Drake 1978. Cf. also Redondi 1987, pp. 31–32; and Drake and O'Malley 1960, pp. xv–xix.

87. Ciampoli in Rome to Galileo in Florence, July 12, 1619, Galileo, *Opere*, XII, pp. 465–466.

88. For the significance of the slight modification, see Drake 1978, p. 494, note 9.

89. Cf. "The Publication of the *Assayer*" and note 104 below.

90. Stelluti in Fabriano to Galileo in Florence, January 27, 1620, in *Carteggio*, no. 560, p. 707.

91. Stelluti in Fabriano to Galileo in Florence, April 4, 1620, in *Carteggio*, no. 568, p. 713.

92. Cesi in Acquasparta to Galileo in Florence, May 18, 1620, in *Carteggio*, no. 570, pp. 715–716 (p. 715).

93. Ciampoli in Acquasparta to Galileo in Florence, May 18, 1620, in *Carteggio*, no. 517, p. 716.

94. For a good account of the extraordinary efforts of Cesarini, Ciampoli, Faber, and Cassiano to prepare and edit the manuscript of the *Saggiatore*, also addressed in the form of a letter, this time to Cesarini, see Redondi 1987, especially pp. 40–51. See also "The Publication of the *Assayer*" in this chapter.

95. *Lettera al M.R.P. Tarquinio Galluzzi* (Florence: 1620).

96. I am immensely grateful to Seth Fagen of Martayan Lan in New York for having let me study item no. 56 of Catalogue 19 (1996) and for having provided me with copies of both it and of Stelluti's *Scandaglio*. The manuscript will henceforward be referred to as MS Martayan Lan 19/56.

97. MS Martayan Lan 19/56, fol. 11. Cf. Stelluti 1622, p. 8.

98. MS Martayan Lan 19/56, fol. 10; cf. Stelluti 1622, fol. 10.

99. MS Martayan Lan 19/56, fols. 23–25; cf. Stelluti 1622, pp. 16–17.

100. Cesi in Rome to Galileo in Florence, November 10, 1612, in *Carteggio*, no. 181, p. 289.

101. Gabrieli 1989, p. 944. Here Gabrieli also suggests that even though the printer's letter to the reader stated that he had received the work from Giovanni Battista, this too might be seen in the light of the various and typical forms of literary deception and pseudonomy of those days.

102. Stelluti in Acquasparta to Galileo in Florence, August 16, 1622, in *Carteggio*, no. 635, p. 771; and also Galileo, *Opere*, XIII, p. 96.

103. For an account of Cesarini's close relationship with the two Florentine cardinals, see now Freedberg 1994c. And see Sluiter 1997, p. 226, for an important and neglected letter by the intelligent and sensitive Bentivoglio (papal nuncio in Brussels between 1607 and 1615), written to Cardinal Scipione Borghese in Rome on April 2, 1609, on the subject of a telescope that Bentivoglio used earlier that year at the court of the archdukes.

104. As translated in Drake and O'Malley 1960, p. 152; Galileo, *Opere*, VI, p. 200. Cf. Galileo's own statement in the *Assayer* that "since it seemed to me that [Grassi] used

too crude a steelyard in his weighing of Sig. Guiducci's propositions, I have elected to employ an assayer's balance precise enough to detect less than the sixtieth part of the grain" (Drake and O'Malley 1960, p. 171; Galileo, *Opere*, VI, p. 220).

105. Drake 1957, pp. 235–236.

106. Galileo, *Opere*, VI, p. 268.

107. On Cesarini's ill-health and death, as well as much else on him, see also Freedberg 1994c.

108. Winther in Acquasparta to Faber in Rome, April 17, 1624, in *Carteggio*, no. 735, pp. 866–868 (p. 866). For more on Winther's trip to Acquasparta in 1624, see Belloni Speciale 1987, especially pp. 65–68, as well as chapter 13 below, note 24.

109. Galileo, *Opere*, XIII, p. 175.

CHAPTER SIX

1. "Invio a V.E. un occhialino per veder da vicino le cose minime, del quale spero che ella sia per prendersi gusto e trattenimento non piccolo, chè chosì accade a me. Ho tardato a mandarlo, perchè non l'ho prima ridotto a perfezzione, havendo hauto dificoltà in trovare il modo di lavorare i cristalli perfettamente. . . . Io ho contemplato moltissimi animalucci con infinita ammirazione: tra i quali la pulce è orribilissima, la zanzara la tigenuola son bellissimi; e con gran contento ho veduto come faccino le mosche et altri animalucci a camminare attaccati a' specchi, et anco di sotto in su." Galileo from Bellosguardo to Cesi in Rome, September 23, 1624, in Galileo, *Opere*, XIII, pp. 208–209 and *Carteggio*, no. 781, pp. 942–943.

2. Ibid.

3. Cesi in Acquasparta to Galileo in Florence, October 26, 1624; Carteggio, no. 791, p. 956.

4. Cf. also Pierre Humbert, "Peiresc et le microscope," in *Revue d'histoire des sciences et de leurs applications* 4 (1951) pp. 154–158, on Peiresc's use of Johann Kuffler's microscope in the Palais de Luxembourg in Paris on May 22, 1622.

5. For a long account of Peiresc's microscopic examinations in 1622, see also Humbert, "Peiresc et le microscope," pp. 154–158; as well as De Waard 1906, pp. 298–300.

6. Pierre Gassendi, *Viri illustris Nicolai Claudii de Peiresc* (Paris, 1641), p. 7.

7. Humbert, "Peiresc et le microscope," pp. 155–158.

8. Peiresc in Paris to Aleandro in Rome, June 7, 1622; transcribed in the splendid and still fundamental article on the early history of the microscope by Govi 1888, p. 20, who also transcribes the further correspondence between Peiresc and Aleandro (until July 1624) on the subject of the new instrument.

9. See, for example, the proposal to combine lenses in the 1589 Naples edition of his *Magia Naturalis*, book XVII, chap. 10, pp. 269–271 and 278–279 (and in the later—but not the earlier—editions), as well as *Carteggio*, p. 114 (Della Porta in Naples to Cesi in Rome, August 28, 1609; cf. also chapter 4 above, under "Della Porta, Galileo, and the Telescope: Early Conflicts"). See also Govi 1888, pp. 13 and 29–30; Singer 1913–14, p. 8; Singer 1915, p. 329; and—much more fully—De Waard 1906, pp. 81–92.

10. Faber in Rome to Cesi in Acquasparta, May 11, 1624, in *Carteggio*, no. 743, pp. 875–876 (p. 875).

11. For further support of the view that this "occhialino" was indeed a compound microscope, see the comments by De Angelis and Lanzara on the distinction between observation by lens and by microscope made by Cesi in his great mycological volumes of 1623–1628 in Paris (De Angelis and Lanzara 1986, p. 255, note 13).

12. For the brief history of the microscope before 1623, see particularly Govi 1888; De Waard 1906; and, inter alia, Singer 1913–14; Singer 1915; and Singer 1953.

13. Faber in Rome to Cesi in Rome, April 13, 1625, in *Carteggio*, no. 841, p. 1038. Here Faber is clearly referring to the passage he had composed on both the telescope and the microscope in the *Tesoro Messicano*, pp. 473–474. In the letter he goes on to ask whether Cesi approves, adding that since the Lincei were the first to use the term *telescope*, they had decided also to find a convenient name for what he was here calling a microscope—deservedly so, Faber declares, because they were the first in Rome to have the instrument (*Carteggio*, no. 841, p. 1039).

14. Fabio Colonna in Naples to Faber in Rome, June 6, 1625, in *Carteggio*, no. 852, pp. 1047–1049. As Gabrieli rightly noted (Gabrieli 1989, p. 361; and *Carteggio*, p. 1047), Colonna's Greek cannot have been very good, since "telescope" clearly derived from

*tele*, "from afar" rather than *téleios*, meaning "complete" or "perfect"; while the etymology of "ponoscope" does not seem to make much sense, unless Colonna meant to write "aposcope" here.

15. Fabio Colonna in Naples to Faber in Rome, June 6, 1625, in *Carteggio*, no. 852, pp. 1047–1048. It is important to note here that already in the 1606 edition of his *Ekphrasis*, Colonna had been using a lens to enlarge details of fungi or "lichens" (Colonna 1606, p. 331); illustrated and briefly discussed in Singer 1915, pp. 320–321.

16. John Wodderburn, *Quatuor problematum quae Martinus Horky contra nuntium sidereum de quatuor planetis novis disputanda propusuit; confutatio* (Padua, 1610), as translated in Singer 1953, p. 197.

17. It is referred to by Cesi in Rome to Galileo in Florence, September 26, 1625, in *Carteggio*, no. 866, p. 1066: "Mi riesce, col mandarle l'accluso foglio, la prima parte; ma la seconda posso solo accennarlene la speranza, della quale la detta espressione ne sia buon auspicio e hieroglifico.

While Kidwell and others have followed Gabrieli in suggesting that Cesi's reference to the "accluso foglio," which formed the "prima parte" of his and the other Linceans' apiarian researches, is to an early draft of the text of the *Apiarium* (see below), it is clear to me, as will become still clearer later on, that this reference must in fact be to the present sheet. Aside from the corroborative evidence offered by the subsequent correspondence on the matter, as well as Rome, BANL, Archivio Linceo MSS 4 and 5 (see below), making it clear that the Linceans were working and reworking the *Apiarium* material even into the New Year, it does seem to me that the words "la detta espressione ne sia un Buon auspicio e hieroglifico" actually refer to something visual and pictorial, rather than to a text *tout court*.

18. Cesi in Rome to Galileo in Florence, September 26, 1625, in *Carteggio*, no. 866, p. 1066: "Questo è fatto per significar tanto più la nostra divotione a' Padroni et esercitar il nostro particolar studio delle naturali osservationi." As observed in the preceding note, this reference seems to me to be indubitably to this sheet, rather than to an early draft of the text of the *Apiarium*.

19. Written some time around 1523–24, but first published in Florence in 1539, and again in 1590, with a commentary by Roberto Titi (Florence, Giunti). The relevant lines are vv. 963–1001. I am grateful to Eileen Reeves for pointing out this critical source of inspiration for the Linceans' bee panegyric.

20. What is altogether unprecedented here is not just the way in which the engraving shows, for very first time, the separation of the various parts of the proboscis into the glossa, labial palps, and galeae, but also the clarity with which it does so. This, in other words, is the complex structure of the "tongues," as Cesi would repeatedly put it in the *Apiarium*. I am extremely grateful to Dr. Jerome Rozen of the American Museum of Natural History for his help with the identification of the parts of the bee, and for his confirmation of just how remarkable and unprecedented these microscopic observations were.

21. This refers to the glossa and the two labial palps and two galeae enclosing it. See the preceding note; cf. also chapter 7 below, under "Colonna, Faber, and the Microscope," as well as note 16 there.

22. On Riquius (*recte* Josse de Rycke) see both Gabrieli 1989, pp. 1133–1176; and especially Van den Bergh, in *Messager des sciences historiques* (Ghent, 1880), pp. 12–32, 189–208; 1881, pp. 160–185, 457–477.

23. For a full bibliography of Riquius's published and unpublished works, see Gabrieli 1989, pp. 1138–1164.

24. Riquius 1625. For the manuscript draft of the *Apes Dianiae*, see Rome, BANL, Archivio Linceo MS 2, fols. 1–9. Riquius's introductory epigram (as he calls it) is dated November 1625 in the published version on sig. a2 and in the manuscript on fol. 7v. Since the first of the two drafts for the poem is signed "Justus Riquius Belga," while the second corrected version and the printed text carry the now predictably proud inscription "Justus Riquius Lyncaeus Belga," the final redaction must have taken place just after he actually became a Lincean, also in 1625.

25. Ever the precise philologist, Riquius notes that although the first had been previously published by Aldrovandi (in his work on insects), and the second by Hubert Goltzius (in one of his then famous numismatic treatises),"they have been so inaccurately represented and explained that they can be regarded as unpublished." The coin

from Megara Hyblaea, on the other hand, had—according to him—never been reproduced before.

26. Riquius 1625, p. 7:

> Hic adyta, & Triviae Virginis esse domum,
> Apparet tota diffusae corpore mammae,
> Nec tamen est ullo foedera passa thoro.
> His alitur mortale genus, vitaeque animantium,
> Vitales succos hinc elementa bibunt.

27. Ibid., p. 6: "Ex sese genita nullo faedata cubili."

28. Ibid.: "Plus tamen est APIBUS tribuit quod virgo pudicis / Gratior & Castae [i.e., Diana] casta volucris adest."

29. Ibid., p. 3: "*Incorrupta tuos servabunt saecula mores / Virgineo castus Praeside Mundus erit.*"

30. This too is why they flocked to the home of the Barberini family, with its motto, *HIC DOMUS*, "here is home." Ibid., pp. 8 and 12.

31. Ibid., p. 12.

32. See Foucault 1973, part 2, especially pp. 264–269, for the basis of this caution.

33. Cesi in Rome to Galileo in Florence, September 26, 1625, in *Carteggio*, no. 866, p. 1066.

34. The coin from Megara was also reproduced and discussed in the *Apes Dianiae*, while that on the left from Ephesus, with a bee and a pasturing stag, was described by Riquius in his note E as belonging to Cesi himself; the other two coins are from Aptera (showing a profile of Diana and a bee) and from Metapontum (a bee alongside the two ears of wheat of the Metapontan mint on the obverse, and a bust of Leucippus with the inscription *HERAKLEION* on the reverse).

35. Colonna in Naples to Cesi in Rome, February 13, 1626, in *Carteggio*, no. 900, pp. 1100–1101.

36. The only extant study of the *Apiarium* in English (Kidwell 1970) is remarkable for its tenacity and for having recognized the significance of this aspect of the Lincean project; but it fails in the end, not just because the translation is inadequate to the difficulty of the text but because the principles of organization of the sheet are not grasped.

37. Cf. Schettini Piazza 1986, p. 239: "A singular work, the *Apiarium*, contradictory, wrongly forgotten, in which truth and legend, theory and experiment are intertwined, where wordplays and the taste for etymology alternate with shows of learning, and an uncritical appeal to classical authority is allied with pride in the new scientific discoveries."

38. For example: *Favus*, or honeycomb, stands at the origin of the noun *favor* (favor) and the verb *favere* (to favor or to be well-disposed); all derive from *favonius*, the gentle west wind that blows at the beginning of spring and thus nourishes the sweetness of flowers. The word for bee itself, *apis*, is repeatedly related to *apex*, summit, or highest point, indicating the loftiness of the bee's prowess and talents. But then the question is posed: could the word not also come from *a-pes*, meaning "without feet," suggesting not that bees actually lack feet but that they are incapable of impurity, and of touching the dirt with their extremities? Moreover, Apis was also the sacred bull worshiped by the Egyptians, and everyone knew that sometimes bees could (it was said) be generated from the putrefying carcasses of dead oxen. In a particularly absurd passage, the words "absis" and "apsis" are punned with *apis*. Stay away—*absis*—from the sting of the bee—*apis*—especially if you are unjust. Next, *apes* is related to *opes*, meaning wealth or richness, just as the name, or *nomen*, of something was constantly said (as always in those days) to contain within it a sign of its deeper meaning, or *omen*. How convenient that in the basic classification of bees, the whole group of social (as opposed to solitary) bees could be divided into *urban* and rural ones! And that Urban's own family name, Barberini, was so closely related to the basic distinction in Pliny the Elder's *Natural History* between *urban* bees and those that are *barbarae*! What this suggests, moreover (from a scientific point of view), is that the Linceans were very aware of the distinction between honeybees—the only social ones—and all the others. But lest the reader make the too-easy equation between *barbara* and barbarian, this panegyric assures one of the virtues of each: the urban bees are distinguished for their customs, virtue, and dignity; the *barbarae* for their movement in swarms, their purity, and their harmony—just like the Barberini, of course.

39. Rome, BANL, Archivio Linceo IV, fols. 416, 460–467.

40. Unfortunately, Kidwell 1970 did not grasp that the central section and these surrounding emblems were separately conceived. The draft designs for the final version (Rome, BANL, Archivio Linceo MSS IV and V) make it clear that while the exact order of the emblems was not essential, those on the right should be generally related to the various capacities of the bee as could be discovered by scientific observation; those on the left to the organs of the body and the appearance of the bee, as well as their symbolic implications; and those at the bottom to the fruit (i.e., honey and wax), use, and character of bees. In practice, needless to say, these divisions turn out to be not quite so strict.

41. For several other sixteenth- and seventeenth-century passages on the good organization of the beehive, such as Olivier de Serre's 1600 description of the hive as "the very model of the well-ordered commonwealth"; or Thomas Moffett's view in his 1638 *Theatre of Insects* that it was a model of "politick and oeconomick virtues" (see Merrick 1988, pp. 15–16).

42. For examples of both the antecedents and the *Nachleben* of the unwillingness of the king bee to use his sting, see Merrick 1988, especially pp. 11–15. Alciati has an emblem of bees and their hive exemplifying the *principis clementia* (e.g., in the Pisan edition of 1621, p. 632). Cf. also Jacques de Bie's "Non utitur aculeo rex cui paremus" in his 1636 *La France métallique*, pp. 49 and 150.

43. For example: The central section begins with a reference to Columella, who like most recent writers says that the leader of the bees does not have a sting; but Aristotle and Ambrose say that he simply does not use his sting; Aelian and Pliny, on the other hand, disagree. And so on, down to modern classical scholars, such as Scaliger and Cardanus. On this tradition, see also Merrick 1988.

44. "*Apice orbis*"—punning, yet once more on *apis*. Cf. note 38 above.

45. For example (to give only the simplest parallels here), they are strong like Jupiter (who derives his strength from his nurture by the honey-bearing nymph Melissa on Mount Hymettus), chaste like Diana, fecund like Venus, and wise as Minerva. They are the ministers of Ceres, conspicuous for their purity. All this is illustrated by ancient coins too.

46. Many other examples of this ultimately Aristotelian insistence on the chastity of the king bee are provided by Merrick 1988. It is worth noting that Aristotle's discussion of these matters occurs in his *De Generatione Animalium*, 759b–760a.

47. Basing himself on ancient writers such as Varro and Columella, Cesi even notes that like the king bee, the pope surely shies away not only from bad odors and from those who are drunk, but also from those who have just engaged in sexual intercourse.

48. See, for example, the long discussion in Aristotle, *De Generatione Animalium*,759a–761a; Pliny, *Natural History*, 3:461; and a host of other Roman writers, many cited by Cesi, ranging from Varro to Aelian.

49. It was precisely in his correspondence with Cesi in the first months of 1626—referring frequently to the *Apiarium* and to their work with the microscope—that Fabio Colonna discussed various possible cures for the continued difficulties Cesi's wife had with her miscarriages and in giving birth to a healthy male offspring (e.g., Colonna in Naples to Cesi in Rome, February 13, 1626, in *Carteggio*, no. 900, pp. 1100–1101; Colonna in Naples to Stelluti in Rome, March 20, 1626, in *Carteggio*, no. 910, pp. 1111–1112).

50. As Merrick 1988, p. 17, has also rightly noted. Though Jan Swammerdam, in his epochal *Histoire générale des insectes* of 1669, mentioned his discoveries about the sexes of the bees (he noted that the queen was incorrectly named the king, and that the "kings" were actually the drones), he did not describe how they mated.

51. Kidwell 1970, p. 110, described the work as a "curious compilation of fact and fancy," which "cannot be classed as a scientific document, because for the greatest part it lacks the objectivity and the observational and theoretical basis which would make it an adequate attempt to explain the natural phenomena which are its subject matter. The treatment of the subject matter, especially in its mythological references and puns, is quite often more literary than scientific, and at times it seems completely uncritical."

52. Cf. also chapter 7, note 7, below.

53. This is an extraordinarily puzzling passage, and Dr. Jerome Rozen has suggested to me that it may in fact simply mean that in order better to view the bee under

examination, Cesi and his colleagues may have placed it in a thin layer of honey itself, as many microscopists have subsequently done.

CHAPTER SEVEN

1. This is prophetically analogous to Salviati's question at the beginning of the second part of Galileo's *Dialogue on the Two Great World Systems* when he asks whether there could possibly be any doubt that Aristotle would have changed his mind about his view of the heavens had he had the telescope at his disposal.

2. Cf. also Colonna's observation, just a few months before the publication of the *Apiarium*, of the "structure of the eye of the fly, with its golden reticulation." See chapter 6 above, under "Microscopes and Mites" and note 15 there.

3. An interesting parallel, in a wholly different field, is perhaps offered by Alexander Marshack's analyses of paleolithic rock markings (which we sometimes might be tempted to call drawings).

4. On this, see chapter 12 below, under "The Evidence of the Senses."

5. Cf. chapter 6 above, under "The *Apiarium*."

6. For this aspect of Ramus's work and all its implications, see the fundamental study by Ong 1958.

7. They are divided and subdivided and include a number of American bees mentioned by writers such as Gregorio de Bolivar and Francisco Hernández, including the degenerate *Xicotli*, the giant and shaggy *Guancoiro*, the numerous *Uruncui*, and the related *Quauxicotli*. What is additionally notable about all of these is that they reveal Cesi's fundamental attentiveness to the native names.

The "middle group," according to the *Apiarium*, also included insects that, as Cesi cheerfully recorded, were hardly known to more recent writers at all, and the subject of much confusion in ancient ones. Pliny, Aelian, and Aristotle all insisted, for example, that the *Sirenes* were not the same as bees, just as in the case of the whole class of *bombyces* or bumble-bees, which do not seem to make honey at all, but which both fly and swim. The latter thus belong to a class that always held much fascination for Cesi, the class of amphibians.

8. Naples, BN, MS XII.E.4, fol. 15 r. See also chapter 11 below, under "Middle Natures," and chapter 13, especially under "The Mirror of Reason and the Theater of Nature" and note 29 there.

9. *Thesaurus*, pp. 460–840. Faber's "Expositio" of Recchi's transcription of Hernández's notes on Mexican animals first appeared in the fragmentary early edition of the *Thesaurus* of 1628, and then, of course, in the final version of 1649–51. See also chapter 10 below.

10. Francis Bacon, *Novum Organum*, II, 39. Since he did not yet have the new term for the instrument, Bacon here simply refers to the use of a *perspicillum* for the enlargement of minute things.

11. *Thesaurus*, p. 473.

12. Stelluti, Faber noted, "has shown us the most remarkable anatomy of all the minute external parts of so small an animal as the bee: the eyes, the tongues, the antennae, the mane, the sting, the digits of the feet and several other parts," which he recently had engraved in copper, "& nuper in aes incidi comisit" (thus clearly referring to the *Melissographia*), *Thesaurus*, p. 757. For Cesi's fundamental observations of ferns, see chapter 8 below, especially under "Ferns, Spores, and Reproduction."

13. Fabio Colonna in Naples to Cesi in Rome, January 9, 1626, in *Carteggio*, no. 887, pp. 1085–1086.

14. Fabio Colonna in Naples to Stelluti in Rome, February 2, 1626, in *Carteggio*, no. 897. pp. 1096–1097; Fabio Colonna in Naples to Cesi in Rome, February 13, 1626, in *Carteggio*, no. 900, pp. 1100–1101.

15. Fabio Colonna in Naples to Cesi in Rome, February 20, 1626, in *Carteggio*, no. 901, p. 1102.

16. It refers very clearly to what in modern terms would be called the galeae, the labial palps, the glossa, and the labellum (the "spoon" at the very tip of the glossa).

17. Fabio Colonna in Naples to Stelluti in Rome, March 20, 1626, in *Carteggio*, no. 910, p. 1111. One cannot pass by in silence another aspect of all these letters from Colonna to Cesi and Stelluti, for they are filled with concern about the recent stillbirth of Cesi's much-wanted son, and with desperation to find reasons and remedies. Had

this calamity to do with the disposition of the stars or, rather, with some more physiological or medical deficiency? Could it have to do with the sperm? Colonna wondered openly. The doctors he consulted seemed divided between astrological and more modern scientific explanations.

18. Fabio Colonna in Naples to Cesi in Acquasparta, July 17, 1626, in *Carteggio*, no. 922, pp. 1123–1124.

19. In his *New Observations of Celestial and Heavenly Things perhaps not published before*. Francesco Fontana, *Novae coelestium terrestriumque rerum observationes, et fortasse hactenus non vulgatae, a Francisco Fontana, specillis a se inventis et ad summam perfectionem perductis, editae* (Naples, 1646).

20. Fabio Colonna in Naples to Cesi in Acquasparta, September 19, 1626, in *Carteggio*, no. 930, pp. 1131–1132 (p. 1131).

21. Fabio Colonna in Naples to Stelluti in Rome, November 15, 1629, in *Carteggio*, no. 999, pp. 1204–1205 (p. 1204).

22. References to Faber and his work include the following: Persius mentions a parrot (Prol. 8); Stelluti refers to Faber's long examination of the various species of parrots in the *Tesoro Messicano* (p. 4). The parrot can say *chaire* (or "hullo"), according to Persius (p. 8); but it can also, adds Stelluti, imitate the sounds of the trumpet, the meowing of cats, the barking of dogs, the howling of babies, whistling, singing, and so on (p. 5). He gives examples from ancient writers and modern experience, and records how Faber transcribed both the words and the music of the popular Flemish song sung by a parrot (*Thesaurus*, p. 717; fig. 10.5 here). When one ornithological allusion follows another in Persius (as in his reference to "raven poets and magpie poetesses," Stelluti brings forth a whole range of classical lore and modern observations on these birds (pp. 5–6, here probably alluding to the work of Cassiano).

"Paint two snakes," declares Persius of the way in which the poetasters hallow their own work: "Don't shit on this, boys; this is sacred ground—don't piss here" (1.112–114). Stelluti elaborates on the ancient apotropaic sign of two serpents and, on the grounds of the word "snake" alone, publicizes Faber's discussion both of Mexican snakes and of a strange serpentlike animal, the *Dracunculus barberinus*, given to him for examination by Francesco Barberini (p. 42); on this problematic beast, see also chapter 12 below, under "The Worm and the Dragon."

References to Colonna's work include these: Persius satirizes the garment of *hyacinthina lana* or "hyacinthine wool" of a Neronian poetaster (I.32); Stelluti remarks that the alternative reading "*iantina*" means violet, from a particular type of seashell; and this then gives him the excuse to praise and discuss Fabio Colonna's work on the murex and other seashells, the *Purpura* of 1616 (pp. 15–16). When Persius speaks of "Calabrian fleece dyed in spoiled purple" (II.65), Stelluti cannot resist recalling the *Purpura* again (p. 73).

An entirely different achievement of Colonna's is referred to in the fifth satire. This was the subtle and sophisticated variation of the ancient *Sambuca* or lyre mentioned in line 95, as well as a modern version of the ancient hydraulic organ; both of these inventions were published in his 1618 book titled the *Sambuca lincea*.

23. Stelluti 1630, pp. 20–22; later on, when Persius refers to ebony (V.136), Stelluti has another note on Cesi's discovery of the metallophytes (pp. 169–170).

24. Then Stelluti describes Francesco Barberini's gift of a lynx to Cesi, and he lauds Faber's discussion in the *Animalia Mexicana* of that animal as well as those of its relatives, the tiger, the leopard, the civet, and the panther. As if this were not enough, he digresses even further, in order to praise and exemplify both the botanical and zoological importance of the contents of the *Tesoro Messicano* as a whole (pp. 36–38). Finally he provides his readers with a charming illustration of the lynx, inserted into the not insubstantial body of the note itself (cf. fig. 7.1).

25. There are of course many straightforwardly classical and antiquarian digressions too, as in Stelluti's comment on and illustration of a strigil (pp. 166–167, referring to V.126). Cesi's great collection of antiquities, he notes, included several examples of this ancient implement, illustrated in several drawings in the corpus of Cassiano's drawings at Windsor Castle. Then, too, there are several campanilistic references that are not entirely without interest: for example, a brief reference to books, parchment, and paper in the third satire (III, 10–11) provokes an immense note on the various kinds of materials used for writing surfaces, including papyrus, parchment, and

above all paper (especially since the famous paper of his own hometown, Fabriano, had been made, Stelluti proudly records, ever since the year 990 (pp. 80–84)).

26. The few writers who have noted this passage (e.g., Gabrieli and Kidwell) have all been perplexed by the fact that Fontana is mentioned here as the draftsman of the sheet, when it was in fact signed by Mattheus Greuter; but of course there is no inconsistency at all. In keeping with the practice of the times, one person (in this case Fontana) would have done the drawing and another (in this case Greuter) would have made the actual engraving.

27. See numbers 4, 11–12, and 7–8 on the print. As Dr. Jerome Rozen has observed to me, the observations of the structure of the proboscis here, especially in 9 and 3—are particularly noteworthy. Here one may clearly see the glossa, the labia, and the galiae folding round the glossa, as well as the way in which the whole proboscis folds up and fits in the ventral groove known as the proboscidial fossa (e.g., in number 8). All this, of course, coincides with Cesi's and Stelluti's constant preoccupation with the "tongues" of bees, both here and in the *Apiarium*.

28. Especially in numbers 9 and 3 on the illustration.

29. Cf. chapter 6 above, under "The *Apiarium*," and note 35 there.

30. Fabio Colonna in Naples to Stelluti in Rome [?], October 2, 1629, in *Carteggio*, no. 995, p. 1200).

31. Stelluti 1630, p. 53.

32. Cf. especially chapter 9, under "A New Botany: The Classical and the Vernacular," and chapter 10, under "Faber's Exposition," on other vernaculars; but also chapter 8, under "Heckius and His Simples."

33. Stelluti 1630, p. 126. That the issue had been on the table for a while is clear from a letter that Colonna wrote to Stelluti on January 29, 1627, not only about their work on bees but also on a curculio beetle. Colonna began with a complicated and implausible etymological justification for the Italian name of the tiny insect: "I have told my friend that you liked the drawing, and he wanted me to tell you that it is called a *gorgoglione* for good reason, because it has its eyes on its throat [*gorga*]; and I replied that this seemed unlikely to me. . . . He has promised me that he will finish the bee as soon as possible" (*Carteggio*, no. 943, p. 1144).

34. For a brief outline and references, see Freedberg 1992a, pp. 41–50.

35. See especially Biagioli 1993.

36. It is important to remember that while one can certainly speak of the separation of physics and metaphysics, to speak of a categorical rift between philosophy and science is somewhat misleading. What was then called "philosophy" included much of what we now called science. Even though metaphysics was shifted away from the domain of physics, both could still be contained under the rubric of philosophy, whether together, as previously, or separately. Blair 1997 contains a full discussion of the meaning of physics at the time.

## CHAPTER EIGHT

1. For Heckius's dates and the dates of his travels to Italy, see Rienstra 1968, especially p. 263.

2. See the good accounts in Carutti 1877; Rienstra 1968 (resuming the earlier literature on Heckius); and Van Kessel 1976. The early iatropharmacological and medical notebooks—that is, those predating his stay in Scandriglia—are bound together in Rome, BANL, MSS Archivio Linceo XVII and XVIII.

3. As he put it in his letter accompanying MS H. 507 in Montpellier: "I send them to you so that you may speculate about them, and thoroughly investigate their powers and effects" (Heckius in Prague to the Linceans in Rome, August 1, 1605; *Carteggio*, no. 32, p. 87).

4. On these, described by Alessandrini (1986, p. 117) as "veri e propri studi di medicina preventiva, che si rivelano di straordinaria attualità," see, for example, Rome, BANL, MSS Archivio Linceo XXII, XXVIII, XXIX.

5. All in Rome, BANL, Archivio Linceo MSS XXII and XXIII.

6. Heckius to Aldrovandi, January 8, 1603, Rome, BANL, MS Archivio Linceo XVIII, fols. 12–15, published in Gabrieli 1989, pp. 1086–1087.

7. For a vivid account of all this, see Carutti 1877, pp. 48–49. For the trial of June 1, 5, 16, and 20, 1603, see Rome, AS, Tribunale Criminale del Governatore di Rome, no.

26, fols. 1210–1231 (*Romana Vulnerum pro Curia et Fisco contra Joannem Ecchium Flandrum*).

8. See notes 10 and 15 below, and the further discussion in the present section; see also chapter 2 above, more briefly, under "Separation and Exile."

9. *Disputatio unica Doctoris Ioannis Heckii Equitis Lyncae Daventriensis. De Peste et quare praecipue grassetur tot ab hinc annis in Belgio* (Deventer: Ioannes Cloppenburch, 1605); Amsterdam, Bibliotheek van het Maatschappij van Geneeskunde; Oxford, Bodleian Library; cf. van Kessel 1976, note 89.

10. Montpellier, BEM, MSS H.506– H.508; the fourth manuscript, MS H.505, carries the subtitle *Mechanica et Naturalia Ioannis Ecchi Lincaei* (see Alessandrini 1978, pp. 288–289 for some further details). As this suggests, it is largely devoted to matters of technical and mechanical interest and contains Heckius's typically rather awkward drawings of machines, instruments, chemical vessels, and a variety of mechanical devices. Introducing the manuscript are thirty pages of medical notes written in Arabic, Syriac, and a strangely hermetic combination of Arabic and the Lincean code, as well as a passionate and moving invocation to the Virgin to assist him in his exploration of the hidden parts of nature. As a whole, the manuscript reminds us not just of the early Linceans' interest in scientific instruments, but also of the close relationships that so often existed in those days between natural history and mechanics, chemistry, astrology and alchemy.

11. Along with the other manuscripts and drawings from Cassiano's collection, they passed into the Albani collections in 1714. The Montpellier manuscripts consist of those confiscated by French revolutionary troops from the Albani collection in 1798 and then eventually deposited in the library of the Medical School there. See Alessandrini 1978, pp. 26–27 and 47–52 on this history. For manuscript numbers, see the preceding note.

12. The earliest edition of Mattioli appeared in 1544, but it was followed by very many others (including at least seventeen sixteenth-century editions in Italian alone); on these see especially Fabiani 1872. For the Linceans' own copy—probably Cesi's—see this chapter, under "The Erbario Miniato."

13. Montpellier, BEM, MS H. 506.

14. Ibid., MS H. 508.

15. Rome, Bibliotheca Vallicelliana, MS R.57. A similar bifurcation between tradition and modernity is revealed by the fourth manuscript in Montpellier, perhaps the earliest of Heckius's *carnets de voyage* there, as discussed in note 10 above.

16. Montpellier, BEM, MS H. 507, fol. VIIIr; Alessandrini 1978, p. 211; Heckius in Prague to the Linceans in Rome, August 1, 1605; *Carteggio*, no. 32, p. 87.

17. Heckius in Prague to the Linceans in Rome, August 1, 1605, in *Carteggio*, no. 32, p. 87.

18. Ibid., in *Carteggio*, no. 33, p. 88.

19. Heckius in Rome to Kepler [in Prague?], April 1, 1606, in *Carteggio*, no. 41, p. 99; Heckius in Rome to Mercuriale in Pisa, April 1, 1606, in *Carteggio*, no. 42, p. 100; Heckius in Rome to Robin in Paris, April 1, 1606, in *Carteggio*, no. 43, pp. 100–101; Heckius in Rome to [Clusius in Leiden?], April 8, 1606, in *Carteggio*, no. 44, pp. 101–103. As I have just suggested, however, these concerns went alongside purely taxonomic and classificatory ones; for these, which emerge in the letters addressed by both Heckius and his fellow Linceans to some of the great botanical and other experts at this time, see also chapter 13, under "The Mockery of Order."

20. Carolus Clusius, *Fungorum in Pannonis Observatorum Brevis Historia*, appended to the *Rariorum Plantorum Historia*, published by Moretus in Antwerp in 1601, and based on material he had collected in the course of a stay in Hungary, and for which he commissioned a beautiful series of watercolor drawings still preserved in the University Library in Leiden (on these, see Aumüller and Jeanplong 1983). For precedents for Clusius' work, and for its relationship to the Lincean project on fungi, see now Pegler and Freedberg, forthcoming.

21. Heckius in Rome to [Clusius in Leiden ?], March 19, 1606, in *Carteggio*, no. 40, pp. 97–98.

22. Heckius in Spoleto to Cesi in Rome, [July 1603?], in *Carteggio*, no. 1, p. 24.

23. Windsor, RL 27691– RL 27901; cf. Gabrieli 1989 pp. 1460–1461 for a basic (as always) early account.

24. Before being bought by James Adam for George III of England in 1762 (thus saving them, as it were, from the depredations of the French revolutionary troops who carried them off to France some thirty-six years or so later; cf. "A Note on the Provenance of Cassiano's Drawings" following chapter 1, above; as well as this chapter, under "The Drawings in Paris"). For the seal on the first page, once identified as that of Cassiano himself, see Solinas 1989b, pp. 52–53. The Index on the first four pages is written in the hand of Cesi's usual scribe (cf. also the Indices to the manuscripts in the Institut de France discussed further in this chapter under "The Drawings in Paris" and also in note 53 below; and the corpus of fossil drawings still in Windsor discussed in chapter 11).

25. That there were at least two more volumes of these drawings may be deduced from the fact that the *first* of the 211 folios in the *Erbario* at Windsor carries the number 499; the related loose drawings of flowers, grasses, leaves, mushrooms, and tubers at Windsor and elsewhere are all numbered between 1 and 498. These, presumably, are from the other volumes of the *Erbario miniato* before they were dismembered by the royal librarian just after the First World War. See McBurney 1989c, pp. 8–9.

26. These are by no means secure identifications. If the hand is that of Cesi, it is far less crabbed than in later years, such as appears in fossil drawings and in the volumes in the Institut de France discussed below. Here the hand seems to me to be more forceful and more consciously elegant (just as one might expect from an enthusiastic young man). If the other hand is that of Heckius, then it is certainly neater than the hand in the Montpellier manuscripts, but these—as we have seen—were prepared in great haste.

There is another possibility altogether, suggested first, I think, by Francesco Solinas; and that is that the main set of inscriptions is actually from the hand of Giovanni Battista Winther, the Bavarian doctor and botanist who participated in many of the Lincean botanical excursions from early on (see also Cortesi 1927 on Winther's botanical trips to Norcia and elsewhere; see chapter 13, note 24 below).

27. "*Dal Matthiolo*," or—even more significantly—"*dal mio Matthiolo*" is one of the most frequently recurring phrases in the manuscript.

28. If this compiler is indeed Cesi, then the *Erbario* offers a spectacular indication of his immersion in the field of botany at an early stage of his life.

29. RL 27749; cf. Mattioli 1585, pp. 1175–1176.

30. The illustrations of the verbascum and angelica (figs. 8.4–5), for example, show just how much closer to the casual scruffiness and informality of nature are these plants than their stiff and uninformative equivalents in Mattioli (figs. 8.6–7).

31. Windsor, RL 27830–27832, *Erbario miniato*, fols. 140–142 (perhaps *Sechium edule* on fols. 27830–27831); cf. Mattioli 1585, p. 1349.

32. Windsor, RL 27788, *Erbario miniato*, fol. 98; cf. Mattioli 1585, p. 1255.

33. Windsor, RL 27809, *Erbario miniato*, fol. 119; cf. Mattioli 1585, p. 1295.

34. RL 19397; cf. Mattioli 1585, p. 945.

35. Windsor, RL 27773, *Erbario miniato*, fol. 83.

36. Windsor, RL 27751, *Erbario miniato*, fol. 61; cf. Mattioli 1585, p. 1180.

37. Compare Mattioli 1585, p. 439, with the remarkable drawings of the aconite on fols. 24–25 of the *Erbario miniato*, Windsor, RL 27714–27715.

38. Cf. Freedberg 1996b and Freedberg 2000.

39. Windsor, RL 19414. The Greek letters AI AI, which form the basis of Ajax's—and Hyacinth's—name, were supposed to have been uttered by him and then formed by the drops of his blood as his died. According to the myth, these are still discernible on petals of the plant.

40. Several species are good for insomnia, such as the various poppies, of course, and the numerous varieties of solanum and atropa (different species of each of these are represented on a number of successive pages, e.g., fols. 15, 16, 19—all described as *sonniferi*). The wild poppy, also serves a similar purpose, especially when cooked up according to the recipe given on fol. 1 (RL 27691). Others work for depression, like the white-flowered buglossum or common borage (*Borago officinalis* L.), which mixed with wine serves to cheer and console the spirit (fol. 93; RL 27783); very many have diuretic or more generally digestive functions; and quite a few are good for excessive menstrual flows and pains.

41. Windsor, RL, *Erbario miniato*, fol. 68, RL 27758 ("*Gallio*").

42. New York, Private Collection.

43. *Erbario miniato*, fol. 58, RL 27748 ("*semprevivo maggiore*"); cf. Mattioli 1585, p. 1175.

44. *Erbario miniato*, fol. 32, RL 27722.

45. Ibid., fol. 52, RL 27742.

46. Ibid., fol. 18, RL 27708.

47. Ibid., fol. 26, RL 27717.

48. Ibid., fol. 31, RL 27721; here identified mistakenly as *brassica canina*, which is in fact *Cynonura erecta* L. Even today, this plant is sometimes called a "*topa*" in Tuscany, meaning a vagina or vulva, with reference to the elliptical shame of the fruit in maturity, with a split in the follicle recalling the shape of the female pudenda.

49. *Erbario miniato*, fol. 196, RL 27886.

50. The first publication to give them at least something of their due was that of Ubrizsy 1980; this was followed in quick and much fuller succession by the important articles by Alessandrini, De Angelis, and Lanzara 1985; and De Angelis and Lanzara 1986.

51. Paris, BIF, MSS 968–970.

52. Ibid., MSS 974–978.

53. The same hand also appears in the indices of the fossil volumes in Windsor Castle (*Natural History of Fossils*, vols. I–V; see chapter 11 below). It should be noted that one of the Paris manuscripts—Paris, BIF, MS 977—lacks the usual scribal index entirely.

54. Alessandrini, De Angelis, and Lanzara 1985, p. 319.

55. The elder Delessert's stamp is still on the inner flysheet of the volumes. Cf. Ubrizsy 1980, p. 132; as well as A. Lasegue, *Musée botanique de M. Benjamin Delessert* (Paris, 1845); and Eugène Fournier, "Note sur le Musée Delessert," in *Actes du Congrès international tenu à Paris en août 1867* (Paris, 1867; reference in Ubriszy 1980, p. 134).

56. Bouteron and Tremblot, *Catalogue général*, p. 233.

57. Paris, BIF, MS 968, preliminary leaf (first endpaper). For a fuller summary of the provenance and adventures of these volumes, see Freedberg in Pegler and Freedberg, forthcoming.

58. Paris, BIF, MS 968, preliminary leaf (first endpaper).

59. Alessandrini 1978, pp. 20–21.

60. There are also many inscriptions in red chalk and in graphite. I believe all these to be Cesi too, though it has been suggested that they may be one or even two further hands. On this problem see the essay by Freedberg in Pegler and Freedberg, forthcoming.

61. On a drawing of a fungus seen at Acquasparta, in Paris, BIF MS 978, fol. 173 (cf. also MS 978 fols. 152, 184, and 195 for other dates around Christmas 1623).

62. Paris, BIF, MS 970, fols. 5 and 6.

63. Paris, BIF, MS 968, fol. 3.

64. As De Angelis and Lanzara 1986, p. 255, note 13, remark, this distinction between observation with a lens (*lente*) and with a microscope may well offer further support to the view that the "occhialino" that Galileo gave to the Linceans in 1624 was in fact a primitive compound microscope. Cf. chapter 6 above, under "Cesi Asks for a Microscope" and "Microscopes and Mites."

65. Clusius 1601; cf. Aumüller and Jeanplong 1983.

66. Pegler in Pegler and Freedberg, forthcoming.

67. Kew, Royal Botanic Gardens, Library. See Pegler and Freedberg, forthcoming.

68. On these see now Pegler and Freedberg, forthcoming. For an excellent survey of the history of mycology before Clusius, see Béla Pozsár, "Das pilzkundliche Wesen vor Clusius," in Aumüller and Jeanplong 1983, pp. 70–72.

69. Particularly noteworthy for garden historians is the appearance of the name of the ubiquitous but altogether mysterious gardener Tranquillo Romauli in MS 976, fols. 92, 119, 121, and passim.

70. For the goose barnacles, see Paris, IDF, MS 977, fol. 38, but see also chapter 14 below, and figs. 14.4–5, as well as notes 14–16 there. For the sea-hare dissection, see Paris, IDF, MS 977, fol. 41.

71. Linnaeus, *Philosophia Botanica*, Vienna 1763 (first ed. Stockholm 1751), p. 92.

72. Prior to the rediscovery of these volumes, the scale of Cesi's contributions to the history of botany had only been recognized by the distinguished Roman botanists Poggioli, Pirotta, and Chiovenda; see the difficult and rare booklet by Poggioli (1865), as

well as Pirotta and Chiovenda 1900–1901 and Pirotta 1904. For the rest, as Gilberto De Angelis and Paola Lanzara have put it, Cesi's discoveries fell victim to "a kind of anti-Cesian and anti-Lincean sourness which in previous times also detracted from the genius of Galileo." Cf. De Angelis and Lanzara 1986, pp. 251–252, who first set out the achievements of Cesi in the Paris volumes, as outlined in the paragraphs that follow here. Recognizing but not detailing their importance, Ubriszy (1980) published a brief notice of these volumes six years before them.

73. Cf. the discussion under "Ferns, Spores, and Reproduction" in this chapter. Here Colonna's own pioneering work in using a lens for the examination of fungoid specimens should be noted; see, for instance, his enlarged figures of "lichens" in the *Ekphrasis* of 1606 (Colonna 1606, p. 331; also illustrated in Singer 1915, p. 320).

74. E.g., BIF, MS 976, fols. 58–65; MS 968, fol. 121.

75. Generally referred to by Cesi as the "sementi fungi." Cf., for example, Paris, BIF, MS 968, fol. 121, where the enlargement is approximately 30X, showing the typical cordlike element or *funiculus* attaching the peridiole to the cup (fig. 8.27).

76. I have taken all this from the excellent account in De Angelis and Lanzara, in Cesi 1986, p. 255.

77. Paris, BIF, MS 976, fol. 137.

78. *Thesaurus*, p. 148; cf. BAV, Stamp. I, 80, fol. 104 (cf. further below in this section, as well as chapter 10 below, under "The Evidence of the Proofs").

79. *Thesaurus*, p. 874.

80. Ibid., p. 875.

81. "Quid enim vetat, quod omnium mater Natura plantis tribuerit, alio vel alio loco uterum & foetum gerere, & excludere magis, minusque conspicuum? Et si Rusco, Hypoglosso, & lauro Alexandrinae flores, & fructus in summo foliorum constuit conspicuos quid mirum: in Filiceis & Capillaribus, si in eorum dorso non tam conspicuos, sed parvos adeo nasci voluit? quos nullo in his diligenti examine, vel anatome praeposita, a iuventute adesse animadvertimus. haec ab Ill.mo nostro Principe Caesio non modo iam diu observata sunt, sed nunc apertius demonstrata parvi illius Telescopij adiumento quod mirum magnumque in modum obiecta, vel minima ob oculos distendit magis quidem, minusque iuxta artificis praestantiam, & lentis exiguitatem, & illius varietatem, imo exigua vel Atomo paria obiecta, ita magna oculis objicit, ut quisque in admirationem incidat, ac Polypodij, & congenerum semina Piso, vel Ciceri paria offert oculis, quae arenae exiguo grano paria simpliciorbius existimantur. Qui autem suo tempore pilosam illam extuberantiam in similibus Telescopio parvo observaverit, & semina, ut diximus, conspicient, & insuper in semine umbilicatam partem etiam quae haerebat, & alimentum, & incrementum capiebat, clarissime observabit, qui absque Telescopio vix nigram arenulam splendidam conspiciet: in hac scilicet pilosa flavescente in folij dorso extuberantia bino ordini dispositos corymbo deprehendet quinquagenos circiter flores sive fructus cum seminibus in singulis. Ex his non imperfectam plantam, sed sua natura dorsiparam existimandam, & sic perfectam, si etiam haec nostra planta ex eo esset genere." *Thesaurus*, p. 875.

82. Windsor, *Erbario miniato*, fol. 143. Windsor, RL 27833.

83. Cf. note 20 above.

84. *Thesaurus*, p. 537.

85. *Thesaurus*, p. 757.

86. Ibid.

87. Windsor, *Erbario miniato*, fol. 143; Windsor, RL 27833.

88. Rome, BANL, Archivio Linceo MS 31; for Cesi's annotation here, see p. 148.

89. On this copy, see chapter 10 below, under "The Evidence of the Proofs."

90. Cf. note 20 above for details.

91. Cf., for example, Paris BIF MS 976, fol. 19, inscribed "Scillae flos 4 septembris 1627 700 circiter floribus onustus."

92. A very typical example is the odd flower and stalk in the shape of the head of a deer in Paris, BIF, MS 977, fol. 110

93. Paris, BIF, MS 976, fol. 47.

94. Ibid., MS 977, fol. 186.

95. For one of the critical and too little known works in the history of the illustration of this species, see Hadrianus Junius's remarkable little treatise titled *Phalli* (Delft: H. Schinckel, 1564).

96. Paris, BIF, MS 968, fols. 110–114. One of the very few parallels to this sort of thing is indeed offered by Junius's treatise of 1564, with its vivid woodcuts by none other than Maerten van Heemskerck (as the text makes clear).

97. In addition to the folios cited in the previous note, see, for example, Paris, BIF, MS 970, fols. 27, 162 ff; MS 974, fol. 26, MS 976, fols. 15, 89, 110, 161, and passim; MS 978, fol. 153.

98. E.g., Paris, BIF, MS 974, unnumbered folios between and including fols. 112–113.

99. Paris, BIF, MS 976, fols. 46–47; cf. fols. 42 and 44.

100. The index is simply headed "*Diversorum*."

101. E.g., Paris, BIF, MS 978, fol. 165.

102. E.g., ibid., fols. 190 ff.

103. E.g., ibid., fols. 135–137.

104. For the Lincean interest in phosphorescent phenomena, especially light-emitting stones, see especially chapter 11 below, under "The Bologna Stone" and "More on Light and Heat"; as well as Redondi 1982; Redondi 1987, especially pp. 20–25 and 298–299; and Reeves 1997.

105. Paris, BIF, MS 978, fol. 224.

106. Cf., for example, the "Tuber Pulverulentus Stellatu duplici basi Aqaspartae 27. Decembri 1623" on fol. 152, and the "Sementi fungus Echinatus Acquaspartae Julij 1624 ad lupariam Sylvan" on fol. 184. Cf. also under "The Drawings in Paris" in this chapter.

107. Paris, BIF, MS fols. 148–50, including the inevitable *pietra fungifera* on fol. 148.

108. In the appendix of illustrations added by Cesi and Faber at some time around 1625, *Thesaurus*, p. 458; Paris, BIF MS 978, fol. 955.

109. Paris, BIF, MS 978, fol. 162.

110. Cf. chapter 3 above, under "Cesi Intervenes" and note 77 there for references.

111. Paris, BIF, MS 978, fols. 227A-C.

112. Cf. chapter 1 above, under "Oranges and Lemons," as well as figs. 1.25–26, for example

113. Cf. Paris, BIF, MS 974, between fols. 112 and 113.

114. *Thesaurus*, p. 547, where Faber—in the course of his long discussion of hermaphroditic humans and animals—records how Cesi was present at the dissection as well, and commissioned a drawing (probably this very one) of it. The skeleton, Faber proudly notes, was prepared by none other than himself. Cf. chapter 12 below, under "The Worm and the Dragon," for Winther's anatomy of the Mexican wolf (*Thesaurus*, pp. 812–814).

115. Cf. especially chapter 11 below, under "Middle Natures," and chapter 13, below, under "The Mockery of Order," but also chapter 7 above, under "Order in the *Apiarium*," for more on Cesi's views on the relationship between imperfect things and the problem of classification, anomaly, and deficiency. On Bodin's view of intermediate classes and their role in classification, see Blair 1997.

## CHAPTER NINE

1. On Imperato and his museum, see his own *Dell'istoria naturale* published in Naples in 1599, along with its famous and often-reproduced illustration of the interior of his museum. See now also Findlen 1994, pp. 31–34, 41, 121, 129, 225–233, and passim; Accordi 1981a; and Stendardo 1992.

2. For the works that Della Porta immediately started dedicating to Cesi, see chapter 2 above, under "separation and Exile," as well as Gabrieli 1989, pp. 719–723.

3. "Plus enim in fronte quam in recessu promittit," Della Porta to Cesi and the Lincei, June 25, 1604; *Carteggio*, no. 14, p. 35.

4. "Napoli paradiso delle delitie, amenità de' piaceri, spasso delle vaghezze, bellissimo, gentilissimo, giocondissimo, stanza di Cerere fertilissima e di Nettuno abbondantissimo e di Venere cortesissima et piacevolissima, ma non più inanzi di gratia che vi è troppo da dire; ma per sbrigarmene in una parola sappia, che s'io havessi havuto i Lyncei, sarebbe per me stato troppo gusto, in questa vita, la stanza di Napoli." Cesi in Rome to Stelluti in Fabriano, July 17, 1604, in *Carteggio*, no. 15, p. 41.

5. Cf. also note 40 below.

6. Cf. note 2 above.

7. "Protomédico general de todas las Indias y tierra firma del Mar Oceano." Hernández 1960–84, I, p. 149.

8. Somolinos d'Ardois 1960, p. 232.

9. Francisco Hernández to Phillip II, March 24, 1576, in Medina 1898–1907, II, p. 285; cf. López Piñero and Pardo Tomás 1994, p. 18.

10. Gabrieli 1989, pp. 373–383; Hernández 1960, I, pp. 296–303 and 409–417; Somolinos d'Ardois 1960, pp. 95–440; Alessandrini 1978, pp. 143–171, with much basic material and relevant bibliographic references; López Piñero and Pardo Tomás 1994, pp. 17–20, 40. Cf. Carutti 1883, p. 53. It is possible that the book or books of dried plants were not sent to Philip in 1576, but were rather brought back with Hernández when he returned to Spain in 1578. There has been much discussion about the language of the volumes sent back by Hernández; but it now seems likely that they were in Latin, even though he himself worked on translations into both Castilian and Nahuatl (cf. the long discussion in Alessandrini 1978, pp. 171–193, with all the relevant evidence, as well as the more firmly assertive pages in López Piñero and Pardo Tomás 1994, p. 40 and passim).

11. Somolinos d'Ardois 1960, pp. 276–283. A number of other manuscript notes and drafts, corrected by Hernández himself, ended up in the Colegio Imperial in Madrid. A century or so later they entered the Convent of San Isidro, where they formed the basis of Casimiro Gómez Ortega's "edition" of Hernández's *Historia Plantarum Novae Hispaniae* published by Ibarra in Madrid in 1790 (they then passed into the Biblioteca del Ministerio de Hacienda; cf. Somolinos d'Ardois, I, p. 396). Of the five planned volumes only three were actually published. The matter was further complicated by the fact that when Hernández himself returned as an old man to Spain in 1578, he brought back with him a quantity of further papers on which the material was based, including several draft manuscripts still preserved in the Biblioteca Nacional in Madrid, MSS 22436–22439. What else he brought back with him is not entirely clear, though I suspect that it may have included one or more volumes of the herbaria of dried plants referred to occasionally in the literature (see, for example, note 35 below).

12. Jose de Siguenza, *Tercera parte de la Historia de la Orden de San Jerónimo* (1605), as in *La fundación del Monasterio de El Escorial* (Madrid: Aguilar, 1963), p. 310.

13. See, for example, the several contemporary references to the work in this chapter under "American Pictures in Naples."

14. Cf. the comment by Stelluti in his afterword to the *Tesoro Messicano*: "opus Mexicanae naturalis historiae, quam Fernandes primum invictissimi Philippi II Hispaniarum Regis in Indias Protomedicus delegatus (etsi tumultuarie satis, ut primis assolet ingenii partus) lustratam oculis, manibusque contrectatum in ipso natali solo contexuerat." *Thesaurus*, p. 951.

15. "No van tan limpios, ni tan limados, o tan par orden (ni ha sido posible) que no deban esperar la última mano antes que se impriman, en especial que van mezcladas muchas figuras que se pintaban como se ofrecían." Francisco Hernández from Mexico to the King in Madrid, March 24, 1576, cited in López Piñero and Pardo Tomás 1994, p. 18.

16. As late as 1649, in his afterword to the great Lincean publication based on the material that forms the subject of this chapter, the *Tesoro Messicano*, Francesco Stelluti would refer to its hugely confused original state; *Thesaurus*, unnumbered page (but 951).

17. On Recchi, see Gabrieli 1989, pp. 377–378 and 732 (as well as many other references to Recchi throughout); and López Piñero and Pardo Tomás 1994, pp. 59–71.

18. "contando que haya de usar y ejercer el oficio de simplicista, teniendo cuidado de hacer plantar y cultivar yerbas medicinales en nuestros jardines y otras partes convinientes y ver lo que truxo escripto de la Nueva Espana el Dr. Francisco Hernández y concertarlo y ponerlo en orden, para que se siga utilidad y provecho . . . ," as cited in Jiménez Muñoz 1977, p. 75.

19. Instituto Valencia de Don Juan, Envío 99, fol. 241r, published in López Piñero and Pardo Tomás 1994, pp. 66 and 82.

20. "He procurado para que esta obra saliese a luz de que se buscasen aquí algunas personas que tallasen y hiciesen las estampas de los dichos simples y yerbas de la forma y tamaño que están en los dichos libros," and continued as in the following note.

21. "Y no hallo cierto cosa en que más bien se haya empelado tal cuantidad de dinero." Instituto Valencia de Don Juan, envío 99, fols. 190v –191r, as published in López Piñero and Pardo Tomás 1994, pp. 66 and 82–84.

22. For an example of these, see López Piñero and Pardo Tomás 1994, p. 86.

23. Recchi's version of Hernández's "Index of the Medicaments of New Spain" was directly translated as a separate section of the first great medical work ever to be published in the New World, Juan Barrios's *Verdadera Medicina*, printed in Mexico City in 1607. Here it was given the expressive heading "On all the plants discovered in this New Spain by Dr Francisco Hernández, *Protomédico*, at the command of His Majesty, and their applications to all ailments" (See López Piñero and Pardo Tomás 1994, pp. 108–104 for both the *Index Medicamentorum Novae Hispaniae*—which they show to be based on Recchi rather than directly on Hernández's manuscript itself—and Barrio's translation). Much more substantially, the bulk of Francisco Ximénez's *Quatro libros de la naturaleza y virtudes de las plantas y animales que estan recevidos en el uso de la Medicina en la Nueva España*, published in Mexico City in 1615, was adapted from a copy of Recchi's manuscript, with many corrections and adjustments by Ximénez himself based on his firsthand knowledge of the hospital gardens at Huaxtepec. (For a brief assessment of the differences between Ximénez's work and the Recchian original (insofar as it can be deduced), see López Piñero and Pardo Tomás 1994, pp. 119–127.) Then, too, that diligent and learned director of the Dutch West Indies Company, Johannes de Laet, edited portions both of Ximénez and of a Recchian manuscript in his well-known series of guides to the New World published in handy pocket formats between 1625 and 1640 (Johannes de Laet, *Nieuwe Wereldt ofte Beschrijvinge van West-Indien*, Leiden: Isaack Elzevier, 1625, reprinted in Dutch in 1630; translated into Latin in 1633 and French in 1640, all published by the Elzeviers in Leiden). Finally, no less than 160 chapters—but only 5 illustrations—of the *History of Mexican Plants* in Books XIV and XV of the Jesuit Juan Eusebio Nieremberg's 1635 *Historia naturae, maxime peregrinae* were based not only on Hernández's manuscripts in the Escorial but also a few remaining in the Colegio Imperial in Madrid, where Nieremberg taught natural history (López Piñero and Pardo Tomás 1994, pp. 129–132). The text and several of the illustrations of the *History of the Animals of New Spain*, also included in the *Historia naturae, maxime peregrinae*, were also based on these manuscripts of Hernández's.

24. In fact, in 1586 Aldrovandi (who had always been interested in American things) had written to the Archduke Francesco de Medici in Florence about "a truly royal book with pictures of various plants, animals and other new things from the Indies" that the archbishop of Piacenza had reportedly seen at Philip's court when he was there at some point before 1582; he was writing now to ask whether the duke could possibly arrange for some copies to be made. Mattirolo 1904, p. 375.

25. Gabrieli 1989, p. 732, citing Bologna, BU, MSS Aldrov. 136, vol. XIII, fol. 294r (= vol. XVII, fol. 53).

26. Della Porta in Naples to Aldrovandi in Bologna, June 7, 1590, in Gabrieli 1989, pp. 731–2. For other letters of summer 1590 from Della Porta to Aldrovandi mentioning the "nitro vero" of Ischia and the remora, as well as Nardo Antonio Recchi, see Gabrieli 1989, pp. 733–734.

27. See especially chapter 4, under "Botany and Astronomy: The Academy Expands," and this chapter, under "Colonna and the Naples Lynceum" below.

28. Colonna 1592, p. 50.

29. On Montano, see Rekers 1972.

30. See the poem he wrote to Montano, in which he also proudly refers to his twenty-four books on the plants of the Americas, as well as to its illustrations (Hernández 1960–84, VI, pp. 28–35). See also ibid., volume 1, especially pp. 280–282.

31. Clusius to Camerarius February 20, 1597; in Hunger 1927–43, II, p. 447.

32. Recchi died in 1595. Cf. also Imperato to Clusius form Naples, August 8 1597, cited in López Piñero and Pardo Tomás 1994, p. 79.

33. Gabrieli 1989, p. 732; see here and p. 733, as well as the reference in the preceding note, for further exchanges between Clusius, Imperato, and Gianvincenzo Pinelli (the Paduan antiquarian and natural historian) on the subject of Recchi and his books of plants.

34. "*pulsitangulus et urinicernulus*" are the self-deprecating words he applies to himself here (in the letter cited in the following note).

35. Heckius in Madrid to Stelluti in Rome, June 2, 1608, in *Carteggio*, no. 48, pp. 110–111.

36. On this episode, in which Markus Welser, Scioppius's distinguished friend from

Augsburg, and probably the philosopher Antonio Persio—all then still supporters of Campanella—were also involved, see Gabrieli 1989, pp. 385–399; De Renzi 1992–93, pp. 49–51; and the letters from Welser to Faber of early 1608 published in Amabile 1887, pp. 33–37; and partly transcribed by De Renzi (1992–93, pp. 52–53).

37. Cf. Ernst 1991, pp. 134–136; and now especially Headley 1997, pp. 26–69.

38. Cf. the letters from Scioppius to Faber of October 31, 1607 and January 16, 1609 cited in Gabrieli 1989, p. 389. Much more on this trip is likely to be found in the surviving correspondence between Scioppius and Faber (most of it was destroyed by Faber himself) in Faber's still underexplored archives formerly in Santo Maria in Aquiro, Archivio degli Orfani, and now housed in Rome, BANL (and not, unfortunately, explored in the otherwise very useful treatment of this whole subject by Headley 1997).

39. Now usefully summarized in De Renzi 1992–93, pp. 49–57. For the complex history of the early German efforts at liberating Campanella, see now also Headley 1997, especially pp. 69–77 and 205.

40. On September 10, 1610, Imperato wrote to Faber saying he was sending Faber and Schreck a packet of 130 seeds of rare plants, because "I have heard from Sig. Fabio Colonna that you are planning to make a fine composition on this subject" (*Carteggio*, no. 58, p. 154). By then the two Germans had already begun working together on a book on American plants. Two months later Imperato sent another 150, "so that Sig. Schreck can complete the work he has begun" (Ferrante Imperato in Naples to Faber in Rome, November 26, 1610; *Carteggio*, no. 59, p. 155).

41. On Petilio, see the typically fascinating biography in Erythraeus 1645, pp. 44–49.

42. Though perhaps it was this material that actually inspired the plans for the expansion of the original Academy in the first place.

43. It should be acknowledged that a whole other scenario is possible here. We know that Petilio spent many years of his later life in Rome. He may even have lived there from before 1610. In this case, the scenario would be that it was only *after* Cesi's trip to Naples—and therefore back in Rome—that he saw the Recchian manuscripts, which Petilio would have brought from his and his uncle's native town.

44. As observed by López Piñero and Pardo Tomás 1994, p. 49.

45. As late as 1624, Faber wrote to Cesi that in making a few adjustments to Schreck's text on the plants, "I was unable to see all the plants in the original. That grumpy old man—*il vecchio moroso* [Petilio]—hardly showed me these ones, saying that it wasn't necessary to see them all." Faber in Rome to Cesi in Acquasparta, September 21, 1624, in *Carteggio*, no. 780, pp. 940–942 (p. 941). Cf. also note 66 below.

46. See note 54 below, as well as chapter 4 above, under "Botany and Astronomy" and note 40 there.

47. *Thesaurus*, p. 788 ("animalium horum et plantarum icones sunt depictae, propriisque ac venustissimis coloribus illustratae").

48. The exceptions are the summary index in Barrios and the extracts in Ximenez, as noted in note 23 above.

49. With the exception of a few engravings made for the epitome of the Recchian material planned by Juan de Herrera (cf. this chapter, under "Recchi Takes Over," above); see López Piñero and Pardo Tomás 1994, pp. 81–86.

50. Indeed, he was constantly working on it. For example, "As soon as I arrived in this quiet place I immediately turned to the Mexican book, which is being printed now," he wrote to Francesco Barberini from Acquasparta on November 25, 1623; *Carteggio*, no. 684, p. 826; Gabrieli 1989, pp. 381–2. Cesi never gave himself a moment's peace.

51. Welser in Augsburg to Faber in Rome, August 19, 1611, in *Carteggio*, no. 72, p. 170–171.

52. For a more detailed history of the printing and other vicissitudes of the book—and its manuscripts—see the rest of this and the next chapter.

53. For a translation of the full and still more cumbersome title, see under "Sponsorship at Last," below.

54. "Adunque dovevo io li giorni passati quando in cassa l'ill.mo et ecc.mo s. Marchese Cesi, mio Signore, veddi le pitture di 500 piante Indiane, affermare, o quella essere una finzione, negando tali piante ritrovarsi al mondo, o vero, se pur fossero, essere frustratorie et superflue, poi che né io né alcuno de i circostanti conosceva le loro

qualità, virtù et effeto?" Galileo, *Opere*, XI, 107–108 (Galileo to Piero Dini in Rome, May 21, 1611: cf. also chapter 4, under "Botany and Astronomy: The Academy Expands").

55. Welser in Augsburg to Faber in Rome, July 29, 1611, in *Carteggio*, no. 69, p. 168.

56. "stampa adesso un libro delle piante nove di Messico," in Gabrieli 1989, p. 1019; citing Rome, BANL, Archivio degli Orfani di Santa Maria in Aquiro, filza 420, fol. 81.

57. Cesi in Rome to Galileo in Florence, September 17, 1611, in *Carteggio*, no. 75, p. 174.

58. Cesi in Tivoli to Galileo in Florence, October 21, 1611, in *Carteggio*, no. 78, pp. 176.

59. Cesi in Rome to Galileo in Florence, December 3, 1611, in *Carteggio*, no. 81, p. 178.

60. In the event, though, Schreck continued to remain close to the Mexican project, even to the extent of visiting the Escorial to see Hernández's originals in 1618.

61. Welser in Augsburg to Faber in Rome, December 10, 1611, in *Carteggio*, no. 82, p. 179. For an important discussion of the *Hortus Eystettensis* in relation to the Lincean work on the *Tesoro* at this time (as well, more generally, on the German Linceans themselves), see Belloni 1983, especially pp. 25–30 (on the *Hortus*).

62. Who as cardinal of Santa Susanna would in 1624 be shown some of the very earliest microscopes by Galileo and Faber; cf. chapter 6, under "Cesi Asks for a Microscope" and "Microscopes and Mites."

63. Cesi to Faber in Rome, June 20, 1612, in *Carteggio*, no. 131, pp. 239–241 (pp. 240–241). For Cobelluzzi's efforts, see Stelluti's application (Stelluti in Rome to Paul V in Rome, June–July 1612, in *Carteggio*, no. 135-a pp. 1272–1273; Paul V in Rome to Stelluti in Rome, July 21, 1612, in *Carteggio*, no. 142-b, pp. 1273–1274) and Cobelluzzi's draft for this privilege (which only appeared in very abbreviated form between the two Medicean privileges of 1627 and 1618 in the printed versions).

64. Gabrieli 1989, p. 532.

65. Cesi in Rome to Galileo in Florence, November 24, 1612, in *Carteggio*, no. 185, 293.

66. For an outline of his still barely explored activities, see Gabrieli 1989, pp. 1011–1052.

67. Cesi in Tivoli or Rome to Faber in Rome, August 21, [1615 ?], in *Carteggio*, no. 409, pp. 509–510.

68. These ten books extend over the first 344 pages of the *Thesaurus* as it was eventually published in 1649–51. The prolegomenon runs from pp. 1–27; and this was followed by Book II on aromatic plants on pp. 27–43; Book III on trees on pp. 44–100; Book IV on *Frutices et suffrutices* on pp. 101–130; Book V on *Acres herbas* on pp. 131–180; Book VI on *Amaras herbas* on pp. 181–219; Book VII on *Salsas et dulces herbas* on pp. 220–258; Book VIII on *Acerbas et acidas herbas* (pp. 259–312); Book IX on animals on pp. 313–334; and Book X on the minerals on pp. 335–344.

69. For the most part the notes on color were based on Schreck's knowledge of the Recchian originals, but he probably made a few later additions on the basis of his examination of Hernández's original illustrations when he visited the Escorial in January 1618 (although at that point he pointedly complained that for four years Cesi had not sent him a copy of his original manuscript in order to do the necessary comparative work. Schreck in Madrid to Faber in Rome, January 18, 1618; *Carteggio*, no. 486, p. 625.

70. Herbals he refers to include those by Monardes 1569; Acosta 1578, and Garcia de Orta (1563), as well as those in earlier and less reliable works of Peter Martyr (1530), Oviedo y Valdés (1526), and Gomara (1554). For the titles and bibliographic details of all these works, see Arber 1986, appendix 1, pp. 271–285, as well, of course, as the authoritative listings in Pritzel and Nissen.

71. *Thesaurus*, pp. 63–65.

72. Ibid., pp. 175–177.

73. Ibid., p. 266.

74. Ibid., pp. 345–455.

75. Ibid., p. 346.

76. Ibid., pp. 450–839.

77. Gabrieli 1989, p. 379. But chapter 10 below, especially under "Faber's Exposition," for further and closer details of Faber's contribution to the *Thesaurus*.

78. Cesi in Acquasparta to Faber in Rome, April 11, 1619, in *Carteggio*, no. 537, pp. 685–686.

79. Cesi in Rome to Galileo in Florence, December 27, 1624, in *Carteggio*, no. 804, pp. 970–971 (p. 971).

80. Faber in Rome to Cesi in Acquasparta, December 20, 1623, in *Carteggio*, no. 699, p. 831; Faber in Rome to Cesi in Acquasparta, September 21, 1624, in *Carteggio*, no. 780, pp. 940–942 (p. 941); Faber in Rome to Cesi in Rome, February 3, 1625, in *Carteggio*, no. 821, p. 1023; Faber in Rome to Cesi in Rome, February 4, 1625, in *Carteggio*, no. 822, p. 1024; Faber in Rome to Cesi in Rome, February 8, 1625, in *Carteggio*, no. 823, p. 1025; Faber in Rome to Cesi in Rome, February 18, 1625, in *Carteggio*, no. 824, pp. 1025–1026; Faber in Rome to Cesi in Rome, January 22, 1626, in *Carteggio*, no. 892, p. 1090; etc.

81. "Verae ac, ut ita dicam, realis philosophiae scientiam" run these pregnant words (*Thesaurus*, p. 464). The passage as a whole comes from the last paragraph of the introduction in the definitive edition of 1648–51. In the edition of 1628 these words occur slightly earlier in Faber's introduction, which at that stage was, of course, addressed to their then patron, Francesco Barberini. In the later edition, that of 1648–51, the introduction as a whole was addressed not to Francesco but simply to the friendly reader, "amico lectori." For similar reasons, too, the final words of the 1628 introduction, "Sedente Urbano VIII. Pont. Opt. Max. Barberino," were suppressed. On all these matters, see this chapter, under "Expunging the Barberini."

82. E.g., the American tiger (506–509), panther (513–514), puma (518), the squirrel with the enormous, allegedly musk-bearing testicles (559, 582–583), the Mexican boar (648–9), the "Peruvian sheep" or llama (661–663), and above all the civet (539–540, 552–553) and various kinds of musk-producing animals (pp. 557–563).

83. E.g., the merganser (*Acitli*, pp. 686–687) and the hummingbird (*Hoitzitziltotl*, pp. 705–706).

84. *Thesaurus*, p. 799. On all these figures, see Gabrieli 1989, pp. 380–381 and 1569–1575 (though it should be noted that on p. 1575 he completely muddles Faber's references to the *Hoitzitziltotl* and the amphisbaena).

85. The valuable *Relazione Diaria* of these travels has long been known but has still not yet been published; it is preserved in two stout volumes in the Vatican Library, Rome, BAV, Barb. Lat. 5688 (the journey to France) and BAV, Barb. Lat. 5689 (to Spain). Two other accounts also survive in the Vatican Library, the first by Cesare Magalotti (Rome, BAV, Cod. Barb. 5686), the second by the Cardinal's secretary, Lorenzo Azzolini (Rome, BAV, Cod. Barb. 5349).

86. Cf. chapter 1, under "Cassiano's Role," as well as McBurney 1992a.

87. Alessandrini 1978, pp. 169–170.

88. Montpellier, BEM, MS H. 101, fol. 12r; transcribed in Alessandrini 1978, pp. 218–219, and discussed by her on pp. 170 and 189.

89. The six treatises and the Index are preserved in the same precious manuscript in Montpellier as de los Reyes's covering letter to Cassiano Montpellier, BEM, MS H. 101, fols. 14r—55v and 56r—77r. For a summary description of this manuscript, see Alessandrini 1978, pp. 234–236, with a good discussion on pp. 189–199. López Piñero and Pardo Tomás have only recently demonstrated just how important a guide the *Index alphabeticus* is to the reconstitution of the original contents of Hernández's works, a reconstitution that would otherwise be impossible (López Piñero and Pardo Tomás 1994, pp. 33–58). But why this important Index was never printed in the *Tesoro Messicano* at all remains a mystery. Perhaps it was simply because such a plethora of other indices (cf. also note 124 below) were printed in the final book; one more large one may simply have seemed too daunting, when there was still so much to do. Only Cesi could really have taken it on, and he, as we shall see, was already engaged in the most critical project of his own scientific life, the *Tabulae Phytosophicae* (cf. "Late Additions" below, but especially chapter 11, under "Cesi's Views," and the whole of chapter 13).

90. "Godo che il *Libro Messicano* pigli ogni dì accrescimento. Di Spagna s'è portato il testo della storia naturale dell Hernando, che servirà a V.S. nel riscontro del testo che lei ha stampato in detto libro Messicano; per il quale s'hebbe un tempo fa il privilegio di Francia" (Cassiano in Castel Gandolfo to Faber in Rome, October 20, 1626, in *Carteggio*, no. 935, p. 1136). Did Cassiano thus also play a role in obtaining the French privilege, perhaps working on the issue as he passed through Paris?

91. *Thesaurus*, pp. 1–90

92. Formerly BAV, MS Barb. Lat. 241; history came full circle when Pope John Paul II offered the Codex Badianus, then of course in the Vatican Library, to the Museo Nacional de Antropologia e Historia de México in Mexico City in 1990.

93. For the text of this foreword, see the facsimile edition with good commentary by Emmart (1940).

94. Solinas 1989b, pp. 78–79, citing Rome, BAV, Barb. Lat. 5689, f. 88, as well as Barb. Lat. 4766, fols 12v-14r.

95. Windsor, RL 27902–27938. Solinas 1989b provides the important documentary evidence, particularly on pp. 77–79. For its acquisition by the Adam brothers acting on behalf of George III of England, along with the other volumes from Cassiano's Museum on Paper, see see chapter 1 above, under "A Brief Note on the Provenance of Cassiano's Drawings."

96. On pp. 846–899 of the *Thesaurus*.

97. On these tables, see especially chapter 11, under "Cesi's Views," and the whole of chapter 13.

98. Cf. chapter 6, note 18.

99. Cesi in Rome to Galileo in Florence, September 26, 1625, in *Carteggio*, no. 866, pp. 1066–1067 (p. 1066).

100. *Thesaurus*, pp. 686–696.

101. Faber in Rome to Thuilius in Padua, October 7, 1626. *Carteggio*, no. 934, p. 1135.

102. *Thesaurus*, pp. 693–695. On the drawings sent by Cassiano from France, cf. also chapter 1 above, under "Cassiano's Role," as well as McBurney 1992a, p. 350 and McBurney 1992b, p. 12.

Faber in Rome to Thuilius in Padua, October 7, 1626. *Carteggio*, no. 934, p. 1135.

*Thesaurus*, pp. 843–899.

105. For some of the details of Colonna's additions, see especially chapter 10, under "Illustration Problems, Yet Again" and "Illustration and Identification."

106. Stelluti in Rome to Galileo in Florence, August 14, 1627, in *Carteggio*, no. 951, pp. 1152–1153 (p. 1152).

107. Cesi in Rome to Galileo in Florence, September 4, 1627, in *Carteggio*, no. 953, pp. 1153–1154.

108. Cesi in Sant'Angelo to Galileo in Florence, January 20, 1628, in *Carteggio*, no. 956, pp. 1157–1158 (p. 1158).

109. *Thesaurus*, p. 839.

110. "Flosculis decisis, quod evenit in finis Septembris, quo tempore hoc anno. 1628 notavimus, dum haec imprimenda erant, rubescunt." *Thesaurus*, p. 899.

111. Stelluti in Rome to Galileo in Florence, December 2, 1628: "From here I have to say that although the projects are very large, nevertheless no time is being lost. The tables on the subject of plants by Prince Cesi are now being printed. They will be added to the Mexican book, and they will follow it to the press, so that one can at least offer the fruit of the first part of this book as soon as possible, which is in any case being much desired" (*Carteggio*, no. 983, p. 1189).

112. Colonna in Naples to Stelluti in Rome, November 10 [or 20?], 1629, in *Carteggio*, no. 998, p. 1204.

113. *Carteggio*, no. 999, p. 1204.

114. *Thesaurus*, p. 901.

115. Not one of the surviving copies of the work looks exactly like another. For some of the differences besides those already mentioned, cf. also "Sponsorship at Last" below; cf. also Guerra (1986), p. 312; Guerra also lists the numbers of surviving copies of each "edition" he examined.

116. For details, see the next section, "Sponsorship at Last."

117. See chapter 10, under "Illustration Problems, Yet Again," citing and translating Johannes de Laet in Leiden to Lucas Holstenius in Rome, October 10, 1636, in *Carteggio*, no. 1238, pp. 1242–1243. See also note 23 above for more on de Laet.

118. For the letter, see Cassiano dal Pozzo in Rome to Nicolas Heinsius probably in Leiden, February 20, 1649, in *Carteggio*, no. 1042, p. 1248. For supplementary information, see also the excellent account in Alessandrini 1978, p. 167.

119. There is not much difference between their designs, except that the lush figure

of an Aztec Ceres on the left of the father's title page of 1628 is replaced by an Aztec prince in the son's; while Matthias's Aztec king on the right becomes a rather different figure, with less extravagantly feathered accoutrements of power, in the engraving by Johann Friedrich (fig. 9.4). In the latter, an alert lynx stands at the feet of the fecund, coral-bearing Indian female on the left.

120. Cf. Alessandrini 1978, p. 149; and Stelluti in Rome to Cassiano in Rome, November 20, 1651, in *Carteggio*, no. 1051, pp. 1262–1263.

121. *Nova Plantarum Animalium et Mineralium Mexicanarum Historia a Francisco Hernández Medico in Indijs primum compilata, dein a Nardo Antonio Reccho in Volumen Digesta, a Io. Terentio, Io. Fabro, et Fabio Columna Lynceis notis & additionibus longe doctissimis illustrata. Cui demum accessere Aliquot ex Principis Federici Caesii Frontispiciis. Theatri Naturalis Phytosophicae Tabulae. Una cum quamplurimis Iconibus, ad octingentes, a quibus singula contemplanda graphice exhibentur* (Rome: Blasio Deversini and Zanobi Masotti, Booksellers, and Vitale Mascardi, Printer, 1651).

122. The earliest one dates from long before Matthias Greuter actually engraved the first title page. It was bestowed by Paul V in 1612 and goes back to those early days when Cesi encouraged Faber to negotiate with Cardinal Cobelluzzi on their behalf (cf. note 63 above). For Stelluti's application and Cobelluzzi's draft for this privilege (which only appeared in very abbreviated form between the two Medicean privileges of 1627 and 1618 in the printed versions), see Stelluti in Rome to Paul V in Rome, June–July 1612, in *Carteggio*, no. 135a, pp. 1272–1273 and Paul V in Rome to Stelluti in Rome, July 21, 1612, in *Carteggio*, no. 142b, pp. 1273–1274. Then come the privileges from Cosimo II, grand duke of Tuscany in 1618, the emperor Ferdinand in 1623, Louis XIII, king of France in 1626, Ferdinand II, grand duke of Tuscany in 1627, and Urban VIII in 1626 and 1627. Finally, after the long wait, came de Las Torres's dedicatory letter to Philip IV in 1650 and the approval of the Jesuit censor, Balthasar de Lagunilla, on August 16, 1651.

123. In the course of my researches I have closely examined more than thirty copies, while the Spanish scholar Francisco Guerra claims to have examined more than one hundred (cf. Guerra in Cesi 1986, p. 312).

124. The number and placement of the many indices is particularly variable. By no means all are present in the earlier printings, but most copies of the final issue contain the general index of plants, animals, and minerals, almost entirely consisting of the Nahuatl names; the index of authors used by Schreck and Faber; a short glossary of Nahuatl botanical terms; the index of Nahuatl names in the *liber unicus* of animals and minerals taken directly from Hernández's notes; and the *Index of the medicaments of New Spain*, arranged according to parts of the body and to different functions. This index Stelluti (that paragon of Lincean virtue) proudly claimed to be his own work, when in fact it was directly taken from the equivalent index in Recchi, long since translated into Castilian by Barrios. (See López Piñero and Pardo Tomás 1994, pp. 108 and 116.) But the great *Index alphabeticus* copied out by Andres de los Reyes for Cassiano (cf. under "Cassiano in Spain" as well as note 89 above) was never printed.

125. For example: in some of the early printings, Giacomo Mascardi's valuable letter to the reader summarizing the contents of the various parts of the work appears on the verso of the title page with a crowned lynx in a wreath as the vignette above it, and a lovely *Flos lincea* to end it; in most of the others, this letter occupies a page and a half. Sometimes Schreck's section on the plants ends with "Finis," sometimes with "Finis Plantarum," and sometimes—in the earliest printings—with no such statement at all. In the early printings only twelve and a half of Cesi's *Tabulae Phytosophicae* appear, as opposed to the full twenty later on (the final seven readily identifiable by the different paper on which they are printed). Occasionally an addition to the eighth of the Tables is attached by pasting to the bottom of the relevant table (pp. 917–918); later this appears separately and more efficiently as an appendix. At some point just before Cesi died, three pages of extra illustrations were added following the end of Schreck's contribution, either updating the illustrations of these plants in Schreck's or Colonna's text, or depicting previously unillustrated plants (such as the *Coccus Maldivae*). In the final issues of the *Tesoro Messicano* the addition of these pages resulted in a double set of pages numbered 457–464. The *liber unicus* appears bound in different places of the book; so too do the many valuable indices, none of which are paginated.

126. It is true that Johann Friedrich Greuter's new title page had already appeared with a date of 1648, without Philip's privilege and with Giacomo still as the typogra-

pher, but for the most part the volumes it introduced clearly constituted a last-ditch attempt, before Cassiano set things up with the Spanish faction, to bind up the earlier material for one last time. Very occasionally it was retrieved to form the frontispiece of factitiously made up copies.

127. Stelluti in Rome to Cassiano in Rome, October 27, 1650, in *Carteggio*, no. 1049, p. 1259.

128. Ibid. Three hundred of these, we learn from his next letter to Cassiano on November 20, 1651, were bought by the booksellers Masotti and Deversini; and fifty remained with Cesi's son-in-law, Duke Salviati (*Carteggio*, no. 1051, p. 1262).

129. Ibid., p. 1263.

130. Stelluti in Rome to Cassiano in Rome, November 20, 1651, in *Carteggio*, no. 1051, pp. 1262–1263.

131. "*cum opportune in Urbe agitarem Hispanensium negotiorum.*" *Thesaurus*, sig. ✲2v.

132. This also served as the frontispiece to the separate printing of Faber's contribution in 1628.

## CHAPTER TEN

1. Gomez Ortega (1790), as cited in De Toni 1901, p. 352.

2. Kurt Polycarp Joachim Sprengel, *Historia rei herbariae* (Amsterdam 1807–1808); *Geschichte der botanik* (Leipzig: Brockhaus, 1817–18).

3. Cf. also note 124 in the previous chapter; see also "A New Botany: The Classical and the Vernacular" there.

4. *Thesaurus*, pp. 462–463.

5. Ibid., pp. 473–474, where he also notes that the word "telescope" was itself invented by none other than Cesi himself. Cf. also chapter 4, under "Rome, 1611," on this episode.

6. *Thesaurus*, p. 473; cf. chapter 7, under "Colonna, Faber, and the Microscope."

7. Cf. ibid. for a translation of the text in *Thesaurus*, pp. 473–474.

8. *Thesaurus*, pp. 532–534.

9. Cf. especially chapter 8, under "Ferns, Spores, and Reproduction"; cf. *Thesaurus*, p. 757.

10. Cf. chapter 6, and chapter 7, under "Stelluti's Persius: Science and the Vernacular."

11. *Thesaurus*, p. 757. All of the passages in Faber on the microscope are to be supplemented by those by Colonna in *his* contributions to the *Thesaurus* on pp. 875 and 894–895.

12. *Thesaurus*, pp. 463 and passim. See also chapter 6, under "Chastity and Insemination," and chapter 11, under "Middle Natives."

13. Cf. Naples, BN, MS XII.E.4, fol. 14v; and Archivio Linceo II, fols. 114–130 ("pioggie prodigiose").

14. *Thesaurus*, p. 475.

15. For Cesarini and his portrait by Van Dyck, see now Freedberg 1994c.

16. Performed in 1608 at the Hospital of Santa Maria della Consolazione, in the presence of the surgeon Prospero Cichino, and recorded, apparently, by a draftsman who was also present; both mother and child had died. *Thesaurus*, pp. 614–615.

17. This particular anatomy was conducted in Faber's own house at the Pantheon, after a morning's discussion in the Vatican gardens in the presence of none other than the pope himself and another famous Roman doctor, Giulio Mancini; *Thesaurus*, pp. 599–615.

18. Ibid., pp. 545–548.

19. Ibid., pp. 462–463; 547.

20. Paris, IDF, MS 978, fol. 357.

21. *Thesaurus*, pp. 545–547.

22. Ibid., pp. 812–814.

23. No wonder, for example, that a lengthy discussion of bearded women is followed by one of several embryological discussions, including the analyses of tortoise, chicken, duck, and lizard eggs (e.g., fig. 10.3).

24. *Thesaurus*, pp. 527 and 630; cf. p. 484, where Faber recalls that he had more than 110 animal skeletons in his museum. On Faber's museum, see now Baldriga 1998.

25. *Liagni scheletra*, [Florence, ca. 1621].

26. *Thesaurus*, p. 694.

27. Montpellier, BEM, MS H. 170, fols. 85r–87v, and 120r–121v; McBurney 1992a, pp. 351–354; cf. *Thesaurus*, pp. 674 and 705–709. Needless to say, Cassiano criticized the illustration of this bird in the *Thesaurus*, on the grounds that it gave a wrong impression of its scale, and that its beak was shown as more curved than it actually was (Montpellier, BEM, MS H. 170, fol. 121v). Cf. also chapter 1, under "Cassiano's Role."

28. For Cassiano's description of the toucan, see Solinas and Nicolò 1988, p. LXIX, note 22; for Carlo Antonio's role in transmitting it in 1628, see *Carteggio* no. 621, pp. 760–761 (there wrongly dated to 1622) and McBurney 1992a, pp. 350 and 359. For the whole passage on t he *Picus Americanus* in the *Thesaurus*, see *Thesaurus*, pp. 697–704; but especially pp. 697–698. See also chapter 1, under "Cassiano's Role" and note 88 there.

29. For the music and words of the Flemish song sung by Boudewijn Breyel's erudite parrot, see p. 717.

30. Prompted, of course, by the sole mention of the word *psittacus* in the Latin poet; Stelluti 1630, pp. 4–5.

31. On Cagnati (d. 1612), see also Ferrari's poignant tribute in the later editions of his *Orationes*.

32. For a few others, see also Gabrieli 1989, pp. 1245–1246.

33. On Corvinus, see also chapter 4, under "Botany and Astronomy" and note 42 there. Davidowicz appears on p. 786 of the *Thesaurus*. The other Roman pharmacists mentioned by Faber include Antonio Belmiseri, Adamo Melfi (who gave him three specimens of *Lyncurium* [cf. chapter 11, under "Illustrating the Treatise," and note 114 there]), and Ludovico Coltri, who gave him some "grey amber." The last two came from the same *rione* of the Pantheon as Faber (*Thesaurus*, pp. 536 and 576 respectively).

34. See chapter 1, under "Cassiano's Role," and notes 73 and 89 there; as well as McBurney 1992a and 1992b.

35. Cf. pp. 680 (pictures of the toothed and untoothed *onocrotalus*; cf. figs. 1.5–6 here); 690 (of a stork—"Avem ad vivum artifice pictoris sui manu coloribus exprimi curavit Cassianus Puteus noster Lynceus"; figs. 1.3–4); 694 (of a merganser; cf. also chapter 9, under "Late Additions"); 697 (of a toucan ["Picus americanus"] sent by Carlo Antonio dal Pozzo to Cassiano, from Fontainebleau). See also the references to Cassiano's and Carlo Antonio's collection of ornithological illustrations on p. 674 (with a further reference to Giovanni Pietro Olina, the author of the *Uccelliera*).

36. For further details of where he got these, see chapter 12, under "Probable and Preposterous Pictures."

37. *Thesaurus*, p. 536. On both phenomena, see chapter 8, under "The Drawings in Paris" and "Similitude Genital and Anthropomorphic," as well as chapter 11, "Illustrating the Treatise" and passim. On lyncurium, see especially chapter 11, note 114.

38. *Thesaurus*, p. 503; Faber's account is translated in chapter 4, under "Botany and Astronomy" and note 43 there.

39. Cf. chapter 4, under "Botany and Astronomy" and notes 44 and 45 there, as well as De Angelis and Lanzara 1985, pp. 127–147; and Gabrieli 1989, p. 1192. Also referred to in chapter 13, under "The Reduction to Order" and notes 23 and 24 there.

40. Cf. chapter 12, under "The Worm and the Dragon," as well as chapter 14.

41. See, for example, the foldout sheet attached to the end of Dati 1664 describing the organization of Cassiano's drawings of antiquities (cf. chapter 1, note 5).

42. *Thesaurus*, pp. 748–749.

43. Ibid., p. 749.

44. Ibid., p. 831.

45. Cf. chapter 9, under "Schreck and Faber at Work," as well as this chapter under "Illustration Problems, Yet Again."

46. On Drebbel and the microscope, see chapter 6 above, under "Cesi Asks for a Microscope," and "Microscopes and Mites."

47. E.g., *Thesaurus*, pp. 543, 551, etc.

48. Gould 1998a, pp. 18–19; and 1998b, p. 10.

49. Cf. Linnaeus, *Genera plantarum* . . . (Leiden, 1737), fols. ¶¶r–¶¶v, §13–§14, and the discussion in Freedberg 1994a, pp. 255–257.

50. I.e., Francisco Ximénes, *Quatro libros de la naturaleza y virtudes de las plantas y animales que estan recevidos en el uso de la medicina en la Nueva España* (Mexico City, 1615).

51. I.e., Nieremberg 1635.

52. Johannes de Laet in Leiden to Lucas Holstenius in Rome, October 10, 1636, in *Carteggio*, no. 1038, pp. 1242–1243.

53. Cf. chapter 9, under "Recchi Takes Over," as well as López Piñero and Pardo Tomás 1994, pp. 81–86.

54. Galileo in Rome to Piero Dini in Rome, May 21, 1611, in *Carteggio*, no. 64, p. 162; cf. chapter 4, under "Botany and Astronomy."

55. That the images were already a consideration at this stage we know from many passing indications in the correspondence, such as Markus Welser's comment to Faber in December 1611 on the difference between the great engravings in the *Hortus Eystettensis* and those in Cesi's book. Welser in Augsburg to Faber in Rome, December 10, 1611, in *Carteggio*, no. 82, p. 179; cf. also chapter 9, under "First Delays."

56. "Retulit idem Bibliothecarius *Mexicani Thesauri* impressionem magis magisque procedere, cum sculptor pictoresque iconibus strenuam operam darent"; Rome, BANL, Archivio Linceo IV [the *Lynceographum*], fol. 281 r.

57. Gabrieli 1989, p. 524. For a more detailed discussion of von Aschhausen's relations with the Linceans, and of the central role of the German Linceans in all the Academic activity (but especially botanical) at this time, see also Belloni 1983.

58. Rome, BAV, Stamp. Barb. N. VI. 175. Cesi sent copies of this book to Della Porta and Colonna in Naples in April 1613, when he sent Stelluti in order to set up the branch of their Academy there (cf. the extraordinary letter in *Carteggio*, no. 236, pp. 343–344). It was probably another copy of the same work that the gardener Tobia Aldini thanks Faber for on December 29, 1616 ("Ho ricevuto le virtù compendiose delle virtuosissime Pitture di V.E. le quali veramente sono grande e singulare"), cited in Cortesi 1908, p. 403.

59. Gabrieli 1989, p. 525.

60. Cf. chapter 9, under "First Delays," as well as note 55 above; and Belloni 1983, especially pp. 26–30.

61. Cf. chapter 9 above, under "Colonna and the Naples Lynceum."

62. For a list of the plants that surround the *Flos Lyncis* or Lincea in many of the headpieces, see De Toni 1901, p. 353.

63. I have generally translated "intagli" in this and the following documents as "engravings," though of course the references are to wood engravings.

64. Baglione 1642, pp. 394–395.

65. From Mainz, and therefore yet another German in this whole group.

66. Rome, BANL, MS Archivio Linceo IV, fols. 374r–376 bis recto. Some of these can be specifically identified, such as the *uccelletti* for which Nuvolstella was paid on October 26 (ibid., fol. 374r), and which surely refer to the small birds illustrated in Schreck's very brief treatment of the fauna of Mexico (*Thesaurus*, pp. 320–321).

67. Solinas 1989b, pp. 58–62.

68. Cesi in Acquasparta to Faber in Rome, February 25, 1619, in *Carteggio*, no. 533, pp. 682–683 (p. 682).

69. Cesi in Acquasparta to Faber in Rome, April 11, 1619, in *Carteggio*, no. 537, p. 685.

70. *Thesaurus*, pp. 313–334.

71. *Carteggio*, p. 685, note 4.

72. *Thesaurus*, pp. 346–355.

73. Faber in Rome to Cesi in Acquasparta, December 20, 1623, in *Carteggio*, no. 699, p. 831.

74. E.g., in the discussion of the *Tepenex comitl* on p. 410.

75. Faber in Rome to Cesi in Acquasparta, December 20, 1623, in *Carteggio*, no. 699, p. 831.

76. Faber in Rome to Cesi in Rome, February 3, 1625, in *Carteggio*, no. 821, p. 1023.

77. Ibid., February 4, 1625, in *Carteggio*, no. 822, p. 1024.

78. Ibid., [1626?], in *Carteggio*, no. 942, p. 1143.

79. Ibid., February 18, 1625, in *Carteggio*, no. 824, p. 1026.

80. Ibid., June 17, 1625, in *Carteggio*, no. 853, p. 1050.

81. Ibid., January 22, 1626, in *Carteggio*, no. 892, p. 1090.

82. The museum of Paludanus (Bernard van den Broeck, 1550–1633); referred to in the *Thesaurus*, p. 742.

83. Faber in Rome to Cesi in Rome, February 6, 1626, in *Carteggio*, no. 898, p. 1098.

84. *Thesaurus*, pp. 626, 587, and 637 respectively.

85. Cf. chapter 1, under "Cassiano's Role," on the *Uccelliera*. On Cassiano's use of plates made by Maggi for Antonio Valli di Todi's *Canto degli Augelli* of 1601 and then reused in the *Uccelliera*, see Solinas 2000a, p. 108.

86. Moreover, even after their apiarian presentation of the Jubilee year, Colonna continued to be vitally concerned with the accuracy of the illustration of their microscopic examination of the bee and worked with his Neapolitan colleague Francesco Fontana—"the friend of the bee"—on the subject. See chapter 7, under "Colonna, Faber, and the Microscope."

87. *Thesaurus*, pp. 580–581; Pliny, *Natural History*, 25, 2; cf. Freedberg 1994a on this passage, as well as chapter 12, under "Probable and Preposterous Pictures."

88. *Thesaurus*, pp. 795–796. But cf. my discussion of this episode in chapter 12, under "The Worm and the Dragon."

89. *Thesaurus*, p. 466.

90. Ibid., p. 517.

91. Along, of course, with the Index prepared by Fra Andres de los Reyes for Cassiano; see chapter 9, under "Cassiano in Spain" and note 89 there.

92. "In the example in Recchi not all of the leaves were painted on the branches. . . for this reason the woodcutter has given a defective image of the plant," he says of the *Cocoquaquilitl* (*Adenophyllum coccineum* Pers.) illustrated by Schreck on p. 171 (*Thesaurus*, p. 877).

93. "The picture of this plant [the *Yolmini Quilizpatli* on p. 197] has been passed over here, but you will find it on fol. 445" [of the manuscript]." *Thesaurus*, p. 878. These pages in Colonna's section are studded with brief comments like these and many others about the accuracy of the illustrations and about the taxonomic issues they raise.

94. Typically, he adds some taxonomic refinement as well: "The plant put forward here, as may be seen from the illustration, seems rather closer to a lesser species, which botanists call the pseudo-Hyoscyamus, with pale flowers; in the other greater and more common species, the flowers are white or pale, and then become ruddy. In this plant, as one can see from the example in Recchi, the flowers are a pale color, and the leaves rather more round at the tip, which in the commoner species are quite narrow around the flowers. See Monardes, chapter 14, and Clusius's *Exoticorum* which puts forward and depicts the plant in two elegant illustrations." *Thesaurus*, p. 877.

95. *Thesaurus*, pp. 877–879.

96. Ibid., pp. 882–883, referring to the *Axochiatl* or "water flower" on p. 252. The plant is in fact *Oenothera laciniata* Hill (Hernández 1942–46, I, pp. 60–61).

97. He would "try, this spring, to find a flowering head, which is a most beautiful thing to see." Fabio Colonna in Naples to Stelluti in Rome, November 28, 1627, in *Carteggio*, no. 954, p. 1155.

98. *Thesaurus*, p. 880.

99. This was the plant described by Hernández as the *Holquahuitl*, or *Arbor holli*, and which Recchi quite mistakenly transcribed as *Arbor chilli*. Colonna gives it its name of *Arbor Stelluta* (now *Castilloa elastica* Cerv.) on p. 865 of the *Annotations*.

100. And, it might be added, many others too, particularly in Naples. From the moment Fra Donato d'Eremita (pharmacist at Santa Caterina at Formello) dedicated his illustrated broadsheet titled *Vera effigie della grandiglia detta fior della passione* to Johannes Faber on December 20, 1619, to the lengthy discussion in Ferrari 1633, it roused the interest of many botanists working in the orbit of the Linceans. D'Eremita also dedicated a similar broadsheet to Colonna on October 30, 1622. Cf. also Gabrieli 1989, pp. 1487–1497, on this interesting figure in this circle. For Schreck and the granadilla, see Gabrieli 1989, p. 1042. For a drawing of the granadilla in the Institut de France manuscripts, see Paris, BIF, MS 178, fol. 288.

For earlier important references to the granadilla, see Clusius, *Exoticarum* (1605), p. 347; Ferdinand Crendel, *Descriptio Floris Granadillae* (1608–9); reprinted in Jacob Gretser, *Hortus Sanctae Crucis* (1610); reprinted in *Gretseri Opera*, III, 2, pp. 27–143 (with propagandistic implications on behalf of the Catholic Church); and, of course, Castelli 1625, pp. 49–56. I am grateful to Eileen Reeves for reminding me of the ways in which the passion fruit could then be used as "papist" propaganda, especially by the Jesuits. See, for example, the hostile English reactions in *Calendar of State Papers: Venetian*, October 3, 1609, pp. 360–361.

101. Cf. especially chapter 12, under "The Worm and the Dragon," on the *Monoceros* or *Dracunculus Barberinus*.

102. Colonna in Naples to Stelluti in Rome, November 28, 1627, in *Carteggio*, no. 954, p. 1156. The *dragoncello* mentioned here is surely a reference to the improbable *Dracunculus monoceros* belonging to Cardinal Barberini that presented such classificatory problems to Faber (*Thesaurus*, pp. 816–821; illustrated on p. 817). As is clear from Colonna's letter to Stelluti on February 17, 1628 (see note 107 below), the reference is certainly to this strange specimen, and not to a plant, as Gabrieli, in *Carteggio* p. 1156, thought might possibly be the case. For the granadilla and the *Narcissus serpentarius*, see notes 100 and 103 here, as well as chapter 9, under "Late Additions" and note 117 there.

103. For the granadilla or passion fruit, see *Thesaurus*, pp. 888–890, as well as note 100 above. For the *Narcissus serpentarius* (perhaps a *Haemanthus coccineus*, also illustrated by Ferrari) see pp. 885–886 and again p. 899.

104. *Thesaurus*, p. 899. For the last-minute printing, see chapter 9, under "Late Additions," and note 110.

105. Fabio Colonna in Naples to Cesi in Rome, November 28, 1627 in *Carteggio*, no. 955, p. 1157. Colonna owed much indeed to Della Valle. Beside the woodcut of the cinnamon leaf he added a short passage in praise of him (*Thesaurus*, p. 862); and no wonder! Della Valle deserved commendation not only for his hardiness of purpose in protracting his journeys through Turkey and Mesopotamia, but also for his helpfulness with the Linceans' favorite project. In addition to the seeds and specimens of the cinnamon and pepper plants, he had also sent a papaya and the plant Colonna named the *Acacia mesopotomica* (*Thesaurus*, pp. 870 and 867–868). In both these cases, as well as in that of the cinnamon on p. 862, the Neapolitan doctor Mario Schipani, friend of both Colonna and Della Valle, is mentioned with gratitude, for having supplied seeds or leaves or branches of these plants grown in his garden. The drawing of the papaya seed sent to Cesi by Della Valle is in Paris, BIF, MS 975, fol. 37.

106. *Thesaurus*, p. 816; cf. especially chapter 12, under "The Worm and the Dragon."

107. Fabio Colonna in Naples to Stelluti in Rome, February 17, 1628, in *Carteggio*, no. 957, pp. 1159–1160.

108. *Thesaurus*, pp. 35 and 163.

109. Ibid., pp. 864 and 876.

110. Ibid., pp. 862–864, also noting the critical but very minor difference between this leaf and that of the *tamalapetra*.

111. Ibid., p. 876.

112. Ibid., pp. 889–890. A whole treatise may be written about the early fortunes of the passion fruit in Naples, allegedly first cultivated by Fra Donato d'Eremita (cf. note 100 above).

113. E.g., those of the *Lysimachia Americana* (*Thesaurus*, p. 882); the *Flos Cardinalis* (a *member*, he says, of the Aztec *Cacauaxochitl* family, but not actually the plant of that name, as noted by De Toni 1901, p. 359, against those who conflate the two species on the basis of a misreading of Colonna in *Thesaurus*, pp. 879–80); the *Acacia Aegyptia*, to be distinguished both from the Mesopotamian and the Mexican acacia, or *Mizquitl* (*Thesaurus*, pp. 866–867); the *Mungo* or *Phaseolus Orthocaulis* (p. 887); and, of course, the *Narcissus Serpentarius*, which he discusses and illustrates twice (the last time in the Appendix, on the final page of the work before Cesi's Tabulae; *Thesaurus*, pp. 885 and 889; cf. also notes 103–104 above, as well as chapter 9, under "Late Additions" and note 110 there).

114. Rome, BAV, R.G. Stampe I, 80–81. These volumes were first published by Giovanni Morello in Morello 1986, no. 75, pp. 83–84.

115. Cf. this chapter, under "Illustration Problems, Yet Again," and note 57 above.

116. The *Atepocapatli* in R.G. Stampe I, 80, fol. 11, for example, is quite different from that reproduced by Schreck on p. 34 of the *Thesaurus* (though it is similar to the plant of that name added in the late addition to Schreck's section on p. 454); only the roots are similar. It is also discussed by Colonna on p. 862 of the *Thesaurus*. The real *Atepocapatli*, as reproduced by Schreck on p. 34, appears in R.G. Stampe I, 81, fol. 7.

117. Rome, BAV, R.G. Stampe I, 81, fol. 103; and *Thesaurus*, p. 309. In his comments on this plant on p. 891, Colonna notes that it can be eaten ("edulem esse accepimus").

118. Eg., strikingly, in the case of the *Coapatli* in Rome, BAV, R.G. Stampe I, 81, fol.

81; and the *Coacihuizpatli*, "dentium medicinae" in ibid., fol. 88. Cf. *Thesaurus*, pp. 90 and 188 respectively, for the corresponding final illustrations.

119. Surprisingly, the hand of Faber (papal *semplicista* after all) does not seem to be present. But perhaps the Linceans always intended him to confine himself to the zoological parts of the *Tesoro*.

120. Just how close this exchange could be appears, for example, in the vigorous annotations on Rome, BAV, R.G. Stampe, 81, fol. 104, illustrating the "*Polipodium tuberosum*," the Aztec *Tuzpatli*, and the Linceans' very own tribute to their leader, the *Planta Caesia*. This, of course, was the same plant that Cesi was the first to examine in the course of his pioneering microscopic investigations (cf. chapter 9, under "Ferns, Spores, and Reproduction"), and which Colonna himself recalled in his own comments on Schreck's discussion of the *Tuzpatli* (*Thesaurus*, pp. 147 and 873–875).

121. A good example is provided by the woodcut of none other than the *Planta lincea* itself, the *Coatzontecoxochitl* (*Stanhopea tigrina* Batem) in R.G. Stampe I, 80, fol. 22 (fig. 10.15).

122. Thus too they even have the printer add the correct Latin name to the already-printed pages of the *Tesoro*, as is very clearly the case with the plant they identify as the *Thalictro Messicano*. This is how they decide the woodcut of the *Cozticpatli Acatlanensi* in Rome, BAV, R.G. Stampe, I, 81, fol. 81 should be described, and they inscribe it thus. See *Thesaurus*, p. 235, for the new letterpress added to the chapter.

123. For the *Atepocapatli*, see note 116 above. As for the difficult case of the "caryophilli," there are no fewer than five folios of them in the second volume (Rome, BAV, Stamp. Barb. I, 81, fols. 36–40), based, according to their inscriptions, on fols. 428–429 of the Recchian manuscripts. These are then weeded out on pp. 154–156 of the *Thesaurus*.

124. Rome, BAV, R.G. Stampe I, 81, fols. 84–85; cf. *Thesaurus*, pp. 196 and 878.

125. *Thesaurus*, p. 56.

126. "Questa e la caragna e non fu messa al suo luogo, perche non risponde bene." Rome, BAV, R.G. Stampe I, 81, fol. 86.

127. *Thesaurus*, p. 865, where Colonna notes that the illustration of the *Caragna* printed on p. 455 was omitted from the discussion of the *Tlahueliloca* on p. 56.

128. Rome, BAV, R.G. Stampe I, 81, fol. 56, the cut also inscribed *481*.

129. Rome, BANL, Corsini 139. H.12. This particular copy of the *Thesaurus* was reproduced as a facsimile edition by the Istituto Poligrafico e Zecco dello Stato in Rome in 1992 along with a brief but useful accompanying book by G.B. Marini Bettòlo, *A Guide for the Reader of the Mexican Treasure, Rerum Medicarum Novae Hispaniae Thesaurus* (Rome: Istituto Poligrafico e Zecco dello Stato, 1992).

130. Cf. the following chapter, note 9.

## CHAPTER ELEVEN

1. Rudwick 1985 remains the best general guide to the history of these problems, and provides essential background to this chapter.

2. For an important discussion of the interpretive issues raised by the problem of resemblance when it came to fossils, see Rudwick 1985, pp. 27–29.

3. I use these terms not in their modern taxonomic senses but rather in the frequently changeable terminology of Cesi and his contemporaries themselves.

4. Bound into four large folio volumes numbered XIV–XVII on their spines and each measuring 608 x 460 mm with the crest of George III blocked in gold on front and back (cf. also chapter 1, under "The Discovery"; but see also "A Brief Note on the Provenance of the Drawings" following chapter 1); Windsor, RL 25536–25734. In the first three of these volumes, the drawings seem to retain their original order (to judge from the numeration on the drawings themselves). The colored drawings of vol. XVII are separately numbered 1–38. A fifth volume in this series is inscribed *Natural History: Fossils* V but is in fact a collection of very beautiful colored drawings of rocks, gems, minerals, colored marles, and other curiosities (such as bottle-imps), though it does contain a few paleontological phenomena such as sharks' teeth, and the like; cf. fig. 1.18.

5. Almost all the colored drawings are found in *Fossils* XVII, the volume consisting predominantly of concretions and baked clays and numbered 1–38 (e.g., figs. 11.17–18). A few other colored drawings are found in the other volumes in the series, such as the site drawing illustrated in fig. 11.10.

6. For Cesi's problematic position on the origins of fossils, see this chapter, "Stelluti and Peiresc on Fossils" and "Cesi's Views."

7. Cf. note 4 above.

8. See below, "Middle Natures," for a further consideration of the implications of such names.

9. Cf. the letter from Stelluti to Galileo of August 2, 1630 (*Carteggio*, no. 1011, p. 1217, and cited in chapter 2, under "The Wanderer of the Heavens"; but see also the important passages in Stelluti 1637a, pp. 5 and 12 (cited below, under "Illustrating the Treatise").

10. These finds—and the correlations with the relevant drawings at Windsor (and in Stelluti's *Trattato* of 1637) were made by Andrew Scott, are discussed at greater length in Scott and Freedberg 2000.

11. On these, see Bromehead 1947b and Schnapper 1988, as well as note 64 below, for a fuller discussion.

12. Besides Bromehead 1947b, cf. also Freedberg 1992b, Freedberg 1996a, and Freedberg and E. Baldini 1997 for the Linceans' interest in so-called pregnant species. See also under "Middle Natures" below, as well as chapter 1, under "The Discovery" and "Oranges and Lemons," and chapter 14 below; and figs. 1.27 and 1.29 and 11.15–16.

13. Ruled around the images either in pen and ink or in black chalk, these frame lines were clearly added at any early stage, before the numbers were written on the sheets.

Another mark to appear on some of the sheets is a small cross in pen and ink. Twelve sheets are marked in this way, one with two crosses by two separate specimens. Although it has been suggested that these crosses indicate images selected for engraving, the drawings thus marked do not relate to any of the drawings with frame lines, nor, indeed, to any of the engravings in Stelluti 1637b. Other possible explanations are that the crosses denoted specimens from a particular part of Cesi's collection, from another collection, or from a particular locality. The fact that these crosses do not appear in Cassiano's indices of the drawings (see chapter 8 above and notes 24 and 53 there) reinforces the possibility that they may have been added by Cesi himself.

14. Stelluti 1637b.

15. On Magini's reservations about the telescope, see, *inter multos alios*, Drake 1978, pp. 159–60; Van Helden 1984; Van Helden 1989, pp. 92–93; Lattis 1994, pp. 177–179.

16. It had been found by a Bolognese alchemist on Monte Paderno near Bologna around 1604, at just about the time Cesi was beginning his own explorations of the fossil woods. See Gomez 1991, and the items listed in the following note.

17. For an early account, see Lagalla 1612, pp. 57–58. On this experiment and its significance see especially Redondi 1984, and Redondi 1987, pp. 5–27; see also Galileo, *Opere*, XX, p. 290; *Carteggio*, p. 215, note 8; De Renzi 1992–1993, pp. 64–65 and 136–137; and now Findlen 1994, pp. 230–232.

18. As in Aristotle *De Anima*, II.2.

19. Galileo, *Opere*, XI, pp. 106–108 (a characteristic anti-Aristotelian position already articulated in the dialogue by "Cecco di Ronchitti" in early 1605; cf., for example, chapter 3 above, under "Cosmos and Countryside."

20. On the nature of light and the atomistic approach, see Redondi 1982 and 1984 and most fully in Redondi 1987, as well as the important material (particularly as it relates to Faber in De Renzi 1992–1993, pp. 148–150.

21. For a clear statement of the argument for Copernicanism from the phases of Venus, see especially Shea 1996. But see also Shea 1977, as well as the articles by Redondi cited in the previous note, for further outlines of these immensely complex and still insufficiently studied and understood, matters. See also the references to Psalm 18:5–6 (Ps. 19 in the King James Version) in the letter to Dini of March 23, 1615 (Galileo, *Opere*, V, p. 289), as well as the discussion in Blackwell 1991, p. 74; and Rossi 1978, pp. 45–46 (cf. also note 42 below)

22. These issues appear in Antonio Persio's hardly known treatise *De natura ignis et caloris* (of which a tiny summary was published in Persio 1613), in an important lecture delivered by Faber in 1622, and in several works by the important but now almost-forgotten friend of the Linceans, Fortunio Liceti. On Persio, whose manuscripts became one of the major unfinished publication projects of the Linceans, see Gabrieli 1989, pp. 442–443 and 865–888. The manuscript of Persio's *De Natura Ignis et Caloris* was ready

for publication but blocked by the hostility of the censors (Rome, BANL, Archivio Linceo VI, fols. 13–347 and VII, 4–370); on this see De Renzi 1992–1993, p. 137, with earlier literature. Faber's lecture was titled *Oratio qua Ignis & Metallorum exemplo, quam parum sciamus demonstratur* (1622), Naples, BN, MS VIII.D.13; described in Gabrieli 1989, pp. 1183–1184; published in full and well-discussed in De Renzi 1992–1993. On Liceti, see also under "More on Light and Heat" and notes 28 and 44 below.

23. Cf. chapter 1, under "The Discovery," for John Evelyn's vivid account of the carbuncle in Cassiano's collection. See also chapter 1, notes 9 and 10 for more on this and other stones in his museum.

24. Pliny, *NH*, XVII.188–190.

25. Ibid. "This chapter," notes its modern editor rightly, "is probably a garbled description of lignite . . . [other] ancient writers mention that moisture causes the spontaneous combustion of lignite" (D. E. Eichholz in Pliny, *Natural History*, vol. V [Cambridge: Harvard University Press, Loeb Classical Library, 1962], p. 242).

26. Pliny, *NH*, XXXVII.92–98.

27. "*Anno 1611 a nobis invent[a] et Romam allat[a]*" he proudly wrote on it; Windsor, RL 25656.

28. In fact, in 1640 Liceti wrote a book with this title, in which he summarized his various positions on the relationship between heat and light. Fortunio Liceti, *Litheosphorus sive de lapide Bononiensi* (Udine, 1640.) For more on Liceti's views on these matters, which even appeared in his more accessible and popular work on Roman lamps (the *De lucernis antiquorum*, first published in Venice in 1622), see also Redondi 1985, pp. 20–25 and 298. Cf. also notes 22 above and 44 below.

29. Imperato in Naples to Faber in Rome, June 10, 1611, in *Carteggio*, no. 65, p. 163. Cf. Della Porta in Naples to Cesi in Rome, April 7, 1612, expressing his frustration about "those small pieces of stone which retain light in the dark" (*Carteggio*, no. 113, p. 215).

30. Pascal, *Oeuvres complètes* (Paris: 1963), p. 589 (Pensée 660).

31. Ferrante Imperato in Naples to Faber in Rome, June 10, 1611; *Carteggio*, no. 165, p. 163.

32. Cf. chapter 4, under "Rome, 1611," as well as note 21 there.

33. Cesi in Tivoli to Galileo in Florence, October 21, 1611; *Carteggio*, no. 78, p. 176. He was clearly referring to the treatise "On the phenomena of the orb of the moon," which together with its annex *De luce et lumine* would only appear in the following year (Lagalla 1612). On Lagalla, whom Cesi rejected as a Lincean on the grounds that the work mentioned in the following note was too Aristotelian (Galileo, *Opere*, XII, p. 18), see Gabrieli 1989, p. 450 and passim.

34. These botanical expeditions to Monte Gennaro in the company of a small group of friends were one of their favorite activities; it was here that they collected many of the specimens they studied, renamed, and classified, and here that they laid their first steps in the binomial system of classification that they started developing. Cf. especially chapter 4, under "Botany and Astronomy," with notes 43 and 44 there; but see also chapter 10, under "Faber's Exposition," and chapter 13, under "The Reduction to Order," with notes 23 and 24 there, as well as De Angelis and Lanzara 1985.

35. Cesi in Tivoli to Galileo in Florence, October 21, 1611; *Carteggio*, no. 78, p. 176.

36. Della Porta in Naples to Cesi in Rome, April 7, 1612; in *Carteggio*, no. 113, p. 215.

37. Cf. also chapter 10, under "Illustration Problems Yet Again," and note 57 there on the prince-bishop's visit to Rome and the Linceans.

38. Häutle 1881, p. 92.

39. Cesi in Rome to Galileo in Florence, February 15, 1613; *Carteggio*, no. 219, p. 327.

40. Cesi in Monticelli to Galileo in Florence, May 30, 1613; *Carteggio*, no. 242, p. 357.

41. Cf. chapter 4, note 16, and chapter 5, under "Nature versus Scripture," for the critical role of this text in the early arguments against heliocentrism.

42. This excellent summary from Blackwell 1991, p. 74, on the basis of Galileo, *Opere*, V, p. 301. Cf. the very similar interpretation, however, in Rossi 1978, pp. 58–60.

43. Galileo, *Opere*, V, pp. 307–348; cf. especially pp. 345–347.

44. Liceti had written much about these problems before too, as, for example, in that extraordinary mixture of antiquarian and physical learning, the popular *De lucernis antiquorum*, first published in Venice in 1621. Cf. also notes 22 and 28 above.

45. Cf. this chapter, under "Middle Natures," and chapter 12, under "The Evidence of the Senses," for the Galilean position; as well as, for example, Freedberg 1994a.

46. Naples, BN, MS XII.E.4; published with a helpful commentary by Gabrieli 1989, pp. 38–60. Cf. also chapter 7, under "Order in the Apiarium," and chapter 13, under "The Mirror of Reason and the Theatre of Nature," as well as note 29 there.

47. Naples, BN, MS XII.E.4, fols. 13r–15v. See also chapter 10, under "Faber's Exposition," and notes 12 and 13 there.

48. Ibid., fol. 14v.

49. Ibid., fol. 15 r. Partly cited in chapter 7, under "Order in the Apiarium."

50. Observations still clearly indebted to Aristotle, especially in *De Generatione Animalium*, 732–734.

51. Naples, BN, MS XII.E.4, fol. 16v; cf. the useful discussion in Gabrieli 1989, p. 45.

52. As in the very curious etching labeled "Mercurius Vegetativus Heckij" on a page headed "an Metallophyta" in Cesi's own hand (Paris, BIF, MS 978, fol. 162), while others show blastoidlike forms and mushrooms seemingly embedded in rocks (or growing directly out of them) (fig. 8.36) (e.g., Paris, BIF, MS 978, fols. 150–151, 154–157). Then too there are the strange sheets with architectural and human fragments (presumably fragments of statues) lying half-embedded in the ground (Paris, BIF, MS 978, fols. 148–149), as if in demonstration of the old use of the term *fossile* to indicate anything dug up from the earth, and where the borderline between natural and artificial so oddly frays (e.g., fig. 8.37). On this important issue, see especially the illuminating work by Bredekamp 1995.

53. No one even began to suspect that they were actually the shells of antique cephalopods.

54. Cf. note 12 above, and especially Freedberg 1996a, on pregnant forms in nature. On "jokes of nature," see also chapter 8, under "Heckius and His Simples," as well as note 105 below.

55. Cf. Freedberg 1994a on these matters, as well as chapter 12 below, under "The Evidence of the Senses."

56. See chapter 1 above and Freedberg and E. Baldini 1997.

57. Cf. also chapter 13, note 24, on Winther in Acquasparta.

58. Winther in Acquasparta to Faber in Rome, March 18, 1624, in *Carteggio*, no. 749, pp. 881–882. The reference to Cesalpino is, of course, to Andrea Cesalpino's *De Plantis Libri XVI*, Florence, 1583.

59. Cesi in Acquasparta to Cassiano in Rome, August 18, 1624, in *Carteggio*, no. 772, pp. 933–934 (p. 934).

60. *Inferriti* is the term Stelluti self-consciously uses here.

61. Stelluti in Acquasparta to Galileo in Florence, August 23, 1624, in *Carteggio*, no. 773, pp. 934–5.

62. Cesi in Acquasparta to Francesco Barberini in Rome, September 29, 1624, in *Carteggio*, no. 785, p. 948. Not surprisingly, it would be in precisely this context—furniture made from fossil woods—that Cassiano referred to the Acquaspartan metallophytes in a letter to Lorenzo Pignoria in Padua (Montpellier, BEM, MS H.267, fol. 34).

63. Cesi in Acquasparta to Francesco Barberini in Rome, December 1, 1624, in *Carteggio*, no. 798, pp. 965–967 (pp. 965–966).

64. On the aetites or eagle stone, which can be defined as "any hollow stone containing loose matter, a smaller stone or sand, which rattles when shaken," see Bromehead 1947b, p. 22. The eagle stone was thus called because it was supposed to be found in eagles' nests and to be used by them as an aid to breeding, as too it was believed to prevent miscarriage in women and help in an easy delivery. This talismanic power no doubt derived from the view that the stones themselves get pregnant and give birth. For a skeptical accounts see also, as so often, Thomas Browne's *Pseudodoxica Epidemica*, book II, chapter 5.

65. Cesi in Acquasparta to Francesco Barberini in Rome, December 1, 1624, in *Carteggio*, no. 798, pp. 965–967.

66. Ibid. Specimens that seem almost perfectly to fit descriptions such as these may still be found in the sites around Dunarobba, Rosaro, and Sismano, such as those recently discovered by Professor Andrew Scott (figs. 11.13–14). The day after his letter to the cardinal, Cesi wrote a brief note to Cassiano, commenting that "I am a sending to Signor Cardinale a little indication of the significance of the things discovered by me

*nella inferior Natura de' terranei*, along with some pieces of such objects" (Cesi in Acquasparta to Cassiano in Rome, December 2, 1624, in *Carteggio*, no. 799, p. 967).

67. Including Rome, BANL, Archivio Linceo 2, fols. 57–60 and 62.

68. Rome, BAV, MS Vat. Lat. 8258; Montpellier, BEM, MSS H. 170 and 173; Paris, BN, MS It. 1684; and Rome, BAV, MS Barb. Lat. 6461, fols. 148–149 (the original); another copy on fols. 157–158.

69. Faber's commentary was actually written between 1625 and 1628.

70. *Thesaurus*, pp. 573–574.

71. Ibid., pp. 502–503.

72. Montpellier, BEM, MS H. 267, fol. 34.

73. "Due calamari di legno-fossile" and "Un tavolino di legno fossile rotto senza piedi, e semplice senza profili" in Rome, BANL, MS Archivio Linceo 32, fol. 87r; also available in Nicolò and Solinas 1986, p. 208.

74. *Thesaurus*, pp. 502–503.

75. Stelluti 1630, p. 21.

76. "Invented" in this case is simply to be taken as meaning "identified and named thus".

77. Stelluti 1630, p. 21; cf. p. 170 as well.

78. Ibid. pp. 169–170.

79. "Ma pare che la natura invidiasse il gran sapere di questo Signore"; Stelluti 1637b, p. 12; but cf. also Stelluti's letter to Galileo the day after Cesi's death on August 1, 1630 (*Carteggio*, no. 1011, p. 1217; cited in chapter 2, under "The Wanderer of the Heavens".

80. Stelluti to Cioli, June 2, 1635, cited in Nicolò 1986, pp. 176–177; Florence, BNC, MS Gal. 102, fol. 18.

81. Cf. the letter written to Cioli by Stelluti cited in the preceding note (Florence, BNC, MS Gal. 102, fol. 18).

82. Stelluti to Galileo, November 3, 1635, in Galileo, *Opere*, XVI, 337.

83. This letter was immediately recognized as central to Cesi's work. In addition to several drafts for it, no fewer than six manuscript copies survive, generally in the company of manuscript versions of Stelluti's *Trattato* (Stelluti 1637b); for these versions, see the following note.

84. Montpellier, BEM, MSS H. 170 and 173, Paris, BN, MS It. 1684, Rome, BAV, MSS Barb. Lat. 4355 and Vat. Lat. 8258. For the contents and relationship of each of these manuscripts, see especially Scott and Freedberg 2000, as well as Alessandrini 1978.

85. The fifth copy was almost certainly owned by Jacques De La Ferrière, discussed below. See Freedberg in Scott and Freedberg 2000, pp. 62–63.

86. As, for example, in the letter of August 2, 1635 (Lhote and Joyal 1989, pp. 198–201).

87. Montpellier, H. 271, fol. 146; Lhote and Joyal 1989, pp. 172–173.

88. Lhote and Joyal 1989, p. 200. Montpellier, BEM, MS H. 271, fol. 167v.

89. Cf. also the remarkable letter from Peiresc to Menestrier dated February 1, 1635. Tamizey de Larroque, *Corrrespondants*, V, pp. 756–759.

90. Montpellier, BEM, MS H.170, fols. 11r–12r. See the full transcription in Alessandrini 1978, pp. 214–217, and the discussion in Scott and Freedberg 2000.

91. Montpellier, BEM, MS 170, fols. 11r–12r; accurately transcribed in Alessandrini 1978, pp. 214–217.

92. Naudé 1650, pp. 667–668 (with an interesting recollection of his own visit to Acquasparta in 1636). For his earlier, almost as acid critique, see Tamizey de Larroque, *Correspondants*, XIII (*Lettres de Naudé . . .*), p. 52 (letter of March 29, 1636). For a strangely parallel position, see Alessandrini's curious assault on Cesi (at least with regard to the fossils) in Alessandrini 1978, p. 133 (in an account otherwise very sympathetic to him).

93. Though no one seems seriously to have discussed the classificatory problems at stake—perhaps because they do not seem to have been much expounded in the *trattatello*.

94. All but the five plates of veined patterns (Stelluti 1637b, plates 3–6, 9) are after drawings in Windsor (cf. fig. 11.21). As the drawn illustrations included in Montpellier, BEM, MS H. 170, fols. 8–10 and Rome, BAV, Vat. Lat. 8258, fol. 26 make clear, a few planned engravings were not made.

95. "Gli occulti parti della natura" (Stelluti 1637b, sig. A2) is, of course, a frequent

phrase at the time, especially in the early correspondence of the Linceans. Generally used in regard to that which is hidden below the surface of things, to the secrets and mysteries of nature, the *arcana naturae*, as these things were so often called, the phrase here has an unusually literal dimension in referring to that which is in fact hidden beneath the earth. For another typical use of this notion, see also chapter 13 below, note 8.

96. Stelluti 1637b, pp. 3–4.

97. As also observed by Gould 1998b.

98. Stelluti 1637b, p. 5, also referring to Pliny, *Natural History*, XIII.15.

99. With its indications of the sites where *metallophita et succensiones* were to be found, it is probable that the map was supplied by Cesi to Stelluti; cf. Gabrieli 1989, pp. 940–941. On the general location shown, see Lippi Boncambi 1960, pp. 67–76; and Biondi 1984, p. 16, citing the important work of R. Almagià, *Documenti cartografici dello Stato Pontificio* (Rome: Città del Vaticano, 1960). See the essay by Scott in Scott and Freedberg 2000 on the geological significance of this map.

100. Stelluti 1637b, p. 6.

101. Cf. the discussion by Scott in Scott and Freedberg 2000, p. 292.

102. Stelluti 1637b, p. 6.

103. A lesson nicely drawn, with regard to just these matters, by Gould 1998b.

104. Stelluti 1637b, p. 6.

105. "*Scherzando cosi la natura per farci forse maravigliare dell'opere sue cosi rare.*" Stelluti 1637b, p. 8. On "jokes of nature," see chapter 8, under "Heckius and His Simples," and chapter 13, under "The Mockery of Order," as well as Findlen 1990; Freedberg 1992b; and Freedberg and E. Baldini 1997.

106. Stelluti 1637b, p. 8.

107. As pointed out to me by Andrew Scott.

108. Stelluti 1637b, plates 3–6; there is in fact a fifth plate of the same kind (plate 9).

109. E.g., Stelluti in Acquasparta to Galileo in Florence, August 23, 1624, in *Carteggio*, no. 773, pp. 934–935; see also Stelluti in Acquasparta to Cassiano in Rome, May 5, 1635, in Rome, BANL, Arch. dal Pozzo, V, fol. 228; cited in Nicolò 1982, pp. 94–95; and again in Nicolò 1986, p. 176 ("one has more stone than wood, which you can tell from its great weight; the other has those cuts and cracks which I refer to in my discourse, and where you can see that white resin"; so cf. Stelluti 1637b p. 9, describing the sample illustrated in his figure 8 (cf. figs. 11.3–4).

110. Stelluti 1637b, p. 7.

111. Stelluti 1637b, plates 10–12. Very similar examples have recently been found in Dunarobba by Andrew Scott (cf. figs. 11.13–14). Cf. also Scott and Freedberg 2000, figs. 21–33, 75–78, and passim.

112. See Gould 1998b.

113. Cf. notes 11 and 64 above.

114. Lyncurium was also known as the "Lapis Lyncis," or stone of the lynx. It was actually a hard-underground sclerotium, often ball-shaped, which could be collected and watered, whereupon it could produce a number of edible fungal fruit bodies over a long period. According to Faber (*Thesaurus*, p. 536), citing Cesalpino and a number of other sources, it resulted from the urine of the Lynx being deposited on the host (whether vegetable or mineral). As its name suggests, it had long been regarded as a kind of concretion of lynxes' urine and therefore a fossil form; but for Cesi and the other Linceans, the term was more often used for stonelike fungi (e.g., Paris, IDF, MS 968, fols. 165–169; cf. fol. 145, depicting a *Polyporus tuberaster* or stone fungus) or for stones, concretions, and sclerotia out of which fungi seemed naturally to grow.

115. I am grateful to Stephen Jay Gould for helping me with the translations of *pietre Astroite* (corals) *pietre Giudaiche* (spines of sea urchins), *la pietra Bucardia* (bucardites or bulls' hearts), *l'ostracite* (fossil oysters), *la vermiculare* (fossil worm tracks), *escara marina* (sea urchins, some of which were thought to be fossil eggs) listed in the remarkable passage on fossil forms that interested Cesi in Stelluti 1637b, pp. 11–12. Several of these are also mentioned in Cesi's Naples manuscript, Naples, BN XII.E.4, fols. 15r–v.

116. Especially in Paris, IDF, MS 978.

117. Inscribed on the spine in gold: *Natural History: Fossils V;* for more on the binding of this volume, see note 4 above.

118. Montpellier, BEM, MS H. 173, fol. 4 r; Paris, BN, MS It. 1684, fol. 3v.

119. On the use of this particular metaphor in the context of Galileo's interpretation

of Scripture (and his opponents' interpretations of data about the natural world!), see chapter 3, under "Cosmos and Countryside," and chapter 5, under "Sunspots."

120. Here probably meaning the movements of the planetary bodies and the effects of these, as defined by Cleomedes's book on the subject, which we know Cesi owned and studied (*Meteora Graece et Latine*, ed. Robert Balfour [Bordeaux, S. Millanges, 1605]). For Cesi's personal copy, with the Lincean stamp, see *100 Books Recently Acquired*, Catalogue 117, New York: Jonathan A. Hill, Bookseller, no. 31 (sold to the John Carter Brown Library, Brown University, Providence, Rhode Island).

121. Colonna in Naples to Stelluti in Rome, November 10, 1628, in *Carteggio*, no. 982, pp. 1187–1188.

122. Colonna in Naples to Stelluti in Rome, November 10, 1628, in *Carteggio*, no. 982, p. 1188.

123. Ibid.

124. In order to do this, Galileo made a rather forced distinction between "fire" and "flame." Cf. Reeves 1991, pp. 3–4 on the rather strained arguments of Galileo here.

125. As always, this example touched on a number of other critical issues too—such as whether any star could be seen through the brilliant light of flames. On all this, see, for example, Galileo, *Opere*, VI, pp. 174–195 passim, as well as pp. 361–365 and 495–497. In his initial use of this instance of transparency of flame, Grassi happened to use the example of semi-burned wood as seen through the flames of a fire (Galileo, *Opere*, VI, p. 174). Cf. also the important discussion in Reeves 1991.

## CHAPTER TWELVE

1. For a survey of the issues at stake, see Freedberg 1994a.

2. "Un ottima scala, et mezzo sicurissimo congionto con gl'altri accidenti cioè odore, sapore, et tatto per venir in cognitione perfettissima de misti o siano perfetti o imperfetti." Bologna, BU, MS Aldrovandi 6, vol. I, fol. 6v. cited in Olmi, 1992, p. 368.

3. Galileo, *Il Saggiatore* (Rome, Mascardi, 1623); G.G. *Opere*, VI, 347–349. For a complex commentary, see also A.C. Crombie, "The Primary and the Secondary Qualities in Galileo Galilei's Natural Philosophy," in *Saggi su Galileo Galilei*, ed. C. Maccagni (Florence, 1972), II, pp. 71–91; as well as Alberto Pasquinelli, *Letture Galileiane*, Bologna: Il Mulino, 1968, pp. 110–120. Cf. also the discussion of "secondary" qualities in chapter 8, under "Heckius and His Simples," and elsewhere throughout this book.

4. The classic survey of this broad topic has become Eisenstein 1979.

5. Fuchs 1542, fol. 7v; cf. Freedberg 1994a on the whole subject.

6. Fuchs 1542, fol. 7v.

7. Linnaeus 1737, fol. ¶¶verso.

8. On Aldrovandi, see above, note 2 as well as the rest of this chapter, but especially chapter 13, under "Precedents."

9. Cesi in Rome to Stelluti and De Filiis, April 10, 1605; *Carteggio*, no. 24, pp. 67–68.

10. As well as *inter multa alia*, Paris, IDF, MS 978, fols. 26, 133, 182, etc.

11. Windsor, RL 32569 (inscribed "388").

12. For more on these works, see the closer discussions in chapter 4, under "Botany and Astronomy," and chapter 9, under "American Pictures in Naples" and "Colonna and the Naples Lynceum."

13. But cf. the discussion in chapter 4, under "Botany and Astronomy," as well as chapter 13 note 21.

14. *Thesaurus*, p. 538 for the illustration of this improbable animal; pp. 538–581 for the discussion. See also chapter 10, under "Illustration Problems, Yet Again."

15. Ibid., p. 580.

16. For more on the shortcomings of the woodcuts, see chapter 9, under "First Delays" and "Sponsorship at Last," as well as chaper 10, "The Evidence of the Proofs."

17. *Thesaurus*, p. 581, citing Pliny, *Natural History*, XXV, 4. On this passage in Pliny and the complexities both of the text and the translation, see Freedberg 1994a, p. 245.

18. Ibid.

19. E.g., on pp. 864 and 876. Cf. the discussion in chapter 10, under "Illustration and Identification."

20. Hence the particular detail of many of the extra woodcuts Colonna provided, including those of the Egyptian and Mesopotamian acacia (a notably more detailed illus-

tration than the *Mizquitl*, or Mexican acacia, which is reproduced alongside it in *Thesaurus*, p. 866; cf. also p. 867), the papaya (p. 870), the *Lobelia Cardinalis* (or, as it is still called here, the *Flos Cardinalis Barberini*, p. 880), the *Lysimachia Americana* (p. 882), the *Mungo* or *Phaseolus orthocaulis* (p. 887), the passion fruit (pp. 889–890), and the *Narcissus serpentarius* (p. 885), on which he added a very last-minute appendix (p. 899). On all of these, see especially chapter 10, under "Illustration and Identification." The passion fruit had been the subject of intense discussion in those days (chapter 10, note 100), and a close-up was provided not just of the fruit but also of the parts of the flower (pp. 889–890); and he was especially proud of the fact that the *Narcissus Indicus Serpentarius* had never been described before, and that *he* had been the one first to do so—on the basis of a specimen grown in the garden of the same Bernardino de Cordoba who owned the civet cat discussed earlier in this chapter. Here too he provided a pair of illustrations, one with a detail of the flower and the other showing a specimen with its bulb and seeds (pp. 885 and 899 respectively).

21. *Thesaurus*, pp. 466 and 538 respectively. The illustrations of an African or Asian civet cat supplied by Fabio Colonna on pp. 580 (cf. fig. 10.7) are indeed more plausible than the absurd version called a "Zibethicum animal americanum" on p. 538 (fig. 10.8).

22. *Llama guanacoe* (ibid., p. 660).

23. In fact a breed of dog, the xolo (ibid., p. 479).

24. In fact a jaguar, *Panthera onca* (ibid., p. 498).

25. Actually an ocelot, *Leopardis pardalis* (ibid., p. 512).

26. That is, a bison (ibid, p. 587).

27. That is, the peccary or *Tayassu Tajacu* (ibid., p. 637)

28. Interestingly, the woodcuts of the birds are much superior both in terms of technical quality and accuracy—perhaps because of the able collaboration of that passionate ornithologist, his friend Cassiano dal Pozzo. Even so, Cassiano himself noticed that an illustration such as that of the hummingbird made the bird seem too large and its beak too curved. Montpellier, BEM, H. 170, fol. 121v, criticizing the illustration in *Thesaurus*, p. 705 (cf. McBurney 1992a, p. 354).

29. *Thesaurus*, p. 498.

30. Ibid., p. 697; on this kind of rhetorical *enargeia*, see the excellent discussion in Ginzburg 1989a.

31. *Thesaurus*, p. 590.

32. Ibid., p. 582.

33. Ibid, p. 637.

34. On the meaning of "conjecture" here, see the discussion in De Renzi 1992–93, pp. 190–193.

35. In art, once more, as in life. What, after all, can be judged about a person on the basis of appearance alone? Conjecture is always permissible; but judgment, in the absence of any further factor enabling one to arrive at the core of being or character, must remain in suspense.

36. *Thesaurus*, p. 790.

37. Cf. part VIII, chapter 10 ("Les vers de l'eau douce") of Charles Bonnet's *Contemplation de la Nature* and the discussion in Dagognet 1970, pp. 73–74.

38. *Thesaurus*, p. 796.

39. Ibid., p. 797.

40. Ibid., p. 797.

41. Ibid., pp. 799–831.

42. Cf. the very splendid drawing in Paris of a bat found in the crypt of a church at Narni, BIF, MS 978 fol. 355 (fig. 8.39).

43. Cf. Paris, IDF, MS 978, fol. 357 (fig. 8.40).

44. *Thesaurus*, pp. 812–814.

45. It should be noted that little help is cast on this puzzling matter by the strange treatise on the Guinea or Medinaworm (*Dracunculus*) published by Welsch [Velschius] in 1674, which also mentions some of the Linceans' researches as well as the *Dracunculus* discussed by Faber—without casting any light on the matter at all.

46. This whole issue may, of course, also be considered in terms of the need to preserve what writers such as Steven Shapin, Mario Biagioli, and Paula Findlen have broadly called "civil conversation." For an illuminating discussion, see especially Shapin 1994, pp. 120–122.

47. De Renzi 1992–93, p. 215.

48. *Thesaurus*, p. 818.

49. Galileo, *Opere*, VIII, pp. 613–614. Cf. Cesi's own words written at the time of the death of Cesarini, on the relationship between philosophy, mathematics, and the study of the world of nature, as cited in the following chapter under "Cesi's Tables: Geometry and the Ambiguity of Classes," and notes 90–91 there.

## CHAPTER THIRTEEN

1. I use the term "species" here in the loose sense it was often used by naturalists of the time: not to indicate a specific class within a hierarchy of classification but rather simply to indicate a group perceived as just that. For good discussions of some of the issues involved in more precise distinctions—between genos and eidos, for example—see the works cited in note 3 below. See also chaper 1, note 61.

2. Windsor, RL 25526. On the same sheet are several other relevant examples, including a bezoar, an aetite, and an apparently petrified mushroom. The drawing, still laid down, has a set of identifying inscriptions on its verso. As the ink has soaked through somewhat over the passage of time, these inscriptions may at least partially be read from the recto with the aid of a mirror: "7. Bezaar orientalis colore palido ad viriden[?] tendente catal. 9. 264; 8. Nux seu cavites cum tegumento substantia durissima colore ferugineo obscuro; 9. Fungite: colore ferugineo; 10. Pseu-corallium [?] vel juncus petreius colore cinereo valde rarus."

3. As set out, for example, in *De Partibus Animalium* 642a–644a. For a very useful summary of the complexity of the Aristotelian use of the term "differences," and its relationship to the even more complex distinction in Aristotle between *genos* and *eidos*, see Peck 1965, pp. lxii–lxvii. For more recent and authoritative discussions of Aristotelian classification see Balme 1987 and Pellegrin 1986, in addition to the important work by Atran 1990 on the later heritage of the Aristotelian positions—as complicated and as apparently inconsistent as they often were.

4. The collocation of "differences and powers" is significant, of course; cf. especially chapter 8, under "Heckius and His Simples," as well as chapter 1, under "Ferrari, Flowers, and the Microscope."

5. Another telling collocation; cf. the preceding note for further references.

6. Heckius (on behalf of the Linceans) in Rome to Clusius in Leiden, March 20, 1604; *Carteggio*, no. 11, p. 32. See also chapter 8, note 19 for the other letters in this group.

7. The original text here is so critical that it deserves citation: "Ludibria naturae ista sunt, et mixtionum miracula. Quis, enim praeter admirationem, tantam colorum varietatem tam miro ordine distinctam et praeter miraculi indicium in his animantibus considerare posset? Quid non molitur natura? Quid non potest? Tentarunt eam imitari quos vides et ego eam scrutans apprehendere: illis lusit illa, eorum mihi ludibria manserunt; coepi tamen et si non ex essentia cognoscere, tamen ex particularibus agendi modum conijcere, invariabilem sane et certum in his etiam, quae nos imperfecta dicimus, reliquis tamen in suo esse longe perfectiora, cum non solum vel in membris non fallant aut morbum ullum multitudinis aut magnitudinis patiantur, sed ne minimus in ijs quidem color mutiletur." Heckius in Prague to the Lincei in Rome, August 1, 1605, Montpellier, BEM, MS 508, fols. 2–3. I have used the transcription in Gabrieli, *Carteggio*, p. 88 rather than that in Alessandrini, p. 212, on the grounds of Gabrieli's greater accuracy and helpful punctuation.

8. Cf. the same letter (cited in the preceding note): "Sic adversis correspondent papilionum aliis maculae et hexagonus pedum numerus ipsis quaternis alis, sic locusta et saltans unum quodque sive aqua sive terra illud nutriat aut pariat semper posteriores pedes anteriorbius longiores et fortiores habet, et perpetuus inter eos firmatus ordo, non nisi specie varians et mutilatum nihil admittens, naturae igitur et haec arcana sunt et neglecta ego ex septentrione collecta." For more on Heckius' view of the *arcana naturae*, see also chapter 8, under "Heckius and His Simples" and note 18 there.

9. Cf., inter alia, the careful exposition in Bremekamp 1953.

10. On Colonna, cf. especially chapter 4, under "Botany and Astronomy," and chapter 9, under "American Pictures in Naples" and "Colonna and the Naples Lynceum."

11. Of course, as already described in the sections enumerated in the preceding note, for Cesi and his fellow Linceans, part of the great attraction of both this work and its second part, the *Ekphrasis*, published in two parts in 1606 and 1616, was the

pioneering use of a new technique for the reproduction of plant illustrations: the etching.

12. For Heckius's letters of November 14, 1602, and January 8, 1603, to Aldrovandi, see Gabrieli 1989, pp. 1084–1086; for his letters of 1604 to the botanists, see this section and note 6 above, as well as chapter 3, under "Cesi Intervenes" and note 76 there. See also chapter 8, under "Heckius and His Simples" and note 19 there.

13. L. Frati, A. Chigi, and A. Sorbelli, *Catalogo dei manoscritti di Ulisse Aldrovandi* (Bologna: N. Zanichelli, 1907), p. 164, note 1.

14. Only two works based on this material were ever published in his lifetime. These were the volumes on ornithology and entomology; all the others, from the books on shells, mollusks, and fishes to those on trees, snakes, quadrupeds, and monsters followed much later. For this reason they present a curiously antiquated "system" of ordering that one might not have expected as late as the second half of the seventeenth century, when the last of Aldrovandi's works appeared in forms much modified by untactful and unsophisticated editors.

15. These range from *aequivoca* (the various meanings of the name of each species) through comparatively straightforward classificatory headings such as *genus* and *differentia* to taxonomic ones such as *cognominata* and *denominata*; from natural and utilitarian headings such as habitat, coitus, birth, gestation, education, voice (or song, in the case of birds), method of locomotion, habitat, intelligence, sympathy, antipathy, to medical, historical, and mythical ones—medical uses, nutritional use, appearance in prophecies, in fables, in mythology, in emblems, in symbols, in proverbs, in hieroglyphs, in sacred images, in coins, in law, in maritime contexts, in sacred ceremonies, in secular ceremonies, and so forth, apparently limitlessly. Cf. Foucault 1973, pp. 38–40.

16. For references to his personal copy of Mattioli, see chapter 8, under "The Erbario Miniato."

17. Fabio Colonna, *Purpura. Hoc est De purpura ab animali testaceo fusa, de hoc ipso aniali, aliisque rarioribus testaceis quibusquam . . .* (Rome: Mascardi, 1616). To this is appended the *De glossopetris dissertatio*, on the fossilized shark's tooth.

18. Interestingly, it was the murex that stood at the origins of Michel Adanson's work on shells. In the *Histoire naturelle du Sénégal: coquillages* (Paris: C. T. Bauche, 1757), p. 8, he discusses the problem raised by the fact that the shape of their shells is different in their youth than in their maturity (cf. Dagognet 1970, p. 43).

19. Thus, at a stroke, he did away with a whole series of widely held commonplaces about them, such as the view that they were produced by bolts of lightning striking the ground, or that they were fossilized serpent's teeth (evidently a less primitive view).

20. Most of these phenomena soon found their way into Cesi's corpus of illustrations (e.g., fig. 14.3).

21. The use of *Ekphrasis* in the title, thus borrowing a term from ancient word for the verbal description of visual works of art, is telling. Cf. chapter 4, under "Botany and Astronomy," and chapter 12, under "Probable and Preposterous Pictures."

22. "Tam in hac, quam in aliis plantis, non enim ex foliis, sed ex flore, seminisque, conceptaculo, et ipso potius semine, plantarum affinitatem dijudicamus." Colonna 1616a, chap. xxvii, p. 62.

23. Gabrieli, *Carteggio*, no. 78, pp. 175–176. On this trip, see also Cortesi in *Annali di Botanica*, VI (1916), pp. 156–160; and now De Angelis and Lanzara 1985, pp. 127–147; and Belloni Speciale 1987, pp. 59–79. Cf. also the following note, as well as chapter 4 under "Botany and Astronomy" and note 43 there for a sense of how affectionately they remembered it.

24. Cf. chapter 4, under "Botany and Astronomy," and notes 44–45 there; Rome, BANL, Archivio degli Orfani di Santa Maria in Aquiro, filza 420, fols. 297–304; De Angelis and Lanzara 1985; and Cortesi 1916, pp. 156–160. Thirteen years later their friend from Bavaria, Cesi's personal doctor and secretary, Giovanni Battista Winther, would himself make a similar list with predominant binomials, following several trips to the mountains and hills around Norcia; cf. Gabrieli, *Carteggio*, pp. 925–926 (*sub* no. 766, Winther to Faber, July 21, 1624); as well as Cortesi 1927, pp. 140–170, especially pp. 146–155. But see especially Belloni Speciale 1987, particularly pp. 65–68 on Winther's role in the botanical work of the Lincei in the region of Acquasparta and Norcia around 1624. Cf. also chapter 11, under "Middle Natures," and note 58 there, for Winther's remarks on the fossil wood of the region.

25. As pointed out by De Angelis and Lanzara 1985, pp. 127–147. Cf. also chapter 8, under "The Drawings in Paris," and note 72 there.

26. Cf. chapter 4, under "Botany and Astronomy," and note 45 there; Cesi in Rome to Galileo in Florence; *Carteggio*, no. 142, p. 252 and Galileo, *Opere*, XI, pp. 365–366 and Galileo, *Opere*, XI, pp. 244–245. See also chapter 5 above, under "Sunspots."

27. Galileo, *Opere*, XI, pp.365–366.

28. Galileo in Florence to Cesi in Rome, June 30, 1612; *Carteggio*, no. 134, pp. 243–4.

29. Naples, BN, MS XII.E.4; reproduced in Gabrieli 1989, pp. 30–58; cf. also chapter 7, under "Order in the Apiarium," and chapter 11, under "Middle Natures." Dictated when he was at the height of his powers, at some point between 1613 and 1615, it gives the fullest conspectus we have of the range of his work, from physics to meteorology, from methodology to astronomy, from the full range of biological sciences to the study of ethnography, archaeology, and philology. It contains notes for the treatises he planned on the constitution of the heavens (the *Coelispicium*), on the atmosphere (*De Aere*), on all kinds of remarkable rains (the *Taumatombria*), on physics, and in various areas of the humanities. As so often, botany and astronomy dominate; but there is also, at various points along the way, a great deal on the different sciences to which Cesi himself contributed so much, such as geology and mycology.

30. "Scilicet Typus Synopticus universalis ad contemplationem omnem et ratiocinationem in plenissimam Encyclopediam methodice conformatus," BNN, MS XII.E.4, fol. 4.

31. Stelluti 1630, p. 21.

32. Allatius 1633, p. 90.

33. "*Primo enim ex rebus ipsis legitimam basim artificiis mentis supposuimus*," Naples, BN, MS XII.E.4, fol. 4r.

34. The same problem as that posed by Galileo's letter to Cesi of June 30, 1612, cited at the end of the preceding section in this chapter and note 28 above.

35. On Ramus, see especially the pathfinding and still fundamental book by Ong 1958.

36. Naples, BN, MS XII.E.4, fol. 4r.

37. If ever there was someone who knew how to turn the terms of the past to good modern use, it was Cesi. No wonder that his actual innovations have largely gone unnoticed!

38. Of course you could always postulate some great chain of being, where everything would be linked to everything else (as Bodin 1597 has already implied in his discussion of intermediate classes); but this was not the way Cesi would go. Cf. also this chapter, under "Cesi's Tables: Geometry and the Ambiguity of Classes," along with notes 98–100 below.

39. As noted in chapter 11, under "Middle Natures," and chapter 7, under "Order in the Apiarium," such "species," according to Cesi, included amphibians and the whole class of oviparous quadrupeds; "zoophytes," "lithophytes," and his beloved "metallophytes"; bats, giraffes, divers, seals, and vipers (many of these were eventually discussed in the pages of the *Tesoro Messicano*); prodigious atmospheric phenomena such as mysterious rains of fire and water (a special interest of Cesi's, to be dealt with at greater length in two other of his projects, the *Taumatombria* and the *De Aere*), and the whole group of imperfect plants (cf., in addition, chapter 8, under "Ferns, Spores and Reproduction").

40. Naples, BN, MS XII.E.4, fol. 15v; cf. also chapter 11, under "Middle Natures," and chapter 7, under "Order in the Apiarium."

41. See Pirotta and Chiovenda 1900–1901, pp. 151–153. Then, after yet another reference to the lithophytes and zoophytes "observed by us," and to "notable plants whose physiological constitution is different from others, and in one way or another confused and irregular" (the Latin term *praepostera* conveys a strong sense of a process that is a reverse of the normal or expected), Cesi proposed a section on "notable forms of fertilization." This would include such strange things as "plantifers corresponding to viviparous animals, not seed-bearing oviparous ones." Then he moves on to "parasitic plants living on other live plants and without roots of their own" (Naples, BN MS XII.E. 4, fol. 15v).

42. See Lepenies 1976 p. 61.

43. Ibid.

44. Cf. chapter 1, under "The Discovery," for Galileo's literary parallel to the dis-

tinction between overarching concepts on the one hand, and piecemeal composition on the other.

45. In their edition of the *Del natural desiderio di sapere*, Altieri Biagi and Basile suggest that "contemplationi" refers to speculative thought and propose "speculazioni" as an equivalent.

46. "Vi sono le raccolte di fiori, di sentenze, d'attione, e theatri e poliantee e giardini et officine varie." These are all extremely common terms for encyclopedic compendia of one or the other kind. "Raccolte di Fiori"—as in *anthology*—here means simply a selection of the highlights of any particular author. I have deliberately avoided using the word that would otherwise seem logical here, viz. "anthology," because this is a better translation for "poliantee" that occurs just below—though of course the sense is almost identical. For "sentenze" and "attioni" together, see, for example, the many Italian translations of Valerius Maximus's *Facta et Dicta Memorabilia* as *Sentenze morali;* for "theatri," very many examples, including Bodin 1597 and the many works listed in Blair 1997 and so forth, often in the spirit of the "teatro del mondo," or the idea that "all the world's a stage . . . ."; for "poliantee," see for example Domenico Nanni Mirabelli's much reprinted *Nova Polyanthea*, first published in 1503 but expanded in the early seventeenth century by Joseph Lang, as in the Venice edition of 1608 (Johannes Guerilius); for "officine," see above all Ravisius Textor's also much reprinted *Officina;* but the term also conveys—importantly in the case of Cesi and Cassiano—the then usual sense of a laboratory or workshop.

47. As I note—and demonstrate—below, to refer this simply to the art of memory of the mid-sixteenth to mid-seventeenth centuries, and the kinds of mnemonic practices and arts derived from ancient writers and found in more recent ones such as Ramon Lull (as does the modern editor of the text, Altieri Biagi, in Altieri Biagi and Basile 1980), is seriously to misjudge both the purpose and the content of the tables and systems actually devised by Cesi and his colleagues.

48. Naples, BN, MS XII.E.4, fol. 24 r. First published in Govi 1888 and now available in Altieri Biagi and Basile 1980. The passage continues with the phrase "both in invention and composition," thus concluding the directly Ramist terminology of the passage as a whole. On the meaning of terms such as these, see especially Ong 1958.

49. These cannot, however, be conceived of solely in terms of facilitating memory, as has been claimed (in Altieri Biagi and Basile 1980, notes 75 and 76). There is more to it than that, just as there is more to the terms "types, conjunctions, divisions, unions, and conditions" than "technical memnotechnic terms." It is true that these are all terms indicating various types of relationship between "the various subjects of the knowable," as the same critic has put it, and that they were engaged to allow "a global comprehension of them all"; but to assimilate these aims to mnemonics seems an oversimplification to me—despite the appeal of setting it in the context of the *ars memorativa* so appealingly and influentially described by Frances Yates and Paolo Rossi, and of the sixteenth- and seventeenth-century desire for better mnemonic systems, as has been done by Altieri Biagi in the notes to her and Bruno Basile's edition of the *Del natural desiderio di sapere* (Altieri Biagi and Basile 1980).

50. But to claim that Cesi's aim was simply to arrive at a "faithful mirror of the harmony of the cosmos" (Altieri Biagi and Basile 1980, notes 75 and 76) is entirely to miss the intensity of Cesi's awareness that the only possibility for science was to order and classify the mass of observational data in ways that would permit analysis of this mass, and encourage the kinds of generalization and theoretical speculation that we now grace with the catch-all term of "hypothesis."

51. Only—and ironically—published on pp. 776–780 of Christoph Scheiner's autodefense, the *Rosa Ursina* of 1630; cf. also chapter 3, under "Scripture and the Fluidity of the Heavens," and chapter 5, under "Cesi versus Bellarmine."

52. This could be a reference to any one of a number of works by Tabernaemontanus (Jakob Theodor Bergzabern, 1520–1588), such as his *Neu volkommen Kreuterbuch*, published in Frankfurt in 1588, or the *Eicones Plantarum* published by Nic. Bassaeus in Frankfurt between 1588 and 1590.

53. The reference is to Pierandrea Mattioli's *Commentari in sex libros Dioscoridis*, edited with editions by Gaspar Bauhin in Basle in 1598.

54. I.e., Mattias Lobelius, *Icones stirpium seu plantarum tam exoticarum quam indigenarum* (Antwerp: Plantin, 1581–91), or possibly the *Plantarum seu stirpium historia* of 1576.

55. Cesi in Acquasparta to Faber in Rome, November 19, 1622; *Carteggio*, no. 643, p. 778.

56. Gaspar Bauhin, ΦΥΤΟΠΙΝΑΧ, *seu enumeratio plantarum* (Basle: Seb. Henricpetri, 1596).

57. Cesi to Faber, from Acquasparta to Rome, January 20, 1621: "Rendo a V.S. le gratie che devo del *Phitopinax* del Bauhino da me tanto desiderato. . . . Desidero anco intendere se vi sia alcuno che habbia ordinatamente distinto et enumerato nelle sue classi li fossili, e particolarmente metallici e mezzi minerali detti; e similmente se alcuno sinopticamente in arbori e tavole habbi ridotto le scienze".

"Mi trovo, sig. Fabri mio, in mezzo alla distinzione e divisione di tutte le cose, e se bene vado con certe mie speculationi e vie particulari, tuttavia è necessario vedo anco quello che hanno fatto gli altri. Nelle pietre particolarmente ho havuto una bella fatiga, che nelli autori non ho trovato pur ombra di ordine che vaglia, e pur ho ritrovato le loro congiuntioni, differenze, divisioni e affinità, con qualche mia sodisfatione, nè credo dispiaceranno a V.S." (*Carteggio*, no. 588, pp. 732–733).

58. Cf. notes 46 and 49 above.

59. Cf. the intensely relevant pages in Ginzburg 1989a, pp. 111–113, on the problem of the inclusion and exclusion of anomaly and monstrosity, and of consequences of the fact that in the "Galilean paradigm," as he calls it, "the more that individual traits were considered pertinent, the more the possibility of attaining exact scientific knowledge diminished" (ibid., p. 111).

60. Which he referred to as "bodies which are most ambiguous, and which show very different combinations: of wood and earth, of wood and stone, of wood that is both stony and metallic, of wood and mineral juices: participating equally in the substance of the body" (*Carteggio*, p. 966; cf. also chapter 11, under "Fossil Woods," and note 63 there).

61. Colonna in Naples to Cesi (or possibly Stelluti, to send on to Cesi) in Rome, October 14, 1628; *Carteggio*, no. 980, p. 1185.

62. "Che di ciò quasi niente da altri è toccato"; Fabio Colonna in Naples to Stelluti in Rome, October 20, 1628; *Carteggio*, no. 981, p. 1186.

63. Ibid.

64. Just how fine Cesi cut things is evident from a letter Stelluti wrote on December 2, 1628, in which the pressure he, Stelluti, and the other Linceans felt to publish the *Tabulae* is also clear: "From here I have to say that although the projects are very large, nevertheless no time is being lost. The tables on the subject of plants by Prince Cesi are now being printed. They will be added to the Mexican book, and they will follow it to the press, so that one can at least offer the fruit of the first part of this book as soon as possible, which is in any case being much desired" (*Carteggio*, no. 983, p. 1189).

65. For his critical early use of the term, see Gabrieli 1989, p. 510, as well as chapter 2, under "Separation and Exile," and note 21 there, for Cesi's early phytosophical lectures.

66. Cf. chapter 11, under "Cesi's Views," and note 121 there.

67. The phrase Colonna uses here is "argento ramoso." This could mean either "branchy silver" which thus makes slightly more sense as a "plant metal"—or simply "coppery silver." Gabrieli rightly notes (in *Carteggio*, p. 1188, note 2) that while this term is not in fact actually used by Imperato 1599, it may conceivably refer to book XV, chapter 25, where Imperato describes the nexus between silver and copper (ibid., p. 447).

68. *Carteggio*, no. 982, pp. 1187–1188.

69. Johan Kentmann, *Nomenclaturae Rerum Fossilium, quae in Misnia praecipue & in aliis quoque regionibus inveniuntur*, printed in Gesner (1565), sigs. a5v–a6r.

70. Lobelius 1576, pp. 666–669.

71. Ramus's "diagrams," of course, are very different, and the spatialization is resolutely horizontal. But even in such systems it was absolutely clear that everything had its proper place.

72. "Ordinata indicatio et expositio illorum omnium, quae in una qualibet planta considerari possunt ac disquiri in tabula Phitologis ob oculos exponenda," Naples, BN MS XII.E.4, fol. 16r.

73. *Thesaurus* 1628, 1630, p. 904.

74. "A PUNCTO, ipsius scilicet primi, summique Plantae nominis CENTRO, per titulo-

rum omnium LINEAS, in capitum superficiem, a quibus corpus deducatur, pertractas, praesigni adumbratione praemissa, qua OBIECTI et TRACTATIONIS circa illud indicia summatim ex omnia materia praesumi possint, unde Plantae proferendae ad omnifariam cognitionem," *Thesaurus*, p. 906.

75. THEORETICIS *praegressionibus illorum omnium, quae ex Planta venire in contemplationem possunt, quibus materiam perlustramus. Thesaurus*, p. 906.

76. The verb used here is *praehendere*, suggesting a preliminary stage of understanding (*Thesaurus*, p. 906).

77. *Dignostica distractione partes sumuntur. Thesaurus*, p. 906.

78. PHYTOTOMIA *imprehendendo ad illa usque continuatur, quae numerum dumtaxat constituunt*. All in *Rerum Medicarum*, p. 906. Cesi's attentiveness to etymological implication, and his masterly sensitivity to the flexibility of language, is apparent throughout this table. Even in this last passage, the prefixes signal a modulation of cognitive understanding: "imprehendendo" clearly represents a deeper stage of grasping than "praehensente," for example. These kinds of subtleties, by which modifications of a process are indicated, are pretty much absent from the de-Latinized and literary English in which much scientific discourse is now popularly presented. There are reasons for the ugliness of scientific discourse.

79. For the process of true apprehension or grasping, Cesi uses the term *praehendere;* to the deep level of correct and thorough knowledge he applies the verb *pernoscere* (*Thesaurus*, p. 906). Once more the skillful use of prefixes is a striking feature of Cesi's inventive work in these tables. Cf. the preceding note.

80. Cf. especially chapter 2, under "Separation and Exile"; chapter 4, under "Della Porta, Galileo, and the Telescope"; chapter 9, under "Naples, 1606" and "Colonna and the Naples Lynceum"; and passim throughout this book.

81. *Thesaurus*, pp. 908–910.

82. "*in recessu magis quam exteriorum pensitatis innotescunt*," as he noted in an almost offhand way in the fourth table. *Thesaurus*, p. 908.

83. Even though it is barely commented upon in Pirotta's otherwise excellent introduction to his edition of the *Tabulae* (Pirotta 1904).

84. Not just form, but the class, as it were, of form. Here divided by Cesi into the notions both of *figure* and of form; *Thesaurus*, p. 908. For an excellent discussion of the relationship between mathematics and physics at the time, see Blair 1997.

85. *Anguli sinus sect. curvit. plan. inaequal.* and *rotund. oblong. teret cylindr. Thesaurus*, p. 910.

86. Ibid., p. 909.

87. Ibid., p. 910.

88. Ibid. D'Arcy Thompson (*On Growth and Form*) would not have been dissatisfied.

89. "La filosofia è scritta in questo grandissimo libro che continuatamente ci sta aperto innanzi a gli occhi (io dico l'universo), ma non si può intendere se prima non s'impara a intender la lingua, e conoscer i caratteri, ne' quali è scritto. Egli è scritto in lingua matematica, e i caratteri son triangoli, cerchi, ed altre figure geometriche, senza i quali mezzi è impossibile a intenderne umanamente parola; senza questi è un aggirarsi vanamente per un oscuro laberinto." Galileo, *Il Saggiatore*, ed. Flora, p. 33, with parallel instances from other parts of Galileo's writings. But cf. also Cesi's reference, in his attempt at a biography of Cesarini, on the book of nature, as cited in the following note.

90. In his attempt at a biography of Cesarini, Cesi wrote that following Galileo's and the Linceans' discoveries, many more pages in this book could be read than the ancients ever saw: "che li caratteri di questo libro," he continued, "erano figure matematiche et esperimenti fisici, e che chiaramente veniva letto dal matematico e dal fisico esperimentatore," with the result that one day Cesarini went to find Prince Cesi, to tell him "che conosceva che la strada da lui propostali era la vera di filosofare e arrivare alla cognitione delle cose della natura, e delle matematiche certezze." Rome, BANL, MS Archivio Linceo IV, fol 311v; also cited in Gabrieli 1989, pp. 783–784.

91. It is extraordinary how commonly one finds this passage (cf. note 89) paraphrased or translated in this rather presumptive way, generally along the lines of "the book of nature is written in characters of geometry" (cf. even Atran 1990, p. 158). When given in English, the passage is invariably said to begin with the words "the book of nature is written in the language [or characters] of geometry" (cf. text above and note 89). But a brief reference to the original text shows that this is not at all what Galileo

said. Of course, a modern interpretation (still under the influence of a very old tradition) may take Galileo to have been referring to "the book of nature"; but what he actually said was that "philosophy is written in this great book (I speak of the universe)," which means something rather different and much more precise than the old cliché. The fact is that almost every commentator on this passage seems to have been unable to avoid thinking of it in terms of the very old *topos* of "the book of nature," as, for example, already discussed in Curtius 1953, pp. 319–326.

92. *Thesaurus*, p. 910.

93. The passage from the imperfect to the anomalous is not as problematic as it may at first seem. Since the norm ("perfect plants") was taken to be the presence of visible seeds, things which grew but which showed no signs of seeds ("imperfect plants") fell easily into the category of the anomalous. It is significant that the last problem Cesi was working on was precisely that of the generation and reproduction of the cryptogams, and that even when evidence of spores and sporangia became overwhelming, he felt that these could only be taken as evidence of the transitional or intermediate signs of organisms that—like sponges, corals, and amber—could not clearly be placed within the system—whether natural or artificial—of nature.

94. The failure to discuss Cesi's work in all these spheres vitiates almost every history of science, especially those that address the progresses of eighteenth-century science (cf. the introduction). The only two works to deal with his contributions to botany—Poggioli 1865 and Pirotta 1904—have remained largely unread.

95. For example, Cesi's inability to free himself from the old doctrines of resemblance, and the frequent indications of the interconnectedess of history, mythology, and morality on the one hand and physical and biological issues on the other.

96. *Thesaurus*, p. 936, curiously misquoted by Pirotta 1904, p. VIII. Cf. "Particularum in figuras [!], indeque figurata membra, compositio: membrorum in plantam coniunctio: spirituum in vegetandi munera conspiratio: hisce omnibus absoluta constitutio; hic consideratur, ut stirps quaelibet suo in genere perfecta" (p. 936) with "Plurium scilicet ipsa ratio syntactica,que ex internis, et externis," here set against judgment of "exterior facies" on the basis of similitude and dissimilitude.

97. For the distinction between internal and external parts, see further the especially dense Table XVI in *Thesaurus*, pp. 942–943. Pirotta 1904, pp. VIII–IX, notes that it is here too that Cesi insists on the importance of avoiding the classificatory errors that result from the grouping together of plants such as the true *Trifolium* and the oxalises solely on the basis of secondary, external characteristics, such as the shape or even number of leaves.

98. "Unaquaeque enim planta alias universas respicit, itidem nulla stirps est, quae in universa coordinatione inter reliquas omnes suum locum praecise non obtineat, quo in ipsarum serie et cathena statuatur." *Thesaurus*, p. 937.

99. And at this point Cesi goes on to give examples: "Quaelibet Planta, praeter propriam Classem in qua pluribus titulis statuitur, degenerans quodammodo [,] propensionem habet, qua peculiaribus quibusdam notis aliam respicit Classem, ut *Monospersum*, *Messagrum*, *Monospermata*, et *Rapaces*." Ibid. Incorrectly and very misleadingly quoted by Poggioli 1865, p. 25, and incompletely quoted by Pirotta 1904, p. IX. For Jean Bodin's less elaborate view of intermediate classes, see Blair 1997.

100. "Sunt et mediae, quae internectere classes ipsas ancipiti quodamodo utrinque affinitate videntur ut . . . ." [ellipsis present in text], *Thesaurus*, p. 937.

101. "L'existence des monstres met en question la vie quant au pouvoir qu'elle a de nous enseigner l'ordre." Canguilhem 1992, p. 171. Cf. also Ginzburg 1989a, p. 112.

102. Canguilhem 1992, p. 174.

103. Ibid., p. 178.

104. Aristotle, *Metaphysics*, *Iota* 4. 1055a6, cited and discussed by Pellegrin 1986, p. 55.

105. Ibid., p. 56. For an excellent account of the ways in which Aristotle viewed the "hitches," as Pellegrin calls them, to the principle of incommunicability of genera, see his note 15 on p. 185, where the Aristotelian distinctions between deviation, intermediate groups, and crossing are discussed at some length.

106. On Ramist systems of structuring knowledge, see especially Ong 1958, and this chapter, under "The Mirror of Reason and the Theatre of Nature" as well as under "Encyclopedias and Frontispieces."

107. Cf. the useful comments in Lepenies 1976 on the later misunderstandings and misuse of passages in Aristotle's *Historia Animalium* about anomaly and hybridity.

108. See Ehrard 1970, p. 113.

109. "Ipsum si vinculum spectes, in individuis tantum habes; si vero aptitudinem ad vincula, quae plures scilicet conjungere valent, jam familias totas connectes." *Thesaurus*, p. 944. Cf. Blair 1997 pp. 131–141 on Bodin's view.

110. Cf. Naples, BN, XII.E.4, fol. 4r: "Primo enim ex rebus ipsis legitimam basim artificiis mentis supposuimus."

111. Already in 1612 Cesi had written to Galileo about Scheiner's refusal to tolerate a sun that was less than perfect by postulating that the *maculae* on the sun's surface was not in fact *on* them, but rather "wandering stars" that moved between our eyes and the sun (cf. Cesi to Galileo, March 3, 1612, in Gabrieli, *Carteggio*, no. 107, p. 207).

112. "non amplius latens sed patens Apelles, omnibus spectandam proponere," Scheiner 1631, n.p. (dedicatory letter to the duke of Albano).

## CHAPTER FOURTEEN

1. I am perfectly aware that in the Gombrichian scheme of things, the schema—or arguably the diagram—lies at the heart of all picturing, and that one could also argue for a continuity between diagram and picture; but to argue for pure continuity would be sophistry. I take the polarity to be a clear, serious, and instructive one.

2. Both in these pages and in works ranging from Agnes Arber's to Elizabeth Eisenstein's. A serious work is Karen Reeds's; a good popularizing one is Wilfred Blunt's.

3. For Faber's magnificent gift of a copy of the *Hortus Eystettensis* to Cesi for Christmas 1615, see *Carteggio*, no. 419, p. 519; for the book on the Farnese gardens titled *Exactissima descriptio rariorum quarumdarum plantarum quae continentur Romae in Horto Farnesiano* attributed to Tobia Aldini see Castelli 1625.

4. Cf. Freedberg 1992b; but see also Foucault 1973 for his brilliant account of the epistemic shift from a culture of similitude to one of what he called "mathesis."

5. A few were by Camillo Cungi; see Freedberg and E. Baldini 1997 for more information on the engravers of the plates in the *Hesperides*.

6. See Freedberg 1992b for a longer discussion of these issues. When Ferrari comes to a digitated or malformed fruit that he feels he cannot account for in biological terms, he comes to a halt, and says "Et hactenus quidem philosophicas inter coniecturas caecutisse sit satis" (Ferrari 1646, p. 299)—before embarking on an extended explanation of the form in a purely poetic vein (cf. Freedberg 1992b, p. 298–299).

7. Both quotations come from a succession of rather disparaging comments (about the tables) in Findlen 1994, pp. 73–75.

8. *Carteggio*, p. 350.

9. Cf. chapter 10, under "Illustration Problems, Yet Again," and notes 58–59.

10. Despite Stelluti's claim that this would be a book in which for once the author would *reason* about physiognomy—that is, present it in a reasoned way, offering a kind of *catalogue raisonné*, one might say, of *all* the equivalents between outward appearance and inner character, he does not seem to have realized just how difficult all this would be.

11. Despite my gendering of the artist of the Institut de France volumes, it is entirely possible that the "draftsman" was a woman.

12. For the inscriptions on the verso of this sheet, see chapter 13, note 2.

13. Cf. the introduction above and chapter 2, under "Four Friends," as well as passim throughout this book. See also, however, the letters Heckius wrote in 1604–1605 on behalf of his fellow Lincei soliciting advice, seeds, and information from learned men all over Europe, as cited in chapter 3, notes 75–76, and chapter 13, note 6 above.

13. London, Collection Cyril Fry.

14. "There is a small Ilande in Lancashire called the Pile of Foulders, wherein are found the broken peeces of old and brused ships, some whereof have been cast thither by shipwracke, and also the trunks or bodies with the branches of old and rotten trees, cast up there likewise; whereon is found a certaine spume or froth, that in time breedeth unto certain shels, in shape like those of the muskle, but sharper pointed and of a whitish color; wherein is conteined a thing in forme like a lace of silke finely woven . . . one ende whereof is fastned unto the inside of the shell, even as the fish of Oisters and Muskles are; the other ende is made fast unto the belly of a rude masse or lumpe,

which in time commeth to the shape and forme of Bird; when it is pefectly formed, the shel gapeth open, and the next thing that appeereth is the foresaid lace or string; next come the legs of the Birde hanging out; and as it groweth greater it openeth the shell by degrees, till at length it is all come foorth . . . and falleth into the sea, where it gathereth feathers, and groweth to a foule, bigger then a Mallard, and lesser than a goose." John Gerard, *The Herball or Generall Historie of Plants* (London: John Norton, 1597), p. 1391. Cf. the earlier account by William Turner, who did not even pretend to base his account on firsthand knowledge; as so often, the appeal to authority (in this case Giraldus Cambrensis) was sufficient, whom he then adapted: "When after a certain time the firwood masts or planks or yard-arms of a ship have rotted on the sea, then fungi, as it were, break out upon them first, in which in course of time one may discern evident forms of birds, which afterwards are clothed with feathers, and last become alive and fly." *Turner on Birds . . . . first published by Dr William Turner, 1544*, ed. A.H. Evans (Cambridge, 1903), p. 27, translated from *Avium praecipuarum . . . per Dn. Guilielmum Turnerum* (Cologne: Gymnicus, 1544).

For another graphic illustration showing a tree of shells spawning birds, see the *Pourtraict des Conches, qui produisent des oyseaux*, in Claude Duret's *Histoire admirable des plantes* of 1605. Significantly (in terms of Cesi's nomenclatures), Duret was particularly interested in what he also termed zoophytes, namely plants with animal properties, or which transgress the boundaries between animal and vegetable.

15. See also Paris, IDF, MS 977, fol. 38, for a remarkable drawing of many goose barnacles attached to a rock.

16. Colonna 1592, in the appendix titled "Piscium Aliquot Plantarum Novarum Historia," pp. 14–19. There pages are headed "Conchas vulgo Anatiferas, non esse fructus terrestres neque ex iis Anates oriri: sed Balani marina speciem," and contain an excellent illustration on p. 17. Could it have been Colonna who supplied or at least inspired such splendid illustrations as that of the civet cat (fig. 1.9) and the goose barnacle (fig. 14.4) to Cassiano? One cannot end the present book without at least raising such a possibility.

17. Though with successive editions of Gerard's *Herball*, the woodcut of the goose barnacle became distinctly more accurate (compare the fanciful illustration in the 1597 edition with the altogether more plausible view of barnacles attached to pieces of wood in the edition of 1631, p. 1578). For Duret, see note 14 above.

18. Cf. Freedberg 1994a for a further discussion of these issues and of these texts (Carolus Linnaeus, *Genera Plantarum, eorumque characteres naturales* (Leiden: 1737), fols. ¶¶r–¶¶v, sec. 13.

19. I should add that I do not myself entirely believe this, as I hope to show in my forthcoming research projects.

20. See now Antonio Damasio, *The Feeling of What Happens: Body and Emotion in the Making of Consciousness* (New York: Harcourt Brace, 1999).

21. Francesco Barberini to Giovanni Muzzarelli, January 25, 1642, *Galileo, Opere*, XVIII, pp. 379–380.

22. Cited in Redondi 1987, p. 24; the first edition was published by Egidio Longo in Naples in 1641; it was edited by Benedetto Croce in 1928 (Bari: Laterza) and again in 1930. Other editions followed. On the whole subject of dissimulation, see especially Ginzburg 1970.

## *Abbreviations*

### MANUSCRIPT LOCATIONS

| | |
|---|---|
| Bologna, BU: | Bologna, Biblioteca Universitaria |
| Florence, BL: | Florence, Biblioteca Laurenziana |
| Florence, BNC: | Florence, Biblioteca Nazionale Centrale |
| London, BL: | London, British Library |
| Montpellier, BEM: | Montpellier, Bibliothèque de l'Ecole de Médecine |
| Naples, BN: | Naples, Biblioteca Nazionale |
| Paris, BN: | Paris, Bibliothèque Nationale |
| Paris, BIF: | Paris, Bibliothèque de l'Institut de France |
| Rome, ARSI: | Rome, Archivium Romanum Societatis Iesu |
| Rome, AS: | Rome, Archivio di Stato |
| Rome, BA: | Rome, Biblioteca Angelica |
| Rome, BANL: | Rome, Biblioteca dell'Accademia Nazionale dei Lincei |
| Rome, BAV: | Rome, Biblioteca Apostolica Vaticana |
| Rome, BV: | Rome, Biblioteca Vallicelliana |
| Siena, AS: | Siena, Archivio di Stato |
| Siena, BC: | Siena, Biblioteca Communale |
| Windsor, RL: | Windsor Castle, Royal Library |

### PRINTED SOURCES

| | |
|---|---|
| DBI: | Dizionario Biografico degli Italiani |
| PL: | Patrologia Latina |

# *Bibliography*


PRIMARY SOURCES

Accetto, Torquato. 1641. *Della dissimulazione onesta*. Naples: Egidio Longo.

Agricola, Georgius. 1546. *De natura fossilium Libri X*. Basle: H. Frobenius et N. Episcopius.

———. 1556. *De re metallica libri XII*. Basle: H. Frobenius et N. Episcopius.

Aguilonius, Franciscus, S.J. 1613. *Opticorum libri sex*. Antwerp: Plantin.

Aldrovandi, Ulisse. 1599–1603. *Ornithologiae hoc est de avibus historiae libri XII*. Bologna: Francesco de Francesco Senese.

———. 1602. *De animalibus insectis libri septum*. Bologna: Giovanni Battista Bellagamba.

———. 1606. *De reliquis animalibus exanguibus libri quatuor, post mortem eius editi: nempe de mollibus, crustaceis, testaceis et zoophytis*. Bologna: Giovanni Battista Bellagamba.

———. 1613. *De piscibus libri V et de cetis lib[er] unus*. Bologna: Giovanni Battista Bellagamba.

———. 1642. *Monstrorum historia*. Bologna: Nicolò Tebaldini.

———. 1648. *Musaeum metallicum in libros IIII distributum*. Bologna: Giovanni Battista Ferrone.

———. 1667–68. *Dendrologiae naturalis scilicet arborum historiae libri duo*, ed. Ovidio Montalbani. Bologna: Giovanni Battista Ferrone.

Allatius, Leo. 1633. *Apes Urbanae, seu de viris illustribus qui ab anno 1630 per totum 1632 Romae adfuerunt*. Rome: Ludovico Grignani.

Alpino, Prospero. 1592. *De plantis Aegypti . . .* Venice: Francesco di Francesco Senese.

Altieri Biagi, Maria, and Bruno Basile, eds. 1980. *Scienziati dal Seicento*, Milan-Naples: Ricciardi.

Aumüller, Stephan A., and József Jeanplong, eds. 1983. *Carolus Clusius in Pannoniis observatorum Brevis Historia et Codex Clusii*, Budapest: Akadèmiai Kiadò & Graz: Akademische Druck- u. Verlagsanstalt.

Baglione, Giovanni. 1642. *Le Vite de Pittori, Scultori et Architetti dal Pontificato di Gregorio XIII sino à tutto quello d'Urbano Ottavo*. Rome: Bartolomeo Fei.

Bauhin, Caspar. 1614. *De Hermaphroditorum Monstrosorumque Partuum Natura*. Oppenheim: Hieronymus Galler/Johann Theodor De Bry.

———. 1623. *Pinax Theatri botanici . . .* Basel: Lud. Rex.

Bellarmine, Robert. 1586. *Disputationes de controversiis de christianae fidei*. 3 vols. Ingolstadt: David Sartorius.

Besler, Basil. 1613. *Hortus Eystettensis*. [Eichstätt and Nuremberg].

Biancani, Giuseppe, S.J. 1635. *Sphaera mundi*, 2nd ed. Modena: Giuliano Cassiano.

Bodin, Jean. 1597. *Universae Naturae Theatrum*. Frankfurt: Wechel.

Borel, Pierre. 1657. *Discours nouveau prouvant la pluralité des mondes, que les astres sont des terres habitées, & la terre une etoile, qu'elle est hors du centre du mode dans le troisiesme ciel, & se tourne devant le soleil qui est fixe, & autres choses très curieuses*. Geneva: N.p..

Brahe, Tycho. 1913–1929. *Astronomiae instauratae progymnasmata*, in J.L.E. Dreyer, *Opera Omnia*. 15 vols. Copenhagen: Gyldendaliana.

Brunfels, Otto von. 1530. *Herbarium vivae eicones . . .* Straßburg: Johann Schotten.
———. 1532. *Contrafayt Kreüterbuch.* Straßburg: Johann Schotten.
Cardanus, Hieronymus. 1550. *De Subtilitate Libri XXI (Liber VII, De Lapidibus).* Nuremberg: I. Petreius.
*Carteggio* [Giuseppe Gabrieli]. 1938–42. "Il carteggio linceo della vecchia accademia di Federico Cesi: 1603–1630." *Atti della Reale Accademia dei Lincei, Memorie della Classe di Scienze morali, storiche e filologiche.* Series 6, vol. 7, fasc. 1–4: 1–1446.
Castelli, Tobia Aldini [recte Pietro Castelli]. 1625. *Exactissima descriptio rariorum quarundam plantarum, quae continentur Romae in Horto Farnesiano.* Rome: Giacomo Mascardi.
———. 1629. *Discorso della differenza tra gli semplici freschi et i secchi con il modo di seccarli.* Rome: Giacomo Mascardi.
Cesalpino, Andrea. 1583. *De plantis libri XVI . . .* Florence: Giorgio Marescotto.
———. 1596. *De metallicis libri tres.* Rome: Zanetti.
Cesi, Federico. 1618. *De caeli unitate, tenuitate fusaque et pervia stellarum motibus natura ex sacris litteris epistola,* in Scheiner 1630, 771–782; Altieri Biagi and Basile 1980, 9–35.
———. [1616?] *Sapere.* "Del naturale desiderio di sapere et instituzione de'Lincei per adempimento di esso." In Altieri Biagi and Basile 1980, 39–70.
Clavius, Christophorus. 1611–12. *Opera Mathematica V tomis distributa ab auctore nunc denuo correcta, et plurimis locis aucta.* Mainz: A. Hierat.
Clusius, Carolus. 1583. *Rariorum aliquot Stirpium, per Pannoniam, Austriam, et vicinas . . . Historia . . .* Antwerp: Plantin.
———. 1601. *Rariorum Plantarum Historia . . .* Antwerp: Johannes Moretus.
Clusius, Istvánffi, ed. 1898. *Caroli Clusii Atrebatis Icones Fungorum in Pannonis Observatorum sive Codex Clusii Lugduno Batavensis . . .* cura et sumptibus Gy. de Istvánffi. 2 volumes. Budapest: Hornyanszkyana.
Colonna, Fabio. 1592. *Phytobasanos.* Naples: Orazio Salviani.
———. 1606. *Minus cognitarum stirpium aliquot ac etiam rariorum nostro coelo orientium* ΕΚΦΡΑΣΙΣ. Rome: Guglielmo Facciotto.
———. 1616a. *Minus cognitarum rariorumque nostro coelo orientium stirpium* ΕΚΦΡΑΣΙΣ. *Qua non paucae ab Antiquioribus Theophrasto, Dioscoride, Plinio, Galeno aliisque descriptae, praeter illos etiam in* ΦΥΤΟΒΑΣΑΝΩ *editas disquiruntur ac declarantur. Item de Aquatilibus aliisque nonnullis animalibus libellus.* Rome: Giacomo Mascardi.
———. 1616b. *Purpura. Hoc est de purpura ab Animali testaceo fusa, de hoc ipso Animali, aliisque rarioribus testaceis quibusdam.* Rome: Giacomo Mascardi.
———. 1618. *La Sambuca Lincea overo dell'Istromento musico perfetto,* libri III. Naples: Constantino Vitale.
———. 1744. *Phytobasanos cui accessit vita Fabi Colonnae et Lynceorum notitia adnotationesque in phytobasanon Iano Planco Ariminensi auctore.* Milan: Pietro Caietano Viviano.
Dati, Carlo. 1664. *Delle lodi del commendatore Cassiano dal Pozzo.* Florence: all'Insegna della Stella.
De Asso [Ignatio Jordan Asso y del Rio]. 1793. *Clariorum Hispaniensium atque exterorum epistolae cum praefatione et notis.* Zaragoza: Typographia Regia.
Decreta. 1830. *Decreta, canones, censurae, et praecepta Congregationum Generalium Societatis Jesu.* Avignon: Sequin.
Della Porta, Giovanni Battista. 1588. *Phytognomica.* Naples: Orazio Salviani.
———. 1589. *Magiae Naturalis libri XX.* Naples: Orazio Salviani
———. 1598. *Della fisonomia dell'huomo.* Naples: Tarquinio Longo.
———. 1610. *Elementorum curvilineorum libri tres. In quibus altera geometriae parte restituta, agitur de circuli quadatura.* Rome: Bartolomeo Zanetti.
———. 1614. *De aeris transmutationibus libri IV.* Rome: Giacomo Mascardi.
Delle Colombe, Lodovico. 1606. *Discorso . . . nel quale si dimostra, che la nuova stella apparita l'ottobre passato 1604 nel Sagittario non è cometa, ne stella generata, ò creata di nuovo, ne apparente: ma una di quelle che furono da principio nel cielo.* Florence: Giunti.
———. 1892. "Contro il moto della terra." In Galileo Galilei, *Le Opere,* II. Florence: Tipografia di G. Barbèra, 253–289.
De Passe, Crispijn Van. 1614. *Hortus floridus.* Arnheim: Johannes Ianssonius.
De Waard, Cornelis, Jr. 1906. *De uitvinding der verrekijkers. Eene bijdrage tot de beschavingsgeschiedenis.* 's-Gravenhage, De Nederlands boek- en steendrukkerij (voorheen H.L. Smits).
*Dialogo de Cecco di Ronchitti da Burzene in Perpuosito De La Stella Nuova . . . Con alcune ot-*

*tave d'Incerto, per la medesima Stella, contra Aristotele*. 1605. Padua: Pietro Paolo Tozzi (2nd ed., Verona: Bartolomeo Merlo, 1605).
Dodonaeus, Rembertus. 1583. *Stirpium historiae pemptades sex. sive libri XXX*. Antwerp: Plantin.
Durante, Castore. 1585. *Herbario nuovo*. Rome: Iacomo Bericchia o e Iacomo Turnierii (also Rome: Bartholomeo Bonfadino et Tita Diani, 1585).
Encelius, Christophorus. 1557. *De re metallica. . . . libri III*. Frankfurt: Hered. Christiani Egenolphi.
Erythraeus, Janus Nicius [Gian Vittorio Rossi]. 1645. *Pinacotheca imaginum illustrium doctrinae vel ingenii laude, virorum*. Cologne: Jodocus Kalcovius [for Amsterdam: Johannes Blau?].
Epistolae. 1635. *Epistolae Praepositorum Generalium ad Patres et Fratres Societatis Jesu*. Antwerp: Johannes Meursius.
Evelyn, John. 1662. *Sculptura or the History and Art of Chalcography and Engraving in Copper*. London: J. C. for G. Beedle and T. Collins.
———. 1955. *Diary: The Diary of John Evelyn*, ed. E.S. de Beer. 2 vols. Oxford: Clarendon Press.
Faber. 1624. See *Praescriptiones*.
Ferrari, Giovanni Battista. 1622. *Nomenclator Syriacus*. Rome: Stephanus Paulinus.
———. 1625. *Orationes*. Lyon: Rouille. Reprints, Rome: Corbelletti, 1627; Rome: P. A. Faciotti, 1634; Rome: P.A. Facciotti, 1635; Venice: Bagioni, 1644; Cologne: Egmont, 1650; London: Roger Daniel, 1657; London: Redmayne, 1668.
———. 1633. *De Florum Cultura Libri IV*. Rome: Stephanus Paulinus.
———. 1638. *Flora ovvero Cultura di fiori*. Rome: P. A. Faciotti.
———. 1646. *Hesperides, sive De Malorum Aureorum Cultura et Usu Libri Quattuor*. Rome: Hermann Scheus.
———. 1652. *Collocutiones*, Siena: Bonetti.
Fontana, Francesco. 1646. *Novae coelestium terrestriumque rerum observationes*. Naples: Apud Gaffarum.
Fuchs, Leonhart. 1542. *De historia stirpium* . . . Basel: Michael Isengrin.
———. 1543. *New Kreüterbuch*. Basel: Michael Isengrin.
Galileo, Galilei. 1633. *Dialogo sopra i due massimi sistemi del mondo*. Reprint, Turin: Einaudi, 1970.
———. 1613. *Istoria e dimostrazioni intorno alle macchie solari e loro accidenti, comprese in tre lettere scritte all'illustrissimo Signor Marco Velseri Linceo*. . . . Rome: Giacomo Mascardi.
———. 1623. *Il Saggiatore*. Ed. Ferdinando Flora. Reprint, Turin: Einaudi, 1977.
———. 1610. *Sidereus Nuncius*. Reprint, translated with introduction, conclusion, and notes by Albert van Helden. Chicago: Chicago University Press, 1989.
———. 1890–1909. *Opere: Le Opere di Galileo Galilei*. Ed. Antonio Favaro. 20 vols. Florence: Tipografia Barbéra .
Gassendi, Pierre. 1658. *Opera omnia*. 6 vols. Lyon: Laurence Anisson & Jean Baptiste Devenet. Reprinted in a facsimile edited by Tullio Gregory (Stuttgart: Friedrich Frommann Verlag, 1964).
———. 1992. *Vie de Peiresc*. Translated by Roger Lassalle and Agnes Bresson. Paris: Belin.
Gesner, Conrad. 1558. *Historiae animalium liber IIII, qui est de piscium & aquatilium animantium natura*. Zurich: Christoph Froschauer.
———. 1565. *De omni rerum fossilium genere, gemmis, lapidibus, metallis, et huiusmodi, libri aliquot*. Zurich: Iacobus Gesner
Grassi, Orazio. 1626. *Ratio ponderum Librae et Simbellae*. Paris: Cramoisy. Reprinted in Galileo, *Opere*, VI: 375–500.
Grew, Nehemiah. 1682. *The Anatomy of Plants. With an idea of a philosophical history of plants, and several other lectures, read before the Royal Society*. London: W. Rawlins.
Gualterotti, Raffaelo. 1605. *Discorso sopra l'apparizione de la nuova stella*. Florence: Cosimo Giunti.
Häutle, Christian, ed. 1881. *Des Bamberger Fürstbischofs Johann Gotffried von Aschhausen Gesandschafts-reise nach Italien und Rom 1612 und 1613*. Tübingen: H. Laupp.
Heckius, Johannes. 1605. *De Nova Stella Disputatio*. Rome: A. Zannetti.
Hernández, Francisco. 1942–46. *Historia de las Plantas de Nueva España*. 3 vols. Mexico:

Imprenta Universitaria (Instituto de Biología de la Universidad Nacional Autónoma de México).
———. 1959. *Historia Natural de Nueva España.* México: Universidad Nacional de México.
———. 1960–1984. *Obras Completas.* I–VI. México: Universidad Nacional de México.
———. *Thesaurus.* See also under *Thesaurus* below.
Hooke, Robert. 1665. *Micrographia: or, Some physiological descriptions of minute bodies made by magnifying glasses.* London: J. Martyn and J. Allestry.
Imperato, Ferrante. 1599. *Dell'historia Naturale . . . libri XXVIII, nella quale ordinatamente si tratta della diversa condition di miniere, e pietre. Con alcune historie di Piante, et Animali.* Naples: C. Vitale.
Imperato, Francesco. 1610. *De fossilibus opusculum.* Naples: D. Roncaioli.
———. 1628. *Discorsi intorno a diverse cose naturali.* Naples: E. Longo.
Inchofer, Melchior. 1633. *Tractatus Syllepticus, in quo, quid de terrae, Solisque motu, vel statione, secundum S. Scripturam, et Sanctos Patres sentiendum, quave certitudine alterutra sententia tenenda sit, . . . ostenditur.* Rome: Lodovico Grignani.
Jouanny, Charles. 1911. *Correspondance de Nicolas Poussin* (Archives de l'Art Francaise, V, 1911). Paris: J. Schemit.
Kepler, Johannes. 1858–1871.*Opera omnia.* Ed. Ch. Frisch. 8 vols. Frankfurt and Erlangen: Heyder and Zimmer.
———. 1937–. *Gesammelte Werke.* Ed. Max Caspar and Walther von Dyck. Munich: Beck.
Lagalla, Giulio Cesare. 1612. *De phoenomenis in orbe Lunae.* Venice: T. Baglioni.
Leeuwenhoek, Antoni van. 1677a. "Concerning little animals." *Philosophical Transactions* XII: 821–833
———. 1677b. "De natis e semine genitali." *Philosophical Transactions* XII: 1040–1046.
———. 1683. "About generation by an animalcule of the male seed." *Philosophical Transactions* XIII: 347–355.
Liceti, Fortunio. 1621. *De lucernis antiquorum reconditis libri quatuor . . .* Venice: Evangelista Deuch.
———. 1640a. *De luminis natura & efficientia libri tres . . .* Udine: Nicola Schiratti.
———. 1640b. *Litheosphorus, sive de lapide Bononiensi lucem . . .* Udine: Nicola Schiratti.
———. 1640c. *De Quaesitis per epistolas a claris viris responsa . . .* 8 vols. Vol. 1, Bologna: Nicolo Tebaldini, 1640; vol. 2, Udine: Nicola Schiratti, 1646.
———. 1641. *De lucidis in sublimi ingenuarum exercitatiunum liber . . .* Padua: Cribelliano.
Lhote, Jean-François, and Danielle Joyal, eds. 1989. Nicholas Claude Fabri de Peiresc, *Lettres à Cassiano dal Pozzo (1626–1637).* Clermont-Ferrand: Adosa.
Lobelius, Mathias. 1576. *Plantarum seu stirpium historia, . . . Cui annexum est Adversariorum volumen.* Antwerp: Plantin.
———. 1581. *Plantarum seu stirpium icones.* Antwerp: Plantin.
Lobelius, Mathias, and Petrus Pena. 1576. *Nova stirpium adversaria . . .* Antwerp: Christophorus Plantinus.
Lorenzini, Antonio [Cesare Cremonini]. 1605. *Discorso. . . . Intorno alla Nuova Stella.* Padua: Pietro Paolo Tozzi.
Magini, Giovanni Antonio. 1589. *Novae Coelestium Orbium Theoricae congruentes cum observationibus N. Copernici.* Venice: Damiano Zenario.
Mattioli, Pierandrea. 1565. *Commentarii in sex libros Pedacii Dioscoridis Anarzabei de Medica materia, . . .* Venice: Valgrisi.
———. 1585. *I discorsi . . . nelli sei libri di Pedacio Dioscoride Anazarbeo, della materia medicinale. . . .* Venice, Valgrisi.
Mercati, Michele. 1719. *Metallotheca. Opus posthumum, auctoritate, & munificentia Clementis Undecimi Pontificis Maximii e tenebris in lucem eductum.* Rome: J. M. Salvioni.
Monardes, Nicolas. 1569. *Dos libros, el veno que trata de todas las cosas que traen de nuestras Indias Occidentales . . .* Sevilla: Hernando Diaz.
———. 1571. *Segunda parte del libro, del las cosas que se traen de nuestras Indias Occidentales . . .* Sevilla: Alonso Escrivano.
Montalbani, Ovidio. 1668. *Ulyssis Aldrovandi patricii bononiensis Dendrologiae naturalis scilicet arborum historiae libri duo.* Bologna: Giovanni Battista Ferrone.
Naudé, Gabriel [Mascurat]. 1650. *Jugement de tout ce qui a esté imprimé contre le cardinal Mazarin, depuis le sixième Janvier, jusques à la declaration du premier Avril mil six cent quarante-neuf.* 2nd ed. Paris.

Nieremberg, Johannes. 1635. *Historia naturae, maxime peregrinae.* Antwerp: Plantin-Moretus.

Olina, Giovanni Pietro. 1622. *Uccelliera overo discorso della natura, e proprietà di diversi uccelli, e in particolare di que' che cantano, con il modo di prendergli, conoscergli, allevargli, e mantenergli.* Rome: Andrea Fei.

Palissy, Bernard. 1580. *Discours admirables, de la nature des eaux et fonteines, tant naturelles qu'artificielles, des metaux, des sels & salines, des pierres, des terres, du feu & des emaux.* Paris: Martin le Jeune.

Panaroli, Domenico. 1645. *Il Camaleonte.* Rome: Francesco Cavalli.

Patrizi, Francesco. 1591. *Nova de universis philosophia.* Ferrara: Benedetto Mamarello.

Pereira, Benito. 1599. *Prior tomus Commentariorum et disputationum in Genesim.* Rome: Giorgio Ferrari, 1599.

Persio, Antonio. 1613. *De Ratione recte philosophandi et De Natura ignis et caloris.* Rome: Giacomo Mascardi.

*Praescriptiones Lynceae Academiae.* 1624. Terni: Tommaso Guerrieri.

Riquius, Justus. 1625. *Apes Dianiae in monimentis veterum noviter observae.* Rome: Giacomo Mascardi.

Scheiner, Christoph. 1630. *Rosa ursina, sive, Sol, ex admirando facularum et macularum suarum phaenomeno varius.* Bracciani: Apud Andream Phoeum.

———. 1631. *Pantographice seu Ars delineandi res quaslibet per parallelogrammum lineare seu cauum, mechanicum mobile.* Rome: Luigi Grignani.

Skippon, Philip. 1752. "An Account of a journey made thro' part of the Low-Countries, Germany, Italy and France." In Awnsham Churchill, ed., *A Collection of Voyages and Travels,* VI. London: H. Lintot and J. Osborn, 359–736.

Stelluti, Francesco. 1622. *Scandaglio sopra la libra astronomica et filosofica di Lotario Sarsi. Nella controversia delle Comete e particolarmente delle tre ultimamente vedute l'Anno 1618.* Terni: Tomasso Guerrieri.

———. 1630. *Persio, tradotto in verso sciolto e dichiarato da Francesco Stelluti, Accademico Linceo da Fabriano.* Rome: Giacomo Mascardi.

———, ed. 1637a. *Della Fisonomia di tutto il corpo humano. . . . Hora brevemente in tavole sinottiche ridotta e ordinata . . .* Rome: Vitale Mascardi.

———. 1637b. *Trattato del Legno Fossile Minerale nuovamente scoperto nel quale brevemente si accenna la varia & mutabil natura di detto Legno rappresentatovi con alcune figure, che mostrano il luogo dove nasce, la diversità dell'onde, che in esso si vedono, e le sue così varie e meravigliose forme.* Rome: Vitale Mascardi.

Tamizey de Larroque, Philippe, ed. 1879–97. *Les Correspondants de Peiresc* [Nicolas Claude Fabri de Peiresc]; lettres inédites et annotés par Philippe Tamizey de Larroque. 21 vols. Paris: L. Techener.

———, ed. 1888–1898. *Lettres de Peiresc* [Nicolas Claude Fabri de Peiresc]. Collection de documents inédits sur l'histoire de France. Série 2. Histoire des lettres et de sciences, 7 vols. Paris: Imprimerie Nationale.

Teti, Girolamo. 1642. *Aedes Barberinae ad Quirinalem a comite Hieronymo Tetio Perusino descriptae.* Rome: Mascardi.

*Thesaurus* [by Francisco Hernández]. 1651. *Rerum Medicarum Novae Hispaniae Thesaurus seu Plantarum Animalium Mineralium Mexicanorum Historia ex Francisci Hernández novi orbis Medici Primarii relationibus in ipsa Mexicana urbe conscriptis a Nardo Antonio Reccho Monte Corvinate Cath. Maiest. Medico et Neap. Regni Archiatro Generali Jussu Philippi II Hisp. Ind. etc. Regis Collecta ac in ordinem digesta a Ioannae Terrentio Lynceo Constantiense Germano Pho. ac Medico Notis Illustrata.* Rome: Vitale Mascardi (also Rome: Giacomo Mascardi, 1628, 1630; and Rome: Vitale Mascardi, 1649).

Welsch, Georg Hieronymus [Velschius]. 1674. *Exercitatio de vena Medinensi. . . . sive de dracunculis veterum.* Augsburg: Theophil Göbel, 1674.

Ximénez. 1615. *Quatro libros. de la naturaleza y virtudes de las plantas . . . en el uso de Medicina en la Nueva España.* Mexico: Diego Lópezá Davalos.

## SECONDARY SOURCES

Accordi, Bruno. 1981a. "Ferrante Imperato (Napoli, 1550–1625) e il suo contributo alla storia della geologia." *Geologia Romana* 20: 43–56.

———. 1981b. "Tentativi di classificazione delle pietre e delle gemme nei secoli XVI e XVII." *Physis* 23: 311–324.

Alessandrini, Ada. 1965. *Documenti lincei e cimeli galileiani*. Rome: Accademia Nazionale dei Lincei.

———. 1976. "Giovanni Heckius Linceo e la sua controversia contro i protestanti." *Rivista di Storia della Chiesa in Italia* XXX/2: 363–404.

———. 1978. *Cimeli lincei a Montpellier*. Rome: Accademia Nazionale dei Lincei (Indici e sussidi della biblioteca 11).

———. 1986. "Originalità dell'Accademia dei Lincei." In *Convegno celebrativo del IV centenario della nascità di Federico Cesi (Atti dei Convegni Lincei, 78)*, Acquasparta, 7–9 ottobre 1985. Rome: Accademia Nazionale dei Lincei, 77–177.

Alessandrini, Ada, Gilberto De Angelis, and Paola Lanzara. 1985. "Il 'Theatrum plantarum' di Federico Cesi nella Biblioteca dell'Institut de France." *Atti della Accademia Nazionale dei Lincei, Rendiconti della Classe di Scienze fisiche, matematiche e naturali*. S. VIII, vol. LXXVIII: 315–325.

Alessandrini, Ada, et al. 1986. *Francesco Stelluti Linceo da Fabriano*. Fabriano: Città e Comune di Fabriano.

Allodi, Federico, ed. 1957. *Studi e Ricerche sui Microscopi Galileiani del Museo di Storia della Scienza*, fasc. I. Florence: Leo S. Olschki Editore.

Alpers, Svetlana. 1983. *The Art of Describing: Dutch Art in the Seventeenth Century*. Chicago: University of Chicago Press.

Altieri Biagi, Maria. 1978. "Lingua della scienza fra Seicento e Settecento." In *Letteratura e scienza nella storia della cultura italiana. Atti del IX Congresso dell' Associazione internazionale per gli studi di lingua e letteratura italiana*. Palermo: Manfredi, 103–162.

Altieri Biagi, Maria, and Bruno Basile, eds. 1980. *Scienziati del Seicento*. Milan-Naples: Ricciardi.

Amabile, Luigi. 1887. *Fra Tommaso Campanella ne' castelli di Napoli, in Roma ed in Parigi*. Naples: A. Morano.

Applebaum, Wilbur, ed. 2000. *Encylopedia of the Scientic Revolution from Copernicus to Newton*. New York: Garland, 2000.

Arber, Agnes. 1986. *Herbals, Their Origin and Evolution: A Chapter in the History of Botany, 1460–1670*. 3rd ed.. Cambridge: Cambridge University Press.

Arrighi, Gino. 1964. "Gli 'Occhiali' di Francesco Fontana in un carteggio inedito di Antonio Santini nella Collezione Galileiana della Biblioteca Nazionale di Firenze." *Physis* 6, 1964: 432–448.

Ashworth, William B., Jr. 1989. "Light of Reason, Light of Nature: Catholic and Protestant Metaphors of Scientific Knowledge." *Science in Context* 3/1: 89–107.

———. 1990. "Natural History and the Emblematic World View." In David C. Lindberg and Robert S. Westmann, eds., *Reappraisals of the Scientific Revolution*. Cambridge: Cambridge University Press, 303–332.

Atran, Scott. 1990. *Cognitive Foundations of Natural History: Towards an Anthropology of Science*. Cambridge: Cambridge University Press.

Baboli, Albino, ed. 1975. *Problemi Religiosi e filosofia*, Padua: La Garangola.

Baccetti, Baccio. 1986. "Il posto di Federico Cesi nella storia della zoologia." In *Convegno celebrativo del IV centenario della nascita di Federico Cesi (Atti dei Convegni Lincei, 78)*, Aquasparta, 7–9 ottobre 1985. Rome: Accademia Nazionale dei Lincei, 225–229.

Baldini, Enrico. 1987. "Simulacri, meraviglie, prodigi e mostruosità nella Dendrologia aldrovandiana e nell'interpretazione scientifica moderna," *Rendiconti dell'Accademia delle Scienze dell'Istituto di Bologna, Classe di Scienze Fisiche* ser. XIV, vol. V, 1987–88: 145–170.

———. 1988–89. "Polimorfismo e teratologia dei frutti nel genere citrus: riscontri storici e attualità biologiche." *Rendiconti della Accademia delle Scienze dell'Istituto di Bologna, Classe di Scienze Fisiche*, ser. XIV, vol. VI: 127–161.

Baldini, Ugo. 1981. "La nova del 1604 e i matematici e filosofi del Collegio Romano." *Annali dell'Istituto e Museo di Storia della Scienza* 6: 63–97.

———. 1984a. "L'astronomia del Cardinale Bellarmino." In Paolo Galluzzi, ed., *Novità celesti e crisi del sapere: Atti del Convegno Internazionale di Studi Galileiani*. Florence: Giunti Barbera, 293–305.

———. 1984b. "Addidamenta Galilaeana: I. Galileo, la nuova astronomia e la critica all' aristotelismo nel dialogo epistolare tra Giuseppe Biancani e i Revisori romani de la Compagnia di Gesù." *Annali dell'Istituto e Museo di storia della scienza di Firenze* 9: 13–43.

———. 1985. "Una fonte poco utilizzata per la storia intellettuale: le 'censurae librorum' e 'opinionum' nell'antica Compagnia de Gesù." *Annali dell'Istituto storico italo-germanico in Trento* 11: 19–67.

———. 1988. "La conoscenza dell'astronomia nell'Italia meridionale anterioramente al Sidereus Nuncius." In P. Nastasi, ed., *Atti del Convegno Il Meridione e le scienze (secoli XVI–XIX)* (Palermo 14–16 maggio 1985). Palermo-Naples: Istituto Gramsci Siciliano-Istituto Italiano per gli studi filosofici, 127–168.

———. 1992. *Legem impone subactis. Studi su filosofia e scienza dei gesuiti in Italia, 1540–1632*. Rome: Bulzoni.

Baldini, Ugo, and George Coyne. 1984. *The Louvain Lectures of Bellarmine and the Autograph Copy of His 1616 Declaration to Galileo*. Vatican Observatory Publications, Special Series, *Studi Galileiani*, I, 2. Vatican City: Specola Vaticana.

Baldriga, Irene. 1998. "Il museo anatomico di Giovanni Faber Linceo." In Sergio Rossi, ed., *Scienza e Miracoli nell'arte del '600. Alle Origini della Medicina Moderna*. Milan: Electa, 1998.

Balme, David M. 1987. "Aristotle's Use of Divisions and Differentiae." In Allan Gotthelf and James G. Lennox, eds., *Philosophical Issues in Aristotle's Biology*. Cambridge: Cambridge University Press, 69–89.

Barker, Peter. 1991. "Stoic Contributions to Early Modern Science." In Margaret Osler, ed., *Atoms, Pneuma and Tranquillity*. Cambridge: Cambridge University Press, 133–154.

Barker, Peter, and Bernard Goldstein. 1988. "The Role of Comets in the Copernican Revolution," *Studies in the History and Philosophy of Science*, XIX: 299–319.

Basso Peressut, ed. 1997. *Stanze della meraviglia. I musei della natura storia e progetto*. Bologna: CLUEB, 1997.

Bell, Janis Callen. 1988. "Cassiano dal Pozzo's Copy of the Zaccolini Manuscripts." *Journal of the Warburg and Courtauld Institutes* 51: 103–125.

———. 1993. "Zaccolini's Theory of Color Perspective." *Art Bulletin* 75: 91–112.

Belli Barsali, Isa. 1981. "Una Fonte per i giardini del Seicento: Il Trattato del Giovan Battista Ferrari." In G. Ragioneri, ed., *Il Giardino Storico Italiano* (Atti del Convegno di Studi Siena-San Quirico d'Orcia, 1981), Florence, 221–234.

Belloni, Luigi. 1972. "Il microscopio applicato alla biologia da Galileo e dalla sua Scuola (1610–1661)." In Carlo Maccagni, ed., *Saggi su Galileo Galilei* (Pubblicazioni del Comitato Nazionale per le Manifestazioni celebrative, vol. III, 2). Florence: G. Barbèra Editore, 689–730.

Belloni, Gabriella. 1983. "Il carteggio italiano-tedesco dei membri dell'*Academia Lynceorum*." *Res Publica Litterarum. Studies in the Classical Tradition* VI, pp. 19–35.

Belloni Speciale, Gabriella. 1987. "La ricerca botanica dei Lincei a Napoli: corrispondenti e luoghi." In F. Lomonaco and M. Torrini, eds., *Galileo e Napoli, Atti del Convegno di Napoli* (12–14 aprile 1984). Naples: Guida, 59–79.

Berlin, Brent. 1992. *Ethnobiological Classification. Principles of Categorization of Plants and Animals in Traditional Sciences*. Princeton: Princeton University Press.

Biagioli, Mario. 1989. "The Social Status of Italian Mathematicians, 1450–1600." *History of Science* 27: 41–95.

———. 1992. "Scientific Revolution, Social Bricolage and Etiquette." In Roy Porter and Mikulas Teich, eds., *The Scientific Revolution in National Context*. Cambridge: Cambridge University Press, 11–54.

———. 1993. *Galileo Courtier: The Practice of Science in the Culture of Absolutism*. Chicago: University of Chicago Press.

———. 1995. "Knowledge, Freedom, and Brotherly Love: Homosociality and the Accademia dei Lincei." *Configurations* II: 139–166.

———. 1996. "Playing with the Evidence." *Early Science and Medicine* 1/1: 70–105.

*Biographie Universelle, ancienne et moderne*. 1811–1862. Joseph François and Louis Gabriel Michaud, eds. 85 vols. Paris.

Biondi, Edoardo, ed. 1984. *Sul 'Trattato del Legno Fossile Minerale' di Francesco Stelluti Accademico Linceo da Fabriano con ristampa anastatica dell'opera*. Fabriano: Casa Editrice Fabrianese.

Blackwell, Richard J. 1991. *Galileo, Bellarmine, and the Bible*. Notre Dame: University of Notre Dame Press.

Blair, Ann. 1997. *The Theater of Nature: Jean Bodin and Renaissance Science*. Princeton: Princeton University Press.

Blunt, Anthony. 1945. *The French Drawings in the Collection of His Majesty the King at Windsor Castle.* Oxford and London: Phaidon.

———. 1966. *The Paintings of Nicolas Poussin. A Critical Catalogue.* London: Phaidon.

———. 1967. *Nicolas Poussin: The A.W. Mellon Lectures in the Fine Arts, 1958.* New York: Bollingen.

Boehm, Laetitia, and Ezio Raimondi, eds. 1981. *Università, accademie e società scientifiche in Italia e in Germania dal Cinquecento al Settecento.* Bologna: Il Mulino.

Bolzoni, Lina. 1995. *La Stanza della Memoria.* Turin: Einaudi.

Boone [sale cat.]. September 16 and 17, 1988. *Property from the Estate of James R. Boone* [sale cat.]. Sotheby's, New York, lots 125–182.

Bouteron, M., and J. Tremblot. 1928. *Catalogue général des manuscrits des bibliothèques publiques de France.* Bibliothèque de l'Institut de France, Paris.

Bredekamp, Horst. 1995. *The Lure of Antiquity and the Cult of the Machine: The Kunstkammer and the Evolution of Nature, Art and Technology.* Princeton: Markus Wiener. (Translated by Allison Brown from *Antikensehnsucht und Maschinenglauben,* 1993.)

———. 200. "Gazing Hands and Blind Spots: Galileo as Draftsman." *Science in Context* 13 (3–4): 423–462.

Bremekamp, C.E.B. 1953. "A Re-Examination of Cesalpino's Classification." *Acta Botanica Neerlandica* I: 580–593.

Brodrick, James. 1928. *The Life and Work of Blessed Robert Francis Cardinal Bellarmine S.J.* 2 vols. London.

Bromehead, Cyril Edward Nowill. 1947a. "A Geological Museum of the Early Seventeenth Century." *Quarterly Journal of the Geological Society* CIII/2: 65–87.

———. 1947b. "Aetites or the Eagle Stone." *Antiquity* XXI: 16–22.

Canguilhem, Georges. 1989. *The Normal and the Pathological.* With an introduction by Georges Foucault. Translated by Carolyn R. Fawcett, with Robert S. Cohen. New York: Zone Books.

———. 1992. *La Connaissance de la Vie.* Deuxieme edition revue et augmentee. Paris: Vrin.

Cantimori, Delio. 1939. *Eretici italiani del cinquecento.* Florence: Sansoni.

Capecchi, Anna Maria. 1986. "Per la ricostruzione di una biblioteca seicentesca: i libri di storia naturale di Federico Cesi 'Lynceorum Princeps.'" *Atti della Accademia Nazionale dei Lincei, Rendiconti della Classe di Scienze morali, storiche e filologiche,* s. VIII, vol. XLI, 145–164.

Capecchi, Anna Maria, et al., eds. 1992. *L'Accademia dei Lincei e la cultura europeo nel XVII secolo. Manoscritti, libri, incisioni, strumenti scientifici.* Cat. Exhib., Accademia dei Lincei, Rome.

Carusi, Enrico. 1929–30. "Lettere di Galeazzo Arconato a Cassiano dal Pozzo per lavori sui manoscritti di Leonardo da Vinci." In *Accademie e Biblioteche d'Italia,* III: 504–518.

Carutti, Domenico. 1877. "Di Giovanni Eckio e della instituzione dell'Accademia dei Lincei, con alcune note inedite intorno a Galileo." *Atti della Reale Accademia dei Lincei, Memorie della Classe di Scienze morali, storiche e filologiche.* S. III, vol. I: 45–77.

———. 1883. *Breve storia dell'Accademia dei Lincei.* Rome: Salviucci.

Casanovas, Juan. 1984. "Il P. Orazio Grassi e le comete dell'anno 1618." In Paolo Galluzzi, ed., *Novità celesti e crisi del sapere: Atti del Convegno Internazionale di Studi Galileiani.* Florence: Giunti Barbera, 307–313.

Casciato, Maristella, Maria Grazia Ianniello, and Maria Vitale, eds. 1986. *Enciclopedismo in Roma barocca: Athanasius Kircher e il Museo del Collegio Romano tra Wunderkammer e museo scientifico,* Venice: Marsili.

Cermenati, Mario. 1906. "Ulisse Aldrovandi e l'America." *Annali di botanica* 4: 3–56.

Cesi, Federico. 1980. "Del natural desiderio di sapere et instituzione de'Lincei per adempimento di esso." In Maria Altieri Biagi and Bruno Basile, eds., *Scienziati del Seicento.* Milan: Ricciardi.

———. 1986. *Convegno celebrativo del IV centenario della nascita di Federico Cesi (Atti dei Convegni Lincei, 78).* Acquasparta, 7–9 ottobre 1985. Rome: Accademia Nazionale dei Lincei.

Chambers, David S., and François Quiviger, eds. 1995. *Italian Academies of the Sixteenth Century.* Warburg Institute Colloquia I. London: The Warburg Institute.

Chiovenda, Emilio. 1936. "Un prezioso esemplare del "Tesoro Messicano dei Lincei." *Atti e Memorie dell'Accademia Modenese di Scienze, Lettere e Arti,* 5: 1–38.

Clark, David H., and F. Richard Stephenson. *The Historical Supernovae.* Oxford: Pergamon Press.

Clericuzio, Antonio, and Silvia De Renzi. 1995. "Medicine, Alchemy and Natural Philosophy in the Early Academia dei Lincei." In David S. Chambers and François Quiviger, eds., *Italian Accademies of the Sixteenth Century.* Warburg Institute Colloquia I. London: The Warburg Institute 1995: 175–194.

Copenhaver, Brian. 1990. "Natural Magic, Hermeticism and Occultism in Early Modern Science." In David C. Lindberg and Robert S. Westmann, eds., *Reappraisals of the Scientific Revolution.* Cambridge: Cambridge University Press: 261–301.

Cortesi, Fabrizio. 1908. "Una lettera inedita da Tobia Aldini a Giovanni Faber." *Annali di Botanica* VI, no. 3: 403–405.

———. 1916. "Per la Storia dei Primi Lincei. I. Il catalogo dell'erbario di uno dei primi Lincei; II. Una escursione botanica dei primi Lincei a Monte Gennaro il 12 ottobre 1611." *Annali di Botanica* VI: 150–160.

———. 1927. "Per la Storia dei Primi Lincei. IV. Lettere di Giovanni Battista Winther a Giovanni Faber." *Annali di Botanica* XVII, no. 4: 140–170.

Cropper, Elizabeth, ed. 1988. *Pietro Testa, 1612–1650, Prints and Drawings.* Cat.exhib., Philadelphia Museum of Art, Philadelphia, and Arthur M. Sackler Museum, Harvard University Art Museums, Cambridge, 1988.

Cropper, Elizabeth, Giovanna Perini, and Francesco Solinas, eds. 1992. *Documentary Culture. Florence and Rome from Grand-Duke Ferdinand I to Pope Alexander VII: Papers from a Colloquium held at the Villa Spelman, Florence 1990.* Villa Spelman Colloquia, 3, Bologna.

Curtius, Ernst Robert. 1953. *European Literature in the Latin Middle Ages.* Translated by Willard R. Trask. London: Routledge.

Dagognet, François. 1970. *Le catalogue de la vie: Étude méthodologique sur la taxinomie.* Paris: PUF.

*Dizionario Biografico degli Italiani.* 1960–. Rome: Istituto della Enciclopedia Italiana.

Daston, Lorraine J. 1991a. "Baconian Facts, Academic Civility and the Prehistory of Objectivity." *Annals of Scholarship* 8: 337–363.

———. 1991b. "Marvelous Facts and Miraculous Evidence in Early Modern Europe." *Critical Inquiry* 18: 93–124.

Daston, Lorraine J., and Katharine Park. 1981. "Unnatural Conceptions: The Study of Monsters in Sixteenth- and Seventeenth-Century France and England." *Past and Present* 92: 20–54.

———. 1998. *Wonders and the Order of Nature, 1150–1750.* New York: Zone Books.

De Angelis, Pietro. 1953. *Giovanni Faber Linceo Primario in Santo Spirito in Saxia (1598–1629): "Il Microscopio."* Rome: Accademia Lancisiana.

De Angelis, Gilberto, and Paola Lanzara. 1985. "Due Elenchi di Piante Osservate e Raccolte dai Primi Lincei a Monte Gennaro (Monti Lucretili, Sabina Meridionale, Lazio)." *Annali di Botanica (Roma)*, suppl. no. 3: 127–147.

———. 1986. "La 'Syntaxis Plantaria' di Federico Cesi nei codici di Parigi: La nascita della microscopia vegetale." In *Convegno celebrativo del IV centenario della nascita di Federico Cesi (Atti dei Convegni Lincei, 78)*, Aquasparta, 7–9 ottobre 1985. Rome: Accademia Nazionale dei Lincei: 251–276.

De Backer, Augustin and Aloys, and Carlos Sommervogel, eds. 1890–1932. *Bibliothèque de la Compagnie de Jésus.* 12 vols. Brussels: Oscar Schepens; Paris: A. Picard.

De Renzi, Silvia. 1989. "Il progetto e il fatto. Nuovi studi sull'Accademia dei Lincei." *Intersezioni. Rivista di storia dele idee* IX/3, 1989: 501–517.

———. 1992–1993. *Storia naturale ed erudizione nella prima età moderna: Johannes Faber (1574–1629) medico linceo.* Università degli Studi di Bari, Tesi di Dottorato in Storia della Scienza. VI ciclo.

———. 1994. "'Fidelissima delineatio': Descrizioni alla prova nelle note di Johan Faber al 'Tesoro Messicano.'" In Andrea Battistini, ed., *Mappe e Letture. Studi in onore di Ezio Raimondi.* Bologna: Il Mulino, 103–120.

———. 1996. "Courts and Conversions: Intellectual Battles and Natural Knowledge in Counter-Reformation Rome." *Studies in the History and Philosophy of Science* 27/4: 429–449.

De Toni, Ettore. 1901. "Le piante LINCEA-CESIA-COLUMNIA-STELLUTA e BARBERINA." *Memorie della Pontificia Accademia dei Nuovi Lincei* XVII: 349–361.

De Toni, Giovanni Battista. 1907. "Spigolature aldrovandiane VI. Le piante dell'antico orto botanico di Pisa ai tempi di Luca Ghini." *Annali di botanica* 5: 421–425.

Dear, Peter. 2001. *Revolutionizing the Sciences: European Knowledge and Its Ambitions, 1500–1700.* Basingstoke: Palgrave.

Donahue, William H. 1981. *The Dissolution of the Celestial Spheres 1595–1650.* New York: Arno Press.

Drake, Stillman. 1957. *Discoveries and Opinions of Galileo.* New York: Anchor Books.

———. 1970. *Galileo Studies. Personality, Tradition, and Revolution.* Ann Arbor: University of Michigan Press, 1970.

———. 1976. *Galileo against the Philosophers.* Los Angeles: Zeitlin and Ver Brugge.

———. 1978. *Galileo at Work: His Intellectual Biography.* Chicago: University of Chicago Press.

———. 1988. "Galileo: A Biographical Sketch." In Ernan McMullin, ed., *Galileo, Man of Science.* 2nd ed. Princeton Junction: The Scholar's Bookshelf, 52–66.

———. 1990. *Galileo: Pioneer Scientist.* Toronto: University of Toronto Press.

Drake, Stillman, and Charles D. O'Malley. 1960. *The Controversy on the Comets of 1618.* Philadelphia: University of Pennsylvania Press.

Eamon, William. 1994. *Science and the Secrets of Nature. Books of Secrets in Medieval and Early Modern Culture.* Princeton: Princeton University Press.

Ehrard, Jean. 1970. *L'Idée de la nature en France a l'aube des lumières.* Paris: Flammarion. (Abridged and unannotated form of *L'Idée de la nature en France dans la première moitié du XVIIIe siècle.* Paris: S.E.V.P.E.N., 1963.)

Eisenstein, Elisabeth. 1979. *The Printing Press as an Agent of Change: Communications and Cultural Transformations in Early Modern Europe.* 2 vols. New York: Cambridge University Press.

Emmart, Emily Walcott. 1940. *The Badianus Manuscript (Codex Barberini, Latin 241) Vatican Library: An Aztec Herbal of 1552.* Baltimore: Johns Hopkins Press.

Ernst, Germana. 1991. *Religione, ragione, e natura: ricerche su Tommaso Campanella e il tardo Rinascimento.* Milan: F. Angeli.

Fabiani, Giuseppe. 1872. *La vita di Pietro Andrea Mattioli raccolta delle sue opere, con aggiunte e annotazioni di Luciano Bianchi.* Siena: Bargellini.

Favaro, Antonio. 1881. *Galileo Galilei ed il "Dialogo de Cecco di Ronchitti da Bruzene in perpuosito de la stella nuova": studi e recherché.* Venice: Antonelli.

———. 1883. *Galileo Galilei e lo Studio di Padova.* 2 vols. Florence: Le Monnier.

———. 1886. *Carteggio inedito di Ticone Brahe, Giovanni Keplero e di altri celebri astronomi e matematici . . . con Giovanni Antonio Magini.* Bologna: Zanichelli.

———. 1894–1914. *Amici e correspondenti di Galileo.* 3 vols. Venice: Officine Grafiche di C. Ferrari. Modern reprint with a foreword by Paolo Galluzzi. Florence: Libreria Editrice Salimbeni, 1993.

———. 1902. *Intorno ai cannocchiali costruiti ed usati da Galileo Galilei.* Venice: Tipografia di Carlo Ferrari.

———. 1968. *Galileo Galilei a Padova.* Padua: Editrice Antenore.

Feldhay, Rivka. 1995. *Galileo and the Church: Political Inquisition or Critical Dialogue?* Cambridge: Cambridge University Press.

Ferrone, Vincenzo. 1984. "Galileo tra Paolo Sarpi e Federico Cesi: Premesse per una ricerca." In Paolo Galluzzi, ed., *Novità celesti e crisi di sapere: Atti del Convegno Internazionale di Studi Galileiani.* Florence: Giunti Barbèra, 239–253.

Ferrone, Vincenzo, and Massimo Firpo. 1985. "Galileo tra inquisitori e microstorici." *Rivista Storica Italiana* 177–238: 957–968.

———. 1986. "From Inquisitors to Microhistorians: A Critique of Redondi's 'Galileo Eretico.'" *Journal of Modern History* 58/2: 285–524.

Feyerabend, Paul. 1988. *Against Method.* Revised edition. London: Verso.

Findlen, Paula. 1990. "Jokes of Nature and Jokes of Knowledge: The Playfulness of Scientific Discourse in Early Modern Europe." *Renaissance Quarterly* 43: 292–331.

———. 1994. *Possessing Nature: Museums, Collecting, and Scientific Culture in Early Modern Italy.* Berkeley and Los Angeles: University of California Press.

Finocchiaro, Maurice A. 1989. *The Galileo Affair: A Documentary History.* Berkeley and Los Angeles: University of California Press.

Fleming, John. 1958. "Cardinal Albani's Drawings at Windsor: Their Purchase by James Adam for George III." *Connoisseur,* CXLII: 164–169.

Foucault, Michel. 1973. *The Order of Things. An Archeology of the Human Sciences* (translation of *Les mots et les choses*, Paris: Gallimard, 1966). New York: Vintage.

Franzoni, Claudio. 1984. "'Rimembranze d'infinite cose': Le collezioni rinascimentali di antichità." In Salvatore Settis, ed., *Memoria dell'antico nell'arte italiana*. Turin: Einaudi, 299–360.

Freedberg, David. 1989a. "From Hebrew and Gardens to Oranges and Lemons: Giovanni Battista Ferrari and Cassiano dal Pozzo." In Francesco Solinas, ed., *Cassiano dal Pozzo, Atti del Seminario Internazionale di Studi*. Rome: De Luca, 37–72.

———. 1989b. "Cassiano, Natural Historian." *Quaderni Puteani* 1: 10–15.

———. 1989c. "Cassiano dal Pozzo's Drawings of Citrus Fruits." *Quaderni Puteani* 1: 16–36.

———. 1992a. "Cassiano and the Jewish Races." *Quaderni Puteani* 3: 41–56.

———. 1992b. "Ferrari on the Classification of Oranges and Lemons." In Cropper, Perini, and Solinas, eds., *Documentary Culture. Florence and Rome from Grand-Duke Ferdinand I to Pope Alexander VII: Papers from a Colloquium held at the Villa Spelman, Florence 1990 (Villa Spelman Colloquia, 3)*. Bologna, 287–306.

———. 1993. "Cassiano and the Art of Natural History." *Quaderni Puteani* 4: 141–222.

———. 1994a. "The Failure of Colour." In John Onians, ed., *Sight and Insight. Essays on Art and Culture in Honour of E. H. Gombrich*. London: Phaidon, 245–262.

———. 1994b. "Poussin et Sienne." *Catalogue Exposition Nicolas Poussin*, ed. Pierre Rosenberg et al. Musée du Louvre, Paris, 62–68.

———. 1994c. "Van Dyck and Virginio Cesarini: A Contribution to the Study of Van Dyck's Roman Sojourns." *Studies in the History of Art*, 46, *Van Dyck 350*, ed. Susan J. Barnes and Arthur K. Wheelock. National Gallery of Art, Washington, D.C., 152–174.

———. 1996a. "Ferrari and the Pregnant Lemons of Pietrasanta." In Alessandro Tagliolini and Margherita Azzi Visentini, eds., *Il Giardino delle Esperidi. Gli agrumi nella storia, nella letteratura e nell'arte*. Florence: Edifir, 41–58.

———. 1996b. "Poussin, Ferrari, Cortone et 'l'Aetas Florea.'" In Alain Mérot, *Nicolas Poussin (1594–1665). Actes du Colloque . . . au Musée du Louvre . . . 19–21 octobre 1994*, I. Paris: La documentation Française, 337–362.

———. 1998. "Iconography between the History of Art and the History of Science: Art, Science, and the Case of the Urban Bee." In Peter L. Galison and Caroline A. Jones, eds., *Picturing Science, Producing Art*. London: Routledge, 272–296.

———. 2000. "Del nominare i fiori: Ferrari, Poussin e la storia della storia naturale." In Luciano Morbiato, ed., *Quaderno di dieci anni, Gruppo Giardino Storico dell'Università di Padova*. Padua: Università degli Studi, 57–84.

———. Forthcoming. "The Failure of Pictures: From Description to Diagram in the Circle of Galileo."

Freedberg, David, and Enrico Baldini. 1997. *Citrus Fruit (The Paper Museum of Cassiano dal Pozzo. A Catalogue Raisonné. Drawings and Prints in the Royal Library at Windsor Castle, the British Museum, the Institut de France and other Collections. Series B. Natural History)*, I. London: Harvey Miller.

Fumaroli, Marc. 1978. "Cicero Pontifex romanus: sul collegio romano e il mecenatismo dei Barberini." In *Mélanges de l'école française de Rome*, XC, 797–835.

Gabrieli, Giuseppe. 1926. "Verbali delle adunanze e cronaca della prima Accademia Lincea (1603–1630)." *Atti della Regia Accademia dei Lincei, Memorie della Classe di Sienze morali, storiche e filologiche*, s. VI, vol. II: 463–512.

———. 1939. "La 'Germania Lincea' ovvero Lincei e Linceabili tedeschi della prima Accademia: in particolare di Teofilo Müller." *Rendiconti della Regia Accademia Nazionale dei Lincei, Memorie della Classe di Scienze morali, storiche e filologiche*, s. VI, vol. XV: 42–66.

———. 1989. *Contributi alla storia della Accademia dei Lincei*. 2 vols. Rome: Accademia Nazionale dei Lincei.

Galison, Peter, and Caroline Jones, eds. 1988. *Picturing Science, Producing Art*. London: Routledge.

Galluzzi, Paolo, ed. 1984. *Novità celesti e crisi del sapere: Atti del Convegno Internazionale di Studi Galileiani*. Florence: Giunti Barbera.

Garboni, Luigi Garboni. 1899. *Un umanista del seicento: Giano Nicio Eritreo*. Città di Castello.

Gardair, Jean-Michel. 1981. "I Lincei: i sogetti, i luoghi, le attività." *Quaderni Storici* (Accademie scientifiche del '600. Professioni borghesi) 48: 763–787.

Gingerich, Owen. 1981. "The Censorship of Copernicus 'De Revolutionibus.'" *Annali dell'Istituto e Museo di Storia della Scienza di Firenze* VI: 45–61.

———. 1985. "Dissertatio cum Professore Righini et Sidereo Nuncio." In Righini Bonelli and Shea 1975, 77–88.

———. 1993. *The Eye of Heaven: Ptolemy, Copernicus, Kepler.* New York: American Institute of Physics.

Ginzburg, Carlo. 1970. *Il Nicodemismo.* Turin: Einaudi.

———. 1986. *Miti emblemi spie: morfologia e storia.* Turin: Einaudi.

———. 1989a. *Clues, Myths, and the Historical Method* (translation of the above by John and Anne C. Tedeschi). Baltimore: Johns Hopkins University Press.

———. 1989b. "Montrer et citer." *Le Débat* 56: 43–54.

Goldman, Jean. 1978. "Aspects of Seicento Patronage: Cassiano dal Pozzo and the Amateur Tradition." Ph.D. diss., University of Chicago.

Golino, Carlo L., ed. 1966. *Galileo Reappraised.* Berkeley and Los Angeles: University of California Press.

Golinski, Jan. 1998. *Making Natural Knowledge: Constructivism and the History of Science.* Cambridge: Cambridge University Press.

Gómez, Susana . 1991. "The Bologna Stone and the Nature of Light." *Nuncius* VI/2: 3–30.

Gorman, Michael John. 1996. "A Matter of Faith? Christoph Scheiner, Jesuit Censorship, and the Trial of Galileo." *Perspectives on Science* 4/3: 283–320.

Gotthelf, Allan, and James G. Lennox, eds. 1987. *Philosophical Issues in Aristotle's Biology.* Cambridge and New York: Cambridge University Press, 1987.

Gould, Stephen Jay. 1997. "[Leonardo's] theory of the living earth." *Natural History* 5.

———. 1998a. "The Sharp-Eyed Lynx, Outfoxed by Nature, Part I." *Natural History* 5.

———. 1998b. "The Sharp-Eyed Lynx, Outfoxed by Nature, Part II." *Natural History* 6.

Govi, Gilberto. 1888. "Il microscopio composto inventato da Galileo." *Atti della Reale Accademia delle Scienze Fisiche e Matematiche.* Naples, II: 1–33.

Grafton, Anthony. 1991. *Defenders of the Text, The Traditions of Scholarship in an Age of Science, 1450–1800.* Cambridge: Harvard University Press.

Greenblatt, Stephen Jay. 1991. *Marvelous Possessions: The Wonder of the New World.* Oxford and New York: Clarendon Press.

Hall, Rupert. 1988. "The Significance of Galileo's Thought for the History of Science." In Ernan McMullin, ed., *Galileo, Man of Science.* 2nd ed. Princeton Junction: The Scholar's Bookshelf, 67–81.

Haskell, Francis. 1980. *Patrons and Painters: A Study of the Relations between Italian Art and Society in the Age of the Baroque.* London, Chatto & Windus: 1963. Revised ed., New Haven and London: Yale University Press.

Haskell, Francis, and Sheila Rinehart. 1960. "The Dal Pozzo Collection: Some New Evidence." *The Burlington Magazine* CII: 318–329.

Headley, John M. 1997. *Tommaso Campanella and the Transformation of the World.* Princeton: Princeton University Press.

Henry, John. 1997. *The Scientific Revolution and the Origins of Modern Science.* London: Macmillan; New York: St Martin's Press.

Herklotz, Ingo. 1999. *Cassiano dal Pozzo und die Archäologie des 17. Jahrhunderts* (Römische Forschungen der Bibliotheca Hertziana, Bd. 28). Munich: Hirmer.

Hoskin, Michael. 1977. *Stellar Astronomy: Historical Studies.* New York: Science History Publications.

Humbert, Pierre. 1951. "Peiresc et le microscope." *Revue d'histoire des sciences et de leurs applications* 4: 154–158.

Hunger, Friedrich Wilhelm Tobias. 1927–43. *Charles de l'Ecluse (Carolus Clusius), Nederlandsch Kruidkunde, 1526–1609.* 2 vols. 's-Gravenhage: M. Nijhoff.

Impey, Oliver, and Arthur MacGregor, eds. 1985. *The Origins of Museums: The Cabinet of Curiosities in Sixteenth- and Seventeenth-Century Europe.* Oxford: Oxford University Press.

Ingegno, Alfonso. 1978. *Cosmologia e filosofia nel pensiero di Giordano Bruno.* Florence: La Nuova Italia.

Jaffé, David. 1989. "The Barberini Circle: Some Exchanges between Peiresc, Rubens and Their Contemporaries," *Journal of the History of Collections* I/2: 119–147.

———. 1990. "Mellan and Peiresc." *Print Quarterly* 7: 168–175.

Jardine, Nicholas. 1988. *The Birth of History and Philosophy of Science: Kepler's "A Defence of Tycho against Ursus," with essays on its Provenance and Significance.* Cambridge: Cambridge University Press.

Jenkins, Ian. 1987. "Cassiano dal Pozzo's Museo Cartaceo: New Discoveries in the British Museum." *Nouvelles de la Républiques des Lettres* II, 29–41.

———. 1989. "Newly Discovered Drawings from the Museo Cartaceo in the British Museum." In Francesco Solinas, ed., *Cassiano dal Pozzo, Atti del Seminario Internazionale di Studi.* Rome: De Luca Edizioni d'Arte, 131–136, 141–175.

Jiménez Muñoz, José Miguel. 1977. *Médicos y cirujanos en Quitaciones de Corte (1435–1715).* Valladolid: Ediciones de la Universidad de Vallodolid.

Kanceff, Emanuele. 1976–77. *Oeuvres de Jean Jacques Bouchard.* 2 vols. Turin: G. Giappichelli.

Kidwell, Clara Sue. 1970. "The Accademia dei Lincei and the 'Apiarium': A Case Study in the Activities of a Seventeenth-Century Scientific Society." Ph.D. diss., University of Oklahoma.

Koyré, Alexandre. 1970. *From the Closed World to the Infinite Universe.* Baltimore: Johns Hopkins University Press.

———. 1978. *Galileo Studies.* Translated from the French by John Mepham. Atlantic Highlands, N.J.: Humanities Press.

Kuhn, Thomas S. 1970. *The Structure of Scientific Revolutions.* 2nd ed. Chicago, University of Chicago Press.

———. 1977. *The Essential Tension. Selected Studies in Scientific Tradition and Change.* Chicago: University of Chicago Press.

Lattis, James. 1994. *Between Copernicus and Galileo: Christoph Clavius and the Collapse of Ptolemaic Cosmology.* Chicago: University of Chicago Press.

Lepenies, Wolf. 1976. *Das Ende der Naturgeschichte: Wandel kultureller Selbstverständlichkeiten in den Wissenschaften des 18. und 19. Jahrhunderts.* Munich: C. Hanser.

Lindberg, David C., and Ronald L. Numbers, eds. 1986. *God and Nature: Historical Essays on the Encounter between Christianity and Science.* Berkeley and Los Angeles: University of California Press.

Lindberg, David C., and Robert S. Westmann, eds. 1990. *Reappraisals of the Scientific Revolution.* Cambridge: Cambridge University Press.

Lippi Boncambi, Cesare. 1960. "Geologia e paleontografia dei bacini lignitiferi dell'Umbria." *Atti del Convegno mostra nazionale delle ligniti, Perugia 7–1 maggio 1959.* Foligno: Poligrafica F. Salviati.

Lomonaco, Fabrizio, and Maurizio Torrini, eds. 1987. *Galileo e Napoli*, Atti del Convegno (12–14 aprile 1984). Naples: Guida.

López Piñero, José María. 1991. *El Codice Pomar (ca. 1590), el interés de Felipe II por la Historia natural y la expedición Hernández a América.* València: Instituto de Estudios Documentales e Históricos sobre la Ciencia Universitat de València-C.S.I.C.

———. 1992. "The Pomar Codex (ca. 1590): Plants and Animals of the Old World and from the Hernández Expedition to America." *Nuntius* 7: 220–279.

López Piñero, José María, and José Pardo Tomás. 1994. *Nuevos materiales y noticias sobre la "Historia de las plantas de Nueva España" de Francisco Hernández.* València: Instituto de Estudios Documentales e Históricos sobre la Ciencia Universitad de València-C.S.I.C.

Lugli, Adalgisa. 1983. *Naturalia et mirabilia: Il collezionismo enciclopedico nelle Wunderkammern d'Europa.* Milan: Mazzotta.

Lumbroso, Giacomo. 1874. "Notizie sulla vita di Cassiano dal Pozzo." *Miscellanea di Storia italiana* XV: 129–388.

Lüthy, Christoph H. 1996. "Atomism, Lynceus, and the Fate of Seventeenth-Century Microscopy." *Early Science and Medicine* 1/1: 1–27.

Maccagni, Carlo, ed. 1972. *Saggi su Galileo Galilei* (Pubblicazioni del Comitato Nazionale per le Manifestazioni celebrative, vol. III, 2). Florence: G. Barbèra Editore.

Magurn, Ruth Saunders. 1955. *The Letters of Rubens.* Cambridge: Harvard University Press.

Mahon, Denis, et al. 1998. *Nicolas Poussin. I Primi Anni Romani.* (Cat. Exhib.) Milan: Electa.

Masson, Georgina. 1961. *Italian Gardens.* London: Thames and Hudson.

Matteoli, Anna. 1959. "Macchie di Sole e Pittura, Carteggio, L. Cigoli-G. Galilei (1609–1613)." *Bolletino della Accademia degli Euteleti della Città di San Miniato* 32: 9–92.

———. 1964–65. "Cinque lettere di Lodovico Cardi Cigoli a Michelangelo Buonarotti il Giovane." *Bolletino della Accademia degli Euteleti della Città di San Miniato* 28: 31–42.

Mattirolo, Oreste. 1904. "Le Lettere di Ulisse Aldrovandi a Francesco I e Ferdinando I." *Memoria della Reale Accademia delle Scienze di Torino*, s. II, vol. 54: 353–401.

Maylender, Michele. 1926–30. *Storie delle accademie d'Italia.* 5 vols. Bologna: L. Cappelli.

Mazzolini, Renato G., ed. 1993. *Non-Verbal Communication in Science prior to 1900.* Florence: Leo S. Olschki Editore.

McBurney, Henrietta. 1989a. "History and Contents of the dal Pozzo Collection in the Royal Library of Windsor Castle." In Francesco Solinas, ed., *Cassiano dal Pozzo, Atti del Seminario Internazionale di Studi.* Rome: De Luca Edizioni d'Arte, 75–93.

———. 1989b. "The Later History of Cassiano dal Pozzo's 'Museo Cartaceo.'" *Burlington Magazine* CXXXI: 549–553.

———. 1989c. "A Brief History of the Museo Cartaceo." *Quaderni Puteani* 1: 5–9.

———. 1989d. "Cassiano dal Pozzo's Drawings of Birds." *Quaderni Puteani* 1: 37– 47.

———. 1992a. "Cassiano dal Pozzo as Scientific Commentator: Ornithological Texts and Images from the 'Museo Cartaceo.'" In Elizabeth Cropper, Giovanna Perini, and Francesco Solinas, eds., *Documentary Culture: Florence and Rome from Grand-Duke Ferdinand I to Pope Alexander VII; Papers from a Colloquium Held at the Villa Spelman, Florence 1990* (Villa Spelman Colloquia, 3), Bologna, 349–363.

———. 1992b. "Cassiano Dal Pozzo as Ornithologist." *Quaderni Puteani* 3: 3–22.

McMullin, Ernan, ed. 1988. *Galileo, Man of Science.* 2nd ed. Princeton Junction: The Scholar's Bookshelf.

Marini Bettólo, G.B. 1992. *A Guide for the Reader of the Mexican Treasure* (printed to accompany the 1992 facsimile of the *Thesaurus*). Rome: *Istituto Poligrafico e Zecca dello Stato.*

Medina, José Toribio. 1898–1907. *Biblioteca Hispano-Americana*, I–VII. Santiago de Chile.

Meinel, Christoph. 1984. *In physicis futurum saeculum respicio: Joachim Jungius und die Naturwissenschaftliche Revolution des 17. Jahrhundert.* Göttingen: Vandenhoeck and Ruprecht.

Mercer, Christia. 1995. "The Vitality and Importance of Early Modern Aristotelianism." In Tom Sorell, ed., *The Rise of Modern Philosophy: The Tension between the New and Traditional Philosophies from Machiavelli to Leibniz.* Oxford: Clarendon Press.

Merrick, Jeffrey. 1988. "Royal Bees: The Gender Politics of the Beehive in Early Modern Europe." *Studies in Eighteenth Century Culture*, XVIII: 7–37.

Merz, Jörg Martin. 1991. *Pietro da Cortona. Der Aufstieg zum führenden Maler im barocken Rom.* Tübingen: E. Wasmuth (Tübinger Studien zur Archeologie und Kunstgeschichte 8).

Morello, Giovanni. 1986. *Federico Cesi e i primi Lincei.* Cat. Exhib. Bibliotheca Apostolica Vaticana, Vatican City.

Morello, Nicoletta. 1977. "Fabio Colonna e gli inizi della paleontologia." *Physis* XIX: 247–279.

———. 1979. *La nascita della paleontologia nel Seicento: Colonna, Stenone e Scilla.* Milan: Angeli.

———. 1981. "*De glossopetris dissertatio:* the demonstration by Fabio Colonna of the true nature of fossils." *Archives internationales d'histoire des sciences* XXXI: 63–71.

Morton, Alan G. 1981. *History of Botanical Science, an Account of the Development of Botany from Ancient Times to the Present Day.* New York: Academic Press.

Moss, Jean Dietz. 1984. "Galileo's Rhetorical Strategies in Defense of Copernicanism." In Paolo Galluzzi, ed., *Novità celesti e crisi del sapere: Atti del Convegno Internazionale di Studi Galileiani.* Florence: Giunti Barbera, 95–103.

———. 1993. *Novelties in the Heavens: Rhetoric and Science in the Copernican Controversy.* Chicago: University of Chicago Press.

Naples. 1988. *Federico Cesi e la fondazione dell'Accademia dei Lincei.* Mostra bibliografica e documentaria. Naples: Accademia Nazionale dei Lincei, Istituto Italiano per gli Studi filosofici, Biblioteca Nazionale Marciana.

Napoleone, Caterina. 1987. "Il gusto dei marmi antichi: descrizioni inedite di marmi e pietre nella relazione diaria del viaggio in Spagna di Cassiano dal Pozzo (1626)." In *Nouvelles de la République des Lettres* II: 43–47.

———. 1989. "Appunti sul 'Natural History of Fossils V' della Royal Library di Windsor." In Francesco Solinas, ed., *Cassiano dal Pozzo, Atti del Seminario Internazionale di Studi*. Rome: De Luca Edizioni d'Arte, 187–198.

Nastasi, Pietro, ed. 1988. *Atti del Convegno Il Meridione e le scienze (secoli XVI-XIX)* (Palermo 14–16 maggio 1985). Palermo and Naples: Istituto Gramsci Siciliano-Istituto Italiano per gli studi filosofici.

Nicolò, Anna. 1982. "Corrispondenza inedita di Francesco Stelluti a Cassiano dal Pozzo nel Carteggio Puteano dell'Accademia dei Lincei." *Atti della Accademia Nazionale dei Lincei, Classe di Scienze morali, storiche e filologiche, Rendiconti*, XXVII, fasc. 3–4: 91–99.

———. 1985. "Francesco Stelluti nelle sue lettere dal 1630 al 1652." *Atti del Convegno Francesco Stelluti*, Fabriano.

———. 1986. "Le Lettere dal 1630 al 1652: Francesco Stelluti e il legno fossile." In Ada Alessandrini et al., *Francesco Stelluti Linceo da Fabriano*. Fabriano: Città e Comune di Fabriano, 176–180.

———. 1991. *Il Carteggio di Cassiano dal Pozzo*. Catalogo. Florence, Olschki.

Nicolò, Anna, and Francesco Solinas. 1986. "Per una analisi del collezionismo Linceo: 'l'Archivio Linceo 32' e il Museo di Federico Cesi." In *Convegno celebrativo del IV centenario della nascita di Federico Cesi (Atti dei Convegni Lincei, 78)*, Acquasparta: 7–9 ottobre 1985. Rome: Accademia Nazionale dei Lincei, 193–212.

———. 1987. "Cassiano dal Pozzo: appunti per una cronologia di documenti e disegni (1612–1630)." In *Nouvelles de la République des Lettres* II: 59–110.

Nissen, Claus. 1951. *Die botanische Buchillustration: ihre Geschichte und Bibliographie*. Stuttgart: Hiersemann.

Odescalchi, Baldassare. 1806. *Memorie istorico-critiche dell'Accademia dei Lincei e del principe Federico Cesi, secondo duca d'Aquasparta, fondatore e principe della medesima*. Rome: Luigi Perego Salvioni.

Olmi, Giuseppe. 1976. *Ulisse Aldrovandi. Scienza e natura nel secondo Cinquecento*. Trent: Libera Università degli Studi di Trento.

———. 1981. "*In essercito universale di contemplazione, e prattica*, Federico Cesi e i Lincei." In Laetitia Boehm and Ezio Raimondi, eds., *Università, Accademie e Società Scientifiche in Italia e in Germania dal Cinquecento al Settecento*. Bologna: Società editrice il Mulino, 169–235.

———. 1982. "Ordine e fama: il museo naturalistico in Italia nei secoli XVI e XVII." *Annali dell'Istituto storico italo-germanico in Trento* 8: 225–274.

———. 1987. "La colonia lincea di Napoli." In Fabrizio Lomonaco and Maurizio Torrini, eds., *Galileo e Napoli. Atti del Convegno di Napoli (12–14 aprile 1984)*. Naples: Guida, 23–58.

———. 1991. "'Molti amici in vari luoghi': Studio della natura e rapporti epistolari nel secolo XVI." *Nuncius* 6: 3–31.

———. 1992. *L'inventario del mondo. Catalogazione della natura e luoghi del sapere nella prima età moderna*. Bologna: Società editrice il Mulino.

———. 1993a. "La bottega artistica di Ulisse Aldrovandi." In Giuseppe Olmi and Lucia Tongiorgi, *De piscibus: La bottega artistica di Ulisse Aldrovandi e l'immagine naturalistica*. Rome: Edizioni dell'Elefante, 9–31.

———. 1993b. "From the Marvellous to the Commonplace: Notes on Natural History Museums (16th–18th Centuries)." In Renato G. Mazzolini, ed., *Non-Verbal Communication in Science prior to 1900*. Florence: Leo S. Olschki Editore, 235–278.

———. 1998a. "*Regiones omnes momento lustrare poteris*: viaggiatori e collezioni nella prima età moderna." In Walter Tega, ed., *Le origini della modernità. I. Linguaggi e sapere tra XV e XVI secolo*. Florence: Olschki: 165–197.

———. 1998b. "Die Accademia dei Lincei." In *Grundriss der Geschichte der Philosophie, Die Philosophie des 17. Jahrhunderts*. Band I. Allgemeine Themen. Iberische Halbinsel Italien, ed. Jean-Pierre Schobinger,

Olmi, Giuseppe, and Lucia Tongiorgi. 1993. *De piscibus: La bottega artistica di Ulisse Aldovrandi e l'immagine naturalistica*. Rome: Edizioni dell'Elefante.

Ong, Walter J., S.J. 1958. *Ramus: Method, and the Decay of Dialogue: From the Art of Discourse to the Art of Reason*. Cambridge: Harvard University Press.

Orbaan, Johannes Albertus Franciscus. 1920. , *Documenti sul Barocco in Roma*. Rome: Nella Sede della Società alla Biblioteca Valicelliana.

Osborne, John, and Amanda Claridge. 1996. *Early Christian and Medieval Antiquities. Vol. I: Mosaics and Wallpaintings in Roman Churches (The Paper Museum of Cassiano dal Pozzo. A Catalogue Raisonné. Drawings and Prints in the Royal Library at Windsor Castle, the British Museum, the Institut de France and other Collections. Series A. Antiquities and Architecture II)*, I. London: Harvey Miller.

Ostrow, Steven F. 1996. "Cigoli's Immaculate Virgin and Galileo's Moon: Astronomy and the Apocalyptic Woman in Early Seicento Rome." *Art Bulletin* 78: 218–235.

Pagano, Sergio M. 1984. *I Documenti del processo di Galileo Galilei*. Vatican City: Pontifica Academia Scientiarum.

Panofsky, Erwin. 1954. *Galileo as a Critic of the Arts*, The Hague: Martinus Nijhoff.

Parker, Gary D. 1985. "Galileo, Planetary Atmospheres, and Prograde Revolution." *Science* 227: 597–600.

Partini, Anna Maria. 1986. "I primi Lincei e l'ermetismo." *Atti della Accademia Nazionale dei Lincei, Rendiconti della Classe di Scienze morali, storiche e filologiche*, s. VIII, vol. XLI: 1–26.

Pastor, Ludwig von. 1866–1938. *Geschichte der Päpste seit dem Ausgang des Mittelalters. Mit Benützung des päpstlichen Geheim-Archives und vieler anderer Archive*. 21 vols. Freiburg im Breisgau: Herder.

Peck, A.L. 1965. Introduction to A.L. Peck, ed., *Aristotle: History of Animals, Books I–III*.Cambridge: Harvard University Press (Loeb Classical Library).

Pegler, David, and David Freedberg. Forthcoming. *Fungi (The Paper Museum of Cassiano dal Pozzo. A Catalogue Raisonné: Drawings and Prints in the Royal Library at Windsor Castle, the British Museum, the Institut de France and other Collections. Series B. Natural History), II*. London: Harvey Miller.

Pellegrin, Pierre. 1986. *Aristotle's Classification of Animals: Biology and the Conceptual Unity of the Aristotelian Corpus*, tr. Anthony Preus. Berkeley and Los Angeles: University of California Press.

Petrucci Nardelli, Franca. 1985. " Il Card. Francesco Barberini Senior e la Stampa a Roma" *Archivio della Società Romana di Storia Patria*. Rome.

Pignatti, Sandro, and Gaspare Mazzolani. 1985. "Federico Cesi Botanico." In *Convegno celebrativo del IV centenario della nascità di Federico Cesi (Atti dei Convegni Lincei, 78)*, Acquasparta, 7–9 ottobre 1985. Rome: Accademia Nazionale dei Lincei, 213–223.

Pintard, Réné. 1943. *Le libertinage erudit dans la première moitié du XVII$^{e}$ siecle*. 2 vols. Paris: Ancienne Librairie Frerne, Boivin et C.ie Éditeurs.

Pirotta, Romualdo. 1904. *Sodales R. Lynceorum Academiae annum CCC. ab eius Institutione concelebrantes Fridirici Cesi auctoris conlegi Opus probatissimum De Plantis ad fidem exemplaris castigatioris studio et cura conlegae Romualdi Pirotta iterum edendum decrevere*. Rome: Accademia dei Lincei.

Pirotta, Romualdo, and Emilio Chiovenda. 1900–1901. "Flora romana." *Annuario del R. Istituto Botanico di Roma* X, fasc. 1, 1900: 1–146; fasc. 2, 1901: 145–304.

Poggioli, Michelangelo. 1865. *De amplitudine doctrinae botanicae qua praestitit Fridericus Caesius*, ed. Giuseppe Poggioli. Rome: Bonae Artes.

Pomian, Krzysztof. 1990. *Collectors and Curiosities: Paris and Venice, 1500–1800*. London: Polity Press.

Pope-Hennessy, John. 1948. *The Drawings of Domenichino in the Collection of His Majesty the King at Windsor Castle*. London: Phaidon.

Poppi, Antonio. 1992. *Cremonini e Galilei Inquisiti a Padova nel 1604*. Padua: Antenore.

Porter, Roy, and Mikulas Teich, eds. 1992. *The Scientific Revolution in National Context*. Cambridge: Cambridge University Press.

Pritzel, Georg August. 1872. *Thesaurus Literaturae Botanicae*. 2nd ed. Leipzig: Brockhaus.

Proja, Salvatore. 1859–60. "Ricerche critico-bibliografiche intorno alla 'Storia naturale del Messico' di Fr. Hernández esposta in dieci libri da N. A. Recchio ed illustrata dagli accademici Lincei." *Atti dell'Accademia Pontificia de'Nuovi Lincei* 1859–60: 441–477.

*Quaderni Puteani* 1: *Il Museo Cartaceo di Cassiano dal Pozzo*. Cassiano Naturalista, Milan: Olivetti, 1989.

*Quaderni Puteani* 2: *Cassiano dal Pozzo's Paper Museum*, vol. I. (Proceedings of a conference held at the British Museum and Warburg Institute, London 14–15 December 1989), Milan: Olivetti: 1992.

*Quaderni Puteani* 3: *Cassiano dal Pozzo's Paper Museum*, vol. II. (Proceedings of a conference held at the British Museum and Warburg Institute, London 14–15 December 1989), Milan: Olivetti: 1992.

*Quaderni Puteani* 4: *The Paper Museum of Cassiano dal Pozzo*. (Cat. Exhib., London: British Museum, 1993), Milan: Olivetti: 1993.

Raimondi, Ezio. 1974. *Il romanzo senza idillo*. Turin: Einaudi.

Rekers, Ben. 1972. *Benito Arias Montano*. London: Warburg Institute and Leiden: Brill.

Redondi, Pietro. 1982. "Galilée aux prises avec les théories aristoteliciennes de la lumière (1610–1640)." *Dix-Septième Siècle* 136: 267–283.

———. 1984. "La luce 'messagio celeste.'" In Paolo Galluzzi, ed., *Novità celesti e crisi del sapere: Atti del Convegno Internazionale di Studi Galileiani*. Florence: Giunti Barbera, 177–186.

———. 1985. "Galileo Eretico: Anatema." *Rivista Storica Italiana* XCVII/1: 934–956.

———. 1987. *Galileo Heretic*. Princeton: Princeton University Press. (Translated by Raymond Rosenthal from *Galileo Eretico*, 1983.)

Reeds, Karen. 1975. "Botany in Medieval and Renaissance Universities." Ph.D. diss., Harvard University.

———. 1976. "Renaissance Humanism and Botany." *Annals of Science* 33: 519–542.

Reeves, Eileen. 1991. "Daniel 5 and the 'Assayer': Galileo Reads the Handwriting on the Wall." *Journal of Medieval and Renaissance Studies* 21/1: 1–27.

———. 1997. *Painting the Heavens*. Princeton: Princeton University Press.

Ricci, Saverio. 1988. "Federico Cesi e la *Nova* del 1604: La teoria della fluidità del cielo e un opusculo dimenticato di Joannes van Heeck." *Atti dell'Accademia Nazionale dei Lincei, Rendiconti morali*. S. VIII, vol. XLIII, fasc. 5–6: 111–133.

Rienstra, Howard. 1968. "Giovanni Ecchio Linceo: Appunti cronologici e bibliografici."*Atti dell'Accademia Nazionale dei Lincei, Rendiconti morali*. s. VIII, vol. XXIII, fasc. 7–12, 255–266.

———. 1971. "Gaetano Marini and the Historiography of the Accademia dei Lincei." *Archivio della Società Romana di Storia Patria* XCIV: 209–233.

Righini Bonelli, Maria Luisa, and William R. Shea. 1975. *Reason, Experiment and Mysticism in the Scientific Revolution*. New York: Science History Publications.

Rinehart, Sheila. 1961. "Cassiano dal Pozzo (1588–1657): Some Unknown Letters." *Italian Studies*: 35–59.

Rivosecchi, Valerio. 1982. *Esotismo in Roma barocca: studi sul Padre Kircher*. Rome: Bulzoni.

Rizza, Cecilia. 1965. *Peiresc e l'Italia*. Turin: Giappichelli.

Ronchi, Vasco. 1958. *Il cannocchiale di Galileo e la scienza del Seicento*. Turin: Einaudi.

Rorty, Richard. 1981. *Philosophy and the Mirror of Nature*. Princeton: Princeton University Press.

Rosen, Edward. 1950. "The Title of Galileo's 'Sidereus Nuncius.'" *Isis* 41: 287–289.

———. 1954. "Did Galileo Claim He Invented the Telescope?" *Proceedings of the American Philosphical Society* 98/5: 304–312.

———. 1985. "The Dissolution of the Solid Celestial Spheres." *Journal of the History of Ideas* XLVI: 13–31.

Rossi, Paolo. 1960. *Clavis Universalis*. Milan-Naples: Ricciardi.

———. 1978. "Galileo Galilei e il libro dei Salmi." *Rivista di Filosofia* 10: 45–71.

———. 1984. *The Dark Abyss of Time: The History of the Earth and the History of the Nations from Hooke to Vico*. Lydia C. Cochrane, trans. Chicago: University of Chicago Press.

Ruelens, Charles Louis, and Max Rooses, eds. 1887–1909. *Correspondance de Rubens et documents epistolaires concernant sa vie et ses oeuvres*. 6 vols. Antwerp: Veuve de Backer.

Rudwick, Martin J. S. 1985. *The Meaning of Fossils: Episodes in the History of Paleontology*. 2nd ed. Chicago: University of Chicago Press.

Ruestow, Edward G. 1983. "Images and Ideas: Leeuwenhoek's Perception of the Spermatozoa," *Journal of the History of Biology* 16: 185–224.

———. 1996. *The Microscope in the Dutch Republic*. Cambridge: Cambridge University Press.

Saccardo, Pier Andrea. 1895. *La botanica in Italia*. Venice: C. Ferrari.

Santillana, Giorgio de. 1955. *The Crime of Galileo*. Chicago: University of Chicago Press. (Midway reprint, 1976.)

Schaffer, Simon, and Steven Shapin. 1985. *Leviathan and the Air-Pump: Hobbes, Boyle and the Experimental Life*. Princeton: Princeton University Press.

Scott, Andrew, and David Freedberg. 2000. *The Fossil Woods and Other Geological Specimens. The Paper Museum of Cassiano dal Pozzo (A Catalogue Raisonné. Drawings and Prints in the Royal Library at Windsor Castle, the British Museum, the Institut de France and Other Collections)*. Series B, Natural History, III. London: Harvey Miller.

Scott, John Beldon. 1991. *Images of Nepotism. The Painted Ceilings of Palazzo Barberini*. Princeton; Princeton University Press.

Schettini Piazza, Enrica. 1980. *Bibliografia storica dell'Accademia Nazionale dei Lincei*. Florence: Leo S. Olschki Editore.

———. 1986. "Teoria e sperimentazione nell' 'Apiario' de Federico Cesi." In *Convegno celebrativo del IV centenario della nascita di Federico Cesi (Atti dei Convegni Lincei, 78)*, Acquasparta, 7–9 ottobre 1985. Rome: Accademia Nazionale dei Lincei, 231–249.

Schlosser, Julius von. 1908. *Die Kunst- und Wunderkammern der Spätrenaissance; ein Beitrag zur Geschichte des Sammelwesens*. Leipzig: Klinkhardt & Biermann.

Schnapper, Antoine. 1988. *Le géant, la licorne, la tulipe: Collections françaises au XVIIe siècle. I. Histoire et histoire naturelle*. Paris: Flammarion.

Schulte van Kessel, Elisja. 1985. "Sapienza, sesso, pietas: i primi lincei e il matrimonio. Un saggi di storia umana." *Mededelingen van het Nederlands Instituut te Rome* XLVI: 121–144.

Segre, Ada. 1995. "Horticultural traditions and the emergence of the flower garden (c.1550–1660)." Ph.D. thesis, University of York, U.K.

Shank, Michael H. 1996. "How Shall We Practise History? The Case of Mario Biagioli's 'Galileo, Courtier.'" *Early Science and Medicine* 1/1: 106–150.

Shapin, Steven. 1994. *A Social History of Truth: Civility and Science in Seventeenth-Century England*. Chicago: University of Chicago Press.

———. 1996. *The Scientific Revolution*. Chicago: Chicago University Press.

Shea, William R. 1970. "Galileo, Scheiner and the Interpretation of Sunspots." *Isis* 61: 498–519.

———. 1975. "La Controriforma e l'esegesi biblica di Galileo Galilei." In Albino Baboli, ed., *Problemi religiosi e filosofia*. Padua: La Garangola, 37–62.

———. 1977. *Galileo's Intellectual Revolution*. New York: Science History Publications.

———. 1986. "Galileo and the Church." In David C. Lindberg and Ronald L. Numbers, eds., *God and Nature: Historical Essays on the Encounter between Christianity and Science*. Berkeley and Los Angeles: University of California Press.

———. 1996. "The Revelations of the Telescope." *Nuncius* XI: 507–526.

Singer, Charles. 1913–14. "Notes on the Early History of Microscopy." *Proceedings of the Royal Society of Medicine* (*Sect. Hist. Med.*), VII: 247–279.

———. 1915. "The Dawn of Microscopical Discovery." *Journal of the Royal Microscopical Society*, XXXV: 317–340.

———. 1953. "The Earliest Figures of Microscopical Objects." *Endeavour* XII/47: 197–201.

Sluiter, Engel. 1997. "The Telescope before Galileo." *Journal for the History of Astronomy* XXVIII: 223–234.

Solinas, Francesco. 1989a. "Percorsi puteani: note naturalistiche ed inediti appunti antiquari." In Francesco Solinas, ed., *Cassiano dal Pozzo, Atti del Seminario Internazionale di Studi*. Rome: De Luca Edizioni d'Arte, 95–129.

———. 1989b. "Il trattato del legno fossile minerale di Francesco Stelluti e i quattro volumi della Natural History of Fossils nelle raccolte della Biblioteca Reale di Windsor." *Quaderni Puteani* 1: 84–94.

———. 1989c. "L'Erbario Miniato e altri fogli di iconografia botanica appartenuti a Cassiano dal Pozzo." *Quaderni Puteani* 1: 52–76.

———. 1989d. "Il primo erbario azteco e la copia romana di Cassiano dal Pozzo." *Quaderni Puteani* 1: 77–83.

———. 1992. "Sull'atelier di Cassiano dal Pozzo: metodi di ricerca e documenti inediti." *Quaderni Puteani* 3: 57–76.

———, ed. 1989. *Cassiano dal Pozzo, Atti del Seminario Internazionale di Studi.* Rome: De Luca Edizioni d'Arte.

———, ed. 2000a. *I Segreti di un Collezionista. Le straordinarie raccolte di Cassiano dal Pozzo 1588–1657.* Cat. Exhib., Rome, Galleria Nazionale di Arte Antica-Palazzo Barberini. Rome: De Luca.

———. 2000b. *L'Uccelliera. Un Libro di Arte e Scienza nella Roma dei primi Lincei.* Florence: Olschki.

Solinas, Francesco, and Anna Nicolò. 1988. "Cassiano dal Pozzo and Pietro Testa: New Documents Concerning the 'Museo Cartaceo'." In Elizabeth Cropper, ed., *Pietro Testa, 1612–1650, Prints and Drawings.* Cat. exhib., Philadelphia Museum of Art, Philadelphia, and Arthur M. Sackler Museum, Harvard University Art Museums, Cambridge (Mass.), LXVI–LXXXVI.

Somolinos d'Ardois, German. 1960. "Vida y obra de Francisco Hernández." In Francisco Hernández, *Obras completas.* México: Universidad Nacional de México, vol I: 95–440.

Sparti, Donatella Livia. 1992. *Le collezioni dal Pozzo. Storia di una famiglia e del suo museo nella Roma seicentesca* (Collezionismo e storia dell'arte, studi e fonti, I, Scuola Normale Superiore di Pisa). Modena: Franco Cosimo Panini.

Spiller, Elizabeth A. "Reading through Galileo's Telescope: Margaret Cavendish and the Experience of Reading." *Renaissance Quarterly* LIII: 192–221.

Standring, Timothy J. 1998. "Some Pictures by Poussin in the Dal Pozzo Collection: Three New Inventories." *The Burlington Magazine*, CXXX: 608–626.

———. 1992. "Observations on the dal Pozzo Library and Its Organization." *Quaderni Puteani* 3: 92–100.

Steiner, George. 1980. "The Cleric of Treason." *The New Yorker*, December 8, 158–195.

Stendardo, Enrica. 1992. "Ferrante Imperato. Il collezionismo naturalistico a Napoli tra '500 e '600, ed alcuni documenti inediti."*Atti e memorie dell'Accademia Clementina*, n.s. 28–29: 43–79.

Stewart, Susan. 1984. *On Longing: Narratives of the Miniature, the Gigantic, the Souvenir, the Collection.* Baltimore: Johns Hopkins University Press.

Stumpo, Enrico. 1986. "Dal Pozzo, Cassiano junior." In *Dizionario Biografico Degli Italiani*, XXXII. Rome: Istituto della Enciclopedia Italiana, 209–213.

Tagliolini, Alessandro, and Margherita Azzi Visentini, eds. 1996. *Il Giardino delle Esperidi: Gli agrumi nella storia, nella letteratura e nell'arte.* Florence: Edifir.

Tarde, Jean. 1984. *A la rencontre de Galilée: Deux voyages en Italie.* Edited and annotated by François Moureau and Marcel Tetel. Geneva: Slatkine.

Thorndike, Lynn. 1923–58. *A History of Magic and Experimental Science.* 8 vols. New York: Columbia University Press.

Thuillier, Jacques. 1960. "Pour un 'Corpus Pussinianum.'" In André Chastel, ed., *Nicolas Poussin, Colloque.* Paris: Centre National de la Recherche Scientifique, II: 49–238.

Tongiorgi Tomasi, Lucia, et al. 1984. *Immagine e Natura. L'immagine naturalistica nei codici e libri a stampa delle Biblioteche Estense e Universitaria. Secoli XV-XVII.* Cat. Exhib., Modena: Biblioteche Estense e Universitaria; Modena: Edizioni Panini.

Tongiorgi Tomasi, Lucia, and Angela Ferrari. 1986. "Botanica Barocca," *Gazzetta del bibliofilo* 17, 1986 (Supplemento fuori commercio al numero 46 di FMR): 3–15.

Turner, Nicholas. 1992. "The Drawings of Pietro Testa after the Antique in Cassiano dal Pozzo's Paper Museum." *Quaderni Puteani* 2: 127–144.

———. 1993. "Some of the Copyists after the Antique Employed by Cassiano." *Quaderni Puteani* 4: 27–37.

Ubrizsy, Andrea. 1980. "Il Codice micologico di Federico Cesi." *Atti della Accademia Nazionale dei Lincei, Rendiconti della Classe di Scienze fisiche, matematiche e naturali.* S. VIII, vol. LXVII: 129–134.

Uhlenbeck, George Eugène. 1924. "Over Johannes Heckius." *Mededeelingen van het Nederlandsch Historisch Instituut te Rome* IV: 217–228.

Van Helden, Albert. 1976. "The 'Astronomical Telescope,' 1611–1650." *Annali dell'Istituto e Museo di Storia della Scienza di Firenze* 1/2: 13–36.

———. 1977. "The Invention of the Telescope." *Transactions of the American Philosophical Society* 67/4: 5–64.

———. 1984. "Galileo and the Telescope." In Paolo Galluzzi, ed., *Novità celesti e crisi del sapere: Atti del Convegno Internazionale di Studi Galileiani.* Florence: Giunti Barbera: 149–158.

———, ed. 1989. *Sidereus Nuncius or the Sidereal Nuncius.* Chicago: University of Chicago Press.

———. 1995. "Galileo and Scheiner on Sunspots: A Case Study in the Visual Language of Astronomy." *Proceedings of the American Philosphical Society* 140/3: 358–396.

Van Kessel, Elisja. 1976. "Joannes van Heeck (1579–?), Co-founder of the Accademia dei Lincei in Rome: A Bio-bibliographical Sketch." *Mededelingen van het Nederlands Instituut te Rome* XXXVIII: 109–134.

Varey, Simon ed. Forthcoming. *The World of Dr. Francisco Hernandez*, Stanford: Stanford University Press.

Vasoli, Cesare. 1978. *L'Enciclopedismo del Seicento.* Naples: Bibliopolis.

———. 1984. "'Tradizione' e 'Nuova Scienza': Note alle lettere a Cristina di Lorena e al P. Castelli." In Paolo Galluzzi, ed., *Novità celesti e crisi del sapere: Atti del Convegno Internazionale di Studi Galileiani.* Florence: Giunti Barbera, 73–94.

Vermeule, Cornelius Clarkson. 1960. "The dal Pozzo-Albani Drawings of Classical Antiquities in the British Museum." *Transactions of the American Philosophical Society*, n.s., I, 5, 5–78.

———. 1966. "The dal Pozzo-Albani Drawings of Classical Antiquities in the Royal Library at Windsor Castle." *Transactions of the American Philosophical Society*, n.s., LVI, 2, 5–77.

Villoslada, Riccardo García, S. J. 1954. *Storia del Collegio Romano dal suo inizio (1551) alla sopressione della Compagnia di Gesù (1773).* Rome: Università Gregoriana.

Volpicelli, Paolo. 1866. "Ritrovamento dell'Inventario degli oggetti appartenenti all'eredità di Federico Cesi." *Atti dell'Accademia dei Lincei* XIV: 203–205.

Wallace, William. 1836. "An Account of the Inventor of the Pantograph." *Transactions of the Royal Society of Edinburgh* 13: 418–439.

———. 1984. *Galileo and His Sources: The Heritage of the Collegio Romano in Galileo's Science.* Princeton: Princeton University Press.

Westfall, Richard S. 1983. "Galileo and the Accademia dei Lincei." In Paolo Galluzzi, ed., *Novità celesti e crisi del sapere: Atti del Convegno internazionale di studi Galileiani,* (Pisa-Venezia-Padova-Firenze, 18–26 marzo 1983). Florence: Giunti Barbera, 189–200.

———. 1985. "Science and Patronage: Galileo and the Telescope." *Isis* 76: 11–30.

———. 1989. *Essays on the Trial of Galileo.* Vatican City: Vatican Observatory.

Westman, Robert S. 1986. "The Copernicans and the Churches." In David C. Lindberg and Ronald L. Numbers, eds., *God and Nature: Historical Essays on the Encounter between Christianity and Science.* Berkeley and Los Angeles: University of California Press, 76–113.

———. 1990. "Proof, Poetics, and Patronage: Copernicus's Preface to 'De revolutionibus'." In David C. Lindberg and Robert S. Westmann, eds., *Reappraisals of the Scientific Revolution.* Cambridge: Cambridge University Press, 166–205.

Wilson, Catherine. 1988. "Visual Surface and Visual Symbol: the Microscope and the Occult in Early Modern Science." *Journal of the History of Ideas* 49: 85–108.

Winkler, Mary G., and Albert Van Helden. 1992. "Representing the Heavens: Galileo and Visual Astronomy." *Isis* 83: 195–217.

White, Lynn, Jr. 1966. "Pumps and Pendula: Galileo and Technology." In Carlo L. Golino, ed., *Galileo Reappraised.* Berkeley and Los Angeles: University of California Press, 96–110.

Yates, Frances. 1964. *Giordano Bruno and the Hermetic Tradition.* Chicago: University of Chicago Press.

———. 1966. *The Art of Memory.* Chicago: University of Chicago Press.

## *Headings*

# Index

DATE DUE